Higher Electrical Principles

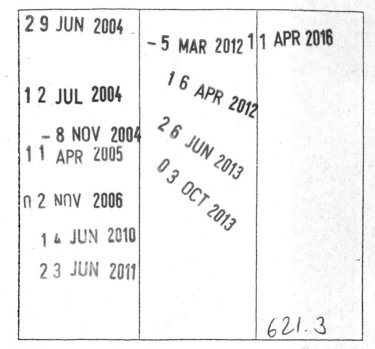

D C Green
MTech, CEng, MIEE

Higher Electrical Principles

Third edition

PEARSON
Longman

Harlow, England • London • New York • Boston • San Francisco • Toronto • Sydney • Singapore • Hong Kong
Tokyo • Seoul • Taipei • New Delhi • Cape Town • Madrid • Mexico City • Amsterdam • Munich • Paris • Milan

Addison Wesley Longman Limited
Edinburgh Gate, Harlow
Essex CM20 2JE, England
and Associated Companies throughout the world

First published by Pitman Publishing Limited as *Electrical Principles IV* 1983
Second edition published by Longman Scientific and Technical 1992
Third edition published by Addison Wesley Longman as *Higher Electrical Principles* 1997

British Library Cataloguing-in-Publication Data
A catalogue entry for this title is available from the British Library

ISBN 0-582-29460-6

Printed and bound by Antony Rowe Ltd, Eastbourne
Transferred to digital print on demand, 2003

Contents

Preface

This book is intended to provide the reader with a good understanding of the fundamental principles of electrical engineering. The study of electrical electronic principles is essential as part of all Higher Certificate/Diploma courses in electrical, electronic, telecommunication and computer engineering, and during the earlier years of many engineering degree courses. The actual topics studied by a student attending a particular institute are selected from the wide-ranging BTEC 'bank'. Some of the contents of this bank seem to be more suited to courses in either control engineering or physical electronics and these topics have been omitted from this book. However, all of the electrical topics are included.

For a student to obtain a good understanding of electrical principles it is necessary to work through a large number of numerical problems of varying complexity. For this reason, extensive use of worked examples is made throughout the book to illustrate various points. Quite often alternative methods have been used to solve the same problem and get the same answer(!); this will hopefully encourage the reader to consider alternatives instead of adopting a blinkered approach to solving problems. Most calculations have been performed to an accuracy of two decimal places (except for currents quoted in amps). Greater accuracy is regarded as somewhat meaningless when most electrical components have tolerances of 5 per cent or more. Each chapter concludes with a large number of exercises, and concise worked solutions to all the low-numbered exercises, and answers only to the exercises with higher numbers, are given at the end of the book.

The first part of the book is concerned with topics that, for most colleges and universities, will probably form level 4 work, while the later chapters deal with what are likely to be level 5 topics. Since each establishment can select topics to form level 4 and level 5 syllabuses as they wish there is very little standardization in course content. However, a recommended sequence of study would be as follows:

- **Level 4**
  Chapter 1 (Complex Algebra, Determinants and Matrices), followed by Chapter 2 (A.C. Circuits) through to Chapter 11 (Iron-cored Transformers) plus, possibly, Chapter 16 (Matched Transmission Lines).
- **Level 5**
  Starting with Chapter 12 (Electric and Magnetic Fields), go straight through to Chapter 20 (Solutions of Circuits using Laplace Transforms).

Of course, many establishments will omit some of these topics in order to satisfy local requirements and so some amendment to this proposed order of work will be necessary for individual readers.

The book assumes that the reader will have studied both electrical/electronic principles and mathematics to BTEC level 3 standard or equivalent. Since the solution of electrical principles problems at this level requires a good understanding of complex numbers Chapter 1 introduces the reader to complex algebra. The use of 'j' notation should be well understood before the later chapters are studied. The work involved in the solution of simultaneous equations is greatly reduced if determinants are employed and an introduction to this useful mathematical tool is also given. It is recommended that determinants are studied before Chapter 7 is read. Lastly, Chapter 1 also gives an introduction to matrices since they are employed in Chapter 16.

D.C.G.

1 Complex algebra, determinants and matrices

The calculation of the currents and voltages at points in an electrical circuit can be carried out using phasor diagrams. It is generally more convenient, however, particularly with the more complex circuits, to employ *complex algebra* to solve problems. Complex numbers are used for the analysis of electric circuits, both series and parallel, for mesh and nodal analysis, and in conjunction with various circuit theorems. Often the analysis of an electric circuit results in a number of simultaneous equations that must be solved to determine the values of two, or more, unknown quantities. The solution of a simultaneous equation is usually made easier if *determinants* are employed and an introduction to this useful mathematical tool is given later in the chapter. Lastly, it is often convenient to write down equations, particularly those describing networks, in matrix form. A matrix is a rectangular array of numbers or equations that are arranged in a number of rows and columns. The numbers of rows and columns need not be the same. The use of matrices is of particular importance when a computer is available which is able to perform matrix calculations.

Complex Numbers

The symbol j represents the square root of minus 1, i.e. $j = \sqrt{-1}$. From this it is evident that $j^2 = -1$, $j^3 = -j$, and, of course, $j^4 = +1$.

j also represents an angle of 90° so that $j = 1\angle 90°$, $j^2 = 1\angle 180°$, $j^3 = 1\angle 270°$, and $j^4 = 1\angle 360°$, $1\angle 0°$.

Example 1.1

Show that $-j = 1/j$.

Solution

$$-j = (-j \times j)/j = -j^2/j = -(-1)/j = 1/j \quad (\textit{Ans.})$$

Example 1.2

Determine the values of (*a*) j^5, (*b*) j^9 and (*c*) j^{11}.

Solution
(*a*) $j^5 = j \times j^4 = j$ (*Ans.*)
(*b*) $j^9 = j^4 \times j^4 \times j = j$ (*Ans.*)
(*c*) $j^{11} = j^9 \times j^2 = j \times -1 = -j$ (*Ans.*)

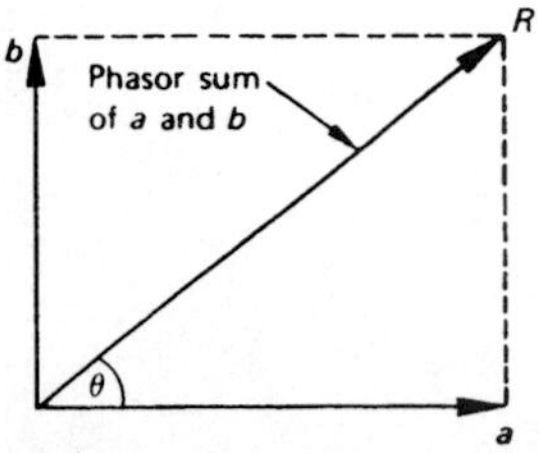

Fig.1.1 Complex number $a + jb$

A quantity $R = a + jb$ is known as a *complex number*, a is the *real part* and b is the *imaginary part* of that number. The term 'imaginary' has its origins in the history of mathematics when the meaning of j was not clearly understood. The *rectangular* number $a + jb$ represents the phasor sum of a distance a in the positive real direction, which is normally taken as being the reference direction, and a distance b in the direction 90° anticlockwise to this reference. This is shown by Fig. 1.1.

The complex number represents a phasor of magnitude R which is at an angle θ to the reference (horizontal) axis. The magnitude of R is $|R| = \sqrt{(a^2 + b^2)}$ and its phase angle θ is $\theta = \tan^{-1}(b/a)$.

Often the magnitude R of a complex number is known as the *modulus*, and the angle θ is known as the *argument*. Examples of complex numbers are 4 + j2, 2 − j4, −6 − j8, and −6 + j10. The real and imaginary parts of a complex number are different kinds of number and so they must be dealt with separately. For example, in the first complex number 4 + j2, the real part 4 cannot be added to the imaginary part 2. Two complex numbers are equal to one another only if their real parts are equal to one another *and* their imaginary parts are also equal. Multiplying a real number by j rotates that number through an angle of 90° in the anticlockwise direction but its magnitude remains unaltered. Multiplying the number by −j rotates it through 90° in the clockwise direction.

Any complex number can be represented on an *Argand diagram* (see Fig. 1.2). Thus (2 + j3) is the point marked as *A* and (−1 − j4) is the point labelled as *B*.

Consider the point *A*:

$$A = 2 + j3 = \sqrt{(2^2 + 3^2)} \angle \tan^{-1} 3/2 = 3.61 \angle 56.3°$$

If the complex number *A* had originally been quoted as 3.61 ∠56.3° it would be possible to obtain the values of its real and imaginary parts. The real part is obtained by multiplying the magnitude of the number by the cosine of its angle, i.e.

$$a = 3.61 \cos 56.3° = 2$$

The imaginary part is obtained by writing

$$b = 3.61 \sin 56.3° = 3$$

Similarly, the point B in Fig. 1.2 is

$$-1 - j4 = \sqrt{(1^2 + 4^2)} \angle \tan^{-1}(-1/-4) = 4.12\angle -104°$$

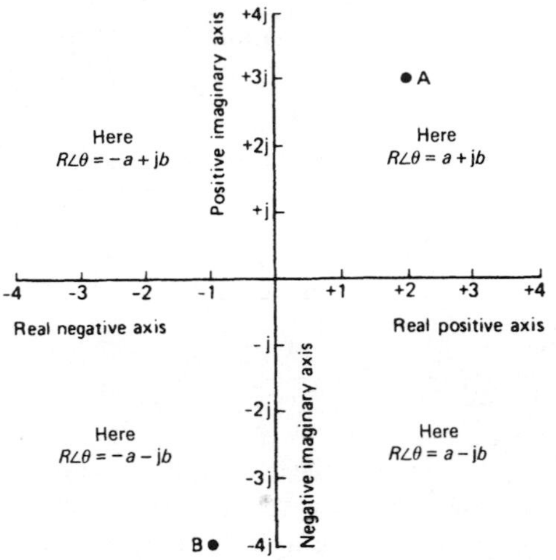

Fig. 1.2 Argand diagram

This process is known as *resolving* the complex number into its real and imaginary parts. A complex number written as $a + jb$ is said to be in its *rectangular form*, and when written as $R\angle\theta°$ it is in its *polar form*. The *conjugate* of a complex number has the same real and imaginary parts but the imaginary part has the opposite sign. Thus the conjugate of $a + jb$ is $a - jb$. In polar form the conjugate of $R\angle\theta$ is $R\angle -\theta$. The real part of a complex number A is often denoted by Re(A) and the imaginary part by Im(A).

Example 1.3

(*a*) Convert the following complex numbers to polar form: (*i*) 4 − j4, (*ii*) −20 + j10 and (*iii*) −10 −j10.

(*b*) Convert the following complex numbers to rectangular form: (*i*) $50\angle -60°$ and (*ii*) $100\angle 30°$

Solution

(*a*) (*i*) $R\angle\theta = \sqrt{(4^2 + 4^2)} \angle \tan^{-1}(-4/4) = 5.66\angle -45°$ (*Ans.*)

(*ii*) $R\angle\theta = \sqrt{(20^2 + 10^2)} \angle \tan^{-1}(10/-20) = 22.36\angle 153.4°$ (*Ans.*)

(*iii*) $R\angle\theta = \sqrt{(10^2 + 10^2)} \angle \tan^{-1}(-10/-10) = 14.14\angle -135°$ (*Ans.*)

(*b*) (*i*) $50\angle -60° = 50\cos(-60°) + j50\sin(-60°) = 25 - j43.3$ (*Ans.*)

(*ii*) $100\angle 30° = 100\cos 30° + j100\sin 30° = 86.6 + j50$ (*Ans.*)

Manipulation of complex numbers

Before complex algebra can be used in the solution of electrical circuit problems it is necessary to know how complex numbers can be added, subtracted, multiplied and divided.

Addition

When two complex numbers are to be added together their real and j parts must be added separately. Thus

$$(a+\mathrm{j}b)+(c+\mathrm{j}d)=a+c+\mathrm{j}(b+d)$$

For example,

$$(6-\mathrm{j}2)+(-2+\mathrm{j}4)=4+\mathrm{j}2 \quad \text{and}$$
$$(4+\mathrm{j}3)+(2+\mathrm{j}5)=6+\mathrm{j}8$$

Two complex numbers given in polar form cannot be directly added together; before addition can take place both of the numbers must be resolved into their real and imaginary parts. Consider the sum of the two numbers $25\angle 60°$ and $40\angle -35°$. Resolving each number into its real and imaginary components gives

$$25\cos 60° + \mathrm{j}25\sin 60° = 12.5+\mathrm{j}21.65$$

and $40\cos(-35°)+\mathrm{j}40\sin(-35°)=32.77-\mathrm{j}22.94$

The sum of the two numbers is

$$12.5+32.77+\mathrm{j}(21.65-22.94) \quad \text{or} \quad 45.27-\mathrm{j}1.29$$

Expressed in polar form this is

$$\sqrt{(45.27^2+1.29^2)}\angle \tan^{-1}(-1.29/45.27)=45.29\angle -1.63°$$

Subtraction

The subtraction of one complex number from another follows the same rules as addition. Thus

$$(a+\mathrm{j}b)-(c+\mathrm{j}d)=a-c+\mathrm{j}(b-d)$$

For example,

$$(6+\mathrm{j}8)-(4-\mathrm{j}10)=2+\mathrm{j}18$$

As for addition, two complex numbers expressed in their polar forms must be changed into their rectangular form before subtraction can be carried out.

Example 1.4

Add, and subtract, the following complex numbers. Give the results in polar form.

(*a*) 40 + j70 and 30 + j20, (*b*) −6 − j10 and −8 + j30; (*c*) $100\angle -30°$ and $50\angle 70°$; (*d*) $50\angle 120°$ and 25 + j30.

Solution

$$(a)\ (40 + j70) + (30 + j20) = 70 + j90$$
$$= \sqrt{(70^2 + 90^2)} \angle \tan^{-1}(90/70)$$
$$= 114.02\angle 52.1° \quad (\textit{Ans.})$$
$$(40 + j70) - (30 + j20) = 10 + j50$$
$$= \sqrt{(10^2 + 50^2)} \angle \tan^{-1}(50/10)$$
$$= 51\angle 78.7° \quad (\textit{Ans.})$$

$$(b)\ (-6 - j10) + (-8 + j30) = -14 + j20$$
$$= \angle(14^2 + 20^2) \angle \tan^{-1}(20/-14)$$
$$= 24.41\angle 125° \quad (\textit{Ans.})$$
$$(-6 - j10) - (-8 + j30) = 2 - j40$$
$$= \sqrt{(2^2 + 40^2)} \angle \tan^{-1}(-40/2)$$
$$= 40.05\angle -87.1° \quad (\textit{Ans.})$$

$$(c)\ 100\angle -30° + 50\angle 70° = 100\cos(-30°) + j100\sin(-30°)$$
$$+ 50\cos 70° + j50\sin 70°$$
$$= (86.6 - j50) + (17.1 + j47) = 103.7 - j3$$
$$= \sqrt{(103.7^2 + 3^2)} \angle \tan^{-1}(-3/103.7)$$
$$= 103.74\angle -1.7° \quad (\textit{Ans.})$$
$$100\angle -30° - 50\angle 70° = (86.6 - j50) - (17.1 + j47) = 69.5 - j97$$
$$= \sqrt{(69.5^2 + 97^2)} \angle \tan^{-1}(-97/69.5)$$
$$= 119.33\angle -54.4° \quad (\textit{Ans.})$$

$$(d)\ 50\angle 120° + (25 + j30) = 50\cos 120° + j50\sin 120° + (25 + j30)$$
$$= (-25 + j43.43) + (25 + j30)$$
$$= j73.3 = 73.3\angle 90° \quad (\textit{Ans.})$$
$$50\angle 120° - (25 + j30) = (-25 + j43.43) - (25 + j30)$$
$$= -50 + j13.3$$
$$= \sqrt{(50^2 + 13.3^2)} \angle \tan^{-1}(13.3/-50)$$
$$= 51.74\angle 165.1° \quad (\textit{Ans.})$$

Multiplication

The multiplication of two complex numbers can be carried out with the numbers expressed in either their polar or their rectangular form. In rectangular form,

$$(a+jb)(c+jd) = ac + jad + jbc + j^2bd$$
$$= ac - bd + j(ad + bc)$$

If the two numbers are in polar form,

$$R\angle\theta^\circ \times S\angle\varphi^\circ = RS\angle(\theta^\circ + \varphi^\circ)$$

Thus the rule is: multiply the magnitudes and add the angles to obtain the product of the two complex numbers.

Example 1.5

Multiply out (*a*) $5\angle 30^\circ \times 10\angle 25^\circ$ and (*b*) $(5+j5)(3-j2)$.

Solution

(*a*) $5\angle 30^\circ \times 10\angle 25^\circ = 5 \times 10\angle(30+25)^\circ = 50\angle 55^\circ$ (*Ans.*)

Alternatively,

$$5\angle 30^\circ = 5\cos 30^\circ + j5\sin 30^\circ = 4.33 + j2.5$$

and $10\angle 25^\circ = 10\cos 25^\circ + j10\sin 25^\circ = 9.06 + j4.23$

Hence,

$$5\angle 30^\circ \times 10\angle 25^\circ = (4.33+j2.5)(9.06+j4.23)$$
$$= 39.23 + j22.65 + j18.32 - 10.58$$
$$= 28.65 + j40.97$$
$$= \sqrt{(28.65^2 + 40.97^2)}\angle \tan^{-1}(40.97/28.65)$$
$$= 50\angle 55^\circ \quad (Ans.)$$

Clearly the first method of solution is much the simpler.

(*b*) $(5+j5)(3-j2) = 15 - j10 + j15 + 10 = 25 + j5$

$$= \sqrt{(25^2 + 5^2)}\angle \tan^{-1}(5/25)$$
$$= 25.5\angle 11.3^\circ \quad (Ans.)$$

Alternatively,

$$5 + j5 = \sqrt{(5^2+5^2)}\angle \tan^{-1}(5/5) = 7.07\angle 45^\circ \quad \text{and} \quad 3 - j2$$
$$= \sqrt{(3^2+2^2)}\angle \tan^{-1}(-2/3)$$
$$= 3.61\angle -33.7 \quad \text{and} \quad 7.07\angle 45^\circ \times 3.61\angle -33.7^\circ$$
$$= 25.5\angle 11.3^\circ \quad (Ans.)$$

Example 1.6

Multiply together the complex numbers $10 - j10$ and $5\angle 30^\circ$.

Solution

(*a*) Working in rectangular form

$$5\angle 30^\circ = 4.33 + j2.5$$

and

$$(10 - j10)(4.33 + j2.5) = 43.3 + j25 - j43.3 - j^2 25$$
$$= 68.3 - j18.3 \quad (\textit{Ans.})$$

In polar form,

$$\sqrt{(68.3^2 + 18.3^2)} \angle \tan^{-1}(-18.3/68.3) = 70.7 \angle -15° \quad (\textit{Ans.})$$

(*b*) $10 - j10 = \sqrt{(10^2 + 10^2)} \angle \tan^{-1}(-10/10) = 14.14 \angle -45°$

and

$$14.14 \angle -45° \times 5 \angle 30° = 70.7 \angle -15° \text{ (as before)} \quad (\textit{Ans.})$$

Complex conjugate

The product of a complex number and its conjugate is always a wholly real number. This property is used in the division of one complex number by another when the numbers are both in the rectangular form. If, $A = 10 + j50$ then

$$A^* = 10 - j50$$

and

$$AA^* = (10 + j50)(10 - j50) = 100 - j500 + j500 + 2500$$
$$= 2600 + j0$$

Division

The division of one complex number by another can be carried out using either the polar or the rectangular forms of the two numbers.

When the rectangular form is used, a procedure known as *rationalization* must be employed. A complex number is rationalized when both its numerator and its denominator are multiplied by the conjugate of the denominator. Thus,

$$(a + jb)/(c + jd) = [(a + jb)(c - jd)]/[(c + jd)(c - jd)]$$
$$= [ac + bd + j(bc - ad)]/[c^2 + d^2 + j(cd - cd)]$$
$$= [ac + bd + j(bc - ad)]/(c^2 + d^2)$$

The reason why the rationalization process is used is that it produces a wholly real denominator. It should be noted that the denominator of a rationalized expression always consists of the square of the real part *plus* the square of the imaginary part of the original denominator.

Example 1.7

Divide $20 + j10$ by $5 + j8$.

Solution

$$(20+j10)/(5+j8) = (20+j10)(5-j8)/(5^2+8^2)$$
$$= (100 - j160 + j50 + 80)/89$$
$$= 2.02 - j1.24 = 2.37\angle -31.5° \quad (\textit{Ans.})$$

When a complex number in polar form is to be divided by another complex number, also in polar form, the division is easily carried out by taking the quotient of their magnitudes and the difference between their angles, i.e.

$$R\angle\theta°/S\angle\varphi° = (R/S)\angle(\theta° - \varphi°)$$

Example 1.8

Convert the complex numbers 20 + j10 and 5 + j8 into their polar forms and then find (20 + j10)/(5 + j8).

Solution

$$20 + j10 = \sqrt{(20^2+10^2)}\angle\tan^{-1}(10/20) = 22.36\angle 26.6°.$$
$$5 + j8 = \sqrt{(5^2+8^2)}\angle\tan^{-1}(8/5) = 9.43\angle 58°$$

Hence,

$$(20+j10)/(5+j8) = (22.36/9.43)\angle 26.6° - 58°$$
$$= 2.37\angle -31.4° \quad (\textit{Ans.})$$

Example 1.9

Divide $10 - j10$ by $5\angle 30°$ by working (*a*) in rectangular form; (*b*) in polar form.

Solution

(*a*) $5\angle 30° = 4.33 + j2.5$. Therefore,

$$(10 - j10)/(4.33 + j2.5)$$
$$= [(10 - j10)(4.33 - j25)]/(4.33^2 + 2.5^2)$$
$$= [43.3 - 25 - j(43.3 + 25)]/(4.33^2 + 2.5^2)$$
$$= (18.3 - j68.3)/25$$
$$= 0.73 - j2.73 \quad (\textit{Ans.})$$

Expressed in polar form,

$$\sqrt{(0.73^2 + 2.73^2)}\angle\tan^{-1}(-2.73/0.73) = 2.83\angle -75° \quad (\textit{Ans.})$$

(*b*) $10 - j10 = 14.14\angle -45°$. Therefore,

$$(14.14\angle -45°)/(5\angle 30°) = 2.83\angle -75° \quad \text{(as before)} \quad (\textit{Ans.})$$

Example 1.10

Calculate the reciprocal of the complex numbers (*a*) $200\angle 20°$ and (*b*) 20 – 40.

Solution

(*a*) $1/(200\angle 20°) = (1\angle 0°)/(200\angle 20°) = 5 \times 10^{-3}\angle -20°$ (*Ans.*)

(*b*) $1/(20 - j40) = (20 + j40)/(20^2 + 40^2) = 0.01 + j0.02$
$= 0.022\angle 63.4°$ (*Ans.*)

Alternatively,

$$20 - j40 = \sqrt{(20^2 + 40^2)}\angle \tan^{-1}(-40/20) = 44.72\angle -63.4°$$

and

$$1/(20 - j40) = 1/(44.72\angle -63.4°) = 0.022\angle 63.4° \quad (Ans.)$$

Example 1.11

Evaluate each of the following equations if $A = 2 + j2$, $B = 3 + j4$ and $C = 4 - j3$:
(*a*) $R = ABC$; (*b*) $R = (A - B)/(A + C)$ and
(*c*) $R = ABC \times [(A - B)/(A + C)]$.

Solution

(*a*) $R = (2 + j2)(3 + j4)(4 - j3) = (6 + j8 + j6 - 8)(4 - j3)$
$= (-2 + j14)(4 - j3) = -8 + j6 + j56 + 42 = 34 + j62$
$= 70.7\angle 61.3°$ (*Ans.*)

Alternatively,

$$A = 2 + j2 = 2.83\angle 45°,$$
$$B = 3 + j4 = 5\angle 53.1° \quad \text{and}$$
$$C = 4 - j3 = 5\angle -36.9°.$$

Thus,

$$R = (2.83 \times 5 \times 5)\angle(45 + 53.1 - 36.9°)$$
$$= 70.7\angle 61.2° \quad (Ans.)$$

[Note the slight difference in angle which is caused by the rounding off of numbers.]

(*b*) $R = [(2 + j2) - (3 + j4)]/[(2 + j2) + (4 - j3)]$
$= (-1 - j2)/(6 - j) = (-1 - j2)(6 + j)/(36 + 1)$
$= (-4 - j13)/37 = -0.11 - j0.35$
$= 0.37\angle - 107.5°$ (*Ans.*)

(*c*) $R = (70.7\angle 61.3°) \times (0.37\angle -107.5°)$
$= 26.16\angle -46.2°$ (*Ans.*)

Example 1.12

The input impedance Z_{in} of an electrical circuit is given by

$$Z_{in} = Z_1 + Z_2(Z_3 + Z_4)/(Z_2 + Z_3 + Z_4)$$

Calculate the input impedance if $Z_1 = 100\angle 45°\ \Omega$, $Z_2 = 25\angle -20°\ \Omega$, $Z_3 = 30\angle -30°\ \Omega$ and $Z_4 = 40\angle 18°\ \Omega$.

Solution

$$Z_1 = 70.71 + j70.71\ \Omega, \quad Z_2 = 23.49 - j8.55\ \Omega,$$
$$Z_3 = 25.98 - j15\ \Omega \quad \text{and} \quad Z_4 = 38.04 + j12.36\ \Omega$$
$$Z_3 + Z_4 = (25.98 - j15) + (38.04 + j12.36) = 64.02 - j2.64$$
$$= \sqrt{(64.02^2 + 2.64^2)}\angle \tan^{-1}(-2.64/64.02)$$
$$= 64.07\angle -2.4°\ \Omega$$
$$Z_2 + Z_3 + Z_4 = (64.02 - j2.64) + (23.49 - j8.55)$$
$$= 87.51 - j11.19$$
$$= 88.22\angle - 7.3°\ \Omega$$

Hence,

$$Z_{in} = (70.71 + j70.71)$$
$$+ (25\angle - 20°) \times [64.07\angle(-2.4°)/88.22\angle -7.3°]$$
$$= (70.71 + j70.71) + 18.16\angle - 15.1°$$
$$= (70.71 + j70.71) + (17.53 - j4.73)$$
$$= 88.24 + j65.98 = 110.18\angle 36.8° \quad (\textit{Ans.})$$

Example 1.13

If $Z_1 = 250 - j120$ and $Z_2 = 120 + j250$, calculate:

(*a*) $1/Z_1 + 1/Z_2$; (*b*) $1/(1/Z_1 + 1/Z_2)$ and (*c*) $Z_1/(1/Z_1 + 1/Z_2)$.

Solution

$$1/Z_1 = 1/(250 - j120) = (250 + j120)/(250^2 + 120^2)$$
$$= (3.26 + j1.56) \times 10^{-3}$$
$$1/Z_2 = 1/(120 + j250) = (120 - j250)/(120^2 + 250^2)$$
$$= (1.56 - j3.25) \times 10^{-3}$$

(*a*) $1/Z_1 + 1/Z_2 = (4.82 - j1.69) \times 10^{-3}$

$$= 5.1 \times 10^{-3}\angle -19.3° \quad (\textit{Ans.})$$

(*b*) $1/(1/Z_1 + 1/Z_2) = 1/(5.11 \times 10^{-3}\angle -19.3°)$

$$= 196.1\angle 19.4° \quad (\textit{Ans.})$$

$$(c)\ Z_1/(1/Z_1+1/Z_2) = (250-\mathrm{j}120)\times 196.1\angle 19.3^\circ$$
$$= 277.31\times 196.1\angle(-25.6+19.3)^\circ$$
$$= 54\,380\angle -6.3^\circ \quad (Ans.)$$

Powers and roots

When two identical complex numbers are multiplied together the result is

$$R\angle\theta \times R\angle\theta = R^2\angle 2\theta$$

If three identical complex numbers are multiplied together their product is

$$R\angle\theta \times R\angle\theta \times R\angle\theta = R^3\angle 3\theta$$

Thus, for n identical complex numbers

$$[R\angle\theta]^n = R^n\angle n\theta \qquad (1.1)$$

A similar result is obtained if the power n is a fractional number. If, for example, $n = 1/2$ (giving the square root of the complex number), then

$$[R\angle\theta]^{1/2} = R^{1/2}\angle\theta/2$$

There will be two roots of equal magnitude but their angles will differ from one another by 180°.

If $n = 1/3$ the cube root of a complex number is obtained as $[R\angle\theta]^{1/3} = R^{1/3}\angle\theta/3$. There will be three roots of equal magnitude but different angle. In general, for the nth root:

$$[R\angle\theta]^{1/n} = R^{1/n}\angle\theta/n \qquad (1.2)$$

Example 1.14

Determine (a) the square and (b) the cube of the complex number $5+\mathrm{j}5$.

Solution

(a) $(5+\mathrm{j}5)^2 = (5+\mathrm{j}5)(5+\mathrm{j}5) = 0+\mathrm{j}50 = 50\angle 90^\circ$ (*Ans.*)

Alternatively, from equation (1.1)

$$5+\mathrm{j}5 = 7.07\angle 45^\circ.\ (7.07\angle 45^\circ)^2 = 7.07^2\angle(2\times 45)^\circ$$
$$= 50\angle 90^\circ \quad (Ans.)$$

$$(b)\ (5+\mathrm{j}5)^3 = (7.07)\angle 45^\circ)^3 = 7.07^3\angle(3\times 45)^\circ$$
$$= 353.4\angle 135^\circ \quad (Ans.)$$

Example 1.15

Calculate the two square roots of the complex number 15 + j25.

Solution

$$15 + j25 = \sqrt{(15^2 + 25^2)} \angle \tan^{-1}(25/15) = 29.16\angle 59°$$

One root is

$$\sqrt{(15 + j25)} = \sqrt{29.16}\angle(59/2)° = 5.4\angle 29.5°$$
$$= 4.7 + j2.66 \quad (\textit{Ans.})$$

The other root is

$$\sqrt{29.16}\angle[(360 + 59)/2]° = 5.4\angle 209.5° = -4.7 - j2.66 \quad (\textit{Ans.})$$

The complex number 15 + j25 and its two square roots are shown on the Argand diagram of Fig. 1.3.

Exponential form of complex numbers

Two important relationships are given in equations (1.3) and (1.4).

$$\angle\theta = e^{j\theta} = \cos\theta + j\sin\theta \qquad (1.3)$$
$$\angle\theta = e^{-j\theta} = \cos\theta - j\sin\theta \qquad (1.4)$$

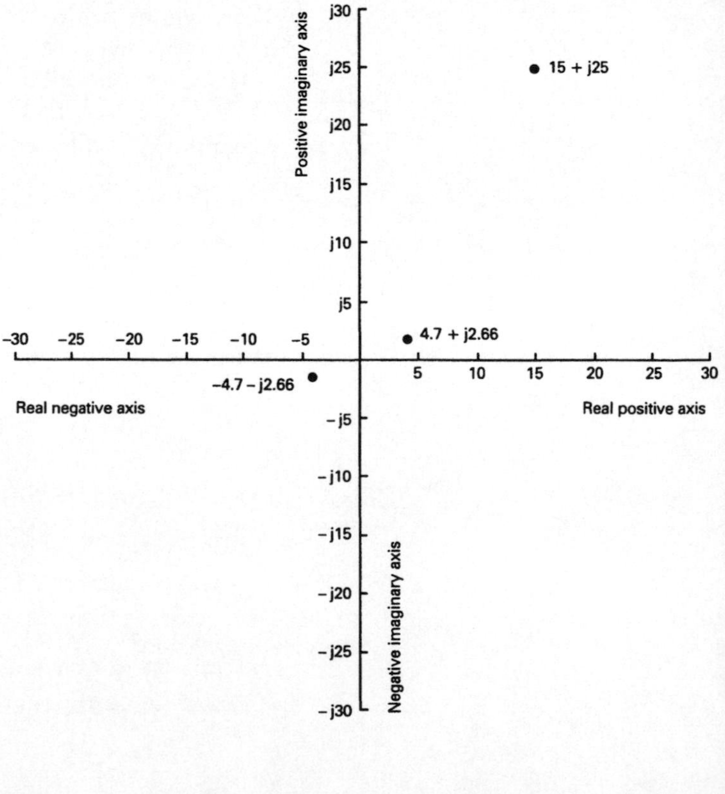

Fig.1.3 Square roots of 15 + j25

Adding equations (1.3) and (1.4)

$$e^{j\theta} + e^{-j\theta} = 2\cos\theta \quad \text{or} \quad \cos\theta = [e^{j\theta} + e^{-j\theta}]/2 \qquad (1.5)$$

Subtracting equation (1.3) from equation (1.4)

$$\sin\theta = (e^{j\theta} - e^{-j\theta})/2j \qquad (1.6)$$

Multiplying both sides of equation (1.3) by R gives

$$R(\cos\theta + j\sin\theta) = R\angle\theta = Re^{j\theta} = e^{p}e^{j\theta} = e^{p+j\theta} \qquad (1.7)$$

Example 1.16

Express the complex number $e^{-(4+j2)}$ in the form $R\angle\theta$.

Solution

$$e^{-(4+j2)} = e^{-4-j2} = 0.018\angle-2 = 0.018\angle-114.6° \quad (\textit{Ans.})$$

Example 1.17

Express the complex number $5 + j10$ in exponential form.

Solution

$$5 + j10 = 11.18\angle1.107(\text{rad}) = e^{2.414+j1.107} \quad (\textit{Ans.})$$

DeMoivre's theorem

If $A = R\angle\theta = Re^{j\theta}$ then $A^2 = R^2\angle2\theta = R^2 e^{j2\theta}$. Also, $A^3 = R^3\angle3\theta = R^3 e^{j3\theta}$, and so on for the higher powers. In general,

$$A^n = R^n\angle n\theta = R^n e^{jn\theta} \qquad (1.8)$$

$e^{jn} = \cos n\theta + j\sin n\theta$, so that

$$A^n = R^n(\cos n\theta = j\sin n\theta) \qquad (1.9)$$

Equations (1.8) and (1.9) are two versions of De Moivre's theorem.

A.C. currents and voltages

The expressions for the instantaneous values of sinusoidal current and voltage waves are

$$i = I_{\max}\sin(\omega t + \theta) \qquad (1.10)$$

and

$$v = V_{\max}\sin(\omega t + \theta) \qquad (1.11)$$

where I_{max} and V_{max} are the peak values of the waveforms, ω is 2π times the frequency, and θ is the angle of the current, or voltage, at time $t = 0$. The voltage, or current, can be written as $V_{max} \angle\theta$, or simply as V_{max} when $\theta = 0°$. Very often the r.m.s. values of the waves are employed and these are simply written down as V or as I. The sum, or the difference, of two sinusoidal voltages at the same frequency gives another sinusoidal wave also at the same frequency. The sum of the sinusoidal voltages $v_1 = 100 \sin \omega t$ volts and $v_2 = 50 \sin(\omega t + 45°)$ volts is

$$\begin{aligned} v = v_1 + v_2 &= 100 \sin \omega t + 50 \sin(\omega t + 45°) \\ &= 100 \angle 0° + 50 \angle 45° \\ &= (100 + j0) + (50 \cos 45° + j50 \sin 45°) = 135.36 + j35.36 \\ &= \sqrt{(135.36^2 + 35.36^2)} \angle \tan^{-1}(35.36/135.36) \\ &= 139.9 \angle 14.6° \text{ V} \end{aligned}$$

Hence, the instantaneous value of the sum voltage waveform is

$$v = 139.9 \sin(\omega t + 14.6°) \text{ V}$$

Example 1.18

Three currents i_1, i_2 and i_3 flowing in a network add to produce a fourth current i_4. If $i_1 = 25 \sin(2000\pi t)$ mA, $i_2 = 20 \sin(2000\pi t + 100°)$ mA and $i_3 = 18 \sin(2000\,\pi t - 20°)$ mA, obtain an expression for i_4.

Solution

$$\begin{aligned} i_4 &= 25 \angle 0° + 20 \angle 100° + 18 \angle -20° \\ &= (25 + j0) + (-3.47 + j19.7) + (16.92 - j6.16) \\ &= 38.45 + j13.54 = 40.76 \angle 19.4° \text{ mA} \quad (\textit{Ans.}) \end{aligned}$$

Example 1.19

The voltage $240 \angle 0°$ V is applied across a circuit which consists of three impedances connected in series. If the voltages across two of the impedances are $v_1 = 100 \angle 62°$ V and $v_2 = 150 \angle -28°$ V, calculate the voltage v_3 across the third impedance.

Solution

$$\begin{aligned} v_3 &= 240 \angle 0° - [(100 \angle 62°) + (150 \angle -28°)] \\ &= (240 + j0) - (46.95 + j88.3 + 132.44 - j70.42) \\ &= 60.61 - j17.88 = 63.19 \angle -16.4° \text{ V} \quad (\textit{Ans.}) \end{aligned}$$

Example 1.20

Obtain an expression for the current supplied to a circuit consisting of three impedances connected in parallel if the currents in the impedances are $i_1 = 5\sin(\omega t + 60°)$ mA, $i_2 = 3\sin(\omega t)$ mA and $i_3 = 10\sin(\omega t - 30°)$ mA.

Solution

$$\begin{aligned} i &= 5\angle 60° + 3\angle 0° + 10\angle -30° \\ &= (2.5 + \text{j}4.33) + (3 + \text{j}0) + (8.66 - \text{j}5) \\ &= 14.16 - \text{j}0.67 = 14.18\angle -2.7° \text{ mA} \end{aligned}$$

Therefore,

$$i = 14.18\sin(\omega t - 2.7°) \text{ mA} \quad (\textit{Ans.})$$

Example 1.21

Solve for r and x in each of the following equations:

(*a*) $(r + \text{j}x)[100/(1 + \text{j}200)] = 400$.
(*b*) $(r + \text{j}x)/\text{j}10 = 22(25 - \text{j}3)$.
(*c*) $520(100 - \text{j}250) = 416(r - \text{j}x)$.

Solution

(*a*) $r + \text{j}x = 400(1 + \text{j}200)/100 = 4 + \text{j}800$. Therefore,

$$r = 4 \quad \text{and} \quad x = 800 \quad (\textit{Ans.})$$

(*b*) $r + \text{j}x = \text{j}220(25 - \text{j}3) = 660 + \text{j}5500$. Therefore,

$$r = 660 \quad \text{and} \quad x = 5500 \quad (\textit{Ans.})$$

(*c*) $r - \text{j}x = 1.25(100 - \text{j}250)$. Therefore,

$$r = 125 \quad \text{and} \quad x = 312.5 \quad (\textit{Ans.})$$

Determinants

A determinant is a square array of numbers that provides a convenient way of solving two, or more, simultaneous equations. Given the two simultaneous equations $Ax + By = P$ and $Cx + Dy = Q$, where x and y are the unknown quantities, the solution is obtained from:

$$x = \frac{\begin{vmatrix} P & B \\ Q & D \end{vmatrix}}{\begin{vmatrix} A & B \\ C & D \end{vmatrix}} = (PD - BQ)/(AD - BC) \quad \text{and}$$

$$y = \frac{\begin{vmatrix} A & P \\ C & Q \end{vmatrix}}{\begin{vmatrix} A & B \\ C & D \end{vmatrix}} = (QA - PC)/(AD - BC)$$

Example 1.22

Solve the equations $6x + 5y = 8$ and $5x + 7y = 3$.

Solution

$$x = \frac{\begin{vmatrix} 8 & 5 \\ 3 & 7 \end{vmatrix}}{\begin{vmatrix} 6 & 5 \\ 5 & 7 \end{vmatrix}} = [(8 \times 7) - (5 \times 3)]/[(6 \times 7) - (5 \times 5)]$$

$$= (56 - 15)/(42 - 25) = 2.41 \quad (\textit{Ans.})$$

$$y = \frac{\begin{vmatrix} 6 & 8 \\ 5 & 3 \end{vmatrix}}{\begin{vmatrix} 6 & 5 \\ 5 & 7 \end{vmatrix}} = [(6 \times 3) - (8 \times 5)]/[(6 \times 7) - (5 \times 5)]$$

$$= (18 - 40)/(42 - 25) = -1.29 \quad (\textit{Ans.})$$

Example 1.23

Solve for I_1 and I_2 the equations

$$(2 + j4)I_1 + (3 - j3)I_2 = 12 \quad \text{and} \quad (4 + j5)I_1 + (5 + j)I_2 = 15$$

Solution

$$I_1 = \frac{\begin{vmatrix} 12 & 3 - j3 \\ 15 & 5 + j \end{vmatrix}}{\begin{vmatrix} 2 + j4 & 3 - j3 \\ 4 + j5 & 5 + j \end{vmatrix}}$$

$$= [12(5 + j) - 15(3 - j3)]/[(2 + j4)(5 + j) - (4 + j5)(3 - j3)]$$

$$= (15 + j57)/(-21 + j19)$$

$$= 0.96 - j1.85 = 2.08 \angle 62.6^\circ \quad (\textit{Ans.})$$

$$I_2 = \frac{\begin{vmatrix} 2 + j4 & 12 \\ 4 + j5 & 15 \end{vmatrix}}{(-21 + j19)}$$

$$= [(30 + j60) - (48 + j60)]/(-21 + j19)$$

$$= 0.47 + j0.43 = 0.64 \angle 42.5^\circ \quad (\textit{Ans.})$$

When there are three unknowns and hence three simultaneous equations, determinants can still be employed although the technique is somewhat more complicated. Suppose the three equations are:

$$Ax + By + Cz = P, \quad Dx + Ey + Fz = Q \quad \text{and}$$
$$Gx + Hy + Iz = R$$

Then the solutions are:

$$x = \frac{\begin{vmatrix} P & B & C \\ Q & E & F \\ R & H & I \end{vmatrix}}{\begin{vmatrix} A & B & C \\ D & E & F \\ G & H & I \end{vmatrix}}$$

The top determinant is evaluated as

$$P\begin{vmatrix} E & F \\ H & I \end{vmatrix} - B\begin{vmatrix} Q & F \\ R & I \end{vmatrix} + C\begin{vmatrix} Q & E \\ R & H \end{vmatrix}$$

and the bottom determinant as

$$A\begin{vmatrix} E & F \\ H & I \end{vmatrix} - B\begin{vmatrix} D & F \\ G & I \end{vmatrix} + C\begin{vmatrix} D & E \\ G & H \end{vmatrix}$$

Also,

$$y = \frac{\begin{vmatrix} A & P & C \\ D & Q & F \\ G & R & I \end{vmatrix}}{\begin{vmatrix} A & B & C \\ D & E & F \\ G & H & I \end{vmatrix}} \quad \text{and} \quad z = \frac{\begin{vmatrix} A & B & P \\ D & E & Q \\ G & H & R \end{vmatrix}}{\begin{vmatrix} A & B & C \\ D & E & F \\ G & H & I \end{vmatrix}}$$

Example 1.24

Solve the equations

$$6x + 5y + 4z = 8, \quad 5x + 7y + 9z = 12 \quad \text{and} \quad x + 2y + 5z = 4$$

Solution

$$x = \frac{\begin{vmatrix} 8 & 5 & 4 \\ 12 & 7 & 9 \\ 4 & 2 & 5 \end{vmatrix}}{\begin{vmatrix} 6 & 5 & 4 \\ 5 & 7 & 9 \\ 1 & 2 & 5 \end{vmatrix}}$$

$$= \frac{8\begin{vmatrix} 7 & 9 \\ 2 & 5 \end{vmatrix} - 5\begin{vmatrix} 12 & 9 \\ 4 & 5 \end{vmatrix} + 4\begin{vmatrix} 12 & 7 \\ 4 & 2 \end{vmatrix}}{6\begin{vmatrix} 7 & 9 \\ 2 & 5 \end{vmatrix} - 5\begin{vmatrix} 5 & 9 \\ 1 & 5 \end{vmatrix} + 4\begin{vmatrix} 5 & 7 \\ 1 & 2 \end{vmatrix}}$$

$$= \frac{[8(35-18)-5(60-36)+4(24-28)]}{[6(35-18)-5(25-9)+4(10-7)]}$$

$$= (136-120-16)/(102-80+12) = 0/34 = 0 \quad (\textit{Ans.})$$

$$y = \frac{\begin{vmatrix} 6 & 8 & 4 \\ 5 & 12 & 9 \\ 1 & 4 & 5 \end{vmatrix}}{34} = \frac{6\begin{vmatrix} 12 & 9 \\ 4 & 5 \end{vmatrix} - 8\begin{vmatrix} 5 & 9 \\ 1 & 5 \end{vmatrix} + 4\begin{vmatrix} 5 & 12 \\ 1 & 4 \end{vmatrix}}{34}$$

$$= [6(60-36)-8(25-9)+4(20-12)]/34$$

$$= (144-128+32)/34 = 48/34 = 1.41 \quad (\textit{Ans.})$$

$$z = \frac{\begin{vmatrix} 6 & 5 & 8 \\ 5 & 7 & 12 \\ 1 & 2 & 4 \end{vmatrix}}{34} = \frac{6\begin{vmatrix} 7 & 12 \\ 2 & 4 \end{vmatrix} - 5\begin{vmatrix} 5 & 12 \\ 1 & 4 \end{vmatrix} + 8\begin{vmatrix} 5 & 7 \\ 1 & 2 \end{vmatrix}}{34}$$

$$= [6(28-24)-5(20-12)+8(10-7)]/34$$

$$= (24-40+24)/34 = 0.24 \quad (\textit{Ans.})$$

Example 1.25

Solve the complex equations

$$I_1 + (1+\mathrm{j})I_2 - \mathrm{j}I_3 = 5, \quad (1+\mathrm{j})I_1 - I_2 + (1-\mathrm{j})I_3 = 6$$

and $(2-\mathrm{j})I_1 + \mathrm{j}4I_2 + 2I_3 = 10$

Solution

$$I_1 = \frac{\begin{vmatrix} 5 & 1+\mathrm{j} & -\mathrm{j} \\ 6 & -1 & 1-\mathrm{j} \\ 10 & \mathrm{j}4 & 2 \end{vmatrix}}{\begin{vmatrix} 1 & 1+\mathrm{j} & -\mathrm{j} \\ 1+\mathrm{j} & -1 & 1-\mathrm{j} \\ 2-\mathrm{j} & \mathrm{j}4 & 2 \end{vmatrix}}$$

The lower (common) determinant is

$$1\begin{vmatrix} -1 & 1-\mathrm{j} \\ \mathrm{j}4 & 2 \end{vmatrix} - (1+\mathrm{j})\begin{vmatrix} 1+\mathrm{j} & 1-\mathrm{j} \\ 2-\mathrm{j} & 2 \end{vmatrix} + (-\mathrm{j})\begin{vmatrix} 1+\mathrm{j} & -1 \\ 2-\mathrm{j} & \mathrm{j}4 \end{vmatrix}$$

$$= 1[-2-(4+\mathrm{j}4)] - (1+\mathrm{j})[2+\mathrm{j}2-(2-\mathrm{j}2-\mathrm{j}-1)]$$

$$-\mathrm{j}(-4+\mathrm{j}4+2-\mathrm{j})$$

$$= -6-\mathrm{j}4-(1+\mathrm{j})(1+\mathrm{j}5)-\mathrm{j}(-2+\mathrm{j}3) = 1-\mathrm{j}8$$

The top determinant is

$$5[-2-j4(1-j)]-(1+j)(12-10+j10)-j(j24+10)$$
$$=5(-2-j4-4)-(1+j)(2+j10)-j10+24$$
$$=2-j42$$

Therefore,

$$I_1=(2-j42)/(1-j8)=5.2-j0.4$$
$$=5.22\angle-4.4^\circ \quad (\mathit{Ans.})$$

$$I_2=\frac{\begin{vmatrix}1 & 5 & -j\\ 1+j & 6 & 1-j\\ 2-j & 10 & 2\end{vmatrix}}{1-j8}$$

$$=1\begin{vmatrix}6 & 1-j\\ 10 & 2\end{vmatrix}-5\begin{vmatrix}1+j & 1-j\\ 2-j & 2\end{vmatrix}-j\begin{vmatrix}1+j & 6\\ 2-j & 10\end{vmatrix}$$
$$=1(12-10+j10)-5(2+j2-1+j3)-j(10+j10-12+j6)$$
$$=(2+j10-10-j10+5-j15-j2+16)/(1-j8)$$
$$=(13-j17)(1-j8)=2.29+j1.34$$
$$=2.65\angle 30.3^\circ \quad (\mathit{Ans.})$$

$$I_3=\frac{\begin{vmatrix}1 & 1+j & 5\\ 1+j & -1 & 6\\ 2-j & j4 & 10\end{vmatrix}}{1-j8}$$

$$=1\begin{vmatrix}-1 & 6\\ j4 & 10\end{vmatrix}-(1+j)\begin{vmatrix}1+j & 6\\ 2-j & 10\end{vmatrix}+5\begin{vmatrix}1+j & -1\\ 2-j & j4\end{vmatrix}$$
$$=1(-10-j24)-(1+j)[10+j10-(12-j6)]$$
$$+5(j4-4+2-j)$$
$$=-10-j24-(1+j)(-2+j16)+5(-2+j3)$$
$$=-2-j23$$

Therefore

$$I_3=(-2-j23)/(1-j8)=2.8-j0.6$$
$$=2.86\angle-12.1^\circ \quad (\mathit{Ans.})$$

Matrices

A *matrix* is a rectangular array of elements arranged in a number of columns and rows that need not be equal to one another. A *square* matrix has equal numbers of columns and rows, a *column* matrix has just a single column and a *row* matrix consists of a single row of elements. A matrix having m rows and n columns is said to be of order $m \times n$. Thus:

$$\begin{bmatrix} 2 & 4 & 6 \\ 3 & 6 & 9 \\ 1 & 2 & 3 \end{bmatrix} \quad \text{is a square matrix,}$$

$$\begin{bmatrix} 2 \\ 3 \\ 1 \end{bmatrix} \quad \text{is a column matrix and}$$

$$\begin{bmatrix} 2 & 4 & 6 \end{bmatrix} \quad \text{is a row matrix.}$$

A matrix provides a convenient way of writing down the equations describing an electric circuit. Thus,

$$\begin{bmatrix} Z_1 & Z_2 \\ Z_3 & Z_4 \end{bmatrix} \begin{bmatrix} I_1 \\ I_2 \end{bmatrix} = \begin{bmatrix} V_1 \\ V_2 \end{bmatrix}$$

is the matrix equivalent of the equations

$$V_1 = I_1 Z_1 + I_2 Z_2 \quad \text{and} \quad V_2 = I_1 Z_3 + I_2 Z_4$$

Addition and subtraction

Two matrices may be added together, or subtracted one from the other, provided they are both of the same order. It is merely necessary to add, or subtract, the corresponding elements in the two matrices.

Example 1.26

Add the matrices $\begin{bmatrix} 10 & 12 \\ 14 & 16 \end{bmatrix}$ and $\begin{bmatrix} 5 & 6 \\ 7 & 8 \end{bmatrix}$.

Solution

$$\begin{bmatrix} 10 & 12 \\ 14 & 16 \end{bmatrix} + \begin{bmatrix} 5 & 6 \\ 7 & 8 \end{bmatrix} = \begin{bmatrix} 15 & 18 \\ 21 & 24 \end{bmatrix} \quad (\textit{Ans.})$$

Example 1.27

Subtract the matrix $\begin{bmatrix} 5 & 6 \\ 7 & 8 \end{bmatrix}$ from the matrix $\begin{bmatrix} 10 & 12 \\ 14 & 16 \end{bmatrix}$.

Solution

$$\begin{bmatrix} 10 & 12 \\ 14 & 16 \end{bmatrix} - \begin{bmatrix} 5 & 6 \\ 7 & 8 \end{bmatrix} = \begin{bmatrix} 5 & 6 \\ 7 & 8 \end{bmatrix} \quad (Ans.)$$

Multiplication

When a matrix is multiplied by a constant all the elements in that matrix are multiplied by that constant, e.g.

$$5\begin{bmatrix} 5 & 6 \\ 7 & 8 \end{bmatrix} = \begin{bmatrix} 25 & 30 \\ 35 & 40 \end{bmatrix}$$

Two matrices can only be multiplied together if the number of columns in the first matrix is equal to the number of rows in the second matrix. The order of multiplication is important since different answers may well be obtained for each order. The procedure is as follows.

For a 2 × 2 square matrix times a column matrix:

$$\begin{bmatrix} A & B \\ C & D \end{bmatrix}\begin{bmatrix} P \\ Q \end{bmatrix} = \begin{bmatrix} AP+BQ \\ CP+DQ \end{bmatrix}$$

For a 3 × 3 square matrix times a column matrix:

$$\begin{bmatrix} A & B & C \\ D & E & F \\ G & H & I \end{bmatrix}\begin{bmatrix} P \\ Q \\ R \end{bmatrix} = \begin{bmatrix} AP+BQ+CR \\ DP+EQ+FR \\ GP+HQ+IR \end{bmatrix}$$

For a 2 × 2 square matrix times another 2 × 2 square matrix:

$$\begin{bmatrix} A & B \\ C & D \end{bmatrix}\begin{bmatrix} E & F \\ G & H \end{bmatrix} = \begin{bmatrix} AE+BG & AF+BH \\ CE+DG & CF+DH \end{bmatrix}$$

Example 1.28

Multiply out $\begin{bmatrix} 1 & 2 \\ 3 & 4 \end{bmatrix}\begin{bmatrix} 5 \\ 6 \end{bmatrix}$.

Solution

$$\begin{bmatrix} 1 & 2 \\ 3 & 4 \end{bmatrix} \times \begin{bmatrix} 5 \\ 6 \end{bmatrix} = \begin{bmatrix} 5+12 \\ 15+24 \end{bmatrix} = \begin{bmatrix} 17 \\ 39 \end{bmatrix} \quad (Ans.)$$

Example 1.29

Multiply $\begin{bmatrix} 1 & 2 \\ 3 & 4 \end{bmatrix} \begin{bmatrix} 5 & 6 \\ 7 & 8 \end{bmatrix}$. Show that if the order of multiplication is reversed a different result is obtained.

Solution

$$\begin{bmatrix} 1 & 2 \\ 3 & 4 \end{bmatrix} \begin{bmatrix} 5 & 6 \\ 7 & 8 \end{bmatrix} = \begin{bmatrix} 5+14 & 6+16 \\ 15+28 & 18+32 \end{bmatrix} = \begin{bmatrix} 19 & 22 \\ 43 & 50 \end{bmatrix} \qquad (Ans.)$$

Reversing the order of multiplication gives:

$$\begin{bmatrix} 5 & 6 \\ 7 & 8 \end{bmatrix} \begin{bmatrix} 1 & 2 \\ 3 & 4 \end{bmatrix} = \begin{bmatrix} 5+18 & 10+24 \\ 7+24 & 14+32 \end{bmatrix} = \begin{bmatrix} 23 & 34 \\ 31 & 46 \end{bmatrix}$$

Example 1.30

Multiply out $\begin{bmatrix} 1 & 2 & 3 \\ 4 & 5 & 6 \\ 7 & 8 & 9 \end{bmatrix} \begin{bmatrix} 2 \\ 2 \\ 2 \end{bmatrix} = \begin{bmatrix} 2+4+6 \\ 8+10+12 \\ 14+16+18 \end{bmatrix} = \begin{bmatrix} 12 \\ 30 \\ 48 \end{bmatrix} \qquad (Ans.)$

Example 1.31

Multiply out $\begin{bmatrix} 1 & 2 & 3 \\ 4 & 5 & 6 \\ 7 & 8 & 9 \end{bmatrix} \begin{bmatrix} 2 & 3 \\ 2 & 3 \\ 2 & 3 \end{bmatrix}$

$$= \begin{bmatrix} 2+4+6 & 3+6+9 \\ 8+10+12 & 12+15+18 \\ 14+16+18 & 21+24+27 \end{bmatrix} = \begin{bmatrix} 12 & 18 \\ 30 & 45 \\ 48 & 72 \end{bmatrix} \qquad (Ans.)$$

Division

Co-factors

The *co-factor* of an element in a square matrix is the determinant of the elements which remain after the column and row that contain that element are removed from the matrix, with the signs alternately positive and negative both horizontally and vertically. Thus for the 2 × 2 matrix $\begin{bmatrix} A & B \\ C & D \end{bmatrix}$, the co-factor of A is D, the co-factor of B is $-C$, of C is $-B$, and of D is A. The array of co-factors is $\begin{bmatrix} D & -C \\ -B & A \end{bmatrix}$.

For the 3×3 square matrix $\begin{bmatrix} A & B & C \\ D & E & F \\ G & H & I \end{bmatrix}$ the co-factors are:

for A, $\begin{vmatrix} E & F \\ H & I \end{vmatrix}$; for B, $-\begin{vmatrix} D & F \\ G & I \end{vmatrix}$; for C, $\begin{vmatrix} D & E \\ G & H \end{vmatrix}$;

for D, $-\begin{vmatrix} B & C \\ H & I \end{vmatrix}$; and so on.

Example 1.32

Determine the co-factor array for the matrix $\begin{bmatrix} 2 & 4 \\ 6 & 8 \end{bmatrix}$.

Solution

The array of co-factors is $\begin{bmatrix} 8 & -6 \\ -4 & 2 \end{bmatrix}$ (*Ans.*)

Example 1.33

Determine the array of co-factors for the matrix $\begin{bmatrix} 2 & 4 & 6 \\ 1 & 3 & 5 \\ 1 & 2 & 3 \end{bmatrix}$.

Solution

$$\begin{aligned}
\text{The co-factors are: upper } 2:&\quad +(3 \times 3 - 2 \times 5) = -1 \\
4:&\quad -(1 \times 3 - 1 \times 5) = +2 \\
6:&\quad +(1 \times 2 - 1 \times 3) = -1 \\
\text{upper } 1:&\quad -(4 \times 3 - 2 \times 6) = 0 \\
3:&\quad +(2 \times 3 - 6 \times 1) = 0 \\
5:&\quad -(2 \times 2 - 4 \times 1) = 0 \\
\text{lower } 1:&\quad +(4 \times 5 - 3 \times 6) = +2 \\
2:&\quad -(2 \times 5 - 6 \times 1) = -4 \\
3:&\quad +(2 \times 3 - 4 \times 1) = +2
\end{aligned}$$

Hence the array of co-factors is $\begin{bmatrix} -1 & 2 & -1 \\ 0 & 0 & 0 \\ 2 & -4 & 2 \end{bmatrix}$ (*Ans.*)

Inverse matrix

In the solution of an electric circuit a matrix equation of the form $[Y][V] = [I]$ is often obtained, where $[Y]$, $[V]$ and $[I]$ represent the admittance, voltage and current matrices respectively. If the $[Y]$ and $[V]$ matrices are known they can be multiplied together to obtain the current matrix $[I]$. If, however, the current and admittance matrices $[I]$ and $[Y]$ are known then $[V]$ must be found from the relationship

$$[V] = [Y]^{-1}[I]$$

where $[Y]^{-1}$ is the *inverse matrix* of $[Y]$.

The product of a square matrix and its inverse matrix is always equal to the unit matrix $\begin{bmatrix} 1 & 0 \\ 0 & 1 \end{bmatrix}$ for a 2×2 matrix or $\begin{bmatrix} 1 & 0 & 0 \\ 0 & 1 & 0 \\ 0 & 0 & 1 \end{bmatrix}$ for a 3×3 matrix.

There are two ways in which an inverse matrix may be determined. Consider the matrix $\begin{bmatrix} 10 & 5 \\ 6 & 2 \end{bmatrix}$.

Method 1

$$\begin{bmatrix} 10 & 5 \\ 6 & 2 \end{bmatrix} \begin{bmatrix} A & B \\ C & D \end{bmatrix} = \begin{bmatrix} 1 & 0 \\ 0 & 1 \end{bmatrix}$$

where $\begin{bmatrix} A & B \\ C & D \end{bmatrix}$ is the inverse matrix of $\begin{bmatrix} 10 & 5 \\ 6 & 2 \end{bmatrix}$.

Multiplying out,

$$\begin{bmatrix} 10A + 5C & 10B + 5D \\ 6A + 2C & 6B + 2D \end{bmatrix} = \begin{bmatrix} 1 & 0 \\ 0 & 1 \end{bmatrix}$$

Therefore,

$$10A + 5C = 1$$
$$6A + 2C = 0$$

giving $A = -0.2$ and $C = 0.6$. Also

$$10B + 5D = 0$$
$$6B + 2D = 1$$

giving $B = 0.5$ and $D = -1$. Hence the inverse matrix is

$$\begin{bmatrix} -0.2 & 0.5 \\ 0.6 & -1 \end{bmatrix}$$

Method 2

(*i*) Interchange the rows and columns of the matrix to obtain the *transpose matrix*. If the transpose matrix is identical to the original matrix then that matrix is *symmetrical*.

(*ii*) Replace each element in the transpose matrix by its co-factor to obtain the *adjoint matrix*.
(*iii*) Divide each element in the adjoint matrix by the determinant of the original matrix.

Applying each of these steps in turn to the matrix $\begin{bmatrix} 10 & 5 \\ 6 & 2 \end{bmatrix}$:

(*i*) Transpose matrix $= \begin{bmatrix} 10 & 6 \\ 5 & 2 \end{bmatrix}$

(*ii*) Adjoint matrix $= \begin{bmatrix} 2 & -5 \\ -6 & 10 \end{bmatrix}$

(*iii*) $\begin{vmatrix} 10 & 5 \\ 6 & 2 \end{vmatrix} = (10 \times 2) - (5 \times 6) = -10$

Therefore, as before, the inverse matrix is:

$$\frac{\begin{bmatrix} 2 & -5 \\ -6 & 10 \end{bmatrix}}{-10} = \begin{bmatrix} -0.2 & 0.5 \\ 0.6 & -1 \end{bmatrix}$$

Example 1.34

Determine the inverse matrix of $\begin{bmatrix} 2 & 2 & 3 \\ 4 & 5 & 6 \\ 7 & 8 & 9 \end{bmatrix}$.

Solution

(*i*) Transpose matrix $= \begin{bmatrix} 2 & 4 & 7 \\ 2 & 5 & 8 \\ 3 & 6 & 9 \end{bmatrix}$

(*ii*) The co-factors are (in order),

2 :	$+(5 \times 9 - 6 \times 8) = -3$	4 :	$-(2 \times 9 - 3 \times 8) = +6$
7 :	$+(2 \times 6 - 3 \times 5) = -3$	2 :	$-(4 \times 9 - 6 \times 7) = +6$
5 :	$+(2 \times 9 - 3 \times 7) = -3$	8 :	$-(2 \times 6 - 4 \times 3) = 0$
3 :	$+(4 \times 8 - 5 \times 7) = -3$	6 :	$-(2 \times 8 - 2 \times 7) = -2$
9 :	$+(2 \times 5 - 2 \times 4) = +2$		

Co-factor matrix $= \begin{bmatrix} -3 & 6 & -3 \\ 6 & -3 & 0 \\ -3 & -2 & 2 \end{bmatrix}$

(*iii*) Determinant of original matrix is:

$2(45 - 48) - 2(36 - 42) + 3(32 - 35) = -3$

Therefore inverse matrix is:

$$\frac{\begin{bmatrix} -3 & 6 & -3 \\ 6 & -3 & 0 \\ -3 & -2 & 2 \end{bmatrix}}{-3} = \begin{bmatrix} 1 & -2 & 1 \\ -2 & 1 & 0 \\ 1 & 2/3 & -2/3 \end{bmatrix} \quad (Ans.)$$

Example 1.35

Determine the inverse matrix of $\begin{bmatrix} 1 & 1+\mathrm{j}1 \\ 1+\mathrm{j}2 & -\mathrm{j}2 \end{bmatrix}$.

Solution

The transpose matrix is $\begin{bmatrix} 1 & 1+\mathrm{j}2 \\ 1+\mathrm{j}1 & -\mathrm{j}2 \end{bmatrix}$.

The co-factor matrix is $\begin{bmatrix} -\mathrm{j}2 & -1-\mathrm{j}1 \\ -1-\mathrm{j}2 & 1 \end{bmatrix}$.

The determinant of the original matrix is

$$-\mathrm{j}2-(1+\mathrm{j}1)(1+\mathrm{j}2) = -\mathrm{j}2-(1+\mathrm{j}2+\mathrm{j}1-2) = 1-\mathrm{j}5$$

Hence the inverse matrix is

$$\frac{\begin{bmatrix} -\mathrm{j}2 & -1-\mathrm{j}1 \\ -1-\mathrm{j}2 & 1 \end{bmatrix}}{1-\mathrm{j}5} = \begin{bmatrix} 0.38-\mathrm{j}0.08 & 0.15-\mathrm{j}0.23 \\ 0.35-\mathrm{j}0.27 & 0.04+\mathrm{j}0.19 \end{bmatrix} \quad (Ans.)$$

Exercises 1

1.1 Write down the real and imaginary parts of each of the following complex numbers:
(*a*) $60+\mathrm{j}20$, (*b*) 25, (*c*) $-\mathrm{j}500$, (*d*) $5\angle 60°$, (*e*) $2\angle -30°$, (*f*) $e^{\mathrm{j}2}$ and (*g*) $e^{-(10+\mathrm{j}2)}$.

1.2 Find the real and imaginary parts of each of the following complex numbers:
(*a*) j^3, (*b*) $1-\mathrm{j}$, (*c*) $10+\mathrm{j}^3 5$, (*d*) $2-\mathrm{j}^5 3$, (*e*) $\mathrm{j}^7-\mathrm{j}^6$, (*f*) $\mathrm{j}^4+\mathrm{j}^5$ and (*g*) $e^{-(1.5-\mathrm{j}3)}$.

1.3 Perform the following additions and give the answers in polar form:
(*a*) $(3+\mathrm{j}5)+(3+\mathrm{j}7)$, (*b*) $(3-\mathrm{j}5)+(-3+\mathrm{j}5)$,
(*c*) $(3+\mathrm{j}5)+(3-\mathrm{j}5)$, (*d*) $(3+\mathrm{j}5)+5\angle 50°$,
(*e*) $(3-\mathrm{j}5)+10\angle -100°$, (*f*) $20\angle 62°+15\angle -150°$ and
(*g*) $1\angle -110°+2\angle 280°$.

1.4 Carry out the following subtractions giving the results in polar form:
(*a*) $(-390+\mathrm{j}150)-(-200-\mathrm{j}50)$,
(*b*) $(0.04-\mathrm{j}0.02)-(0.02-\mathrm{j}0.02)$, (*c*) $14\angle 50°-18\angle -37°$,
(*d*) $106\angle -35°-106\angle -250°$, (*e*) $e^{-(1.5+\mathrm{j}1.5)}-e^{-(1.4-\mathrm{j}0.8)}$,
(*f*) $e^{-(0.5+\mathrm{j}0.6)}-22\angle 60°$ and (*g*) $1200\angle -30°-(1000-\mathrm{j}520)$.

1.5 Multiply out the following products and express the results in polar form: (*a*) $0.3\angle 10° \times 14\angle 96°$, (*b*) $(22+\text{j}10)(-16-\text{j}5)$, (*c*) $(5-\text{j}5)(6-\text{j}6)$, (*d*) $\text{e}^{-(0.4+\text{j}0.8)} \times \text{e}^{-(0.8+\text{j}0.4)}$, (*e*) $\text{e}^{-(1+\text{j}0.6)} \times 0.5\angle -35°$, (*f*) $0.607\angle -0.6$ rad $\times\ 22\angle 0.8$ rad and (*g*) $100\angle 10° \times (14-\text{j}6)$.

1.6 Perform the following divisions and give the answers in polar form: (*a*) $(12\angle -12°)/(6+\text{j}2.8)$, (*b*) $(140-\text{j}60)/(5+\text{j}10)$, (*c*) $(38\angle 93°)/(19\angle -90°)$, (*d*) $\text{e}^{\text{j}2}/\text{e}^{\text{j}}$, (*e*) $\text{e}^{-(0.2+\text{j}0.3)}/\text{e}^{-(0.18+\text{j}0.7)}$, (*f*) $(20\angle 15° \times 5\angle 30°)/(200\angle 45°)$ and (*g*) $(25-\text{j}40)/(40\angle 0°)$.

1.7 Evaluate:
(*a*) $\text{e}^{-(0.228+\text{j}0.225)}$, (*b*) $\sqrt{(40+\text{j}200)/[(4+\text{j}100) \times 10^{-3}]}$,
(*c*) $(6+\text{j}10)^2/4$,
(*d*) $\log_e[1+(2+\text{j}6)/2(3+\text{j}5)+(10+\text{j}8)/(3+\text{j}5)]$ and
(*e*) $\sqrt{[(40+\text{j}200)((4+\text{j}100) \times 10^{-3})]}$.

1.8 Evaluate the following equations if $A = 4+\text{j}10$, $B = 10-\text{j}20$ and $C = -5-\text{j}10$. Give the answers in polar form.
(*a*) $Z = 1/(A+B)$, (*b*) $Z = 1/A + 1/B$,
(*c*) $Z = \sqrt{[(A-B)/(B-C)]}$ and (*d*) $Z = \sqrt{[(AB)/(A+B+C)]}$.

1.9 Solve the following equations giving the results in polar form if $A = 50-\text{j}60$, $B = 60+\text{j}40$ and $C = 20+\text{j}10$.
(*a*) $Z = C + AB/(A+B)$, (*b*) $Z = AB/(A+B+C)$,
(*c*) $Z = (AB+BC+AC)/B$.

1.10 Evaluate:
(*a*) $Z = (AB+C)/(A+C)$, (*b*) $Z = (A/B)/(1/B+1/C)$ and
(*c*) $Z = (1/A+1/B)C$ if $A = 10-\text{j}16$, $B = 20+\text{j}20$ and $C = 50+\text{j}15$.

1.11 Calculate (*a*) the square and (*b*) the square root of each of the following complex numbers: (*i*) $-4-\text{j}6$, (*ii*) $4+\text{j}6$, (*iii*) $20\angle 32°$ and (*iv*) $\text{e}^{-(0.4+\text{j}0.2)}$.

1.12 Write down the phasor form of each of the following waveforms:
(*a*) $v_1 = 50\sin(\omega t + 35°)$ V, (*b*) $v_2 = 100\sin(\omega t - 60°)$ V,
(*c*) $v_3 = 5\cos(\omega t)$ V, (*d*) $v_4 = 10\cos(\omega t + 35°)$ V, and
(*e*) $v_5 = 40\cos(\omega t - 60°)$ V.

1.13 A current $i_1 = 15\sin(\omega t + 20°)$ mA flows into a circuit. It divides into two currents i_2 and i_3. If $i_2 = 8\sin(\omega t - 40°)$ mA, calculate i_3.

1.14 The current $i_1 = 25\sin(2000\pi t)$ mA flows into a circuit. The circuit splits into three branches. If two of the branches carry currents $i_2 = 15\sin(2000\pi t - 30°)$ mA and $i_3 = 10\sin(2000\pi t + 130°)$ mA, calculate the current i_4 in the third branch.

1.15 (*a*) Determine an expression for the instantaneous voltage applied to a circuit consisting of two impedances in series if the voltage drop across the impedances are (*i*) $v_1 = 12.5\sin(1000\pi t + 20°)$ V and (*ii*) $v_2 = 7.5\sin(1000\pi t - 60°)$ V.

1.16 Four voltage sources are connected in series. The sources are $v_1 = 20\sin \omega t$, $v_2 = 25\sin(\omega t - 60°)$, $v_3 = 30\sin(9\omega t + 90°)$ and $v_4 = 40\sin(\omega t + 150°)$. Obtain an expression for the resultant sum voltage.

1.17 Solve, using determinants, the equations $605x - 600y = 10$, $-600x + 2200y - 600z = 0$ and $-600y + 800z = 0$.

1.18 Solve, using determinants, the equations $5x + 10y - 4z = 8$, $-x + 7y + 2z = 0$ and $3x - 7y + 9z = 0$.

1.19 Solve, using matrix algebra, $\begin{bmatrix} 4 & 3 & 5 \\ 3 & 4 & 6 \end{bmatrix} \begin{bmatrix} 2 & 4 \\ 3 & 7 \\ 1 & 1 \end{bmatrix}$.

1.20 Multiply out $\begin{bmatrix} 2+j3 & -j6 \\ -2+j2 & 4+j2 \end{bmatrix} \begin{bmatrix} 4 & -j4 \\ 1-j & 2+j3 \end{bmatrix}$.

1.21 Solve for A and B $\begin{bmatrix} 10 & 2 \\ 4 & 6 \end{bmatrix} \begin{bmatrix} A \\ B \end{bmatrix} = \begin{bmatrix} 8 \\ 6 \end{bmatrix}$.

1.22 Multiply out $\begin{bmatrix} 1 & j \\ 2 & -j \end{bmatrix} \begin{bmatrix} 4 & j2 \\ -3 & -j \end{bmatrix}$.

2 A.C. circuits

When resistances, inductors and capacitors are connected in series and/or in parallel with one another to form an electric circuit the current flowing in the circuit, and the voltage across each component, can be determined using Ohm's law and complex algebra.

Components

Resistance

When a sinusoidal voltage $V \sin \omega t$ is applied across a linear resistance R the current that flows is given by Ohm's law:

$$i = (V/R) \sin \omega t = I \sin \omega t$$

The current is in phase with the voltage and so it does not possess an imaginary component. The resistance R of the circuit is

$$R = (V \sin \omega t)/(I \sin \omega t) = (V\angle 0°)/(I\angle 0°) = R + \mathrm{j}0 = R \qquad (2.1)$$

Inductance

The self-induced voltage in an inductance is equal to the inductance times the rate of change of the current: $V = -L(\mathrm{d}i/\mathrm{d}t)$. For a sinusoidal current $i = I \sin \omega t$ and $\mathrm{d}i/\mathrm{d}t = \omega I \cos \omega t = \omega I \sin(\omega t + 90°)$. The voltage across the inductor, $v = \omega L I \sin(\omega t + 90°)$ leads the current by 90°. The reactance X_L of the inductance is the ratio of the peak (or r.m.s.) values of the voltage and the current. Hence

$$X_L = [V \sin(\omega t + 90°)]/(I \sin \omega t) = (V\angle 90°)/(I\angle 0°) = \omega L\angle 90° = \mathrm{j}\omega L \qquad (2.2)$$

Clearly, inductive reactance is directly proportional to frequency.

Example 2.1

The voltage developed across a 80 mH inductance is $v = 20\sin(2000t + 60°)$ volts. Calculate the current flowing in the inductance.

Solution
The reactance of the inductance is $X_L = 2000 \times 80 \times 10^{-3} = \text{j}160\ \Omega$. Hence, the current flowing is

$$I = (20\angle 60°)/(160\angle 90°) = 0.125\angle{-30°}\ \text{A} \quad (\textit{Ans.})$$

Or $i = 125\sin(2000t - 30°)$ mA (*Ans.*)

Example 2.2

When a 240 V, 50 Hz sinusoidal voltage is applied across an inductance the current which flows is to be no larger than 2 A. Calculate the minimum value for the inductance.

Solution
Minimum inductive reactance $X_{L(min)} = 240/2 = 120\ \Omega$. Therefore,

$$L_{min} = 120/(2\pi \times 50) = 382\ \text{mH} \quad (\textit{Ans.})$$

Capacitance

When a sinusoidal voltage $v = V\sin\omega t$ is applied across a capacitance C a charge $q = CV\sin\omega t$ will be supplied to the capacitance. Current is the rate of change of charge and so $i = \text{d}q/\text{d}t = \omega CV\cos\omega t = \omega CV\sin(\omega t + 90°)$. The reactance X_C of the capacitance is the ratio of the peak (or r.m.s.) values of the voltage and the current, and hence,

$$\begin{aligned} X_C &= V/I = (V\angle 0°)/(\omega CV\angle 90°) = (1/\omega C)\angle 90° \\ &= 1/\text{j}\omega C = -\text{j}/\omega C \end{aligned} \qquad (2.3)$$

This means that capacitive reactance is inversely proportional to frequency.

Example 2.3

The current flowing through a 0.22 μF capacitor is $i = 5\sin(2000t + 18°)$ mA. Determine the voltage across the capacitor.

Solution

$$X_C = -\text{j}/\omega C = -\text{j}(2000 \times 0.22 \times 10^{-6}) = -\text{j}2273\ \Omega$$

$$V_C = IX_C = 5\angle 18° \times 2273\angle -90° = 11.37\angle -72°\ \text{V} \quad (\textit{Ans.})$$

Or $v_C = 11.37\sin(2000t - 72°)$ V (*Ans.*)

Series Circuits

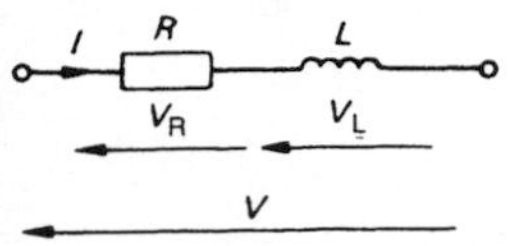

Fig. 2.1 Resistance and inductance in series

Resistance and inductance in series

Referring to Fig. 2.1 $V_R = IR$ and $V_L = Ij\omega L$. The applied voltage V is the phasor sum of V_R and V_L and is written as $V = V_R + jV_L$, $V_R = IR$ and $V_L = Ij\omega L$ and hence $V = I(R + j\omega L)$.

The impedance Z of the circuit is the ratio of the applied voltage to the circuit current,

$$Z = V/I = R + j\omega L \tag{2.4}$$

$$= \sqrt{(R^2 + \omega^2 L^2)} \angle \tan^{-1}(\omega L/R) \tag{2.5}$$

Since $I = V/Z$, the current flowing in the circuit lags the applied voltage by angle $\theta = \tan^{-1}(\omega L/R)$.

Example 2.4

Calculate the resistive and inductive components of the circuit whose impedance at 796 Hz is $1000 + j1000\ \Omega$.

Solution

$$R = 1000\ \Omega \quad (Ans.)$$

$$X_L = 1000\ \Omega, \text{ so } L = 1000/(2\pi \times 796) = 0.2\ \text{H} \quad (Ans.)$$

Example 2.5

A current of 500 mA flows in an inductor of 60 mH inductance and 10 Ω resistance. Calculate the voltage across the inductor at a frequency of 1000 Hz.

Solution

The reactance of the inductor is

$$X_L = 2\pi \times 1000 \times 60 \times 10^{-3} = 377\ \Omega$$

The impedance of the circuit is

$$Z = 10 + j377\ \Omega$$

Voltage across inductor is

$$IZ = 0.5(10 + j377) = 5 + j188.5$$

$$= 188.56 \angle 88.5°\ \text{V} \quad (Ans)$$

Example 2.6

A circuit with a resistance of 330 Ω and an inductance of 0.1 H has a voltage $v = 20 \sin(2000t + 20°)$ volts applied across it. Calculate (*a*) the impedance of the circuit, (*b*) the current that flows, (*c*) the voltage across each component and (*d*) the r.m.s. voltages in the circuit.

Solution

(*a*) $X_L = j2000 \times 0.1 = j200\ \Omega$

$Z = 330 + j200 = \sqrt{(330^2 + 200^2)} \angle \tan^{-1}(200/330)$
$= 385.9 \angle 31.2°$ (*Ans.*)

(*b*) $I = (20 \angle 20°)/(385.9 \angle 31.2°) = 51.83 \angle -11.2°$ mA (*Ans.*)

Or $i = 51.83 \sin(2000t - 11.2°)$ mA (*Ans.*)

(*c*) Voltage across resistance: $v_R = IR = (51.83 \times 10^{-3}) \angle -11.2° \times 330$

$= 17.1 \angle -11.2°$ V (*Ans.*)

Or $v_R = 17.1 \sin(2000t - 11.2°)$ V (*Ans.*)

Voltage across inductance: $V_L = I j X_L$

$= (51.83 \times 10^{-3}) \angle -11.2° \times 200 \angle 90°$

$= 10.37 \angle 78.8°$ V (*Ans.*)

Or $v_L = 10.37 \sin(2000t + 78.8°)$ V (*Ans.*)

(*d*) $V = 20/\sqrt{2} = 14.14$ V, $V_R = 17.1/\sqrt{2} = 12.1$ V,

$V_L = 10.37/\sqrt{2} = 7.33$ V (*Ans.*)

Example 2.7

Determine the resistance required in series with a 100 mH inductor to cause the current to lag the applied voltage by 20° at a frequency of 318 Hz.

Solution

(*a*) $X_L = \omega L = 2\pi \times 318 = 1998\ \Omega$

$Z = R + j1998 = \sqrt{(R^2 + 1998^2)} \angle \tan^{-1}(1998/R)$

$1998/R = \tan 20° = 0.364$

$R = 1998/0.364 = 5489\ \Omega$ (*Ans.*)

Example 2.8

An inductive reactance of 2000 Ω is connected in series with a 1000 Ω resistor and a $15 \angle 0°$ V voltage is applied across the circuit. Calculate (*a*) the current that flows into the circuit and (*b*) the voltage dropped across each component. (*c*) Verify Kirchhoff's voltage law. (*d*) Give expressions for the instantaneous value of each voltage in the circuit. [Note: Kirchhoff's laws are stated in Chapter 5.]

Solution

(*a*) Impedance of circuit: $Z = 1000 + j2000\ \Omega$

Current: $I = V/Z = 15/(1000 + j2000) = 15/(2236 \angle 63.4°)$

$= 6.71 \angle -63.4°$ mA (*Ans.*)

(*b*) $V_R = IR = 6.71 \times 10^{-3} \angle -63.4° \times 1000 \angle 0°$

$= 6.71 \angle -63.4°$ V (*Ans.*)

$V_L = IX_L = 6.71 \times 10^{-3} \angle{-63.4°} \times 2000 \angle 90°$
$= 13.42 \angle 26.6°$ V (*Ans.*)

(*c*) Applied voltage: $V = V_R + V_L = 6.71 \angle -63.4° + 13.42 \angle 26.6°$
$= (3 - j6) + (12 + j6) = 15 + j0$
$= 15 \angle 0°$ V (*Ans.*)

(*d*) $v = 15 \sin \omega t$ V, $v_R = 6.71 \sin(\omega t - 63.4°)$ V
$V_L = 13.42 \sin(\omega t + 26.6°)$ V (*Ans.*)

Resistance and capacitance in series

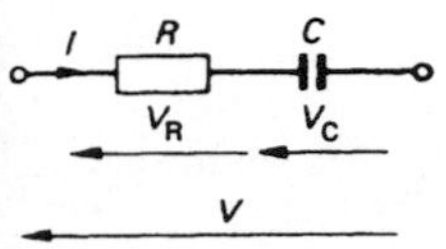

Fig. 2.2 Resistance and capacitance in series

From Fig. 2.2 the applied voltage is equal to the phasor sum of the voltages across R and C. Hence,

$$V = V_R + V_C = I(R + 1/j\omega C)$$
$$Z = V/I = R + 1/j\omega C$$
$$= \sqrt{(R^2 + 1/\omega^2 C^2)} \angle \tan^{-1}(1/\omega CR) \qquad (2.6)$$

Example 2.9

A circuit has an impedance of $560 - j220\ \Omega$. Calculate (*a*) its resistance and capacitive reactance and (*b*) the magnitude and phase angle of the impedance.

Solution
(*a*) $R = 560\ \Omega$, $X_C = 220 \angle{-90°}\ \Omega$ (*Ans.*)
(*b*) $Z = \sqrt{(560^2 + 220^2)} \angle \tan^{-1}(-220/560) = 601.7 \angle{-21.5°}$ (*Ans.*)

Example 2.10

The voltage $v = 12 \sin(314t + 45°)$ volts is applied across a circuit that consists of a 3.3 kΩ resistor connected in series with a 2.2 μF capacitor. Calculate (*a*) the impedance of the circuit, (*b*) the current flowing in the circuit, (*c*) the voltage across each component, (*d*) verify Kirchhoff's voltage law and (*e*) give expressions for the instantaneous voltages across each component.

Solution
(*a*) $X_C = -j/\omega C = -j/(314 \times 2.2 \times 10^{-6}) = -j1448\ \Omega$
Impedance of circuit: $Z = \sqrt{(3300^2 + 1448^2)} \angle \tan^{-1}(-1448/3300)$
$= 3603.7 \angle{-23.7°}$ (*Ans.*)

(*b*) Current: $I = V/Z = (12 \angle 45°)/(3603.7 \angle{-23.7°})$
$= 3.33 \angle 68.7°$ mA (*Ans.*)

(*c*) $V_R = 3.33 \times 10^{-3} \angle 68.7° \times 3.3 \times 10^3 = 11 \angle 68.7°$ V (*Ans.*)
$V_C = 3.33 \times 10^{-3} \angle 68.7° \times 1448 \angle{-90°}$
$= 4.82 \angle{-21.3°}$ V (*Ans.*)

(*d*) $V = V_R + V_C = 11\angle 68.7° + 4.82\angle -21.3°$
$= (4 + j10.25) + (4.49 - j1.75)$
$= 8.49 + j8.5 = 12\angle 45°$ V (*Ans.*)

(*e*) $v_R = 11\sin(314t + 68.7°)$ V (*Ans.*)
$v_C = 4.82\sin(314t - 21.3°)$ V (*Ans.*)

Series RLC circuit

When a voltage is applied to a series *RLC* circuit the phasor sum of the voltages across the three components is equal to the applied voltage. Thus,

$$V = V_R + V_L + V_C = IR + IX_L + IX_C$$
$$= I(R + X_L + X_C) = I(R + \omega L - j/\omega C)$$

The impedance Z of the circuit is

$$Z = V/I = R + j(\omega L - 1/\omega C) \quad (2.7)$$
$$= \sqrt{[R^2 + (\omega L - 1/\omega C)^2]}\angle \tan^{-1}[(\omega L - 1/\omega C)/R] \quad (2.8)$$

Example 2.11

Calculate the voltage applied across the circuit given in Fig. 2.3 if, at a frequency of 2000 Hz, 2 V are dropped across the 1000 Ω resistance.

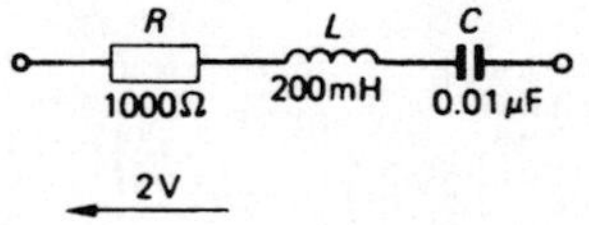

Fig. 2.3

Solution
The current flowing in the circuit is 2 / 1000 = 2 mA.

$X_L = 2\pi \times 2 \times 10^3 \times 200 \times 10^{-3} = j2513\ \Omega$
$X_C = 1/(2\pi \times 2 \times 10^3 \times 10^{-8}) = -j7958\ \Omega$
Impedance of circuit: $Z = 1000 + j(2513 - 7958) = 1000 - j5445$
$= 5536\angle 79.6°\ \Omega$

$V = IZ = 2 \times 10^{-3} \times 5536\angle 79.6° = 11\angle -79.6°$ V (*Ans.*)

(Angle relative to the current).

Example 2.12

A 1 kΩ resistor is connected in series with a 100 mH inductance and a 1 μF capacitor. If the applied voltage is $5\sin(4000t)$ V calculate (*a*) the impedance of the circuit and (*b*) the current flowing. (*c*) Verify Kirchhoff's voltage law.

Solution
(*a*) $X_L = 4000 \times 100 \times 10^{-3} = j400\ \Omega$
$X_C = 1/(4000 \times 1 \times 10^{-6}) = -j250\ \Omega$

Impedance: $Z = 1000 + j400 - j250 = 1000 + j150\ \Omega$

$= \sqrt{(1000^2 + 150^2)}\ \angle \tan^{-1}(150/1000)$

$= 1011.2\angle 8.5°\ \Omega$ (*Ans.*)

(*b*) $I = V/Z = 5/(1011.2\angle 8.5°) = 4.95\angle -8.5°$ mA (*Ans.*)

Or $i = 4.95 \sin(4000t - 8.5°)$ mA (*Ans.*)

(*c*) $V = IR + IX_L + IX_C$

$= 4.95 \times 10^{-3}\angle -8.5° \times 1000 + 4.95 \times 10^{-3}\angle -8.5°$

$\times 400\angle 90° + 4.95 \times 10^{-3}\angle -8.5° \times 250\angle -90°$

$= 4.95\angle -8.5° + 1.98\angle 81.5° + 1.238\angle -98.5°$

$= (4.90 - j0.73) + (0.29 + j1.96) + (-0.18 - j1.22)$

$\simeq 5 + j0 = 5\angle 0°$ V (*Ans.*)

Example 2.13

An inductive reactance of 100 Ω, a capacitive reactance of 80 Ω and a resistance R are connected in series and the voltage $120\angle 0°$ V is applied across the circuit. The output voltage of the circuit is to be taken from the resistor R. Calculate (*a*) the required value of R for the output voltage to be 80 V, (*b*) the phase angle then between the input and output voltages of the circuit.

Solution

(*a*) Impedance of circuit: $Z = R + j100 - j80 = R + j20\ \Omega$

$I = 120/(R + j20), \quad |I| = 120/\sqrt{(R^2 + 20^2)}$

$V_{out} = |I|R = 80 = 120R/\sqrt{(R^2 + 20^2)}$, or

$120/80 = 1.5 = \sqrt{(R^2 + 20^2)}/R$

Squaring both sides of the equation,

$2.25 = (R^2 + 400)/R^2$ and $R = \sqrt{(400/1.25)} = 17.9\ \Omega$ (*Ans.*)

(*b*) $Z = 17.9 + j20\ \Omega, \quad I = 120/(17.9 + j20)$

$V_{out} = (120 \times 17.9)/(17.9 + j20)$

$= 2148/\sqrt{(17.9^2 + 20^2)}\ \angle \tan^{-1}(-20/17.9)$

$= 80\angle -48.2°$ V

Hence,

Phase angle $= -48.2°$ (*Ans.*)

Example 2.14

$240\angle 0°$ volts at 50 Hz are applied across the series connection of a 15 Ω resistor, a 100 mH inductor, and a 220 μF capacitor. Calculate the voltage across the capacitor.

Solution

$$X_L = 2\pi \times 100 \times 10^{-3} \times 50 = 31.4\ \Omega$$
$$X_C = 1/(2\pi \times 220 \times 10^{-6} \times 50) = 14.5\ \Omega$$
$$Z = 15 + j(31.4 - 14.5) = 15 + j16.9 = 22.6\angle 48.4^\circ\ \Omega$$
$$I = (240\angle 0^\circ)/(22.6\angle 48.4^\circ) = 10.62\angle -48.4^\circ\ \text{A}$$
$$V_C = IX_C = 10.62\angle -48.4^\circ \times 14.5\angle -90^\circ$$
$$= 154\angle -138.4^\circ\ \text{V} \quad (\textit{Ans.})$$

Alternatively,

$$V_C = (240 \times 14.5\angle -90^\circ)/(22.6\angle 48.4^\circ) = 154\angle -138.4\ \text{V} \quad (\textit{Ans.})$$

Parallel circuits

Resistance and inductance in parallel

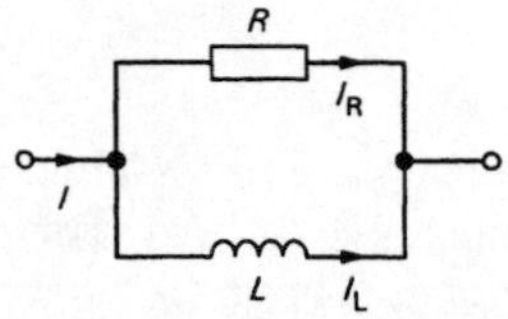

Fig. 2.4 Resistance and inductance in parallel

When a resistance R is connected in parallel with a pure inductance L across a sinusoidal voltage of V volts, the current I that flows into the circuit is equal to the phasor sum of the currents I_R and I_L that flow in the two components, see Fig. 2.4.

$$I = I_R + I_L = V/R + V/j\omega L = V/R - jV/\omega L$$

The impedance Z of the circuit is

$$Z = V/I = 1/R - j/\omega L = (\omega L - jR)/\omega LR \qquad (2.9)$$

Example 2.15

A 120 Ω resistor is connected in parallel with a 30 mH pure inductor. The circuit is supplied with the current $i = 1\sin(2\pi \times 10^3 t)$ mA. Calculate the current flowing in each component and verify Kirchhoff's current law.

Solution

$$X_L = 2\pi \times 10^3 \times 30 \times 10^{-3} = 188.5\ \Omega$$
$$I_R = (1 \times j188.5)/(120 + j188.5) = (188.5\angle 90^\circ)/(223.46\angle 57.5^\circ)$$
$$= 0.84\angle 32.5^\circ\ \text{mA} \quad (\textit{Ans.})$$
$$I_L = (1 \times 120)/(120 + j188.5) = (120\angle 0^\circ)/(223.46\angle 57.5^\circ)$$
$$= 0.54\angle -57.5^\circ\ \text{mA} \quad (\textit{Ans.})$$
$$I_R + I_L = 0.84\angle 32.5^\circ + 0.54\angle -57.5^\circ$$
$$= (0.71 + j0.46) + (0.29 - j0.46) = 1\ \text{mA} \quad (\textit{Ans.})$$

Resistance and capacitance in parallel

The total current I supplied to the parallel combination of a resistance R and a capacitance C in parallel (Fig. 2.5) is the

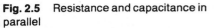

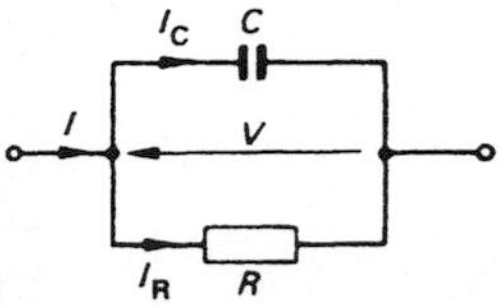

Fig. 2.5 Resistance and capacitance in parallel

phasor sum of the currents I_C and I_R flowing in each component. Since the components are connected in parallel they will have the same voltage developed across them and so the voltage is taken as the reference. Therefore

$$I = I_R + I_C = V/R + [V/(1/j\omega C)] = V/R + jV\omega C$$

The impedance Z of the circuit is the ratio voltage/current and hence,

$$Z = V/I = 1/[(1/R) + j\omega C] = R/(1 + j\omega CR) \qquad (2.10)$$

Example 2.16

A 130 Ω resistor is connected in parallel with a capacitor of reactance 250 Ω at the frequency of the voltage source. The current flowing into the network is 20 mA. Calculate the current in each component.

Solution

$$I = I_R + I_C = V/R + jV\omega C$$
$$Y = I/V = 1/Z = 1/R + j\omega C = 1/130 + j/250$$
$$= (7.69 + j4) \times 10^{-3}\ S$$

Hence,

$$Z = 102.33 - j53.21 = \sqrt{(102.33^2 + 53.21^2)}$$
$$\angle \tan^{-1}(-53.21/102.33)$$
$$= 115.37\angle -27.5°\ \Omega$$
$$V = IZ = 20 \times 10^{-3} \times 115.37\angle -27.5° = 2.31\angle -27.5°\ \text{V}$$
$$I_R = V/R = (2.31\angle -27.5°)/130 = 17.77\angle -27.5\ \text{mA} \quad (\textit{Ans.})$$
$$I_C = V/X_C = (2.31\angle -27.5°)/(250\angle -90°)$$
$$= 9.24\angle 62.5°\ \text{mA} \quad (\textit{Ans.})$$

Alternatively, from equation (2.10),

$$Z = 130/(1 + j130/250) = (130\angle 0°)/(1.13\angle 27.5°)$$
$$= 115.04\angle -27.5°\ \Omega$$
$$I_C = (20 \times 130)/(130 - j250) = (2600\angle 0°)/(281.78\angle -62.53°)$$
$$= 9.23\angle 62.5°\ \text{mA} \quad (\textit{Ans.})$$
$$I_R = 20\angle 0° - 9.23\angle 62.5° = 20 - (4.26 + j8.19) = (15.74 - j8.19)$$
$$= 17.74\angle -27.5°\ \text{mA} \quad (\textit{Ans.})$$

Example 2.17

A 100 Ω resistor is connected in parallel with a 2 nF capacitor and the circuit is connected to a 6 V supply at a frequency of $\omega = 5 \times 10^6$ radians. Calculate (*a*) the impedance of the circuit, (*b*) the current supplied to the circuit, (*c*) the capacitor current, and (*d*) the current in the resistor.

Solution

(*a*) $Z = 100/(1 + j5 \times 10^6 \times 2 \times 10^{-9} \times 100) = 100/(1 + j)$
$= 70.7\angle{-45°}\ \Omega$ (*Ans.*)

(*b*) Supply current: $V/Z = 6/(70.7\angle{-45°})$
$= 84.87\angle 45°$ mA (*Ans.*)

(*c*) $I_C = V/X_C = V\omega C = 6 \times 5 \times 10^6 \times 2 \times 10^{-9}$
$= 60\angle 90°$ mA (*Ans.*)

(*d*) $I_R = V/R = 6/100 = 60\angle 0°$ mA (*Ans.*)

Admittance, conductance and susceptance

Admittance is the reciprocal of impedance. The input admittance of a circuit is equal to the ratio (current flowing into circuit)/(applied voltage). The unit of admittance is the *siemen* S and its symbol is Y. Thus,

$$Y = I/V = 1/Z \qquad (2.11)$$

Usually, the admittance of a circuit is a complex quantity. Its real part is known as the *conductance* G and its imaginary part is known as the *susceptance* B. Hence,

$$Y = G \pm jB \qquad (2.12)$$

Resistance

For a pure resistance $Z = R + j0$ and $Y = 1/Z = 1/(R + j0) = 1/R$. This means that the conductance of a resistance is $G = 1/R$ while its susceptance $B = 0$.

Inductance

For a pure inductance $Z = 0 + jX_L$ and $Y = 1/Z = 1/(0 + jX_L) = -j/X_L$. Thus the conductance of a pure inductor is zero and its susceptance $B_L = -j/X_L = 1/j\omega L$.

Capacitance

For a pure capacitance $Z = 0 - jX_C$ and $Y = 1/Z = 1/(0 - jX_C) = j/X_C$. Now the conductance of a capacitor is zero and its susceptance $B_C = j/X_C = j\omega C$.

Each susceptance has the opposite sign to its corresponding reactance, i.e. inductive susceptance is negative and capacitive susceptance is positive. When a voltage V is applied across an admittance Y the current that flows is $I = VY$.

Example 2.18

Determine (*a*) the conductance of a 1 kΩ resistor, (*b*) the susceptance of (*i*) a 10 mH pure inductance, (*ii*) a 1 μF capacitor, at a frequency of 796 Hz.

Solution

(*a*) $G = 1/1000 - 1$ mS (*Ans.*)

(*b*) $X_L = 2\pi \times 796 \times 10 \times 10^{-3} = j50\ \Omega$

$B_L = 1/j50 = -j20$ mS (*Ans.*)

(*c*) $X_C = 1/(2\pi \times 796 \times 1 \times 10^{-6}) = -j200\ \Omega$

$B_C = 1/-j200 = j5$ mS (*Ans.*)

Series Circuit

The impedance of a reactance X connected in series with a resistor R is $Z = R \pm jX$. The admittance of the circuit is

$$Y = 1/Z = 1/(R \pm jX) = (R \pm jX)/(R^2 + X^2) \qquad (2.13)$$

The conductance G is the real part of the admittance, i.e.

$$G = R/(R^2 + X^2) \qquad (2.14)$$

[Note that G is *not* equal to the reciprocal of R.]

The susceptance B is given by the imaginary part of the admittance, i.e.

$$B = \mp jX/(R^2 + X^2) \qquad (2.15)$$

Example 2.19

A 1 kΩ resistor is connected in series with an inductive reactance of 600 Ω. Calculate (*a*) the impedance, (*b*) the admittance, (*c*) the conductance, and (*d*) the susceptance of the circuit.

Solution

(*a*) $Z = 1000 + j600 = \sqrt{(1000^2 + 600^2)} \angle \tan^{-1}(600/1000)$

$= 1166\angle 31°\ \Omega$ (*Ans.*)

(*b*) $Y = 1/Z = 1/(1166\angle 31°) = 8.58 \times 10^{-4}\angle -31° = 858\ \mu S$ (*Ans.*)

(*c*) $G = 8.58 \cos(-31°) = 7.35 \times 10^{-4} = 735\ \mu S$ (*Ans.*)

(*d*) $B = 8.58 \sin(-31°) = -4.42 \times 10^{-4}$ S $= -442\ \mu S$ (*Ans.*)

[Note: $1/R = 1/1000 = 10 \times 10^{-4} \neq G$.]

Parallel Circuits

When a resistance R is connected in parallel with a reactance X the impedance of the circuit is $Z = \pm jXR/(R \pm jX)$ and the admittance is

$$Y = (R \pm jX)/(\pm jXR) = 1/R \pm j(1/X) \qquad (2.16)$$

Now the conductance, again the real part of the admittance, is equal to the reciprocal of the resistance R, i.e.

$$G = 1/R \qquad (2.17)$$

The susceptance B is equal to the reciprocal of the reactance with the opposite sign, i.e.

$$B = \mp j(1/X) \qquad (2.18)$$

Example 2.20

An admittance $Y = 20 + j50$ mS consists of two components connected (*a*) in series and (*b*) in parallel. Calculate the values of the components if the frequency is 1 kHz.

Solution

(*a*) Impedance is:

$$\begin{aligned} Z &= 1/Y = 1/[(20 + j50) \times 10^{-3}] \\ &= [(20 - j50) \times 10^{-3}]/[(20^2 + 50^2) \times 10^{-6}] \\ &= (20 - j50)/2.9 = 6.9 - j17.24\ \Omega \end{aligned}$$

Therefore,

$R_s = 6.9\ \Omega$ (*Ans.*)

$C_s = 1/(2\pi \times 1000 \times 17.24) = 9.23\ \mu F$ (*Ans.*)

(*b*) $R_p = 1/G_p = 1/(20 \times 10^{-3}) = 50\ \Omega$ (*Ans.*)

$X_p = 1/B_p = 1/(j50 \times 10^{-3}) = -j20\ \Omega$

$C_p = 1/(2\pi \times 1000 \times 20) = 7.96\ \mu F$ (*Ans.*)

Example 2.21

A 39 kΩ resistor is connected in parallel with a 4.7 nF capacitor. Calculate (*a*) the impedance, (*b*) the admittance, (*c*) the conductance and (*d*) the susceptance of the circuit at 1 kHz.

Solution

(*a*) $Z = jX_C R/(R + jX_C) = R/(1 + j\omega CR) = (39 \times 10^3)/(1 + j1.152)$

$= 16759 - j19306 = 25565\angle{-49^\circ}\ \Omega$ (*Ans.*)

(*b*) $Y = 1/Z = 1/(25565\angle 49^\circ) = 39.12\angle{+49^\circ}\ \mu S$ (*Ans.*)

(*c*) $G = 39.12 \times 10^{-6} \cos(49^\circ) = 25.67\ \mu S$ (*Ans.*)

(*d*) $B = 39.12 \times 10^{-6} \sin(49^\circ) = 29.52\ \mu S$ (*Ans.*)

Alternatively,

(*c*) $G = 1/(39 \times 10^3) = 25.64\ \mu S$ (*Ans.*)

(*d*) $B = j2\pi \times 1000 \times 4.7 \times 10^{-9} = 29.53\ \mu S$ (*Ans.*)

(*b*) $Y = G + jB = (25.64 + j29.53) \times 10^{-6}\ S = 39.11\angle 49^\circ\ \mu S$ (*Ans.*)

(*a*) $Z = 1/Y = 1/(39.11\angle 49^\circ) = 25570\angle -49^\circ\ \Omega$ (*Ans.*)

Example 2.22

An inductive reactance of 10 Ω, a 20 Ω resistor, and a capacitive reactance of 40 Ω are connected in parallel. Calculate (*a*) the admittance, conductance, and susceptance, (*b*) the impedance, resistance and reactance of the circuit and (*c*) the input current and the current in each component when the applied voltage is $10\angle 0^\circ$ V.

Solution

(*a*) $Y = 1/20 + 1/j10 + 1/-j40 = 0.05 - j0.1 + j0.025$

$= 0.05 - j0.075 = 0.09\angle -56.3^\circ\ S$ (*Ans.*)

$G = 0.09\cos(-56.3^\circ) = 0.05\ S$ (*Ans.*)

$B = 0.09\sin(-56.3^\circ) = -j0.075\ S$ (*Ans.*)

(*b*) $Z = 1/Y = 1/(0.09\angle -56.3^\circ) = 11.11\angle -56.3^\circ\ \Omega$ (*Ans.*)

$R = 11.11\cos 56.3^\circ = 6.16\ \Omega$ (*Ans.*)

$X = 11.11\sin 56.3^\circ = j9.24\ \Omega$ (*Ans.*)

(*c*) $I = YV = 10\angle 0^\circ \times 0.09\angle -56.3^\circ = 0.9\angle -56.3^\circ\ A$ (*Ans.*)

$I_L = B_L V = [(1/10)\angle 90^\circ] \times 10\angle 0^\circ = 1\angle -90^\circ\ A$ (*Ans.*)

$I_R = GV = (1/20) \times 10 = 0.5\angle 0^\circ\ A$ (*Ans.*)

$I_C = B_C V = [(1/40)\angle 90^\circ] \times 10\angle 0^\circ = 0.25\angle 90^\circ\ A$ (*Ans.*)

Example 2.23

For the circuit shown in Fig. 2.6 calculate (*a*) the admittance, (*b*) the conductance, (*c*) the susceptance, (*d*) the impedance, (*e*) the resistance, and (*f*) the reactance.

Solution

(*a*) $Y_A = 1/(56 - j50) = 9.94 + j8.87\ mS$

$Y_B = 1/(56 + j50) = 9.94 - j8.87\ mS$

$Y_C = 1/(22 + j20) = 24.88 - j22.62\ mS$

$Y_D = 1/100 = 10 + j0\ mS$

$Y_E = 1/-j100 = 0 + j10\ mS$

$Y_T = Y_A + Y_B + Y_C + Y_D + Y_E$

$= 54.76 - j12.62 = 56.2\angle 13^\circ\ mS$ (*Ans.*)

(*b*) $G = 56.2\cos 13^\circ = 54.76\ mS$ (*Ans.*)

(*c*) $B = 56.2\sin 13^\circ = -j12.62\ mS$ (*Ans.*)

(*d*) $Z = 1/Y_T = 1/(56.2 \times 10^{-3}\angle 13^\circ) = 17.79\angle -13^\circ\ \Omega$ (*Ans.*)

(*e*) $R = 17.79\cos(-13^\circ) = 17.33\ \Omega$ (*Ans.*)

(*f*) $X = 17.79\sin(-13^\circ) = -j4\ \Omega$ (*Ans.*)

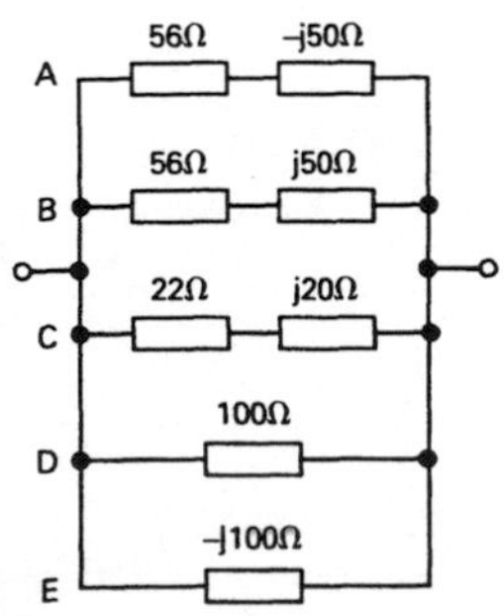

Fig. 2.6

Equivalent series and parallel circuits

It is often convenient in design or analysis to be able to replace, on paper, a series circuit with its equivalent parallel circuit or, alternatively, replace a parallel circuit with its equivalent series circuit. Two circuits are said to be equivalent if, at one particular frequency, their impedances (or admittances) are equal to one another in both magnitude and angle.

Series-to-parallel conversion

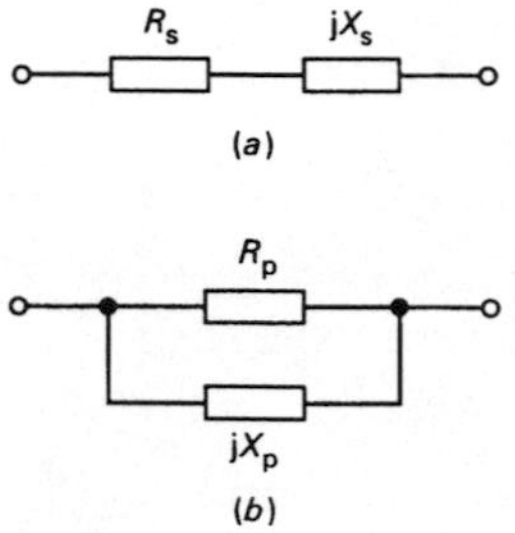

Fig. 2.7

Figure 2.7(*a*) shows a resistance R_s connected in series with a reactance jX_s (X_s may be either an inductive or a capacitive reactance). The impedance of the series circuit is $Z_s = R_s + jX_s$, and hence its admittance is

$$Y_s = 1/Z_s = 1/(R_s + jX_s) = (R_s - jX_s)/(R_s^2 + X_s^2)$$
$$= R_s/(R_s^2 + X_s^2) - jX_s/(R_s^2 + X_s^2)$$

The first term represents the conductance G of a resistance $R_p = R_s/(R_s^2 + X_s^2)$ which is connected in parallel with a susceptance $B_p = -jX_s/(R_s^2 + X_s^2)$. Therefore the equivalent parallel resistance is

$$R_p = (R_s^2 + X_s^2)/R_s \qquad (2.19)$$

and the equivalent parallel reactance X_p is

$$X_p = (R_s^2 + X_s^2)/X_s \qquad (2.20)$$

Example 2.24

Determine the equivalent parallel circuit for a 10 Ω resistance connected in series with a 15.92 μH inductance (*a*) at 100 kHz and (*b*) at 10 MHz.

Solution

(*a*) $\omega L_s = 2\pi \times 100 \times 10^3 \times 15.92 \times 10^{-6} = 10\ \Omega$

$R_p = (10^2 + 10^2)/10 = 20\ \Omega$ (*Ans.*)

$X_p = j(10^2 + 10^2)/10 = j20\ \Omega$

$L_p = 20/(2\pi \times 100 \times 10^3) = 31.8\ \mu H$ (*Ans.*)

(*b*) $\omega L_s = 10(10^7/10^5) = 1000\ \Omega$

$R_p = (10^2 + 1000^2)/10 = 100.01\ k\Omega$ (*Ans.*)

$X_p = j(10^2 + 1000^2)/1000 = j1000\ \Omega$

$L_p = 1000/(2\pi \times 10 \times 10^6) = 15.92\ \mu H$ (*Ans.*)

Clearly, as the frequency increases L_p approaches the same value, as L_s and R_p gets very large.

Parallel-to-series conversion

The impedance of the circuit shown in Fig. 2.7(b) is

$$\begin{aligned} Z_p &= jR_pX_p/(R_p + jX_p) \\ &= [jR_pX_p(R_p - jX_p)]/(R_p^2 + X_p^2) \\ &= R_pX_p^2/(R_p^2 + X_p^2) + jR_p^2X_p/(R_p^2 + X_p^2) \end{aligned} \tag{2.21}$$

The equivalent series resistance, R_s, is

$$\begin{aligned} R_pX_p^2/(R_p^2 + X_p^2) &= (X_p^2/R_p)/[1 + (X_p^2/R_p^2) \\ &= (1/R_p)/[(1/X_p^2) + (1/R_p^2) \end{aligned} \tag{2.22}$$

or

$$R_s = G_p/(G_p^2 + B_p^2) \tag{2.23}$$

The equivalent series reactance, X_s is the imaginary part of equation (2.21), i.e.

$$\begin{aligned} X_s &= jR_p^2X_p/(R_p^2 + X_p^2) = jX_p/[1 + (X_p^2/R_p^2)] \\ &= j(1/X_p)/[(1/X_p^2) + (1/R_p^2)] \end{aligned}$$

or

$$X_s = jB_p/(G_p^2 + B_p^2) \tag{2.24}$$

Example 2.25

A 0.1 μF capacitor is connected in parallel with a 470 Ω resistor. Determine the equivalent series circuit at a frequency of 10 kHz.

Solution

$$G_p = 1/470 = 2.13 \text{ mS}$$
$$B_p = j\omega C = 2\pi \times 10^4 \times 0.1 \times 10^{-6} = 6.28 \text{ mS}$$
$$R_s = (2.13 \times 10^{-3})/[(2.13^2 + 6.28^2) \times 10^{-6}] = 48.44\ \Omega \quad (\textit{Ans.})$$
$$X_s = 48.44 \times (j6.28/2.13) = j142.82\ \Omega$$
$$C_s = 1/(2\pi \times 10^4 \times 142.82) = 111.5 \text{ n}F \quad (\textit{Ans.})$$

Example 2.26

A 31.8 μH pure inductance is connected in parallel with a 20 Ω resistor. Determine its equivalent series circuit at 100 kHz.

Solution

$$G_p = 1/20 = 0.05 \text{ S}$$
$$B_p = 1/j\omega L = -j/(2\pi \times 31.8 \times 10^{-6} \times 10^5) = -j0.05 \text{ S}$$
$$R_s = 0.05/(0.05^2 + 0.05^2) = 10\ \Omega \quad (\textit{Ans.})$$

$$X_s = j0.05/(0.05^2 + 0.05^2) = j10\ \Omega$$
$$L_s = 10/(2\pi \times 10^5) = 15.92\ \mu H \quad (Ans.)$$

Series-parallel circuits

Many of the circuits which require to be analysed are neither series nor parallel circuits but a combination of both. To analyse such a circuit a suitable combination of the techniques used so far must be employed.

Example 2.27

For the series-parallel circuit given in Fig. 2.8 calculate the current flowing in each component.

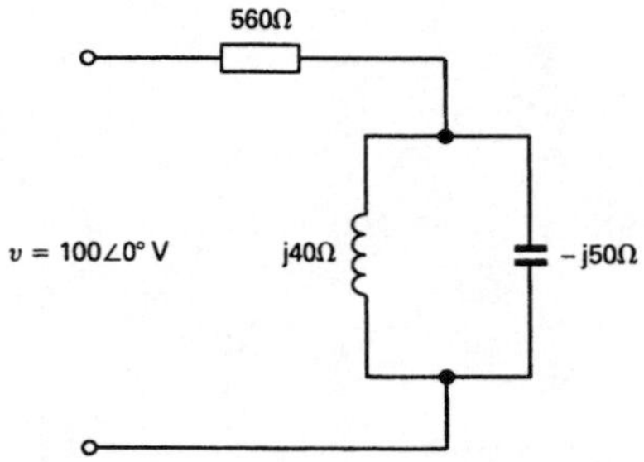

Fig. 2.8

Solution
The admittance of the LC combination is

$$Y_{LC} = 1/j40 + 1/-j50 = -j0.025 + j0.02 = -j5 \times 10^{-3}\ S$$
$$Z_{LC} = 1/Y_{LC} = 1/-j(5 \times 10^{-3}) = j200\ \Omega$$

Impedance of circuit:

$$Z = 560 + j200 = 594.6\angle 19.7°\ \Omega$$

Current in 560 Ω resistor $= (100\angle 0°)/(594.6\angle 19.7°)$
$= 0.168\angle -19.7°$ A (*Ans.*)

Current in capacitor $= (0.168\angle -19.7° \times j40)/(j40 - j50)$
$= (6.72\angle 70.3°)/(10\angle -90°)$
$= 0.67\angle 160.3°$ A (*Ans.*)

Current in inductor $= (0.168\angle -19.7° \times -j50)/-j10$
$= (8.4\angle -109.7°)/(10\angle -90°)$
$= 0.84\angle -19.7°$ A (*Ans.*)

Example 2.28

For the circuit shown in Fig. 2.9 calculate (*a*) the impedance, (*b*) the voltage across each component, (*c*) the current in each component, (*d*) the total current and (*e*) confirm Kirchhoff's voltage law.

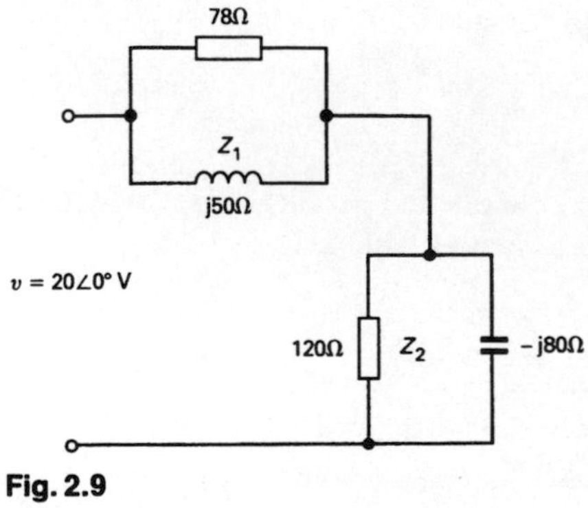

Fig. 2.9

Solution
(*a*) The impedance Z_1 of 78 Ω in parallel with j50 Ω is

$$(j78 \times 50)/(50 + j78) = (3900\angle 90°)/(92.65\angle 57.3°) = 42.1\angle 32.7°$$
$$= 35.43 + j22.74\ \Omega$$

The impedance Z_2 of 120 Ω in parallel with −j80 Ω is

$$(-j80 \times 120)/(120 - j80) = 36.92 - j55.39\ \Omega = 66.57\angle -56.3°\ \Omega$$

Therefore,

$$\text{Impedance of circuit } Z = (35.43 + j22.74) + (36.92 - j55.39)$$
$$= 72.35 - j32.65$$
$$= 79.38\angle{-24.3°}\ \Omega \quad (Ans.)$$

(*b*) $\text{Voltage across } Z_1 = (20\angle 0° \times Z_1)/(Z_1 + Z_2)$
$$= (20\angle 0° \times 42.1\angle 32.7°)/(79.38\angle{-24.3°})$$
$$= 10.61\angle 57°\ \text{V} \quad (Ans.)$$

$$\text{Voltage across } Z_2 = (20\angle 0° \times 66.57\angle{-56.3°})/(79.38\angle{-24.3°})$$
$$= 16.77\angle{-32°}\ \text{V} \quad (Ans.)$$

(*c*) Current in 78 Ω $= (10.61\angle 57°)/78 = 0.14\angle 57°$ A (*Ans.*)
Current in j50 Ω $= (10.61\angle 57°)/(50\angle 90°) = 0.21\angle{-33°}$ A (*Ans.*)
Current in 120 Ω $= (16.77\angle{-32°})/120$
$= 0.14\angle{-32°}$ A (*Ans.*)
Current in −j80 Ω $= (16.77\angle{-32°})/(80\angle{-90°})$
$= 0.21\angle 58°$ A (*Ans.*)

(*d*) Total current $= V/Z = (20\angle 0°)/(79.38\angle{-24.3°})$
$= 0.25\angle 24.3°$ A (*Ans.*)

Alternatively,

$$I = I_{120} + I_{j50} = 0.14\angle{-32°} + 0.21\angle 58°$$
$$= 0.12 - j0.07 + 0.11 + j0.18 = 0.23 - j0.11$$
$$= 0.26\angle 25.6°\ \text{A} \quad (Ans.)$$

(*d*) $V = 10.61\angle 57° + 16.77\angle{-32°} = 5.78 + j8.9 + 14.22 - j8.89$
$= 20\angle 0°$ V (*Ans.*)

Example 2.29

For the circuit shown in Fig. 2.10 calculate (*a*) the impedance and (*b*) the current in the 220 Ω resistor.

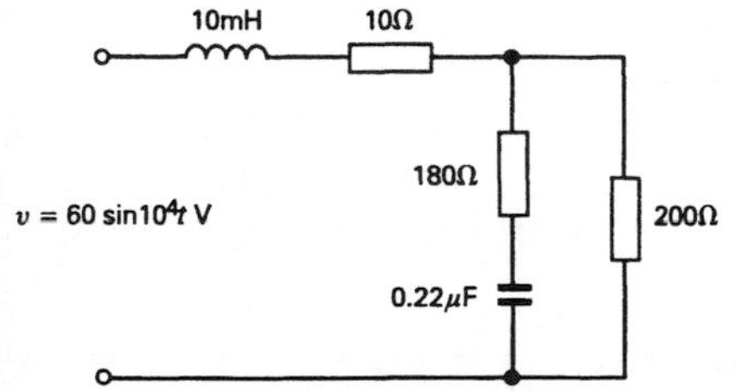

Fig. 2.10

Solution

(*a*) $X_L = 10^4 \times 10^{-3} = j10\ \Omega$
$X_C = 1/(10^4 \times 0.22 \times 10^{-6}) = -j454.5\ \Omega$
Impedance of 180 Ω in series with −j454.5 Ω = 180 − j454.5 Ω
Admittance $= 1/Z = 1/(180 - j454.5)$
$= 7.53 \times 10^{-4} + j1.9 \times 10^{-3}$ S

Conductance of 200 Ω $= 1/200 = 5 \times 10^{-3}$ S.
Admittance of parallel branch $= (5.753 + j1.9) \times 10^{-3}$ S
Impedance of parallel branch $= 10^3/(5.753 + j1.9)$
$= 156.6 - j51.7\ \Omega$
Impedance of circuit $Z = (10 + j100) + (156.6 - j51.7)$
$= 166.6 - j48.3\ \Omega$
$= 173.46\angle 16.2°\ \Omega$ (*Ans.*)

(b) Voltage across parallel branch

$$= [60 \times (156.6 - j51.7)] / (173.46 \angle 16.2°)$$
$$= (60 \times 164.9 \angle 18.3°) / (173.46 \angle 16.2°)$$
$$= 57.21 \angle 2.1° \text{ V}$$

Current in 200 Ω resistor $= (57.21 \angle 2.1°)/200$
$= 0.29 \angle 2.1° \text{ A}$ (*Ans.*)

Example 2.30

For the circuit given in Fig. 2.11 calculate (a) the impedance and (b) the current taken from the supply.

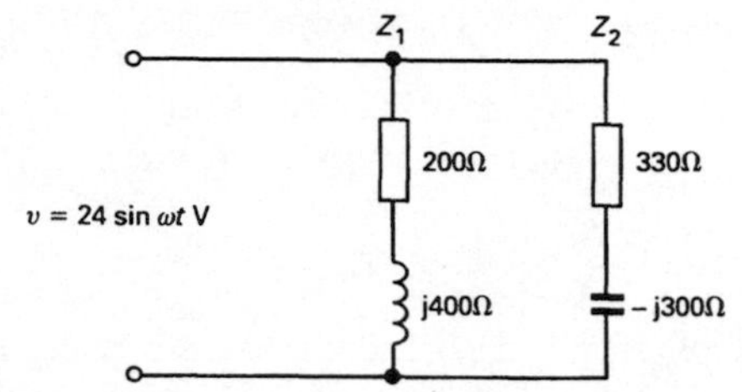

Fig. 2.11

Solution

(a) $Z_1 = 200 + j400$,

$Y_1 = 1/Z_1 = 1/(200 + j400) = (1 - j2) \times 10^{-3}$ S

$Z_2 = 330 - j300$,

$Y_2 = 1/Z_2 = 1/(330 - j300) = (1.66 + j1.51) \times 10^{-3}$ S

$Y_T = Y_1 + Y_2 = (2.66 - j0.49) \times 10^{-3}$ S

$Z_T = 1/Y_T = 1/[(2.66 - j0.49) \times 10^{-3}]$

$= 363.6 + j66.98$ Ω

$= 369.7 \angle 10.4°$ Ω (*Ans.*)

(b) $I = (24 \angle 0°)/(369.7 \angle 10.4°) = 64.9 \angle -10.4°$ mA (*Ans.*)

Exercises 2

2.1 Calculate the resistance and the inductance or capacitance of each of the following impedances if the frequency is 50 Hz: (a) $10 + j120$ Ω, (b) $25 - j80$ Ω, (c) $50 \angle -20°$ Ω and (d) $120 \angle 60°$ Ω.

2.2 A 270 Ω resistor is connected in series with an 0.2 H inductor of self-resistance 30 Ω. Calculate (a) the impedance of the circuit at a frequency of $1000/\pi$ Hz and (b) the current flowing into the circuit when the applied voltage is $17 \angle 0°$ volts.

2.3 (a) Calculate the impedance of each of the following circuits: (i) a 0.3 H inductor in series with a 100 Ω resistor, (ii) a 0.47 μF capacitor in series with a 100 Ω resistor and (iii) a 0.3 H inductor, a 0.47 μF capacitor and a 100 Ω resistor connected in series, if the applied voltage is $v - 15 \sin(800\pi t)$ volts. (b) For each circuit write down an expression for the current.

2.4 The voltage $v = 120 \sin(2000\pi t + 30°)$ V is applied to a circuit that consists of a 60 Ω resistance connected in series with a 120 Ω inductive reactance. (a) Obtain an expression for the current flowing in the circuit. (b) Calculate the r.m.s. voltages dropped across the two components. (c) Confirm Kirchhoff's voltage law.

2.5 A circuit consists of two capacitive reactances, −j20 Ω and −j100 Ω, connected in series with one another and also with a resistor of 560 Ω. (a) Calculate the impedance of the circuit. (b) Calculate the current in, and the r.m.s. voltage across, each component when the voltage $v = 5\cos\omega t$ is applied to the circuit.

2.6 An inductor with a reactance of 60 Ω and a self-resistance of 10 Ω at a particular frequency is connected in parallel with a 100 Ω resistor. The combination is connected in series with a capacitor of 10 Ω reactance. Calculate the current in each component when the applied voltage is 150 V.

2.7 The voltage $v = 100\sin(100\pi t)$ V is applied to a series RC circuit. The voltage across the 1 μF capacitor is to lag the applied voltage by 40°. Calculate (*a*) the necessary value for the resistor R, (*b*) an expression for the current then flowing in the circuit.

2.8 Two inductors are connected in parallel with one another and the combination is connected in series with a 10 Ω resistor and a capacitive reactance of 10 Ω. One inductor has a reactance of 12 Ω and a resistance of 8 Ω while the other has a reactance of 10 Ω and a resistance of 5 Ω. The voltage $100\angle 0°$ V is applied across the circuit. Calculate the current (*a*) into the circuit, (*b*) in each inductor.

2.9 An inductor, $R = 10$ Ω and $X_L = 82$ Ω, is connected in parallel with a capacitor of reactance −j44 Ω. A 47 Ω resistor is connected in series with the combination and a voltage of $12\angle 0°$ V is applied to the circuit. Calculate the current that flows in the capacitor.

2.10 A series circuit consists of a capacitive reactance of −j2600 Ω, a 3300 Ω resistor, another capacitive reactance of −j5200 Ω and another resistor of 2200 Ω. The voltage $v = 12\angle -40°$ V is applied across the circuit. (*a*) Calculate the voltage across the 2200 Ω resistor. (*b*) What will this voltage become if the frequency of the applied voltage is (*i*) doubled, and (*ii*) halved?

2.11 A 270 Ω resistor is connected in series with an inductor of reactance 250 Ω and self-resistance 30 Ω and a capacitor C, and a $50\angle 0°$ V voltage is applied to the circuit. Determine the value of C which will make the magnitude of the output voltage equal to 25 V when the frequency of the applied voltage is (*a*) 50 Hz and (*b*) 5 kHz.

2.12 A 100 Ω resistor is connected in parallel with a 160 Ω capacitive reactance. $60\angle 0°$ volts are applied across the circuit. Calculate (*a*) the admittance of the circuit, (*b*) the current supplied to the circuit and (*c*) the current in each component.

2.13 A reactance of −j100 Ω, a reactance of j200 Ω, a resistance of 50 Ω and a reactance of −j200 Ω are connected in parallel with one another. Calculate (*a*) the admittance, (*b*) the resistance and (*c*) the reactance of the circuit.

2.14 A parallel circuit has two branches: one is the series connection of a 40 Ω resistor and a −j60 Ω reactance, and the other consists of a 52 Ω resistance in series with a 30 Ω inductive reactance. The voltage $100\angle 0°$ V is applied to the circuit. Calculate (*a*) the admittance and (*b*) the impedance of the circuit. Also calculate (*c*) the current supplied to the circuit and (*d*) the current in each component.

2.15 A 10 kΩ resistor is connected in parallel with both a 10 kΩ inductive reactance and a 8 kΩ capacitive reactance. Determine the equivalent series network.

2.16 The voltage $v = 50\sin(\omega t - 45°)$ V is applied to a circuit that consists of an inductor, whose reactance is 10 Ω and whose self-resistance is 1 Ω, connected in series with the parallel combination of a 5 Ω resistor and a capacitive reactance of 6 Ω. (*a*) Calculate the impedance of the circuit and (*b*) obtain an expression for the current which flows into the circuit.

2.17 A 20 kΩ resistor is connected in series with the parallel combination of a 39 kΩ resistor and a 20 kΩ inductive reactance. (*a*) Calculate the impedance and admittance of the circuit. (*b*) Calculate the resistance, reactance, conductance and susceptance of the circuit. If a voltage of $18\angle 0°$ V is applied across the circuit calculate (*c*) the supply current and (*d*) the current in each branch.

2.18 The current $i = 0.5\angle -30°$ A flows into a circuit that consists of an inductor of reactance j50 Ω and resistance 5 Ω in series with the parallel combination of a 100 Ω resistor, and a 47 Ω resistor in series with a reactance of −j30 Ω. Determine the voltages across (*a*) the inductor, (*b*) the 47 Ω resistor and (*c*) the 100 Ω resistor.

2.19 An inductor of reactance 32 Ω and resistance 5 Ω is connected in parallel with a 20 Ω resistor. This circuit is connected in series with a capacitor of reactance −j20 Ω and a $100\angle 0°$ V voltage source. Calculate (*a*) the impedance and admittance of the circuit, (*b*) the effective resistance and reactance of the circuit, (*c*) the effective conductance and susceptance of the circuit, (*d*) the current taken from the supply and (*e*) the current flowing in each resistance.

2.20 Calculate the impedance of the circuit shown in Fig. 2.12. Also find the current that flows when a voltage of 120 V is applied across the circuit.

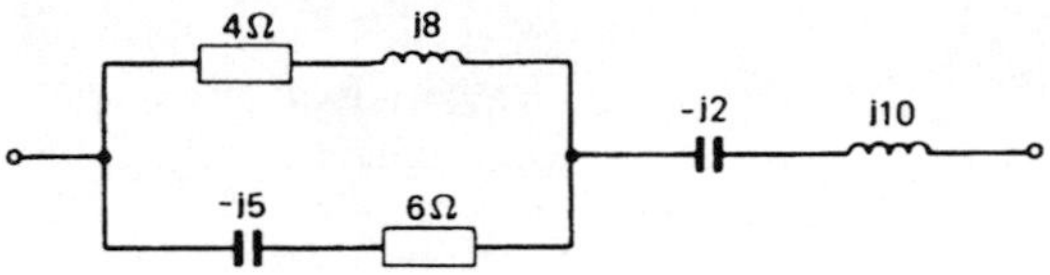

Fig. 2.12

2.21 An inductor has a resistance of 500 Ω and an inductance of 250 mH at a frequency of 1.59 kHz. Calculate the current that flows in the circuit when 25 V at 1.59 kHz are applied across the circuit.

2.22 Calculate the current taken from the supply in the circuit of Fig. 2.13.

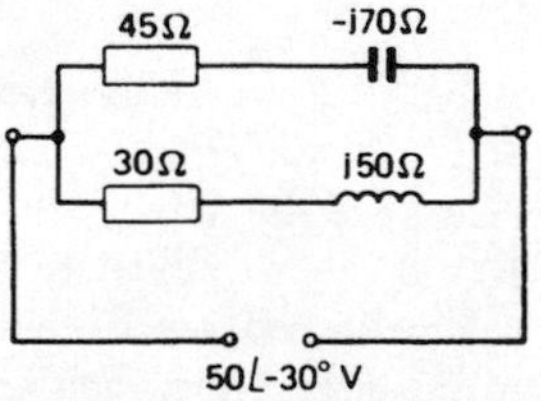

Fig. 2.13

2.23 For the circuit given in Fig. 2.14 calculate the impedance.

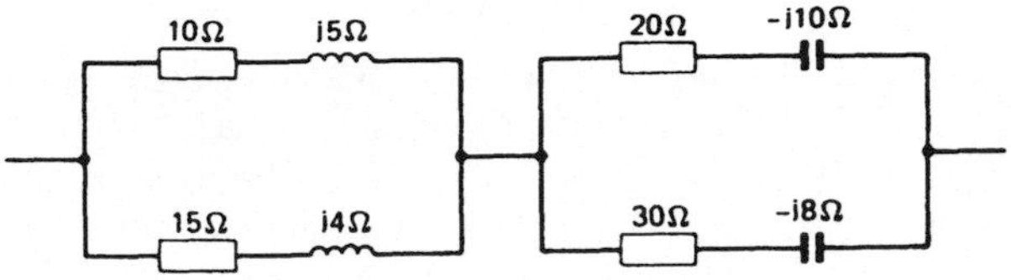

Fig. 2.14

2.24 Calculate the impedance at 796 Hz of a 0.2 H inductor with a self-resistance of 108 Ω.

2.25 A current of $40\angle 37°$ mA flows in a 40 mH inductor of self-resistance 8 Ω at a frequency of 100 Hz. Calculate the applied voltage.

2.26 A circuit with an admittance of $22 + j12$ mS has 240 V applied across it. Calculate the current flowing into the circuit.

2.27 A 1 μF capacitor is connected in parallel with a 100 Ω resistor and the voltage $v = 100 \sin 5000t$ V is applied across the circuit. Obtain expressions for (*a*) the current in each component and (*b*) the total current.

2.28 A 20 mH inductor of self-resistance 10 Ω is connected in parallel with both a 22 μF capacitor and a 12 Ω resistor. (*a*) Determine the instantaneous voltage across the circuit when a current of $7.17\angle 0°$ A is supplied at a frequency of 50 Hz. (*b*) Calculate the resistance and the inductance, or capacitance, for each of the following admittances if the frequency is 50 Hz: (*i*) $0.02 + j0.05$ S, (*ii*) $0.008 - j0.002$ S, (*iii*) $4 \times 10^{-3}\angle 37°$ S.

2.29 A 40 mH inductor of self-resistance 10 Ω is connected in series with a capacitor *C*. Calculate the required value of *C* for the current to be 750 mA when the applied voltage is 24 V at 50 Hz.

2.30 An inductor having a resistance of 50 Ω and a reactance of 45 Ω at a particular frequency is connected in parallel with a capacitor whose reactance is 62 Ω. Calculate the impedance of the circuit.

2.31 Calculate the admittance and the impedance of (*a*) a 20 Ω resistor in parallel with an inductive reactance of 20 Ω, (*b*) a resistor of 1250 Ω in parallel with both an inductive reactance of 2000 Ω and a capacitive reactance of 1000 Ω.

2.32 The current $2.4\angle 0°$ A flows into the parallel combination of a 100 Ω resistor and a capacitive reactance of 72.3 Ω. Determine the current flowing in each component.

2.33 Find the equivalent parallel circuit for the series connection of j80 Ω, −j20 Ω and 50 Ω.

2.34 A circuit has an admittance of $0.04 - j0.06$ S. Calculate its resistance and inductive reactance if the components are connected (*a*) in parallel and (*b*) in series.

2.35 The voltage 240 V, 50 Hz is applied across (*a*) a 20 Ω resistance in series with a 100 mH inductance, (*b*) a 30 Ω resistance in series with a 22 μF capacitance, (c) circuits (a) and (b) connected in series, (d) circuits (a) and (b) connected in parallel. For each circuit calculate the supply current.

2.36 A circuit consists of the series connection of, in turn, a 150 Ω resistor, reactances of j1250 Ω, –j900 Ω, j300 Ω, –j250 Ω and another resistor of 750 Ω. The voltage $v = 40\sin(200t)$ V is applied to the circuit. Calculate (a) the impedance of the circuit, (b) an expression for the instantaneous current flowing and (c) the voltage across each component.

2.37 A 120 Ω resistor is connected in series with the reactances j45 Ω, –j112 Ω, –j48 Ω, j56 Ω and j22 Ω. Calculate the effective capacitance of the circuit at a frequency of 2000 rad/s.

2.38 A T-network has series arms of –j130 Ω and 780 Ω and a shunt arm of –j220 Ω. A 120 Ω resistor is connected between the 780 Ω resistor and the common line. Determine the input impedance of the network.

2.39 (a) Obtain the impedance in polar form at 50 Hz of each of the following: (i) a 25 Ω resistor in series with a 100 mH inductor, (ii) a 50 Ω resistor in series with a 22 μF capacitor, (b) Calculate the impedance when the two circuits are connected (i) in series and (ii) in parallel with one another. (c) If the applied voltage is $v = 340\sin 100\pi t$ V, obtain an expression for the current in both ($b(i)$) and ($b(ii)$).

2.40 Determine the voltage which when applied across a circuit consisting of $R = 200$ Ω in series with $X_C = -\text{j}300$ Ω causes a current of 0.5 A to flow. Also find the voltage across each component.

2.41 Three impedances, $Z_1 = 50 + \text{j}50$ Ω, $Z_2 = 100 + \text{j}150$ Ω and $Z_3 = 100 - \text{j}90$ Ω, are connected in parallel across a 240 V supply. Calculate the current taken from the supply.

2.42 Determine the parallel RC circuit which has the same impedance at a frequency of 1 kHz as the impedance 18 – j20 Ω.

2.43 A circuit has five branches in parallel. The impedance of each branch is: $A = 22 - \text{j}20$ Ω, $B = 30 + \text{j}10$ Ω, $C = 47$ Ω, $D = -\text{j}50$ Ω, $E = 2 + \text{j}48 - \text{j}50$ Ω. Calculate (a) the admittance, (b) the conductance and (c) the susceptance of the circuit. The current flowing in the 47 Ω resistor is to be 1.5 A. Determine (d) the required applied voltage and (e) the current that flows in the 30 Ω resistance.

2.44 A parallel circuit contains three separate branches. Branch A consists of an inductor of reactance 30 Ω and resistance 2 Ω, branch B is a 10 Ω resistor in series with a capacitor of reactance 20 Ω and branch C is another capacitor of reactance 50 Ω. The parallel circuit is connected in series with the impedance 1 + j15 Ω. Calculate (a) the impedance of the circuit and (b) the current in each branch when the applied voltage is $100\angle 0°$ V.

2.45 A 5 mH inductor of self-resistance 5 Ω is connected in parallel with a 2.2 μF capacitor and a 100 Ω resistor. The voltage $v = 10\sin(1000t)$ V is applied to the circuit. Calculate (a) the current supplied to the circuit, (b) the current in the 100 Ω resistor, (c) the current in the inductor and (d) the impedance of the circuit.

3 Power in A.C. circuits

The instantaneous power dissipated in a circuit is the product of the instantaneous values of the current flowing in the circuit and the voltage applied across it. In general, the current and the voltage will not be in phase with one another. If their phase difference is θ then the instantaneous power is given by

$$p = V_m \sin(\omega t \times I_m) \times \sin(\omega t + \omega) \qquad (3.1)$$

Example 3.1

The voltage $v = 340\sin(100\pi t)$ V is applied across a circuit whose impedance is $50\angle -20°\ \Omega$. Calculate the instantaneous power dissipated in the circuit when the time t is (a) 0.5 ms and (b) 1 ms.

Solution

$I_{max} = 340/50 = 6.8$ A. Hence, $I = 6.8\sin(100\pi t + 20°)$ A.

(a) $p = 340\sin(100\pi \times 0.5 \times 10^{-3})$
$\times 6.8\sin(100\pi \times 0.5 \times 10^{-3} + 20°)$
$= 340\sin 9° \times 6.8\sin 29° = 53.19 \times 3.3 = 175.5$ W (*Ans.*)

(b) $p = 340\sin(100\pi \times 10^{-3}) \times 6.8\sin(100\pi \times 10^{-3} + 20°)$
$= 340\sin 18° \times 6.8\sin 38° = 105.1 \times 4.19 = 440.4$ W (*Ans.*)

Mean power

The instantaneous power is not a very useful concept and more useful is the *average* or *mean power* that is dissipated in a circuit.

A trigonometric identity is

$$2\sin A\sin B = \cos(A - B) - \cos(A + B)$$

Using this, with $A = \omega t + \theta$ and $B = \omega t$, equation (3.1) becomes

$$p = V_m I_m\,[\cos\theta - \cos(2\omega t + \theta)]/2$$

The average power dissipated over a complete cycle is

$$P = \frac{1}{2\pi}\int_0^{2\pi} vi \, \mathrm{d}\omega t$$

$$= \frac{V_m I_m}{4\pi}\int_0^{2\pi} [\cos\theta - \cos(2\omega t + \theta)] \, \mathrm{d}\omega t$$

$$= \frac{V_m I_m}{4\pi}\left[\cos\theta \cdot \omega t - \frac{\sin(2\omega t + \theta)}{2}\right]_0^{2\pi}$$

$$= \frac{V_m I_m}{4\pi}[2\pi\cos\theta] = \frac{V_m I_m}{2}\cos\theta$$

$$P = VI\cos\theta \qquad (3.2)$$

$$= |I|^2 \times \text{real part of impedance } V/I \qquad (3.3)$$

Thus, the *mean* power dissipated is the product of the r.m.s. values of the current and the voltage and a term, $\cos\theta$, known as the *power factor* of the circuit. Note carefully that the power factor is *only* equal to $\cos\theta$ when both the current and the voltage are of sinusoidal waveform. The general definition of power factor, which is true for any waveform, is

$$\text{Power factor} = \text{Power/volt-amps} \qquad (3.4)$$

Example 3.2

Calculate the power dissipated when a voltage $V = (5 + j2)$ volts is developed across an impedance carrying a current of $(20 + j15)$ mA.

Solution

$$V = 5 + j2 = 5.39\angle 21.8° \text{ V} \qquad I = 20 + j15 = 25\angle 36.9° \text{ mA}$$

From equation (3.3),

$$P = 5.39 \times 25 \times 10^{-3}\cos 15.1° = 130.1 \text{ mW} \quad (\textit{Ans.})$$

Example 3.3

An inductor of reactance 20 Ω and resistance 12 Ω is connected in parallel with a capacitor whose reactance is 6 Ω, and the combination is connected in series with a 10 Ω resistor and another capacitor whose reactance is 4 Ω. Calculate the power dissipated in the circuit when a voltage of $50\angle 0°$ is applied to the circuit.

Solution

The impedance Z of the circuit is

$$Z = 10 - j4 + [-j6(12 + j20)]/(12 + j20 - j6)$$

$$= 11.27 - j11.48 = 16.1\angle -45.5° \ \Omega$$

The current I flowing into the circuit is

$$I = V/Z = (50\angle 0°)/(16.1\angle -45.5°) = 3.11\angle 45.5° \text{ A}$$

The power dissipated $= 50 \times 3.11 \cos 45.5° = 109$ W (*Ans.*)

Alternatively,

The current in the 12 Ω resistance

$$= (3.11\angle 45.5° \times 6\angle -90°)/(12 + j14)$$

$$= (18.66\angle -44.5°)/18.44\angle 49.4°$$

$$= 1.01\angle 3.9° \text{ A}$$

Power in 12 Ω resistance $= 1.01^2 \times 12 = 12.24$ W

Power in 10 Ω resistance $= 3.11^2 \times 10 = 96.72$ W

Therefore,

Total power dissipated $= 12.24 + 96.72 \simeq 109$ W (*Ans.*)

Reactive power and apparent power

Although power cannot be dissipated in a pure resistance it is often convenient to introduce a quantity known as *reactive power*. Reactive power is given the symbol Q and is calculated as

$$Q = I^2X = V^2/X = VI \sin\theta \text{ vars} \tag{3.5}$$

Just as with reactance, inductive vars are taken as being positive, while capacitive vars are considered to be negative. The total reactive power in a circuit is equal to the algebraic difference between the inductive and the capacitive vars. The reactive power has zero contribution to make to the transfer of energy from a source to its load but it still takes just as much current as though it did make a contribution. Reactive power is of little, if any, significance in light-current engineering but it is of importance in the generation, distribution, and use of electrical power.

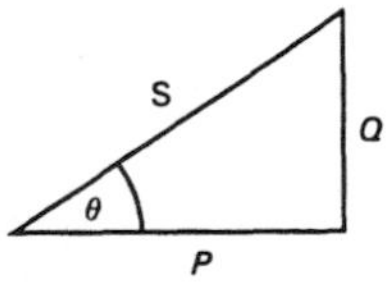

Fig. 3.1 Power triangle

Figure 3.1 shows the power triangle for an inductive circuit. The *apparent power* S is the product of the applied voltage V and the resulting current I. Thus:

$$S = VI \text{ volt-amps} \tag{3.6}$$

Clearly,

$$S = \sqrt{(P^2 + Q^2)} \tag{3.7}$$

Complex power

The total apparent power in circuit is *not* equal to the sum of the individual apparent powers in that circuit. This means that the total apparent power must be calculated using equation (3.7), where P and Q are the total real and total reactive powers. An alternator may be specified as having an apparent power of 10 kVA. The phase angle between the generated voltage and the current taken from the alternator will be determined by the impedance of the load. If the phase angle of the load is 0° then the power delivered to the load is equal to the apparent power, i.e. 10 kW. If the phase angle is, say, 30° the power delivered is $10\cos 30° = 8.66$ kW. In the extreme case in which the phase angle is 90° there would be zero power delivered to the load The factor $\cos\theta$ is known as the *power factor*, and from Fig. 3.1,

$$\text{Power factor} = (\text{Real power})/(\text{Apparent power}) \qquad (3.8)$$

Use of Complex Numbers

If the supply voltage is $v = V\angle\theta = V(\cos\theta + j\sin\theta) = a + jb$ volts, and the resulting current is $I\angle\varphi = I(\cos\varphi + j\varphi) = c + jd$ amps, then the power supplied to the load is

$$\begin{aligned} P &= VI\cos(\theta - \varphi) = VI(\cos\theta\cos\varphi + \sin\theta\sin\varphi) \\ &= V\cos\theta I\cos\varphi + V\sin\theta I\sin\varphi \end{aligned}$$

or

$$P = ac + jbd \text{ W} \qquad (3.9)$$

Also, the reactive power is

$$\begin{aligned} Q &= VI\sin(\theta - \varphi) = VI(\sin\theta\cos\varphi - \cos\theta\sin\varphi) \\ &= V\sin\theta I\cos\varphi - V\cos\varphi I\sin\theta \\ &= bc - ad \quad \text{vars} \qquad (3.10) \end{aligned}$$

The product VI^*, where I^* is the *conjugate* of I is often known as the *complex power*. The real part of the complex power is the true power and the imaginary part is the reactive power.

Example 3.4

Determine (a) the real power; (b) the reactive power and (c) the apparent power dissipated by the voltage $100\angle 60°$ V and the current $8\angle 6.9°$ A .

Solution

(a) $P = 100 \times 8\cos(60° - 6.9°) = 480.34$ W (*Ans.*)

(b) $Q = 100 \times 8\sin(60° - 6.9°) = 639.75$ vars (*Ans.*)

(c) $S = \surd(480.34^2 + 639.75^2) = 800$ VA (*Ans.*)

Alternatively, $100\angle 60°$ V $= 50 + j86.6$ V and $8\angle 6.9°$ A $= 7.94 + j0.96$ A.

Therefore,

(a) $P = 50 \times 7.94 + 86.6 \times 0.96 = 480.2$ W (*Ans.*)

(b) $Q = 86.6 \times 7.94 - 50 \times 0.96 = 639.6$ vars (*Ans.*)

Example 3.5

A circuit dissipates 10 kW at a lagging power factor of 0.82. If the applied voltage is 240 V calculate (a) the reactive and apparent powers and (b) the current flowing into the circuit.

Solution

(a) $10000 = S \times 0.82$. Hence $S = 10000/0.82 = 12.2$ kVA (*Ans.*)

(b) $\theta = \cos^{-1} 0.82 = 34.9°$ and so the reactive power is

$Q = 12.2 \times 10^3 \sin 34.9° = 6.98$ kvars (inductive) (*Ans.*)

(c) $I = (6.988 \times 10^3)/240 = 29.08$ A (*Ans.*)

Example 3.6

Determine the real and reactive powers in the circuit of Fig. 3.2, (a) by adding the powers dissipated in the resistors and adding the reactive powers; (b) from the product VI. (c) Calculate the apparent power.

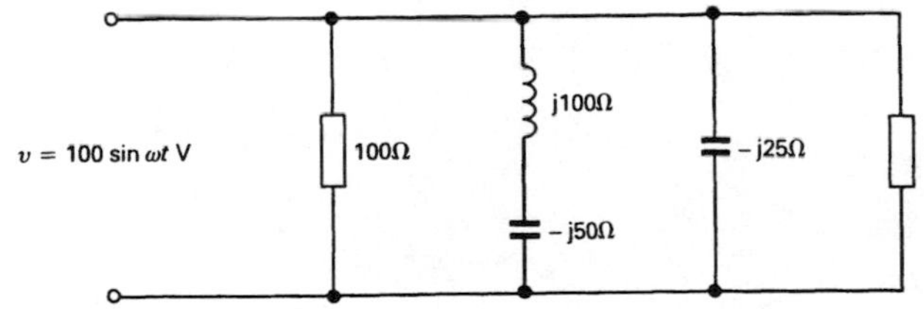

Fig. 3.2

Solution

(a) Power in 100 Ω $= (100/\sqrt{2})^2/100 = 49.99$ W.

Power in 120 Ω $= (100/\sqrt{2})^2/120 = 41.65$ W.

Total real power $= 49.99 + 41.65 = 91.64$ W (*Ans.*)

Reactive power in j50 Ω $= (100/\sqrt{2})^2/50 = 99.97$ vars.

Reactive power in $-$j25 Ω $= (100/\sqrt{2})^2/25 = -199.94$ vars.

Total reactive power $= 99.97 - 199.94 = -99.97$ vars (*Ans.*)

(b) Admittance of circuit is

$Y = 1/100 + 1/\text{j}50 + 1/-\text{j}25 + 1/120$

$= 0.018 + \text{j}0.02\ S = 0.027\angle 48°\ S$

$I = VY = 100\angle 0° \times 0.027\angle 47.5° = 2.7\angle 47.5°$ A

Total real power $= (100/\sqrt{2}) \times (2.7/\sqrt{2}) \cos 47.5° = 90.3$ W (*Ans.*)

Total reactive power $= 50 \times 2.7 \sin 48° = 100.3$ vars (*Ans.*)

(c) Apparent power $S = \sqrt{(90.3^2 + 100.3^2)} = 135$ VA (*Ans.*)

Loads in parallel

When two loads having different apparent powers and power factors are connected in parallel to an a.c. supply the total apparent power and the total power factor are not equal to the algebraic sum of their individual values. The problem is illustrated by the following example.

Example 3.7

For the system shown in Fig. 3.3 calculate (*a*) the total apparent power and (*b*) the power factor.

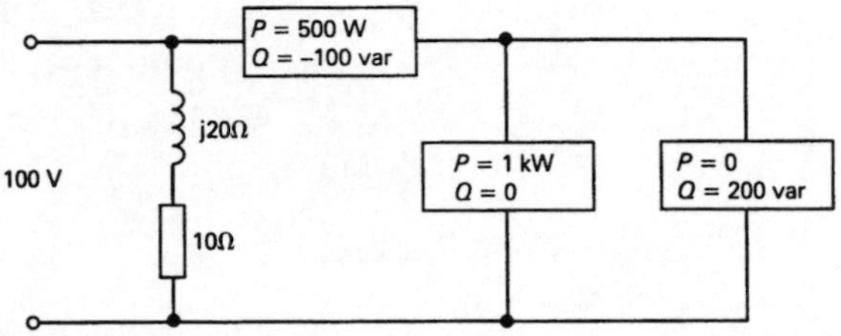

Fig. 3.3 Loads in parallel

Solution

(*a*) The current in the inductive branch of the circuit is

$$I_L = 100/(10 + j20) = 2 - j4\ \text{A} = 4.47\angle{-63.4°}\ \text{A}$$

Real power $P = 4.47^2 \times 10 \approx 200$ W;

Reactive power $Q = 4.47^2 \times 20 = 400$ vars.

Load A: $P = 500$ W, $Q = -100$ vars;

Load B: $P = 1000$ W, $Q = 0$;

Load C: $P = 0$ W, $Q = 200$ vars.

Total real power $= 200 + 500 + 1000 = 1700$ W.

Total reactive power $= -100 + 400 + 200 = 500$ vars.

Therefore,

Total apparent power $S = \sqrt{(1700^2 + 500^2)} = 1772$ VA (*Ans.*)

Example 3.8

For the circuit shown in Fig. 3.4 calculate (*a*) the true power dissipated, (*b*) the power factor and (*c*) the apparent power.

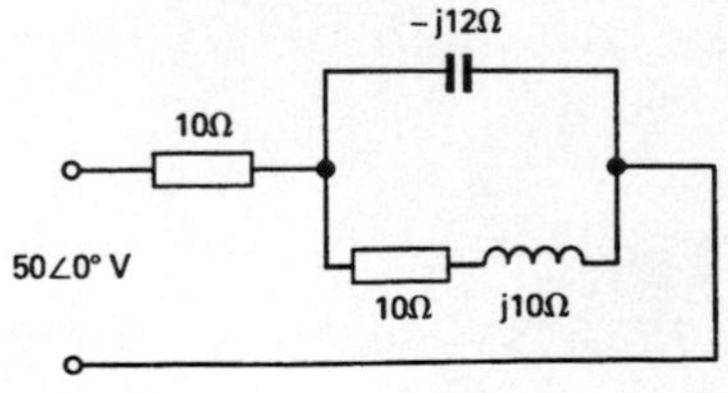

Fig. 3.4

Solution

(*a*) Impedance of circuit

$$Z = 10 + [(10 + j10)(-j12)]/(10 + j10 - j12) = 23.85 - j9.23\ \Omega$$
$$= 25.57\angle{-21.2°}\ \Omega$$

The current I flowing into the circuit is

$$I = (50\angle 0°)/(25.57\angle{-21.2°}) = 1.96\angle 21.2°\ \text{A}$$

$$\text{Power dissipated} = VI\cos\theta = 50 \times 1.96\cos 21.2°$$

$$= 91.37\ \text{W} \quad (\textit{Ans.})$$

(*b*) Power factor $= \cos 21.2° = 0.93$ (*Ans.*)

(*c*) Reactive power $= 50 \times 1.96\sin 21.2° = 35.44$ vars.

Apparent power $= \sqrt{(91.37^2 + 35.44^2)} = 98$ VA (*Ans.*)

Alternatively,

Power dissipated in the series 10 Ω resistor $= 1.96^2 \times 10 = 38.42$ W.

$I = 1.83 + \text{j}0.71$ A

Current in inductor $= (1.83 + \text{j}0.71)(-\text{j}12)/(10 + \text{j}10 - \text{j}12)$

$= 1.24 - \text{j}1.94 = 2.3\angle{-57.4°}$ A.

Power dissipated in 10 Ω parallel resistance $= 2.3^2 \times 10 = 52.9$ W.

Hence

The total power dissipated $= 38.42 + 52.9 = 91.32$ W (*Ans.*)

Reactive power in inductance $= 2.3^2 \times 10 = 52.9$ vars.

Current in capacitor $= (1.83 + \text{j}0.71) - (1.24 - \text{j}1.94)$

$= 0.59 + \text{j}2.65 = 2.72\angle 77.5°$ A.

Reactive power in capacitor $= 2.72^2 \times 12 = -88.78$ vars.

Total reactive power $= 52.9 - 88.78 = -35.9$ vars (*Ans.*)

Example 3.9

For the circuit given in Fig. 3.5 calculate (*a*) the impedance, (*b*) the power factor, (*c*) the real power dissipated, (*d*) the apparent power dissipated and (*e*) the real power dissipated in each resistor.

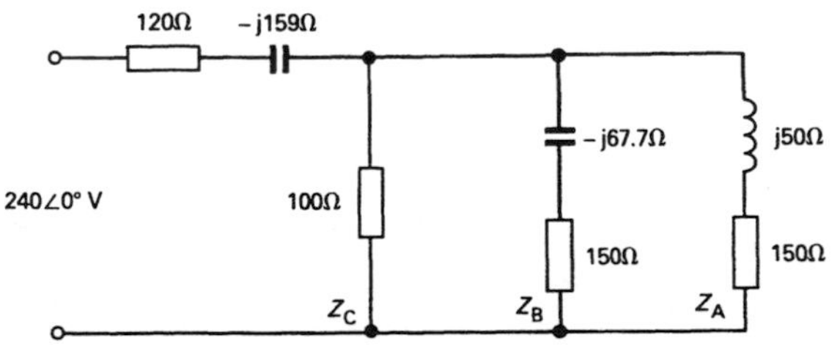

Fig. 3.5

Solution

$Z_A = 150 + \text{j}50\ \Omega$, $Y_A = 1/Z_A = 6 - \text{j}2$ mS

$Z_B = 150 - \text{j}67.7\ \Omega$, $Y_B = 5.54 + \text{j}2.5$ mS

$Z_C = 100\ \Omega$, $Y_C = 10$ mS

$$Y_T = Y_A + Y_B + Y_C = 21.54 + j0.5 \text{ mS}$$
$$Z_T = 1/Y_T = 46.4 - j1.1\ \Omega$$
$$Z = 120 - j159 + 46.4 - j1.1 = 166.4 - j160.1\ \Omega$$
$$= 230.9\angle{-43.9^\circ}\ \Omega \quad (Ans.)$$

(*b*) Power factor $= \cos(-43.9^\circ) = 0.72$ (*Ans.*)

(*c*) $I_{in} = 240/(230.9\angle{-43.9^\circ}) = 1.04\angle 43.9^\circ$ A

Real power dissipated $= VI\cos\theta = 240 \times 1.04\cos 43.9^\circ$

$= 179.85$ W (*Ans.*)

(*d*) Reactive power $= VI\sin\theta = 240 \times 1.04\sin 43.9^\circ = 173.07$ vars.

Apparent power $= \sqrt{(179.85^2 + 173.07^2)} = 249.6$ VA (*Ans.*)

Alternatively,

Reactive power $= VI^* = 240\angle 0^\circ \times 1.04\angle -43.9^\circ = 173.07$ vars,

(*e*) Power in 120 $\Omega = 1.04^2 \times 120 = 129.8$ W.

Voltage drop across $(120 - j159)\ \Omega = 1.04\angle 43.9^\circ \times 199.2\angle{-53^\circ}$

$= 207.2\angle{-9.1^\circ}$ V.

Voltage across parallel part of the circuit is

$= 240\angle 0^\circ - 207.2\angle{-9.1^\circ} = 48.24\angle 42.8^\circ$ V

$I_A = (48.24\angle 42.8^\circ)/(158.1\angle 18.4^\circ) = 0.31\angle 24.4^\circ$ A and

$P_{150} = 0.31^2 \times 150 = 14.42$ W (*Ans.*)

$I_B = (48.24\angle 42.8^\circ)/(169\angle{-24.3^\circ}) = 0.29\angle 67.1^\circ$ A and

$P_{150} = 0.29^2 \times 150 = 12.62$ W (*Ans.*)

$I_C = (48.24\angle 42.8^\circ)/100 = 0.48\angle 42.8^\circ$ and

$P_{100} = 0.48^2 \times 100 = 23.04$ W

Total power $= 14.42 + 12.62 + 23.04 + 129.8 = 179.9$ W

Power factor correction

When a particular real power is supplied to a load the nearer the power factor is to unity the smaller will be the reactive power. This, in turn, will reduce the current taken by the load and so minimize the cost of the cables, transformers and generating equipment. The loads on most electrical power systems are predominately inductive and so they have a lagging power factor. The provision of energy to such systems is uneconomical unless the load power factors can be made to be as near to unity as possible. Capacitive loads are often added in parallel with the load to cancel out some of the inductive vars with capacitive vars and hence increase the power factor and reduce the apparent power. This practice is known as *power factor correction* and it ensures that a power distribution system operates with greater efficiency.

The basic concept of power factor correction is illustrated by Fig. 3.6. In the absence of the power factor correction capacitor

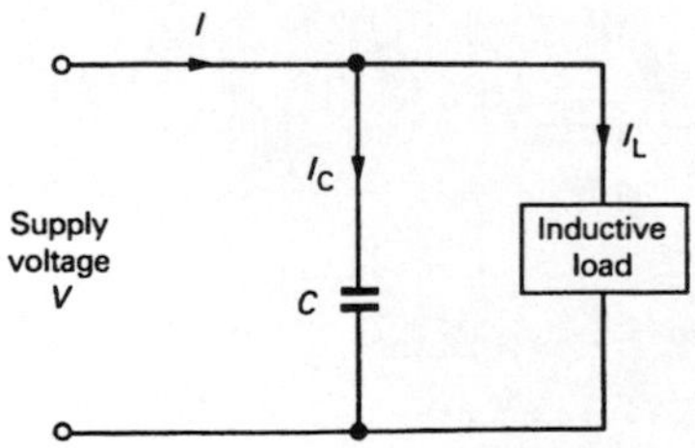

Fig. 3.6 Power factor correction

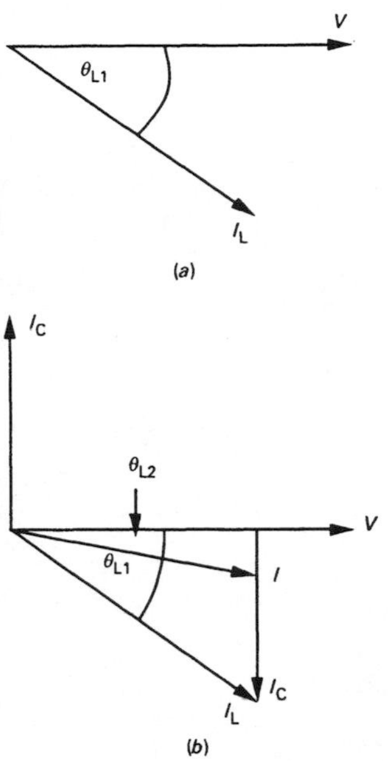

Fig. 3.7 Phasor diagram of power factor correction

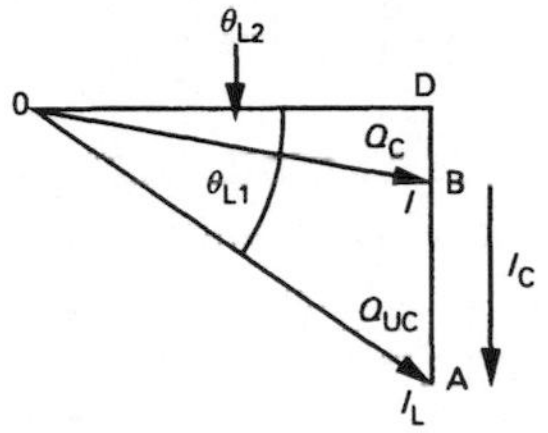

Fig. 3.8

C the load takes an inductive current I_L from the supply. The phasor diagram is shown by Fig. 3.7(a). The current I_L lags the applied voltage V by angle θ_{L1}. When the power factor correction capacitor C is fitted it takes a current I_C from the supply which leads the applied voltage by 90°. The total current I taken from the supply is the phasor sum of I_L and I_C and it has a smaller phase angle θ_{L2}. Before power factor correction was carried out the reactive power was, from Fig. 3.7(b), $VI\sin\theta_{L1}$. After correction the reactive power is $VI\sin\theta_{L2}$.

The capacitor current $I_C = I_L\sin\theta_{L1} - I\sin\theta_{L2}$. The power triangle is shown by Fig. 3.8; where Q_C is the reactive power with the correction capacitor fitted, and Q_{UC} is the reactive power without the correction capacitor. Thus, $Q_{UC} - Q_C$ is the change in the reactive power that is caused by power factor correction.

Referring to Fig. 3.8,

$$I_C = V\omega C = AB = AD - DB = OD\tan\phi_{L2} - OD\tan\phi_{L2},$$
$$OD = OA\cos\phi_{L1} = I_L\cos\phi_{L1}$$

Therefore,

$$C = [I_L\cos\phi_{L1}(\tan\phi_{L1} - \tan\phi_{L2})]/V\omega$$
$$I_L\cos\phi_{L1} = P/V = (\text{kilowatts} \times 10^3)/V$$
$$\text{and } C = [\text{kilowatts } (\tan\phi_{L1} - \tan\phi_{L2}) \times 10^3]/V^2\omega \text{ F}$$
$$= [\text{kilowatts } (\tan\phi_{L1} - \tan\phi_{L2}) \times 10^9]V^2\omega\ \mu\text{F} \quad (3.11)$$

Example 3.10

For the power system shown in Fig. 3.9: (*a*) calculate the total apparent power, the power factor and the current taken from the supply before the power factor correction capacitor is fitted: (*b*) calculate the capacitive vars needed to obtain a power factor of unity: (*c*) determine the necessary capacitance value; and (*d*) find the total apparent power and the current supplied to the load after power factor correction.

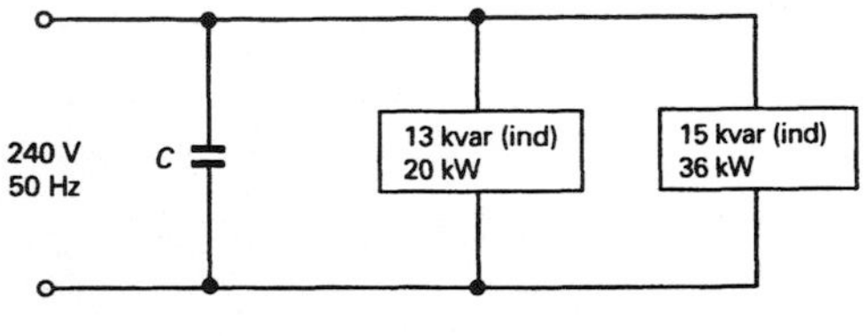

Fig. 3.9

Solution

(*a*) $Q = 13 + 15 = 28$ kvar; $P = 20 + 36 = 56$ kW;

$S = \surd(28^2 + 56^2) = 62.61$ kVA (*Ans.*)

Power factor $= \cos\theta = P/S = 56/62.61 = 0.89$ lagging (*Ans.*)

Current taken from supply is

(Apparent power)/(Supply voltage) $= (62.61 \times 10^3)/240$

$= 260.9$ A (*Ans.*)

(*b*) For unity power factor the capacitive vars must be equal to 28 kvar (*Ans.*)

(*c*) Reactive power in correction capacitor $= 240^2/X_C = 28 \times 10^3$, or $X_C = 2.06\ \Omega$.

Therefore,

$C = 1/(2\pi \times 50 \times 2.06) = 1545\ \mu F$ (*Ans.*)

(*d*) $Q = 0$ and $S = P = 56 \times 10^3$.

Therefore,

$I = (56 \times 10^3)/240 = 233.3$ A (*Ans.*)

Example 3.11

A 240 V, 50 Hz, 20 kVA transformer is operated at full load with a power factor of 0.8 lagging. It is required to improve the power factor to (*i*) unity and (*ii*) 0.9 lagging. (*a*) Calculate the capacitive vars needed in each case. (*b*) Calculate the value of the power factor correction capacitor in each case.

Solution

(*a*) Real power $P = 20 \times 10^3 \times 0.8 = 16$ kW. Phase angle between the applied voltage and the supply current $= \cos^{-1} 0.8 = 36.9°$. Reactive power $Q = 20 \times 10^3 \sin 36.9° = 12$ kvar lagging. The power triangle is shown in Fig. 3.10.

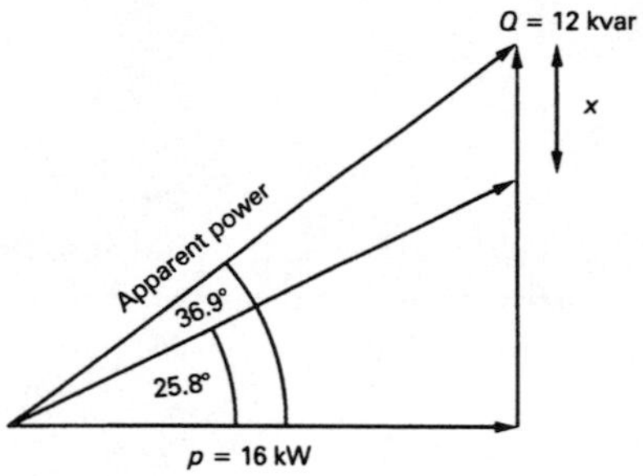

Fig. 3.10

(*i*) When the power factor is improved to unity, $\theta = \cos^{-1} 1 = 0°$ and then the capacitive vars required are equal to 12 kvar (*Ans.*)

(*ii*) When the new power factor is 0.9,

$\theta = \cos^{-1} 0.9 = 25.8°; \tan 25.8° = 0.483 = (12 - x)/16$, or

$x = 12 - 7.735 = 4.265$ kvar capacitive (*Ans.*)

Note the large difference between the capacitive vars needed to improve the power factor to unity or to 0.9. This illustrates why the power factor of a system is rarely improved to unity.

(*b*) (*i*) Power factor = 1: Capacitor current $= (12 \times 10^3)/240 = 50$ A

$X_C = 240/50 = 4.8\ \Omega$

$C = 1/(2\pi \times 50 \times 4.8) = 663\ \mu F$ (*Ans.*)

(*ii*) Power factor = 0.9 : $I_C = (4.265 \times 10^3)/240 = 17.77$ A

$X_C = 240/17.77 = 13.51\ \Omega$

$C = 1/(2\pi \times 50 \times 13.51) = 236\ \mu F$ (*Ans.*)

Example 3.12

A load takes 125 kW at a lagging power factor of 0.6 from a 415 V, 50 Hz supply. Calculate the value of a power factor correction capacitor that will improve the power factor to (*a*) 0.9 lagging and (*b*) 1. For each case calculate the reduction in the apparent power supplied to the load.

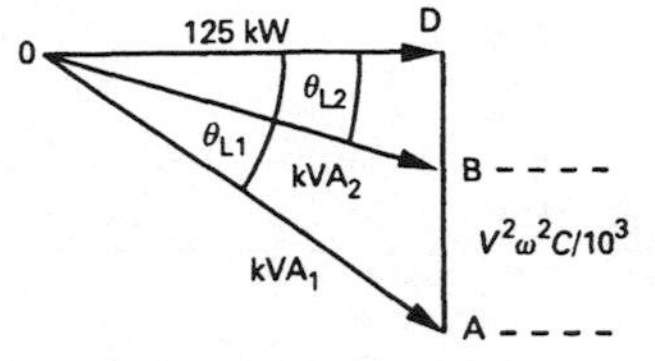

Fig. 3.11

Solution

The phasor diagram is shown in Fig. 3.11. The initial phase $\phi_{L1} = \cos^{-1} 0.6 = 53.13°$, the initial kVA = 125/0.6 = 208.3 kVA, and initial kvar = 125 tan 53.13° = 166.66 kvar.

(*a*) Phase angle $= \cos^{-1} 0.9 = 25.84°$; kvar $= DB = 125 \tan 25.84° =$ 60.54 kvar

$$\text{Capacitive kvar} = AB = DA - DB = 166.66 - 60.54 = 106.12 \text{ kvar}$$

Hence,

$$V^2\omega C = 106.12 \times 10^3 \quad \text{or} \quad C = 509\ \mu\text{F} \quad (\textit{Ans.})$$

$$\text{Reduction in apparent power} = 208.3 - 125/0.9 = 69.54 \text{ kVA } (\textit{Ans.})$$

(*b*) Total kvar = 0 and capacitive kvar $= AD = 166.66$ kvar.

Therefore,

$$C = (166.66 \times 10^3 \times 10^6)/(415^2 \times 2\pi \times 50) = 3080\ \mu\text{F} \quad (\textit{Ans.})$$

kVA = kW = 125, so reduction = 208.3 − 125 = 83.3 kVA (*Ans.*)

Example 3.13

A load of 75 kW at 0.75 lagging power factor is supplied by a 1100 V, 50 Hz source. The real power is to be increased to 85 kW without any increase in the apparent power. Calculate (*a*) the required power factor and (*b*) the power factor correction capacitor that is required.

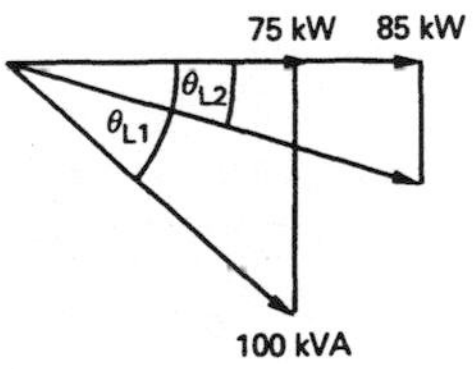

Fig. 3.12

Solution

(*a*) The phasor diagram of the system is shown in Fig. 3.12. Initial kVA = 75/0.75 = 100 kVA.

If ϕ_{L2} is the final phase angle then

$$\cos \phi_{L2} = \text{power factor} = 85/100 = 0.85 \quad (\textit{Ans.})$$

(*b*) $\phi_{L1} = \cos^{-1} 0.75 = 41.4°$, and $\phi_{L2} = \cos^{-1} 0.85 = 31.8°$

Capacitive kvar = 100 sin 41.4° − 100 sin 31.8° = 13.43 kvar.

Hence,

$$V^2\omega C = 13.43 \times 10^3 \quad \text{or} \quad C = (13.43 \times 10^3)/(1100^2 \times 2\pi \times 50) = 35.3\ \mu\text{F} \quad (\textit{Ans.})$$

Exercises 3

3.1 The voltage $v = 12\sin(314.2t)$ V is applied across a 50 Ω resistor. Calculate (*a*) the average power and (*b*) the peak power dissipated in the resistor.

3.2 An inductor with 9 Ω resistance and 18 Ω reactance is connected to a 240 V, 50 Hz supply. Calculate the average, apparent and reactive powers in the circuit.

3.3 A generator supplies a mixed load of 200 kW at unity power factor, 800 kVA at 0.72 lagging power factor, and 400 kVA at 0.8 leading power factor. Calculate the current flowing in the load.

3.4 A load of 12 kW at 0.8 lagging power factor is supplied by a 240 V mains supply. Calculate (*a*) the impedance of the circuit, (*b*) the power factor and (*c*) the average power dissipated.

3.5 A circuit operated from a 400 V, 50 Hz supply takes a current of 30 A at a power factor of 0.72 lagging. Calculate (*a*) the apparent power, (*b*) the real power and (*c*) the reactive power in the circuit.

3.6 An inductor with a reactance of 80 Ω and a resistance of 20 Ω is connected in parallel with a 56 Ω resistor. A voltage of $50\angle 0°$ is applied to the circuit. Calculate the power dissipated in the circuit.

3.7 A circuit consists of the series connection of a 30 Ω resistor, an inductor of reactance 20 Ω, and a capacitor of reactance 10 Ω. A voltage of $50\angle 0°$ V is applied across the circuit. Calculate (*a*) the impedance of the circuit, (*b*) the power factor and (*c*) the average power dissipated.

3.8 The voltage $v = 120\sin 314t$ is applied across a 100 mH inductor of self-resistance 5 Ω. (*a*). Calculate the value of the capacitance which must be connected in parallel with the inductor to make the power factor unity. (*b*) Calculate the average power then supplied to the circuit.

3.9 A voltage of $100\angle 0°$ V is applied across (*a*) a 56 Ω resistor, (*b*) an inductor of reactance 150 Ω and negligible resistance in series with a capacitor of 50 Ω reactance and (*c*) a capacitor of 200 Ω reactance that is connected in series with a 100 Ω resistor. Calculate the real, reactive, and apparent powers in each case.

3.10 For the system shown in Fig. 3.13 calculate (*a*) the total apparent power, (*b*) the total real power and (*c*) the overall power factor.

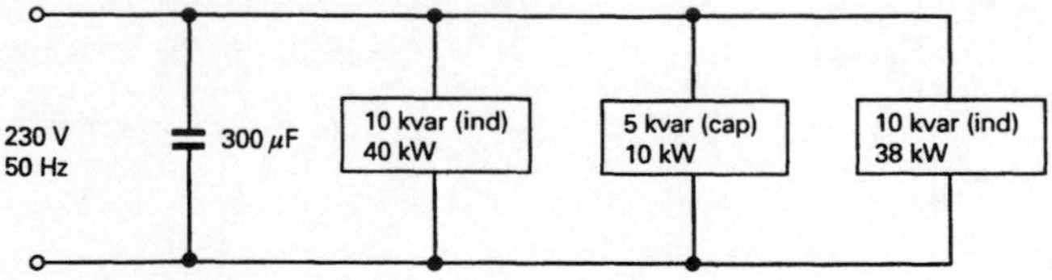

Fig. 3.13

3.11 The voltage $30 + j20$ V is applied across a circuit when a current of $1 - j0.5$ A flows. Calculate (*a*) the impedance of the circuit, (*b*) the real power dissipated, (*c*) the reactive power, and (*d*) the apparent power.

3.12 A circuit dissipates 22 kW power at a lagging power factor of 0.883 when a voltage of 240 V is applied. Calculate (*a*) the apparent power and (*b*) the current that flows into the circuit.

3.13 A 240 V, 50 Hz circuit operates at its full load of 25 kVA with a lagging power factor of 0.81. The power factor is to be improved to 0.93. Determine the required value of the power factor correction capacitor.

3.14 An a.c. source operates at 1 kV and 50 Hz to supply a load of 50 Ω resistance in series with a 200 mH inductance. Calculate (*a*) the apparent power, (*b*) the power factor, (*c*) the real power dissipated, (*d*) the reactive power and (*e*) the value of the power factor correction capacitor needed to improve the power factor to 0.92 lagging.

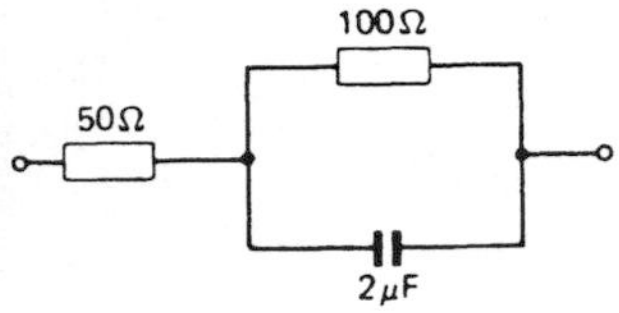

Fig. 3.14

3.15 Explain what is meant by the power factor of a circuit. Determine the voltage across the circuit shown in Fig. 3.14, the current supplied to it, and the phase difference between them at a frequency of $5000/2\pi$ R/S, if the current in the 100 Ω resistor is 10 mA. Also find the total power dissipated in the network and its power factor.

3.16 A sinusoidal voltage of 100 V r.m.s. is applied across a circuit whose impedance is 300 + j200 Ω. Calculate (*a*) the current flowing in the circuit, (*b*) the power dissipated and (*c*) the conductance and susceptance of the circuit.

3.17 A sinusoidal signal of 12 V r.m.s. is applied across a circuit whose conductance is 100 μS and whose susceptance is +22 μS. Calculate (*a*) the current that flows, (*b*) the power dissipated, (*c*) the power factor and (*d*) the resistance and reactance of the circuit.

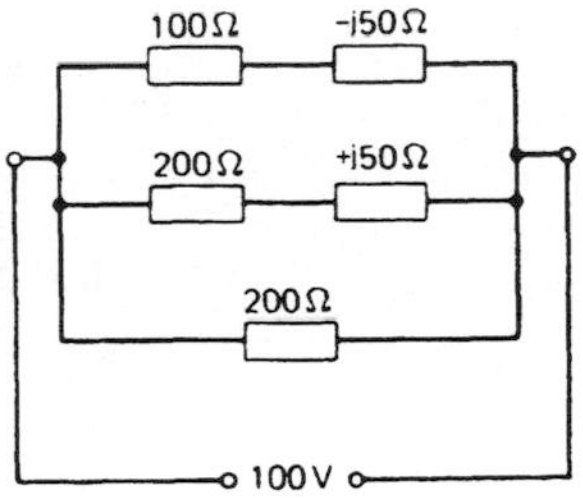

Fig. 3.15

3.18 For the circuit given in Fig. 3.15 calculate (*a*) the impedance and the admittance, (*b*) the current taken from the supply and (*c*) the power dissipated.

3.19 At a frequency of 200 kHz a coil has a resistance of 50 Ω and a reactance of 1200 Ω. Calculate its impedance and its admittance. Also calculate the power dissipated when a 12 V signal is applied across the circuit.

3.20 The admittance of a circuit at 120 kHz *is* (200 − j250) μS. Calculate (*a*) its resistance and reactance, (*b*) the power dissipated when 50 V are applied across the circuit, and (*c*) the impedance of the circuit when a capacitance of 400 pF is connected in parallel.

3.21 A circuit has a resistance of 3500 Ω and a reactance of +j5000 Ω. Calculate (*a*) its admittance and (*b*) the capacitance that should be connected in parallel with the circuit in order to make the power factor unity. $\omega = 5000$ R/S.

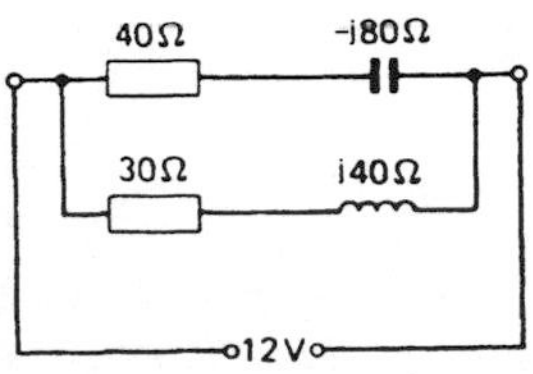

Fig. 3.16

3.22 An inductor has a resistance of 500 Ω and an inductance of 250 mH at a frequency of 1.59 kHz. Calculate the current that flows in the circuit when 25 V at 1.59 kHz are applied across the circuit.

3.23 For the circuit shown in Fig. 3.16 calculate (*a*) the admittance (*b*) the current taken from the supply and (*c*) the power dissipated.

3.24 A circuit dissipates a real power of 50 kW and an apparent power of 68 kVA. Calculate the reactive power and the power factor.

3.25 A circuit is supplied with 250 kW power at a lagging power factor of 0.64. If the kVA is kept at a constant value calculate the increase in the dissipated power if the power factor is increased to unity.

3.26 When a voltage of $240\angle 0°$ V is applied across a circuit 4 kW power is dissipated at a lagging power factor of 0.81. Calculate (*a*) the complex power, (*b*) the reactive power and (*c*) the impedance of the circuit.

3.27 The voltage $v = 340\sin(100\pi)$ V is applied across a 50 mH inductor of 12 Ω resistance. Calculate (*a*) the apparent power, (*b*) the real power and (*c*) the reactive power.

3.28 A load takes a current of 2.5 A at a lagging power factor of 0.73 when it is connected to the mains supply voltage of 240 V. Determine the value of a capacitor connected in parallel with the load that will increase the power factor to 0.92.

3.29 Calculate the mean power dissipated in the circuit shown in Fig. 3.17.

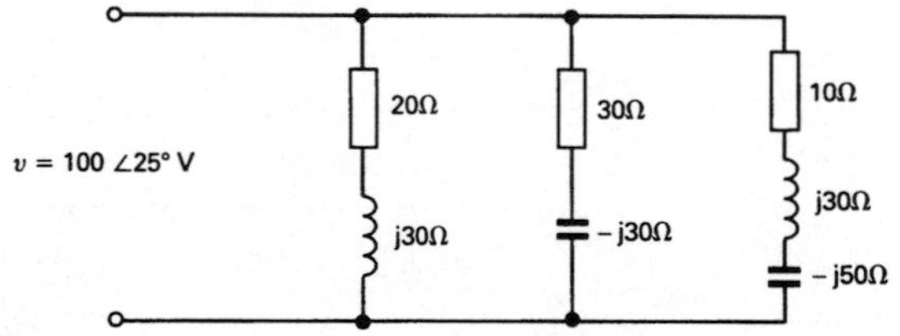

Fig. 3.17

3.30 An inductor of reactance 5 Ω and resistance 5 Ω has a voltage $10\angle 45°$ V applied across it. Calculate the real power dissipated in the inductor.

3.31 The average power dissipated in a 1000 Ω resistor is 200 mW. Calculate (*a*) the r.m.s. current and (*b*) the peak voltage across the resistor.

3.32 For the circuit given in Fig. 3.18 calculate (*a*) its impedance, (*b*) the power factor, (*c*) the true power dissipated, (*d*) the reactive power and (*e*) the apparent power.

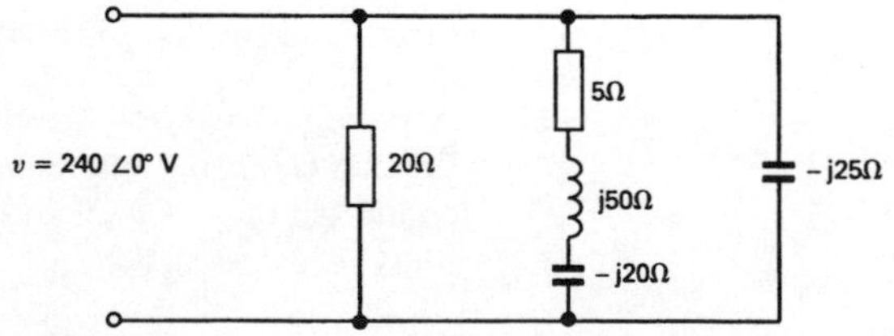

Fig. 3.18

3.33 The peak value of the voltage applied across a circuit is 200 V. The average power dissipated in the circuit is 1.5 kW and the power factor is 0.9. Calculate (*a*) the r.m.s. current flowing into the circuit, and (*b*) the phase angle between the applied voltage and the current.

3.34 An inductor of reactance 50 Ω and resistance 2.5 Ω is connected in parallel with a capacitor of reactance 15 Ω. Another capacitor of reactance 20 Ω is connected in series and $100\angle 0°$ V is applied across the circuit. Calculate the real power dissipated in the circuit.

3.35 The voltage $v = 20\sin(\omega t + 20°)$ is applied to a circuit whose impedance is $Z = 100 + \mathrm{j}60$ Ω. (*a*) Calculate the r.m.s. current flowing into the circuit. (*b*) Calculate the complex power.

3.36 A 50 mH inductor of 100 Ω resistance is connected in parallel with a 200 Ω resistor. (*a*) Calculate the conductance, susceptance, and impedance of the circuit at $\omega = 20 \times 10^3$ rad/s. (*b*) Calculate the value of capacitance which when connected in parallel will give unity power factor.

3.37 100 V at 8 kHz is applied across a circuit of impedance $100 + \mathrm{j}300$ Ω. Calculate (*a*) the current and (*b*) the power dissipated.

3.38 Two admittances, (*i*) 0.01 – 0.01 mS and (*ii*) $0.03 + \mathrm{j}0.04$ mS, are connected in series, and a 100 Ω resistor is connected in parallel with them. Determine the capacitance value which when connected in series will give the circuit unity power factor.

3.39 A circuit having a conductance of 39 μS and a susceptance of –50 μS has a voltage of 30 V applied across it. Calculate the power dissipated in the circuit.

4 Resonant circuits

When an a.c. voltage is applied to a circuit which contains both inductance and capacitance the amplitude and phase of the current that flows will depend upon the frequency of the applied voltage. At one particular frequency, known as the *phase resonant frequency*, the circuit will be purely resistive and so the current will be in phase with the applied voltage. The circuit is then said to be *phase resonant*. For a series circuit the impedance at resonance is at its minimum value, equal to the resistance when the current is at its maximum value. When a parallel circuit is phase resonant its impedance is high and purely resistive and known as the *dynamic resistance*, but it is not quite at its maximum possible value. The maximum impedance occurs at a slightly higher frequency that is known as the *amplitude resonant frequency*. The *Q factor* of a circuit is a measure of its 'quality' and has a magnifying effect upon either the applied voltage or the input current. If the Q factor of the circuit is greater than about 10 the difference between the phase and amplitude resonant frequencies is small and it is generally neglected. When a parallel circuit is said to be resonant, phase resonance is normally assumed.

A *resonant circuit* consists of an inductor and a capacitor connected either in series or in parallel with one another. Some resistance is always present because of the unavoidable resistance of the inductor windings, but often the resistance of the capacitor is negligibly small. Resonant circuits are widely used, particularly in radio engineering, because of their ability to be tuned to select a required narrow band of frequencies from a range of applied frequencies.

The series resonant circuit

A *series resonant circuit* consists of an inductor and a capacitor connected in series as shown by Fig. 4.1. The impedance of the

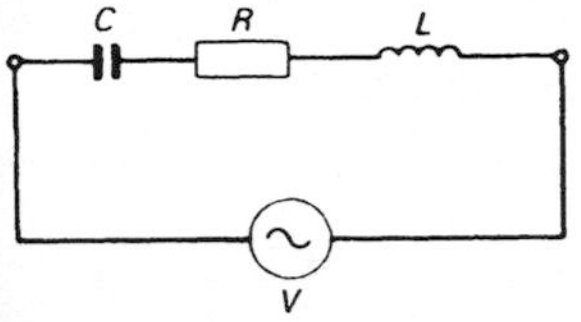

Fig. 4.1 Series resonant circuit

circuit at a particular frequency $\omega/2\pi$ Hz is

$$Z = R + \mathrm{j}[\omega L - (1/\omega C)] \quad (4.1)$$

$$= \sqrt{[R^2 + (\omega L - 1/\omega C)^2]} \angle \tan^{-1}[(\omega L - 1/\omega C)/R] \quad (4.2)$$

The impedance of the circuit varies with change in frequency because of the $[\omega L - (1/\omega C)]$ term. Figure 4.2 shows how the inductive reactance ωL and the capacitive reactance X_C vary with frequency. At one particular frequency, known as the *resonant frequency* f_0, the inductive and capacitive reactances are equal to one another and cancel out, i.e. $\omega_0 L = 1/\omega_0 C$. Then,

$$\omega_0^2 = 1/LC, \quad \omega_0 = 1/\sqrt{(LC)}$$

or $$f_0 = 1/[2\pi\sqrt{(LC)}] \quad (4.3)$$

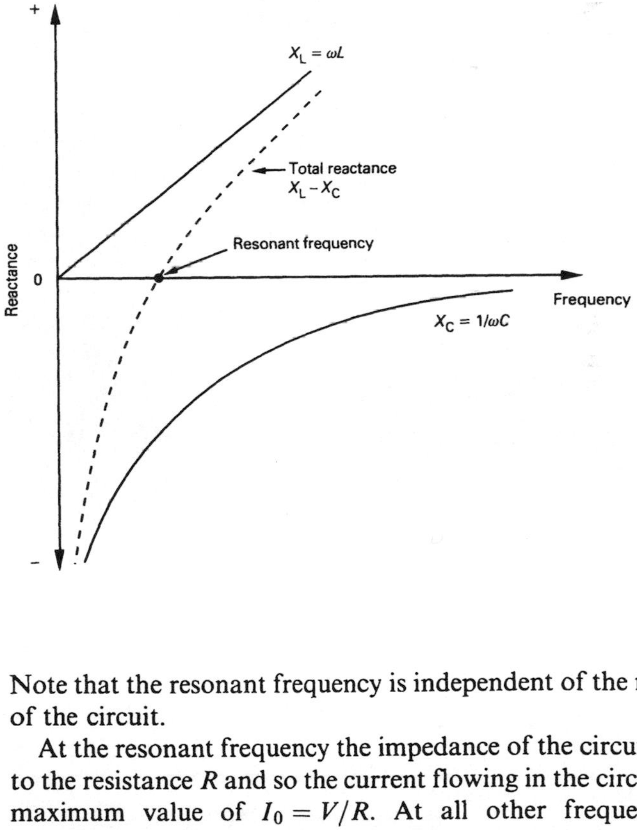

Fig. 4.2 Reactance-frequency curves

Note that the resonant frequency is independent of the resistance of the circuit.

At the resonant frequency the impedance of the circuit is equal to the resistance R and so the current flowing in the circuit has its maximum value of $I_0 = V/R$. At all other frequencies the impedance of the circuit is greater than the resonant value since $[\omega L - (1/\omega C)] \neq 0$. The impedance at any frequency *can* be obtained with the aid of equation (4.2) but a more convenient method involves the use of the Q factor (p. 80). However, it should be evident that at frequencies above resonance the impedance of the circuit will be inductive, while the impedance will be capacitive at frequencies below resonance.

Example 4.1

A 200 mH inductor of resistance 15 Ω is connected in series with a 0.22 μF capacitor. Calculate (*a*) the resonant frequency, (*b*) the inductive reactance, the capacitive reactance and the impedance at resonance and (*c*) the resonant current when 5 V is applied across the circuit.

Solution

(*a*) $f_0 = 1/[2\pi\sqrt{(0.2 \times 0.22 \times 10^6)}] \approx 759$ Hz (*Ans.*)

(*b*) At the resonant frequency,

$X_L = 2\pi \times 759 \times 200 \times 10^{-3} = \text{j}953.8\ \Omega$ (*Ans.*)

$X_C = 1/(2\pi \times 759 \times 0.22 \times 10^{-6} = -\text{j}953.1\ \Omega$

$Z_0 = R = 15\ \Omega$ (*Ans.*)

(*c*) Resonant current $I_0 = 5/15 = 0.33$ A (*Ans.*)

The parallel resonant circuit

Figure 4.3 shows a resistance R connected in parallel with a pure inductance L and a pure capacitance C to form an 'ideal' parallel resonant circuit.

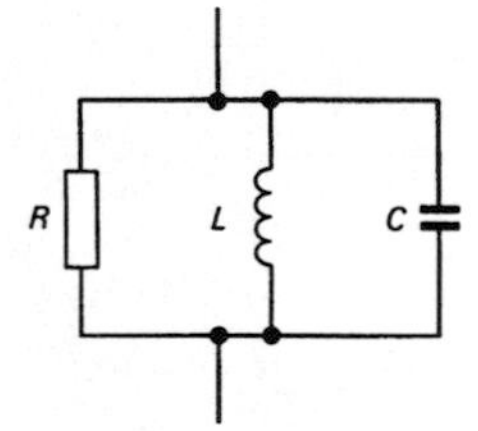

Fig. 4.3 Ideal parallel resonant circuit

The admittance Y of the circuit is

$$Y = G + \text{j}(B_C - B_L) = 1/R + \text{j}(\omega C - 1/\omega L)$$

The admittance is at its minimum value when its imaginary part is equal to zero. Then, $\omega_0 C = 1/\omega_0 L$, $\omega_0^2 = 1/LC$ and the resonant frequency f_0 is

$$f_0 = 1/[2\pi\sqrt{(LC)}] \qquad (4.4)$$

At this frequency the admittance of the circuit is $Y = 1/R$ and hence its impedance at resonance is $Z = 1/Y = R$. The ideal parallel resonant circuit has its maximum impedance of $R\ \Omega$ at the resonant frequency.

A practical parallel resonant circuit consists of an inductor and a capacitor connected in parallel as shown in Fig. 4.4. R is the resistance of the inductor and the capacitor is assumed to have a negligible resistance. The total current supplied to the circuit is the phasor sum of the inductive current I_L and the capacitive current I_C. At low frequencies the inductive branch takes the larger current because the inductive reactance is low while the capacitive reactance is high. This means that the impedance of the circuit is inductive. At high frequencies the converse is true: the current in the capacitive branch is larger than the current in the inductive branch and so the impedance of the circuit is capacitive.

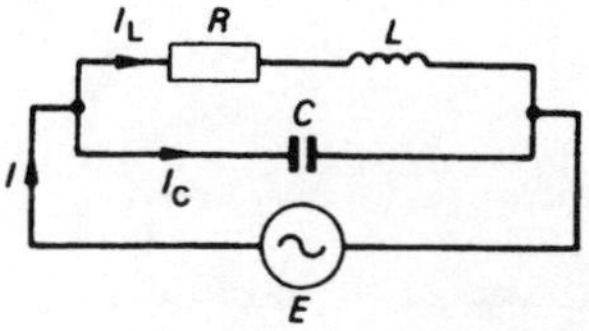

Fig. 4.4 Parallel resonant circuit

At a particular frequency, known as the *resonant frequency*, the impedance of a parallel tuned circuit is purely resistive and it is then generally known as the *dynamic resistance* R_d.

Resonant frequency

The admittances of the inductive and capacitive branches are, respectively, given by

$$Y_L = 1/(R + j\omega L) = (R - j\omega L)/(R^2 + \omega^2 L^2) \quad \text{and}$$
$$Y_C = j\omega C$$

The total admittance of the circuit is $Y_T = Y_L + Y_C$, or

$$Y_T = R/(R^2 + \omega^2 L^2) + j[(\omega C - \omega L)/(R^2 + \omega^2 L^2)] \qquad (4.5)$$

At the resonant frequency $\omega_0/2\pi$ the admittance of the circuit is wholly real; therefore,

$$\omega_0 C - \omega_0 L/(R^2 + \omega_0^2 L^2) = 0, \quad R^2 C + \omega_0^2 L^2 C = L$$
$$\omega_0^2 = L/L^2 C - R^2 C/L^2 C = 1/LC - R^2/L^2 \quad \text{and}$$
$$f_0 = (1/2\pi)\sqrt{(1/LC - R^2/L^2)} \qquad (4.6)$$

This expression can be written as

$$f_0 = [1/(2\pi\sqrt{LC})]\sqrt{(1 - R^2 C/L)} \qquad (4.7)$$

Very often the second term, R^2/L^2, is negligibly small compared with the first term and then equation (4.6) reduces to the same form as that derived for the series resonant frequency, i.e.

$$f_o = 1/(2\pi\sqrt{(LC)})$$

Resonant impedance

Substituting for ω_0^2 in the real part of equation (4.5) gives

$$Y_T = G_T = R/\{R^2 + [(1/LC) - (R^2/L^2)]L^2\}$$
$$= R/[R^2 + (L/C) - R^2] = CR/L$$

Therefore, the dynamic resistance R_d of a parallel tuned circuit is

$$R_d = 1/G_T = L/CR \ \Omega \qquad (4.8)$$

When a series or a parallel tuned circuit is at resonance its impedance is purely resistive and so the circuit then has unity power factor.

Amplitude resonant frequency

Although the impedance R_d at phase resonance is purely resistive it is *not* the maximum value of the impedance. The maximum value of the impedance occurs at the amplitude-resonant frequency f_a. The amplitude resonant frequency f_a is the frequency at which the impedance of a parallel tuned circuit reaches its maximum value of impedance:

$$f_a = [1/(2\pi\sqrt{(LC)})]\sqrt{(1 - R^2C/4L)} \qquad (4.9)$$

Example 4.2

A 10 mH inductor of 50 Ω resistance is connected in parallel with a 22 nF capacitor. Calculate (*a*) the phase resonant frequency, (*b*) the amplitude resonant frequency, (*c*) the dynamic resistance and (*d*) the maximum value of the impedance.

Solution

$$(a)\ f_0 = \left[\frac{1}{2\pi\sqrt{(0.01 \times 22 \times 10^{-9})}}\right]\sqrt{\left[1 - \frac{(50^2 \times 22 \times 10^{-9})}{0.01}\right]}$$
$$= 10700 \text{ Hz} \quad (Ans.)$$

$$(b)\ f_a = \left[\frac{1}{2\pi\sqrt{(0.01 \times 22 \times 10^{-9})}}\right]\sqrt{\left[1 - \frac{(50^2 \times 22 \times 10^{-9})}{4 \times 0.01}\right]}$$
$$= 10723 \text{ Hz} \quad (Ans.)$$

(*c*) $R_d = L/CR = (10 \times 10^{-3})/(50 \times 22 \times 10^{-9}) = 9091\ \Omega$ (*Ans.*)

(*d*) $jX_L = 2\pi \times 10723 \times 0.01 = j673.75\ \Omega$

$X_C = 1/(2\pi \times 10723 \times 22 \times 10^{-9}) = -j674.65\ \Omega$

$Z_{max} = [(50 + j673.75) \times (-j674.65)]/(50 + j673.75 - j674.65)$

$= 9100 - j510.6 = 9115\angle -3.2°\ \Omega$ (*Ans.*)

Example 4.3

An inductor has a series resistance of 100 Ω and an inductance of 40 mH. (*a*) Calculate its conductance and susceptance at a frequency of 5000/π Hz. (*b*) Calculate the value of capacitance to be connected (*i*) in series and (*ii*) in parallel with the inductor for the circuit to have unity power factor. For each case calculate the circuit's impedance.

Solution

(*a*) $Y_L = 1/(100 + j400) = (100 - j400)/(170 \times 10^3)$
$= (0.588 - j2.353) \times 10^{-3}$ S (*Ans.*)

(*b*) (*i*) The series connected capacitor must have a reactance of −j400 Ω. Therefore,

$400 = 1/10^4C_s$ or $C_s = 1/(4 \times 10^6) = 0.25\ \mu F$ (*Ans.*)

(*ii*) The parallel-connected capacitor must have a susceptance of $+j2.353 \times 10^{-3}$ S. Therefore,

$$2.353 \times 10^{-3} = 10^4 C_p \quad \text{or} \quad C_p = 0.235\ \mu\text{F} \quad (Ans.)$$

Resistance in both branches

If an inductor L of resistance R_1 is connected in parallel with a capacitor C and a resistor R_2 the admittance Y_T of the circuit is equal to the sum of the admittances of each branch. The admittance of the inductor is

$$Y_L = 1/(R_1 + jX_L) = (R_1 - jX_L)/(R_1^2 + X_L^2)$$

and the admittance of the capacitive branch is

$$Y_C = 1/(R_2 - jX_C) = (R_2 + jX_C)/(R_2^2 + X_C^2)$$

Hence the total admittance of the circuit is

$$Y_T = R_1/(R_1^2 + X_L^2) + R_2/(R_2^2 + X_C^2) + j[X_C/(R_2^2 + X_C^2) - X_L/(R_1^2 + X_L^2)]$$

At resonance the j terms sum to zero, and hence

$$(1/\omega_0 C)/[R_2^2 + (1/\omega_0 C)^2] = \omega_0 L/(R_1^2 + \omega_0^2 L^2),$$

$$\omega_0 L[R_2^2 + (1/\omega_0 C)^2] = (1/\omega_0 C)(R_1^2 + \omega_0^2 L^2),$$

$$\omega_0^2 LC[R_2^2 + (1/\omega_0 C)^2] = R_1^2 + \omega_0^2 L^2$$

$$\omega_0^2 = (R_1^2 - L/C)/(LCR_2^2 - L^2)$$

$$= (R_1^2 - L/C)/[LC(R_2^2 - L/C)], \quad \text{and}$$

$$f_0 = [1/(2\pi\sqrt{(LC)})]\sqrt{[(R_1^2 - L/C)/(R_2^2 - L/C)]} \qquad (4.10)$$

At this frequency the admittance of the circuit is

$$Y_0 = R_1/(R_1^2 + \omega_0^2 L^2) + R_2/[R_2^2 + (1/\omega_0 C)^2] \qquad (4.11)$$

The impedance of the circuit is purely resistive.

Example 4.4

A 2 mH inductor of 6 Ω resistance is connected in parallel with a 20 μF capacitor that is in series with a 4 Ω resistor. Calculate (*a*) the resonant frequency of the circuit and (*b*) its impedance at resonance.

Solution

(*a*) $f_0 = [1/(2\pi\sqrt{(2 \times 10^{-3} \times 20 \times 10^{-6})})]\sqrt{[(36 - 100)/(16 - 100)]}$

≈ 695 Hz (*Ans.*)

(*b*) At 695 Hz,

$X_L = 2\pi \times 695 \times 2 \times 10^{-3} = 8.73\ \Omega$ and

$X_C = 1/(2\pi \times 695 \times 20 \times 10^{-6}) = 11.45\ \Omega$

$Y = 4/(4^2 + 11.45^2) + 6/(6^2 + 8.73^2) = 0.081$ S and

$Z = 1/0.081 = 12.4\ \Omega$ (*Ans.*)

Q factor

Ideally, components such as inductors and capacitors would have zero resistance and hence dissipate zero power. In practice, of course, all components possess some self-resistance although the resistance of a capacitor is usually small enough to be ignored. Losses in inductors and capacitors will be discussed in Chapter 9.

The losses in a component, or in a circuit, can be expressed in terms of its *Q factor*. Q factor is defined as

$$Q = \frac{2\pi \times \text{Maximum energy stored}}{\text{Energy dissipated per cycle}} \tag{4.12}$$

(at resonance if a tuned circuit)

Inductor

A practical inductor has an inductance *L* and a resistance *R*. The maximum energy stored in the magnetic field due to the inductance is

$$LI^2_{max}/2 \quad \text{or} \quad L(I\sqrt{2})^2/2 = LI^2 \text{ joules}$$

The energy dissipated per cycle in the resistance is I^2R/f joules. Substituting into equation (4.12),

$$Q = (2\pi LI^2)/(I^2R/f) = 2\pi fL/R = \omega L/R \tag{4.13}$$

i.e. the *Q* factor is the ratio reactance/resistance.

Capacitor

The losses in a capacitor *C* can be represented by a resistance *R* connected in series with the capacitor. The maximum energy stored in the electric field due to the capacitance is

$$CV^2_{C\,(max)}/2 = C(V_C\sqrt{2})^2/2 = CV^2_C \text{ joules}$$

The voltage V_C developed across the capacitance *C* is equal to the product of the current *I* and the reactance $1/\omega C$ of the capacitor. Thus the maximum energy stored may be written as

$$CI^2/(\omega C)^2 \quad \text{or} \quad I^2/\omega^2 C \text{ joules}$$

The energy dissipated per cycle in the resistance is I^2R/f and so

$$Q = [(2\pi I^2)/(\omega^2 C)]/(I^2R/f) = (2\pi f)/(\omega^2 CR) = 1/\omega CR \qquad (4.14)$$

This equation may also be written as the ratio reactance/resistance.

Resistance in parallel with inductance

Maximum energy stored is

$$E = L(\sqrt{2}\, I_L)^2/2 = LI_L^2 \text{ J}$$

$$I_L = V/\omega L \quad \text{so} \quad E = LV^2/\omega^2 L^2 = V^2/\omega^2 L$$

Power dissipated per cycle $= I_R^2 R/f$ W

$$I_R = V/R \quad \text{so} \quad P = (V^2/R)(R/f) = V^2/Rf \text{ W}$$

Therefore,

$$Q = 2\pi(V^2/\omega^2 L)/(V^2/R)f_0 = R/\omega_0 L \qquad (4.15)$$

Thus the Q factor is the ratio (resistance/reactance).

Resistance in parallel with a capacitor

The maximum energy stored is $E = CV^2$ J. The power dissipated per cycle is V^2/Rf W.

Therefore,

$$Q = 2\pi(CV^2)/(V^2/Rf) = \omega CR \qquad (4.16)$$

Again, Q = resistance/reactance

Series resonant circuit

In a series resonant circuit energy is continually being transferred between the electric field of the capacitance and the magnetic field of the inductor. Only at resonance are the two fields equal to one another and so it is customary to consider the Q factor of a resonant circuit at its resonant frequency. The maximum energy stored in a series resonant circuit is then LI^2 joules and the energy dissipated per cycle is I^2R/f_0 joules.

Therefore,

$$Q = (2\pi LI^2)/(I^2R/f_0) = \omega_0 L/R = 1/\omega_0 CR$$

$$(\text{since } \omega_0 L = 1/\omega_0 C) \qquad (4.17)$$

The Q factor of a series tuned circuit is the same as that of the inductor alone if the capacitor is lossless and there is no added resistance. $\omega_0 = 1/\sqrt{(LC)}$ and substituting into equation (4.17),

$$Q = L/[R\sqrt{(LC)}] = (1/R)\sqrt{(L/C)} \qquad (4.18)$$

Equation (4.18) demonstrates clearly that for the maximum Q factor the resistance must be as small as possible *and* the L/C ratio should be as high as possible.

The current I_0 flowing in a series tuned circuit at resonance is $I_0 = V/R$, where V is the applied voltage. This current flows in both the inductance and the capacitance and develops voltages, I_0X_L and I_0X_C, respectively, across them. Therefore,

$$V_C = I_0/\omega_0 C = V/R\omega_0 C = QV \qquad (4.19)$$

Example 4.5

A series resonant circuit is connected across a 10 V, 2 MHz supply having an internal resistance of 5 Ω. Calculate the values of inductance and capacitance required to give a capacitor voltage of 250 V at the resonant frequency. The resistance of the inductor is 5 Ω.

Solution
The required Q factor is 250/10 = 25 and hence

$$25 = 0.1\sqrt{(L/C)} \quad \text{or} \quad \sqrt{L} = 10 \times 25 \times \sqrt{C}$$

Thus $2\pi \times 2 \times 10^6 = 1/250C$. Hence,

$$C = 1/(1000\pi \times 10^6) = 318 \text{ pF} \quad (Ans.)$$

Also

$$L = 1/(4\pi^2 \times 4 \times 10^{12} \times 318 \times 10^{-12}) = 19.9\ \mu\text{H} \quad (Ans.)$$

Example 4.6

A 10 mH inductor of resistance 5 Ω is connected in series with a 0.1 μF capacitor. A voltage of 1 V at the resonant frequency is applied to the circuit. Calculate (*a*) the resonant frequency, (*b*) the Q factor and (*c*) the voltages across the capacitor and the inductor.

Solution
(*a*) $f_0 = 1/[2\pi\sqrt{(10 \times 10^{-3} \times 0.1 \times 10^{-6})}] = 5033$ Hz (*Ans.*)
(*b*) $Q = \omega L/R = (2\pi \times 5033 \times 10 \times 10^{-3})/5 = 63.25$ (*Ans.*)
(*c*) $V_C = QV = 63.25$ V (*Ans.*)
$V_L = QV = 63.25$ V, $V_R = I_0 r = (1/5) \times 5 = 1$ V
Voltage across inductor $= 1 + j63.25 \approx 63.26$ V (*Ans.*)

Example 4.7

The Q factor of the circuit in Example 4.6 is to be reduced to 20. What series resistor is required?

Solution

$$R = (2\pi \times 5033 \times 10 \times 10^{-3})/20 \approx 15.8\ \Omega$$

Therefore,

The extra resistance = 10.8 Ω (*Ans.*)

Reactive power

Q reactance/resistance. Multiplying both numerator and denominator by I^2 gives

$$Q = \text{(reactive power)/(real power)} \quad (4.20)$$

Example 4.8

A series tuned circuit dissipates a reactive power of 500 var and a real power of 25 W at the resonant frequency. Calculate the Q factor of the circuit.

Solution

$Q = 500/25 = 20$ (*Ans.*)

Parallel resonant circuit

The maximum energy stored in a parallel tuned circuit is LI_L^2 joules, where I_L is the current flowing in the inductive branch. The energy dissipated per cycle at resonance is I_L^2R/f_0 joules. Therefore,

$$Q = (2\pi LI_L^2)/(I_L^2R/f_0) = \omega_0L/R \quad (4.21)$$

This is the same expression as obtained for the series tuned circuit and it supposes that the capacitor possesses negligible resistance.

The dynamic resistance of a parallel tuned circuit is $L/CR\ \Omega$. Multiplying both the numerator and the denominator by ω_0 gives

$$R_d = \omega_0L/\omega_0CR = Q\omega_0L = Q/\omega_0C \quad (4.22)$$

Also, $R_d = Q\omega_0L = (\omega_0L/R)\omega_0L = \omega_0^2L^2/R$. Hence

$$X_L = \omega_0L = \sqrt{(RR_d)} \quad (4.23)$$

When a parallel tuned circuit has a current I supplied to it a voltage IR_d will be developed across its terminals. The current I_C flowing in the capacitive branch is equal to this voltage divided by the reactance of the capacitor, i.e.

$$I_C = V/X_C = IR_d/(1/\omega_0 C) = IL\omega_0 C/CR$$
$$= I\omega_0 L/R \quad \text{or} \quad I_C = QI \tag{4.24}$$

Similarly, the current I_L flowing in the inductive branch is

$$I_L = V/\surd(R^2 + \omega_0^2 L^2) = V/\{\omega_0 L\surd[1 + R^2/(\omega_0 L)^2]\}$$
$$= (IL/CR)/\{\omega_0 L\surd[1 + (1/Q^2)]\}$$
$$I_L = QI/\surd[1 + (1/Q^2)] \approx QI \tag{4.25}$$

Equations (4.24) and (4.25) show that the current flowing in either branch of a parallel tuned circuit is Q times greater than the total current supplied to the circuit.

Equation (4.6) for the phase resonant frequency of a parallel tuned circuit can be written in terms of the Q factor,

$$f_0 = [1/(2\pi\surd(LC))]\surd[1 - (1/Q^2)] \tag{4.26}$$

Example 4.9

Calculate the Q factor of a 120 μH inductor of 10 Ω resistance connected in parallel with a 100 pF capacitor.

Solution

$$R_d = (120 \times 10^{-6})/(10 \times 100 \times 10^{-12}) = 120 \text{ k}\Omega$$
$$X_L = \surd(120 \times 10^3 \times 10) = 1095.45 \ \Omega$$
$$120 \times 10^3 = Q \times 1095.45, \text{ therefore, } Q = 109.6 \quad (\textit{Ans.})$$

Example 4.10

A parallel tuned circuit consists of an inductance L of 50 μH and 10 Ω connected in parallel with a loss-free capacitor at a frequency of $10^7/2\pi$ Hz. The circuit is connected in series with a 1 mA constant current source. Calculate (a) the current in the inductor, (b) the dynamic resistance and (c) the voltage across the circuit.

Solution

$$Q = \omega_0 L/R = (10^7 \times 50 \times 10^{-6})/10 = 50$$

(a) $I_L = QI = 50$ mA (*Ans.*)

(b) $R_d = Q\omega_0 L = 50 \times 500 = 25$ kΩ (*Ans.*)

(c) $V = IR_d = 25$ V (*Ans.*)

Example 4.11

A 200 mH inductor of Q factor 63.6 is connected in parallel with a 0.22 μF capacitor. Calculate its resonant frequency.

Solution

$$f_0 = [1/(2\pi\sqrt{(0.2 \times 0.22 \times 10^{-6})})]\sqrt{(1 - 63.6^2)} = 759 \text{ Hz} \quad (Ans.)$$

Example 4.12

A parallel tuned circuit is to be resonant at 1 MHz with a Q factor of 30. When 1 V at 1 MHz is applied the resonant current is to be 50 μA. Calculate the required values of the inductance, capacitance and resistance in the circuit.

Solution

$$1 \times 10^6 = 1/(2\pi\sqrt{(LC)}), \qquad 30 = (2\pi \times 1 \times 10^6 L)/R$$
$$R_d = L/CR = 1/(50 \times 10^{-6}) = 20 \times 10^3$$

Therefore,

$$10^{12} = 1/(4\pi^2 LC) \quad \text{and} \quad C = 1/(4\pi^2 L \times 10^{12})$$
$$20 \times 10^3 = (4\pi^2 L^2 \times 10^{12})R, \qquad R = (4\pi^2 L^2 \times 10^9)/20$$
$$30 = (2\pi \times 1 \times 10^6 L \times 20)/(4\pi^2 L^2 \times 10^9) \quad \text{or}$$
$$L = 106\ \mu\text{H} \quad (Ans.)$$

Therefore

$$R = (2\pi \times 1 \times 10^6 \times 106 \times 10^{-6})/30 = 22.2\ \Omega \quad (Ans.)$$

Lastly,

$$C = (106 \times 10^{-6})/(22.2 \times 20 \times 10^3) = 239 \text{ pF} \quad (Ans.)$$

Effective Q factor

In most cases the voltage source applied across a series resonant circuit will not have an internal resistance of very nearly zero ohms. The source impedance will then be in series with a series tuned circuit or in parallel with a parallel tuned circuit. The Q factor of the series circuit will be reduced from $Q = \omega_0 L/R$ to $Q_{eff} = \omega_0 L/(R + R_s)$ or

$$Q_{eff} = (\omega_0 L/R)[1/(1 + R_s/R)] = Q/(1 + R_s/R) \qquad (4.27)$$

For a parallel circuit consider Fig. 4.5(a) which shows such a circuit shunted by a resistor R_p. At the resonant frequency the tuned circuit can be represented by its dynamic resistance R_d (see Fig. 4.5(b)). The two resistances are in parallel and their total resistance, the *effective dynamic resistance*, $R_{d(eff)}$, is

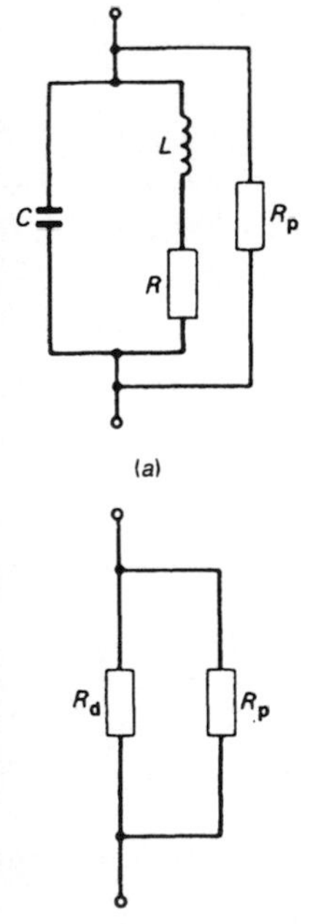

Fig. 4.5 Effective Q factor

$R_{d(eff)} = R_d R_p/(R_d + R_p)$. Therefore,

$$Q_{eff} = R_{d(eff)}/\omega_0 L = Q\omega_0 L R_p/(Q\omega_0 L + R_p)\omega_0 L$$
$$= QR_p/(R_d + R_p), \text{ or}$$
$$Q_{eff} = Q/(1 + R_d/R_p) \quad (4.28)$$

Example 4.13

A parallel tuned circuit has a 100 pF capacitor connected in parallel with an inductor of Q factor 100. The circuit is resonant at 1 MHz. Calculate its effective Q factor when a resistance of 20 kΩ is connected in parallel.

Solution
From equation 4.22

$$R_d = 100/(2\pi \times 1 \times 10^6 \times 100 \times 10^{-12}) = 10^6/2\pi \ \Omega$$

Hence

$$Q_{eff} = 100/[1 + 10^6/(2\pi \times 20 \times 10^3)] = 11.2 \quad (Ans.)$$

Example 4.14

For the circuit shown in Fig. 4.6 calculate (*a*) the resonant frequency, (*b*) the Q factor and (*c*) the percentage error in the resonant frequency if the presence of the 10 Ω resistor is neglected.

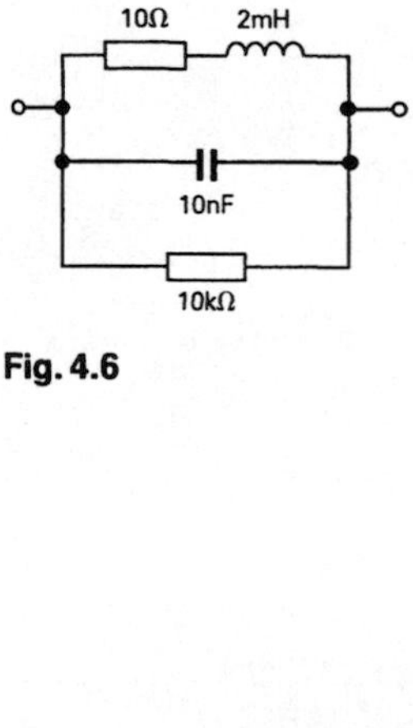

Fig. 4.6

Solution

$$(a)\ f_0 = \left[\frac{1}{2\pi\sqrt{(2 \times 10^{-3} \times 10 \times 10^{-9}}}\right]\sqrt{\left[1 - \frac{(10^2 \times 10 \times 10^{-9})}{(2 \times 10^{-3})}\right]}$$
$$= 35.579 \text{ kHz} \quad (Ans.)$$

$$(b)\ Q = \omega_0 L/R = (2\pi \times 35.579 \times 10^3 \times 2 \times 10^{-3})/10 = 44.7$$
$$Q_{eff} = 44.7/(1 + 20/10) = 14.9 \quad (Ans.)$$
$$R = QX_L$$
$$= 44.7 \times 2\pi \times 35.579 \times 10^3 \times 2 \times 10^{-3}$$
$$\approx 20 \text{ k } \Omega$$

$$(c)\ f_0 = 1/(2\pi)\sqrt{(2 \times 10^{-3} \times 10 \times 10^{-9})} = 35.588 \text{ kHz}$$

$$\% \text{ error} = [(35.579 - 35.588)/35.579] \times 100 = -0.025\% \quad (Ans.)$$

Q Factors in series and in parallel

When the losses of a capacitor are not negligibly small, the overall Q factor of the circuit will depend upon the Q factors of the individual components.

Suppose that the Q factors of the inductor and the capacitor are, respectively, Q_L and Q_C. The overall factor Q_T is

$$Q_T = (1/R_T)\sqrt{(L/C)} \quad \text{where} \quad R_T = R_L + R_C$$

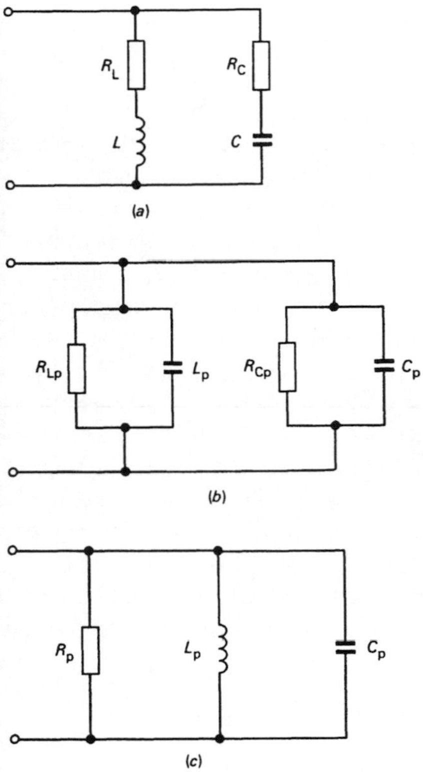

Fig. 4.7 *Q* factors in parallel

Hence

$$Q_T = \{1/[(\omega_0 L/Q_L) + (1/\omega_0 C Q_C)]\}[\sqrt{(L/C)}]$$
$$= \{1/[(1/Q_L)\sqrt{(L/C)} + (1/Q_C)\sqrt{(L/C)}]\}[\sqrt{(L/C)}]$$
$$Q_T = 1/[(1/Q_L) + (1/Q_C)] = Q_L Q_C/(Q_L + Q_C) \qquad (4.29)$$

Equation (4.29) can also be applied to find the overall Q factor of two Q factors in *parallel*. Figure 4.7(a) shows an inductor L of Q factor Q_L in parallel with a capacitor C of Q factor Q_C. Converting both components to their equivalent parallel circuits gives Fig. 4.7(b). This circuit can be simplified to give Fig. 4.7(c), where $R_p = R_{Lp}R_{Cp}/(R_{Lp} + R_{Cp})$. Now $Q_L = R_{Lp}/\omega L_p$ and $Q_C = \omega C_p R_{Cp}$. The Q factor of figure (c) is $R_p\sqrt{(C_p/L_p)}$, and $\omega_0 = 1/\sqrt{(L_p C_p)}$, or $\sqrt{C_p} = 1/(\omega_0\sqrt{L_p})$. Therefore,

$$1/Q = (1/R_p)\sqrt{(L_p/C_p)} = \omega_0 L_p/R_p$$
$$= \omega_0 L_p/R_{Lp} + \omega_0 L_p/R_{Cp} = 1/Q_L + 1/Q_C \qquad (4.30)$$

Example 4.15

An inductor of Q factor 80 is connected in parallel with a capacitor having a Q factor of 500. Calculate the Q factor of the combination.

Solution
From equation (4.30)

$$Q = (80 \times 500)/(80 \times 500) = 69 \quad (Ans.)$$

Selectivity of a resonant circuit

The *selectivity* of a resonant circuit is its ability to discriminate between signals at different frequencies. If a constant voltage source is applied to a series resonant circuit the current flowing in the circuit will vary with frequency in the manner shown in Fig. 4.8(a). The current flowing at the resonant frequency

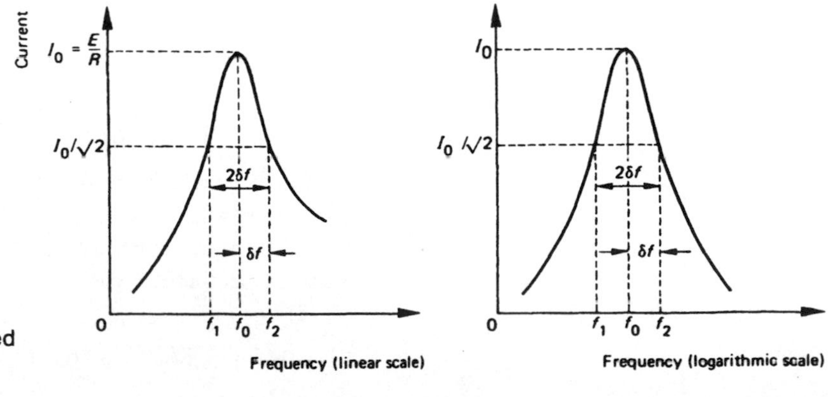

Fig. 4.8 Selectivity of a series tuned circuit

depends solely upon the resistance of the circuit ($I_0 = E/R$), and the shape of the curve depends upon the Q factor of the circuit. Reducing the circuit resistance increases the peak value of the curve, and by increasing the Q factor, makes the circuit more selective. The selectivity of a resonant circuit is usually expressed in terms of its 3 dB bandwidth. This is the bandwidth $f_2 - f_1$ over which the current is not less than $1/\sqrt{2}$ times the resonant current. The current-frequency curve is not symmetrical about the resonant frequency if a linear frequency base is employed, since, for example, the current at $0.8f_0$ is equal to the current at $1.25f_0$. To obtain a symmetrical curve the current should be plotted to a logarithmic frequency base, Fig. 4.8(*b*), although if the Q factor is fairly high the difference between the two curves is small and is often neglected.

If a current is supplied to a parallel resonant circuit, then the voltage developed across the circuit will vary with change in frequency in the same way as the current in a series resonant circuit. Thus, Fig. 4.6 will show the voltage-frequency characteristic of a parallel resonant circuit if the vertical axis is changed to read $V_0 = IR_d$ at resonance, where I is the constant current and R_d is the dynamic impedance of the circuit.

Series tuned circuit

The impedance at any frequency $\omega/2\pi$ of a series tuned circuit is $Z = R + j[\omega L - (1/\omega C)]$.

The current i flowing in the circuit is $i = V/Z$ or

$$i = V/[R + j(\omega L - 1/\omega C)]$$

$$= (V/R)/\{1 + (j\omega_0 L/R)[\omega/\omega_0 - 1/(\omega_0 \omega LC)]\}$$

$$= I_0/[1 + jQ(\omega/\omega_0 - \omega_0/\omega)] = I_0/[1 + jQ(\omega^2 - \omega_0^2)/\omega_0\omega]$$

Let $\omega = \omega_0 + \delta\omega_0$. Then $\omega^2 = \omega_0^2 + 2\omega_0\,\delta\omega_0 + (\delta\omega_0)^2$. The last term is negligibly small and hence

$$i \approx I_0/[1 + jQ(2\omega_0\,\delta\omega_0/\omega_0^2)] = I_0/[1 + jQ(2\delta\omega_0/\omega_0)] \quad (4.31)$$

The term $2\delta\omega_0$ is the bandwidth of the circuit for any required number of dB reduction on the maximum current value. Hence equation (4.31) can be re-written as

$$i = I_0/[1 + (jQB/f_0)] \quad (4.32)$$

where B is the bandwidth and f_0 is the resonant frequency. Equation (4.32) can be used to obtain a very useful equation for the 3 dB bandwidth of a series tuned circuit in terms of its resonant frequency and its Q factor. The magnitude of equation (4.32) is

$$|i/I_0| = 1/\sqrt{[1 + (Q^2B^2/f_0^2)]} \quad (4.33)$$

3 dB bandwidth

3 dB is a current ratio of $1/\sqrt{2}$. Thus for 3 dB reduction,

$$1/\sqrt{2} = 1/\sqrt{[1 + (Q^2 B_{3dB}^2/f_0^2)]}$$

Hence $1 = Q^2 B_{3dB}^2/f_0^2 = QB_{3dB}/f_0$ or

$$B_{3dB} = f_0/Q \tag{4.34}$$

The lower 3 dB frequency f_1 is equal to

$$f_0 - (B_{3dB}/2) = f_0 - (f_0/2Q) = f_0[1 - (1/2Q)] \tag{4.35}$$

The upper 3 dB frequency f_2 is equal to

$$f_0[1 + (1/2Q)] \tag{4.36}$$

The product $f_1 f_2$ of the two 3 dB frequencies is

$$f_0^2(1 - 1/2Q)(1 + 1/2Q) = f_0^2 \tag{4.37}$$

Therefore,

$$f_0 = \sqrt{(f_1 f_2)} \tag{4.38}$$

This means that the resonant frequency is equal to the geometric mean of the upper and lower 3 dB frequencies.

A useful expression for the current flowing in a series tuned circuit can be obtained in terms of the resonant frequency, the Q factor and the 3 dB bandwidth:

$$|i/I_0| = 1/\sqrt{\{1 + Q^2[(\omega^2 - \omega_0^2)/\omega_0\omega]\}} \tag{4.39}$$

At the lower and upper 3 dB frequencies f_1 and f_2, this ratio is equal to $1/\sqrt{2}$. Hence at the lower 3 dB frequency,

$$\sqrt{2} = \sqrt{[1 + Q^2[(f_0^2 - f_1^2)/f_0 f_1]^2]}$$
$$1 = Q^2[(f_0^2 - f_1^2)/f_1 f_0]^2$$

Taking the positive root gives

$$1 = Q[(f_0^2 - f_1^2)/f_0 f_1] \quad \text{and hence} \quad f_0 f_1 = Q(f_0^2 - f_1^2)$$

From this,

$$f_1 = [f_0\sqrt{(1/Q + 4)} - f_0/Q]/2 \tag{4.40}$$

Similarly at the upper 3 dB frequency, f_2,

$$\sqrt{2} = \sqrt{\{1 + Q^2[(f_2^2 - f_0^2)/f_0 f_2]^2\}}, \text{and}$$
$$f_2 = [f_0\sqrt{(1/Q + 4)} + f_0/Q]/2 \tag{4.41}$$

The 3 dB bandwidth is $f_2 - f_1 = f_0/Q$ as before.

Example 4.16

A 15 mH inductor has negligible self-resistance and is connected in series with a 75 Ω resistor and a 22 nF capacitor. Calculate (*a*) the resonant frequency, (*b*) the Q factor, (*c*) the lower and upper 3 dB frequencies and (*d*) the 3 dB bandwidth.

Solution

(*a*) $f_0 = 1/[2\pi\sqrt{(LC)}] = [1/(2\pi\sqrt{(15 \times 10^{-3} \times 22 \times 10^{-9})})]$
$= 8761$ Hz (*Ans.*)

(*b*) $Q = (1/R)\sqrt{(L/C)} = (1/75)\sqrt{[(15 \times 10^{-3})/(22 \times 10^{-9})]}$
$= 11$ (*Ans.*)

(*c*) $f_1 = [(8761\sqrt{(1/121 + 4)}) - 8761/11]/2 = 8372$ Hz (*Ans.*)
$f_2 = [(8761\sqrt{(1/121 + 4)}) + 8761/11]/2 = 9168$ Hz (*Ans.*)

(*d*) 3 dB bandwidth $= 9168 - 8372 = 796$ Hz (*Ans.*)

Note that the two 3 dB frequencies are not symmetrically situated either side of the resonant frequency f_0.

Other bandwidths

Equation (4.33) can also be used to obtain the 6 dB, 10 dB and other bandwidths of a tuned circuit.

6 dB bandwidth

6 dB down is a voltage or current ratio of 1/2. Hence, $1/2 = 1/\sqrt{(1 + Q^2B^2_{6dB}/f_0^2)}$,

$$3 = Q^2B^2_{6dB}/f_0^2 = B^2_{6dB}/B^2_{3dB} \quad \text{or} \quad B_{6dB} = \sqrt{3}\,B_{3dB} \qquad (4.42)$$

10 dB bandwidth

10 dB down is a voltage or current ratio of $1/\sqrt{10}$ and hence $10 = 1 + Q^2B^2_{10dB}/f_0^2$, and

$$B_{10dB} = 3B_{3dB} \qquad (4.43)$$

20 dB bandwidth

20 dB down is a voltage or current ratio of 1/10 and hence $100 = 1 + Q^2B^2_{20dB}/f_0^2$, and

$$B_{20dB} = \sqrt{99}\,B_{3dB} \qquad (4.44)$$

Example 4.17

A series tuned circuit is resonant at 400 kHz and has a Q factor of 50. Calculate (*a*) its 3 dB bandwidth, (*b*) its 6 dB bandwidth, (*c*) its 10 dB bandwidth and (*d*) its 20 dB bandwidth.

Solution

(*a*) $B_{3dB} = f_0/Q = (400 \times 10^3)/50 = 8$ kHz (*Ans.*)

(*b*) $B_{6dB} = \sqrt{3} \times 8 = 13.86$ kHz (*Ans.*)

(*c*) $B_{10dB} = 3 \times 8 = 24$ kHz (*Ans.*)

(*d*) $B_{20dB} = \sqrt{99} \times 8 = 75.6$ kHz (*Ans.*)

Example 4.18

A series tuned circuit is resonant at 15 kHz. Calculate its lower and upper 3 dB frequencies (*a*) using equations (4.40) and (4.41) and (*b*) equations (4.35) and (4.36) if the Q factor is 8.

Solution

(*a*) (*i*) $f_1 = \{(15 \times 10^3)\sqrt{[(1/8^2 + 4)} - (15 \times 10^3)/8]\}/2$
$= 14.09$ kHz (*Ans.*)

$f_2 = \{(15 \times 10^3)\sqrt{[(1/8^2 + 4)} + (15 \times 10^3)/8]\}/2$
$= 15.967$ kHz (*Ans.*)

(*ii*) $f_1 = (15 \times 10^3) - [(15 \times 10^3)/8]/2 = 14.063$ kHz (*Ans.*)

$f_2 = (15 \times 10^3) + [(15 \times 10^3)/8]/2 = 15.938$ kHz (*Ans.*)

Equation (4.1) can be written as

$$Z = R + j\omega L(1 - 1/\omega^2 LC) = R[1 + jQ(1 - \omega_0^2/\omega^2)] \quad (4.45)$$

Parallel tuned circuit

The impedance of a parallel tuned circuit can be written as the product of the impedances of the two arms divided by their sum, i.e.

$$\begin{aligned} Z &= [(R + j\omega L)(1/j\omega C)]/[R + j(\omega L - 1/\omega C)] \\ &\approx (L/C)/[R + j(\omega L - 1/\omega C)] \\ &= (L/CR)/[1 + j(1/R)(\omega L - 1/\omega C)] \\ &= R_d/[1 + j(1/R)(\omega L - 1/\omega C)] \end{aligned}$$

This equation is of the same form as that obtained for the series circuit and following the same steps as before, will lead to a similar result, i.e.

$$Z = R_d/[1 + j(QB/f_0)] \quad (4.46)$$

Example 4.19

A parallel tuned circuit is resonant at 120 kHz and has a 3 dB bandwidth of 6 kHz. Calculate the bandwidth over which its impedance is (*a*) 6 dB and (*b*) −10 dB less than the dynamic resistance.

Solution

$$Q = f_0/B_{3dB} = (120 \times 10^3)/(6 \times 10^3) = 20$$

(*a*) 6 dB is a voltage ratio of 2 : 1, hence

$$|Z/R_d| = 1/2 = 1/\sqrt{[1 + (Q^2 B_{6dB}^2/f_0^2)]}$$
$$= 1/\sqrt{[1 + (400 \times B_{6dB}^2/(144 \times 10^8))]}$$
$$3 = 400B_{6dB}^2/(144 \times 10^8)$$
$$B_{6dB} = \sqrt{[(3 \times 144 \times 10^8)/400]} = 10.39 \text{ kHz} \quad (\textit{Ans.})$$

(*b*) 10 dB is a voltage ratio of 3.162:1, hence

$$|Z/R_d| = 1/3.162 = 1/\sqrt{[1 + (400 \times B_{10dB}^2/(144 \times 10^8))]}$$
$$9 = 400B_{10dB}^2/(144 \times 10^8)$$

and therefore

$$B_{10dB} = 3B_{3dB} = 18 \text{ kHz} \quad (\textit{Ans.})$$

Series-parallel circuit

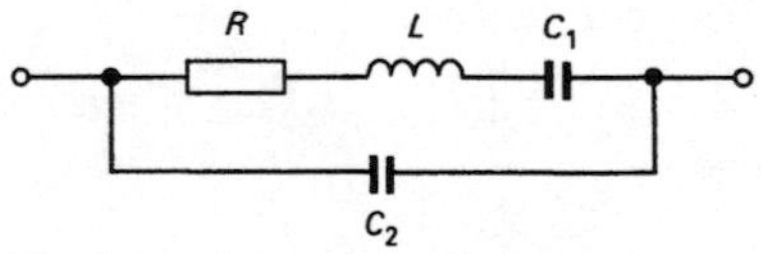

Fig. 4.9. Series-parallel circuit

Figure 4.9 shows a circuit in which the series connection of an inductance L, a resistance R and a capacitor C_1 is connected in parallel with another capacitor C_2. The series branch of the circuit will be series resonant at a frequency f_s at which $\omega_s L = 1/\omega_s C_1$, or $\omega_s^2 = 1/LC_1$, and hence

$$f_s = 1/(2\pi\sqrt{LC_1}) \tag{4.47}$$

At frequencies higher than f_s the inductive reactance of the circuit will become greater than the reactance of the series capacitor C_1 and so the effective reactance of the series arm is inductive. At some particular frequency the effective inductive reactance of the series branch will become equal to the reactance of the shunt capacitor C_2. At this frequency parallel resonance takes place. Then,

$$\omega_p C_2 = 1/(\omega_p L - 1/\omega_p C_1)$$
$$\omega_p C_2(\omega_p L - 1/\omega_p C_1) = 1$$
$$\omega_p^2 C_2 L - C_2/C_1 = 1,$$

or

$$\omega_p^2 = (1 + C_2/C_1)/C_2 L = 1/C_2 L + 1/C_1 L$$
$$= (1/LC_1)(1 + C_1/C_2) = \omega_s^2(1 + C_1/C_2)$$

and

$$f_p = f_s\sqrt{(1 + C_1/C_2)} \tag{4.48}$$

Example 4.20

A 50 mH inductor of 20 Ω resistance is connected in series with a 180 Ω resistor and a 100 nF capacitor. A 20 nF capacitor is then connected in parallel with the series circuit. Determine (*a*) the series resonant frequency and (*b*) the parallel resonant frequency of the circuit.

Solution

(*a*) $f_s = 1/[2\pi\sqrt{(50 \times 10^{-3} \times 100 \times 10^{-9})}] = 2251$ Hz (*Ans.*)

(*b*) $f_p = 2251\sqrt{(1 + 100/20)} = 5514$ Hz (*Ans.*)

This kind of series-parallel circuit occurs in the electrical equivalent circuit of piezo-electric crystals that are commonly employed in electronic circuits. Typical figures for such a crystal are: $L = 1$ H, $R = 1000$ Ω, $C_1 = 0.5$ pF and $C_2 = 10$ pF.

Exercises 4

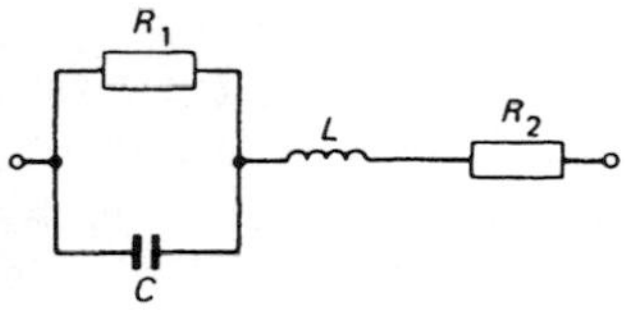

Fig. 4.10

4.1 Derive an expression for the resonant frequency of the circuit shown in Fig. 4.10. If $R_1 = 2000$ Ω, $R_2 = 30$ Ω, $L = 20$ mH and $C = 0.3$ μF, calculate (*a*) the resonant frequency and (*b*) the impedance of the circuit at resonance.

4.2 A parallel tuned circuit is resonant at 100 kHz and has a 3 dB bandwidth of 5 kHz. Calculate the bandwidth over which its impedance is (*a*) −6 dB, (*b*) −10 dB and (*c*) −20 dB relative to the resonant value.

4.3 Three series tuned circuits are connected in series with one another. The data for the circuits are
Circuit A $R = 10$ Ω, LC product $= 2 \times 10^{-10}$, L/C ratio $= 5000$
Circuit B $R = 20$ Ω, LC product $= 2 \times 10^{-10}$, L/C ratio $= 5000$
Circuit C $R = 10$ Ω, LC product $= 2 \times 10^{-10}$, L/C ratio $= 1250$
For each of these circuits find its capacitance, its inductance, its resonant frequency and its Q factor. Also find the 3 dB bandwidth of each circuit, and the overall 3 dB bandwidth and resonant frequency.

4.4 A coil of inductance 50 mH and Q factor 80 is connected in parallel with a 600 pF loss-free capacitor. This circuit is supplied by a constant current source at the resonant frequency. If the supply current is I mA and the voltage developed across the circuit is 25 V calculate (*a*) the resonant frequency, (*b*) the capacitor current, (*c*) the Q factor of the coil and (*d*) the voltage across the circuit when a resistor of 250 kΩ is connected in parallel.

4.5 A series tuned circuit is resonant at 5 MHz and has a Q factor of 63. Calculate its lower and upper 3 dB frequencies.

4.6 An inductor of $Q = 200$ has a dynamic resistance of 1 MΩ. A 4.7 kΩ resistor is coupled via a 1 : 10 tapping on a transformer. Calculate the effective Q factor of the inductor.

4.7 A series tuned circuit uses a 10 mH inductor of self-resistance 28 Ω and it is to be resonant at 24 kHz with a 3 dB bandwidth of 3 kHz. Calculate (*a*) the required capacitor value, (*b*) the maximum resistor value across which the output voltage can be developed and (*c*) if the nearest preferred value of capacitance is employed what is the percentage error in the resonant frequency?

4.8 A 15 mH inductor, a 470 pF capacitor and a 22 Ω resistor are connected in series. (*a*) Calculate the resonant frequency. (*b*) Calculate the impedance (*i*) at resonance, (*ii*) 25 per cent below resonance, (*iii*) 5 per cent above resonance. (*c*) Calculate the Q factor. (*d*) If the voltage across the capacitor is 12 V at the resonant frequency, determine the applied voltage.

4.9 A parallel circuit has three branches. Branch A contains a 100 Ω resistor, branch B contains a 4 mH inductor, and branch C a 2 μF capacitor. The applied voltage is $15\angle 0°$ V. Calculate (*a*) the resonant frequency of the circuit, (*b*) the current in each component at resonance, (*c*) the total current into the circuit and (*d*) the magnitude of the circuit's impedance at frequencies of $0.1f_0$, $0.5f_0$, f_0, $2f_0$, and $10f_0$.

4.10 A 15 mH inductor of 50 Ω resistance is connected in parallel with a 15 nF capacitor. Calculate (*a*) the phase resonant frequency, (*b*) the amplitude resonant frequency, (*c*) the Q factor and (*d*) the impedance of the circuit of each resonant frequency.

4.11 A 60 mH inductor of 12 Ω resistance is connected in series with a 47 nF capacitor. A 24 V voltage is applied to the circuit. Calculate (*a*) the resonant frequency, (*b*) the resonant current, (*c*) the Q factor and (*d*) the voltage across the capacitor. (*e*) Give an expression for the instantaneous value of the resonant current.

4.12 A series circuit has $Q = 45$ and is connected to a $2\sin(5000\pi t)$ V supply and is tuned to resonance by a 47 nF capacitor. Calculate (*a*) the voltage across the capacitor, (*b*) the values of the resistance and inductance of the circuit and (*c*) the new Q factor when a 50 Ω resistor is connected in series with the circuit.

4.13 A 1 H inductor has a resistance of 100 Ω and is connected in series with a 0.1 μF capacitor and a 560 Ω resistor. Calculate (*a*) the resonant frequency, (*b*) the Q factor and (*c*) the upper and lower 3 dB frequencies.

4.14 An inductor having $L = 100$ μH and $Q = 80$ is connected in parallel with a 220 pF capacitor. Calculate the magnitude and phase of the circuit's impedance at a frequency of 99.2 per cent of the resonant frequency. Also calculate the 3 dB bandwidth of the circuit.

4.15 A 50 mH inductor has a resistance of 14.3 Ω. Calculate (*a*) the resonant frequency, (*b*) the Q factor and (*c*) the impedance when a 68 nF capacitor of (*i*) zero resistance, (*ii*) 10 Ω resistance is connected in series.

4.16 A 20 μH inductor of Q factor 100 is connected in parallel with a capacitor. If the circuit is resonant at 15 MHz calculate (*a*) the capacitance, (*b*) the dynamic resistance and (*c*) its 3 dB, 6 dB and 10 dB bandwidths.

4.17 A 250 mH inductor is connected in series with a resistor so that the total resistance is 40 Ω. A 1 μF capacitor is then connected in parallel. Determine (*a*) the phase-resonant frequency, (*b*) the amplitude-resonant frequency, (*c*) the lower and upper 3 dB frequencies, (*d*) the dynamic resistance and (*e*) the maximum value of the impedance.

4.18 A 1 kΩ resistor is connected in series with a 0.5 H inductor and a 1 pF capacitor. An 8 pF capacitor is connected across the series circuit. Calculate (*a*) the series resonant frequency, (*b*) the parallel resonant frequency and (*c*) the Q factor at the series resonant frequency.

4.19 A series tuned circuit is resonant at 2 kHz and has a Q factor of 25. When 1 V at this frequency is applied across the circuit the current that flows is 100 mA. Calculate (*a*) the inductance, (*b*) the capacitance, (*c*) the 3 dB bandwidth, (*d*) the lower and upper 3 dB frequencies and (*e*) the current that flows at the upper 3 dB frequency.

4.20 A parallel tuned circuit has lower and upper 3 dB frequencies of 15.26 kHz and 13.48 kHz. Calculate (*a*) the resonant frequency and (*b*) the Q factor of the circuit.

4.21 A series-tuned circuit is resonant at 5000 rad/s when it has its minimum impedance of 42 Ω and a 3 dB bandwidth of 200 rad/s. Calculate (*a*) the inductance, (*b*) the capacitance and (*c*) the two frequencies at which the impedance of the circuit has a phase angle of $\pm\pi/3$ rad.

4.22 An 0.4 H inductor of resistance 20 Ω is connected in parallel with a 22 μF capacitor. Calculate (*a*) its resonant frequency, (*b*) its Q factor, (*c*) the current that flows in each component when 100 V at the resonant frequency is applied to the circuit.

4.23 Calculate the resonant frequency and the impedance at resonance of a 50 mH inductance of 10 Ω resistance that is connected in parallel with a 1 μF capacitor and a 12 Ω resistor in series.

4.24 Three capacitors or respective values 10 nF and $Q = 120$, 47 nF and $Q = 100$, and 22 nF and $Q = 80$ are connected (*a*) in series and (*b*) in parallel. Calculate the capacitance and total Q factor in each case.

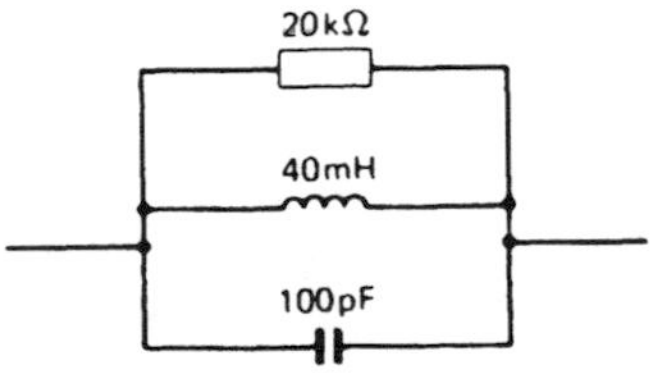

Fig. 4.11

4.25 For the circuit shown in Fig. 4.11 calculate (*a*) the dynamic resistance, (*b*) the resonant frequency and (*c*) the Q factor.

4.26 A coil of inductance 200 mH and Q factor 25 is connected in series with a loss-free 0.1 μF capacitor. Calculate (*a*) the resonant frequency and (*b*) the impedance at resonance. Also find (*c*) the Q factor and (*d*) the resonant frequency if the Q factor of the capacitor is 380.

4.27 An inductor was measured and found to have an inductance of 7.5 mH and a Q factor of 80 at 600 kHz. At 1450 kHz the Q factor was only 30. Calculate the effective resistance of the inductor at each frequency.

4.28 A coil of inductance 150 mH and Q factor 25 is connected in series with a capacitor of 0.03 μF. Calculate (*a*) the resonant frequency, (*b*) the impedance at resonance, (*c*) the Q factor and (*d*) the impedance at resonance if the capacitor has a Q factor of 800.

4.29 An inductance of 2 mH and $Q = 50$ is connected in parallel with a 620 pF loss-free capacitor. Calculate (*a*) the resonant frequency, (*b*) the capacitor current and (*c*) the *Q* factor of the circuit when a constant current of 1 mA is supplied to the circuit.

4.30 An inductance of 250 μH is connected in series with a capacitor and the circuit formed is resonant at 500 kHz. A voltage at this frequency is then applied across the circuit with the result that 300 mV appears across the capacitor. A 12 Ω resistor connected in series with the circuit causes the capacitor voltage to fall to 200 mV. Calculate (*a*) the capacitance, (*b*) the resistance and (*c*) the *Q* factor of the circuit.

4.31 A 10 mH coil has a *Q* factor of 100 and is connected in parallel with a 3 nF capacitor. A variable frequency 6 V source is connected across the circuit. Calculate (*a*) the resonant frequency, (*b*) the minimum current taken from the source and (*c*) the current in the capacitor at the resonant frequency.

4.32 A 20 kΩ resistor, a 25 μH inductance of $Q = 100$ at 2 MHz, and a capacitor are connected in parallel. Calculate (*a*) the capacitance to resonate the circuit, (*b*) the dynamic impedance of the circuit and (*c*) the *Q* factor of the circuit.

5 Mesh and nodal analysis

In many circuits, the components are connected neither in parallel nor in series with one another and then analysis of the circuit will require the use of Kirchhoff's laws. Kirchhoff's laws are:

- *Kirchhoff's first law* The algebraic sum of the currents entering any node in a circuit is zero.
- *Kirchhoff's second law* The e.m.f. applied to a loop in a circuit is equal to the sum of the potential differences around that loop.

However, some simplification of the work involved can usually be obtained if the method of circulating currents due to Maxwell is used. With this method a current is assumed to circulate around *each* loop in the circuit in a clockwise direction. Then Kirchhoff's second law is applied. If the solution of the problem yields any current with a negative sign, this merely means that the assumed clockwise direction was in error and the current really flows in the opposite direction. The actual current flowing in any component is then the algebraic sum of the circulating currents in that component.

Kirchhoff's laws may be applied to the solution of a circuit using either *mesh analysis* or *nodal analysis*. When mesh (loop) analysis is employed, either Kirchhoff's current law can be applied to label the currents flowing in each branch of the circuit, or the method of circulating currents, due to Maxwell, may be used. Kirchhoff's voltage law is then written down around two or more loops in the circuit and the resulting equations are solved simultaneously. The unknowns are called the *loop currents*. Once all the loop currents are known the actual currents flowing in each component and the voltages across the components can easily be found. Nodal analysis chooses certain points, or *nodes*, in the circuit and then applies Kirchhoff's current law to each node in turn to determine the voltage at each node relative to a reference node. A node is a point in the circuit at which two or more components are connected together. Usually the reference node that is selected is earth or the common line in the circuit.

When the number of *nodes* or junctions in the circuit is less than the number of loops, it will generally prove to be easier to employ *nodal analysis*. For a nodal analysis of a circuit the currents entering each node are summed to produce a number of simultaneous equations which may be solved to obtain the required currents and/or voltages.

Mesh analysis

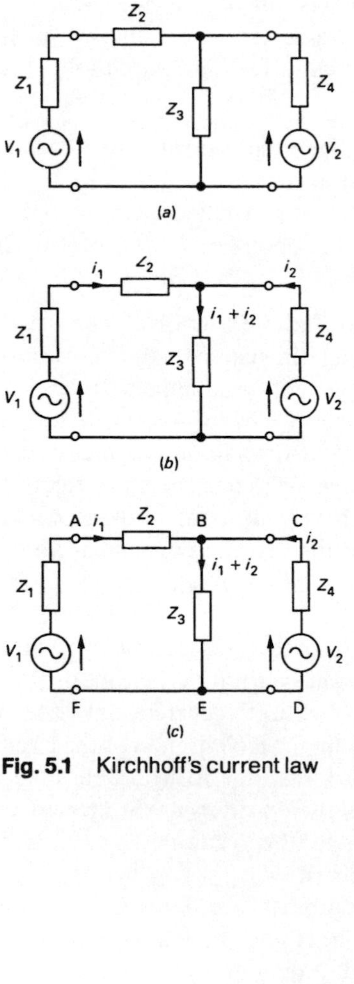

Fig. 5.1 Kirchhoff's current law

A *loop* is a closed path in a circuit that commences and ends at the same node and only passes through any other node once. A *mesh* contains two or more loops. Figure 5.1(a) shows a purely resistive network in which there are two voltage sources V_1 and V_2, six nodes, and three loops. Suppose that the currents flowing in the circuit are to be determined. The procedure to be followed is then:

- Use Kirchhoff's current law to label the network with the currents flowing in each branch assuming that current flows from the positive terminal of each voltage source. If the direction of an arbitrarily chosen current turns out to be incorrect, this will be indicated by its calculated value being negative. This step in the procedure is illustrated by Fig 5.1(*b*).
- Label all the nodes in the circuit as A, B, C, etc. as shown by Fig 5.1(*c*).
- Apply Kirchhoff's voltage law to each loop in the circuit. Three loops are possible in Fig. 5.1(*c*); they are loop ABEF, loop BCDE and loop ABCDEF.

Loop ABEF:

$$V_1 = i_1(Z_1 + Z_2) + Z_3(i_1 + i_2) \quad \text{or}$$
$$V_1 = i_1(Z_1 + Z_2 + Z_3) + i_2Z_3 \qquad (5.1)$$

Loop BCDE:

$$V_2 = i_2Z_4 + Z_3(i_1 + i_2) \quad \text{or}$$
$$V_2 = I_1Z_3 + i_2(Z_3 + Z_4) \qquad (5.2)$$

Loop ABCDEF:

$$V_1 - V_2 = i_1(Z_1 + Z_2) - i_2Z_4 \qquad (5.3)$$

Since there are only two unknowns, I_1 and I_2, just two of these equations are required. The chosen pair of equations must be solved simultaneously to determine the unknown current(s).

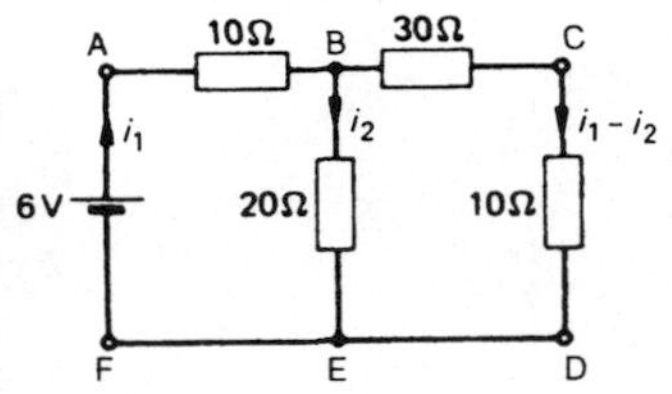

Fig. 5.2

Example 5.1

Calculate the current flowing in the 20 Ω resistor in the circuit of Fig. 5.2.

Solution

Referring to Fig. 5.2, Loop ABEF:

$$6 = 10i_1 + 20i_2 \tag{5.4}$$

Loop ABCDEF:

$$6 = 10i_1 + 40(i_1 - i_2) = 50i_1 - 40i_2 \tag{5.5}$$

Loop BCDE:

$$0 = 40(i_1 - i_2) - 20i_2 = 40i_1 - 60i_2 \tag{5.6}$$

Multiply equation (5.4) by 2 and add the result to equation (5.5) to obtain $18 = 70I_1$. Hence,

$$i_1 = 18/70 = 0.257 \text{ A}$$

Therefore, from equation (5.6),

$$60i_2 = 40i_1 \quad \text{and} \quad i_2 = (40 \times 0.257)/60 = 0.171 \text{ A} \quad (\textit{Ans.})$$

Using determinants,

$$i_2 = \frac{\begin{vmatrix} 50 & 6 \\ 40 & 0 \end{vmatrix}}{\begin{vmatrix} 50 & -40 \\ 40 & -60 \end{vmatrix}} = (0 - 240)/(-1400) = 0.171 \text{ A} \quad (\textit{Ans.})$$

Using matrices,

$$\begin{bmatrix} 50 & -40 \\ 40 & -60 \end{bmatrix} \begin{bmatrix} A & B \\ C & D \end{bmatrix} \begin{bmatrix} 1 & 0 \\ 0 & 1 \end{bmatrix}$$

$$= \begin{bmatrix} 50A - 40C & 50B - 40D \\ 40A - 60C & 40B - 60D \end{bmatrix} \begin{bmatrix} 1 & 0 \\ 0 & 1 \end{bmatrix}$$

Solving for A, B, C and D gives $A = 0.043$, $B = -0.029$, $C = 0.029$ and $D = -0.036$.

Hence,

$$\begin{bmatrix} 0.043 & -0.029 \\ 0.029 & -0.036 \end{bmatrix} \begin{bmatrix} 6 \\ 0 \end{bmatrix} = \begin{bmatrix} I_1 \\ I_2 \end{bmatrix}$$

$$I_1 = 6 \times 0.043 = 0.258 \text{ A} \quad (\textit{Ans.})$$

$$I_2 = 0.029 \times 6 = 0.174 \text{ A} \quad (\textit{Ans.})$$

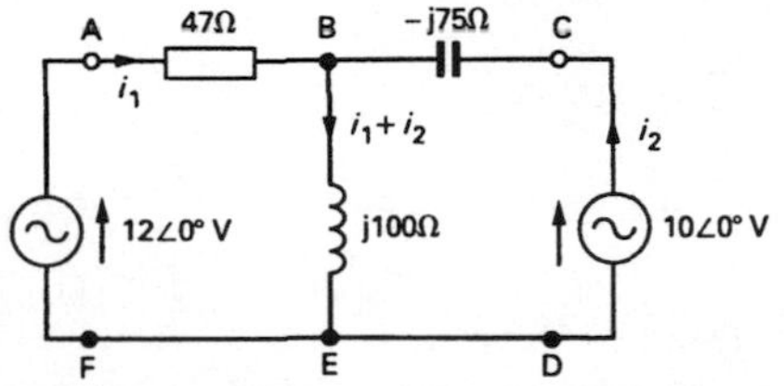

Fig. 5.3

Example 5.2

Calculate the currents flowing in the capacitor and in the resistor in Fig. 5.3.

Solution

Loop ABEF:

$$12 = 47i_1 + j100(i_1 + i_2) = (47 + j100)i_1 + j100i_2 \quad (5.7)$$

Loop BCDE:

$$10 = -j75i_2 + j100(i_1 + i_2) = j100i_1 + j25i_2 \quad (5.8)$$

From equation (5.8), $i_1 = (10 - j25i_2)/j100$. Substituting into equation (5.7) gives

$$12 = [(47 + j100)(10 - j25i_2)/j100] + j100i_2$$
$$2 + j4.7 = i_2(-11.75 + j75) \quad \text{and}$$
$$i_2 = (2 + j4.7)/(-11.75 + j75) = 0.057 - j0.036$$
$$= 0.067\surd - 32.3° \text{ A} \quad (\textit{Ans.})$$
$$i_1 = [(10 - j25)(0.057 - j0.036)]/j100 = -0.014 - j0.091$$
$$= 0.092\angle -98.9° \text{ A} \quad (\textit{Ans.})$$

Using determinants,

$$i_2 = \frac{\begin{vmatrix} 47 + j100 & 12 \\ j100 & 10 \end{vmatrix}}{\begin{vmatrix} 47 + j100 & j100 \\ j100 & j25 \end{vmatrix}}$$
$$= [(470 + j1000) - j1200]/[(-2500 + j1175) + (10 \times 10^3)]$$
$$= (470 - j200)/(7500 + j1175) = 0.057 - j0.036$$
$$= 0.067\angle - 33.3° \text{ A} \quad (\textit{Ans.})$$

$$i_1 = \frac{\begin{vmatrix} 12 & j100 \\ 10 & j25 \end{vmatrix}}{7500 + j1175}$$
$$= (j300 - j1000)/(7500 + j1175) = -0.0143 - j0.091$$
$$= 0.092\angle -98.9° \text{ A} \quad (\textit{Ans.})$$

Example 5.3

Calculate the current flowing in the capacitor in the circuit shown in Fig. 5.4.

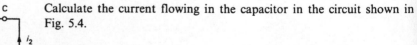

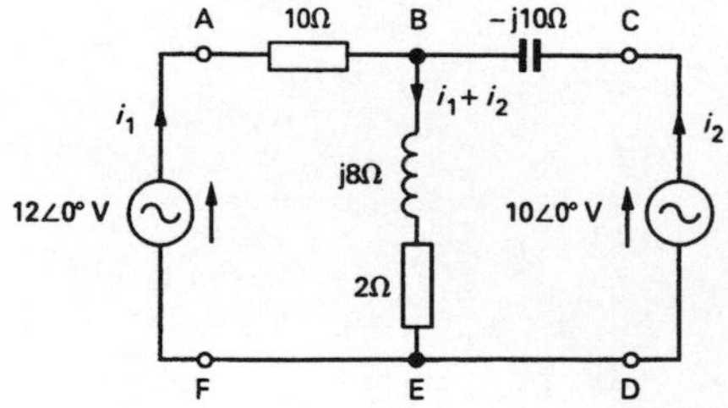

Fig. 5.4

Solution

Loop ABEF:

$$12 = (12 + j8)i_1 + (2 + j8)i_2 \quad (5.9)$$

Loop BCDE:

$$10 = (2 + j8)i_1 + (2 - j2)i_2 \quad (5.10)$$

Subtracting equation (5.10) from (5.9) gives,

$$2 = 10i_1 + j10i_2, \quad i_1 = (2 - j10i_2)/10 = 0.2 - ji_2$$

Substituting into equation (5.10) gives,

$$10 = (2 + \mathrm{j}8)(0.2 - \mathrm{j}i_2) + (2 - \mathrm{j}2)i_2$$

$$i_2 = (9.6 - \mathrm{j}1.6)/(10 - \mathrm{j}4) = 0.88 + \mathrm{j}0.19$$

$$= 0.90\angle 12.3^\circ \text{ A} \quad (\textit{Ans.})$$

Using determinants,

$$i_2 = \frac{\begin{vmatrix} 12 + \mathrm{j}8 & 12 \\ 2 + \mathrm{j}8 & 10 \end{vmatrix}}{\begin{vmatrix} 12 + \mathrm{j}8 & 2 + \mathrm{j}8 \\ 2 + \mathrm{j}8 & 2 - \mathrm{j}2 \end{vmatrix}}$$

$$= [(120 + \mathrm{j}80) - (24 + \mathrm{j}96)]/[(40 - \mathrm{j}8) - (-60 + \mathrm{j}23)]$$

$$= (96 - \mathrm{j}16)/(100 - \mathrm{j}40)$$

$$= 0.88 + \mathrm{j}0.19 = 0.90\angle 12.3^\circ \quad (\textit{Ans.})$$

Example 5.4

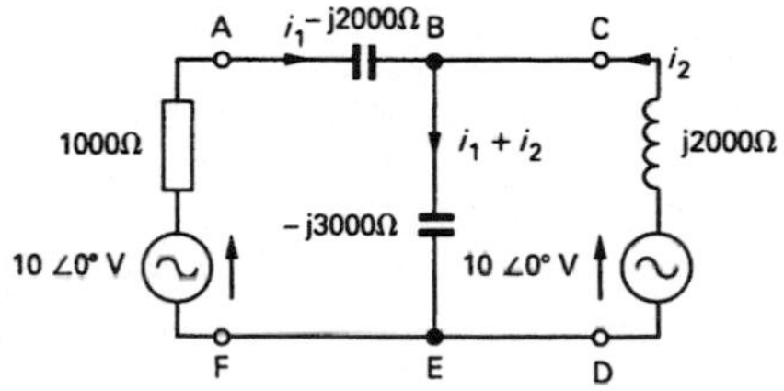

Fig. 5.5

Calculate the current in the inductor in the circuit of Fig. 5.5.

Solution

Loop ABEF:

$$10 = i_1(1 - \mathrm{j}2) - \mathrm{j}3(i_1 + i_2) = i_1(1 - \mathrm{j}5) - \mathrm{j}3i_2 \quad (5.11)$$

Loop BCDE:

$$10 = \mathrm{j}2i_2 - \mathrm{j}3(i_1 + i_2) = -\mathrm{j}3i_1 - \mathrm{j}i_2 \quad (5.12)$$

Multiply equation (5.12) by 3,

$$30 = -\mathrm{j}9i_1 - \mathrm{j}3i_2 \quad (5.13)$$

Subtract equation (5.11) from (5.13) to get,

$$20 = (-1 - \mathrm{j}4)i_1, \quad i_1 = 20/(-1 + \mathrm{j}4) = -1.18 + \mathrm{j}4.71$$

Substitute this value of i_1 into equation (5.12), then $10 = -\mathrm{j}3(-1.18 - \mathrm{j}4.71) + \mathrm{j}i_2$,

or $i_2 = 3.53 - \mathrm{j}4.12 = 5.43\angle{-49.4^\circ}$ mA (*Ans.*)

Or, using determinants,

$$i_2 = \frac{\begin{vmatrix} 1 - \mathrm{j}5 & 10 \\ -\mathrm{j}3 & 10 \end{vmatrix}}{\begin{vmatrix} 1 - \mathrm{j}5 & -\mathrm{j}3 \\ -\mathrm{j}3 & -\mathrm{j}1 \end{vmatrix}}$$

$$= [(10 - \mathrm{j}50) - (-\mathrm{j}30)]/[(-5 - \mathrm{j}) - (-9)] = 3.53 - \mathrm{j}4.12$$

$$= 5.43\angle{-49.4^\circ} \text{ A} \quad (\textit{Ans.})$$

Maxwell's circulating currents

It may often prove easier to assume that all the currents in a network flow around each loop in a clockwise direction. If any of the assumed directions turns out to be incorrect it will be indicated by a minus sign in front of the calculated current value. The concept of circulating currents is illustrated by the following example in which it is assumed that the two currents i_1 and i_2 flow clockwise around the two loops in the circuit.

Example 5.5

Find the current i_1 and i_2 in the circuit shown in Fig. 5.6. [Note that this is the same circuit as the one in example 5.2.]

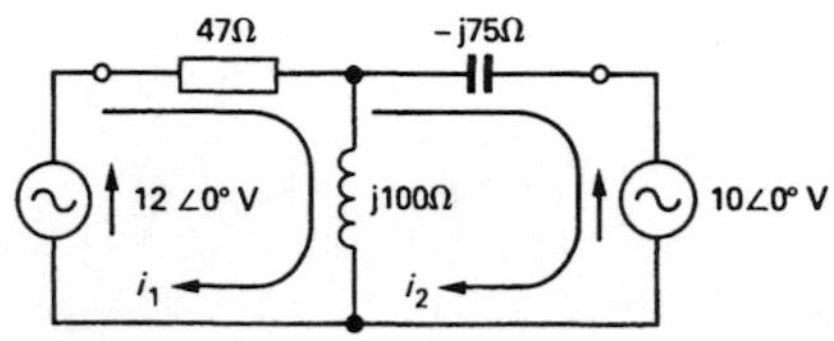

Fig. 5.6

Solution

$$12 = i_1(47 + j100) - j100 i_2 \qquad (5.14)$$

$$-10 = -j100 i_1 + i_2(j100 - j75) = -j100 i_1 + j25 i_2 \qquad (5.15)$$

These are the same equations as were obtained earlier for example 5.2, except that some of the signs are reversed. Using determinants,

$$i_1 = \frac{\begin{vmatrix} 12 & -j100 \\ -10 & j25 \end{vmatrix}}{\begin{vmatrix} 47 + j100 & -j100 \\ j100 & j25 \end{vmatrix}}$$

$$= (j300 - j1000)/[(-2500 - j1175) - (-1 \times 10^4)]$$

$$= -0.014 - j0.091 = 0.092\angle{-98.9^\circ}\ \text{A} \quad (\textit{Ans.})$$

$$i_2 = \frac{\begin{vmatrix} 47 + j100 & 12 \\ j100 & 10 \end{vmatrix}}{7500 + j1175}$$

$$= [(470 + j1000) - j1200]/(7500 + j1175) = 0.057 - j0.036$$

$$= 0.065\angle - 32.2^\circ\ \text{A} \quad (\textit{Ans.})$$

Example 5.6

Use mesh analysis to calculate the power dissipated in the 600 Ω resistor in the circuit shown in Fig 5.7.

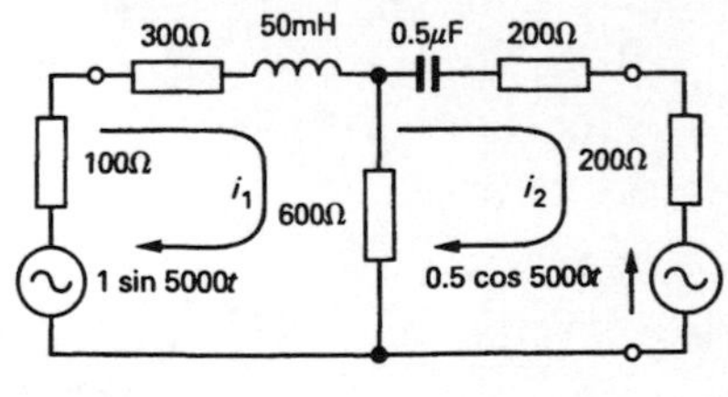

Fig. 5.7

Solution

$$0.707 = (1000 + j250)i_1 - 600 i_2$$

$$j0.707/2 = 600 i_1 - (1000 - j400) i_2$$

If the currents are in milliamps, then,

$$j0.707/2 \times 0.6 = 0.36 i_1 - 0.6(1 - j0.4) i_2$$

and

$$(1-j0.4)\times 0.707=(1+j0.25)(1-j0.4)i_1-0.6(1-j0.4)i_2$$

$$0.707(1-j0.707)=(0.74-j0.15)i_1$$

$$i_1=(0.865\angle{-35^\circ})/(0.755\angle{-11.5^\circ})=1.145\angle{-23.5^\circ}$$

$$=1.05-j0.457$$

$$i_2=[(1+j0.25)(1.05-j0.457)-0.707]/0.6$$

$$=0.76-j0.325=0.826\angle{-23.2^\circ}$$

Current in 600 Ω resistor is

$$i_1-i_2=-0.29-j0.132=0.319\angle{-155.5^\circ}\ \text{mA}$$

Therefore,

$$\text{Power dissipated}=(0.319\times 10^{-3})^2\times 600=61.1\ \mu\text{W}\quad (\textit{Ans.})$$

Nodal analysis

First choose one node in the circuit to be the reference; this node is usually the common or earth line. All other nodes are labelled A, B, C, etc. Kirchhoff's current law is then applied to each node in turn, other than the reference node. A current flowing into a node is taken as being positive while a current flowing out of a node is assumed to be negative. The sum of the currents entering and leaving a node is, of course, equal to zero. Each current is expressed in terms of the voltage between the node and the reference node. The steps to be taken for the nodal analysis of a circuit are as follows:

- Label all the nodes except the reference node.
- Assume that the voltage at each node is greater than the voltage (usually zero) at the reference node.
- Write down Kirchhoff's current law at each node and equate the total currents entering each node to the total current that leaves that node.
- Solve the simultaneous equations thus obtained for the unknown node voltage(s).

Example 5.7

Repeat example 5.3 using nodal analysis.

Solution

$$(12-V_B)/10=V_A/(2+j8)+(V_B-10)/-j10$$

$$1.2-0.1V_A=0.0294V_B-j0.1177V_B+j0.1V_B$$

$$V_B=8.066+j8.831\ \text{V}$$

$$I_C=(V_B-10)/-j10=-0.8831-j0.1934=0.904\angle{-167.7^\circ}\ \text{A}\ (\textit{Ans.})$$

But the current actually flows in the opposite direction, hence

$$I_C=0.904\angle(-167.7+180)^\circ=0.904\angle 12.3^\circ\ \text{A}\quad (\textit{Ans.})$$

Example 5.8

Repeat example 5.4 using nodal analysis.

Solution

$$(10 - V_B)/(1000 - j2000) + (10 - V_B)/j2000 = V_B/-j3000$$

Multiply throughout by 1000,

$$2 + j4 - V_B(0.2 + j0.4) + (-j5) + j0.5V_B = j0.33V_B$$
$$2 + j4 - j5 = V_B(0.2 + j0.33 + j0.4 - j0.5)$$
$$2 - j1 = V_B(0.2 + j0.23)$$
$$V_B = (2.236\angle{-26.6°})/(0.305\angle 49°) = 7.331\angle{-75.6°}\ V$$
$$I_L = (10 - 7.331\angle{-75.6°})/(2000\angle 90°)$$
$$= (10.829\angle 41°)/(2000\angle 90°)$$
$$= 5.42\angle{-49°}\ mA \quad (Ans.)$$

Example 5.9

Use nodal analysis to determine the value of each current flowing in the circuit given in Fig 5.8

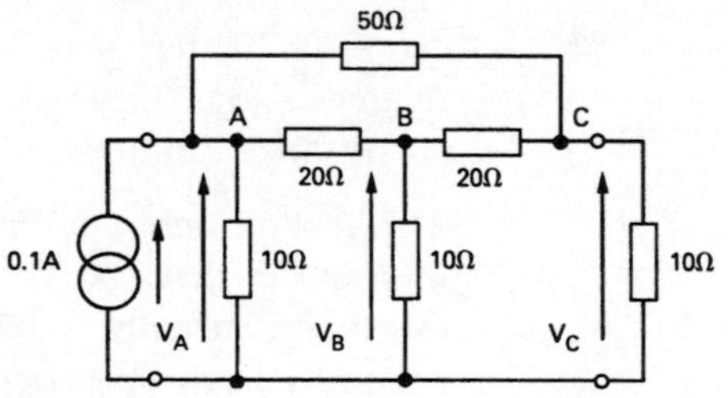

Fig. 5.8

Solution
Node A:

$$0.1 = V_A/10 + (V_A - V_B)/20 + (V_A - V_C)/50$$
$$= V_A(0.1 + 0.05 + 0.02) - 0.05V_B - 0.02V_C$$
$$0.1 = 0.17V_A - 0.05V_B - 0.02V_C \quad (5.16)$$

Node B:

$$(V_A - V_B)/20 = V_B/10 + (V_B - V_C)/20$$
$$0 = -0.05V_A + V_B(0.05 + 0.1 + 0.05) - 0.05V_C$$
$$0 = -0.05V_A + 0.2V_B - 0.05V_C \quad (5.17)$$

Node C:

$$(V_B - V_C)/20 + (V_A - V_C)/50 = V_C/10$$
$$0 = -0.02V_A - 0.05V_B + V_C(0.1 + 0.05 + 0.02)$$
$$0 = -0.02V_A - 0.05V_B + 0.17V_C \quad (5.18)$$

$$V_A = \frac{\begin{vmatrix} 0.1 & -0.05 & -0.02 \\ 0 & 0.2 & -0.05 \\ 0 & -0.05 & 0.17 \end{vmatrix}}{\begin{vmatrix} 0.17 & -0.05 & -0.02 \\ 0.05 & 0.2 & -0.05 \\ -0.02 & -0.05 & 0.17 \end{vmatrix}}$$

$$= 0.1[0.2 \times 0.17 - (-0.05)^2]/[0.17(0.2 \times 0.17 - 0.05^2)$$
$$+ 0.05(8.5 \times 10^{-3} - 1 \times 10^{-3})$$
$$- 0.02(-25 \times 10^{-4} + 4 \times 10^{-3})]$$
$$= (3.15 \times 10^{-3})/(5.57 \times 10^{-3}) = 0.55 \text{ V}$$

$$V_B = \frac{\begin{vmatrix} 0.17 & 0.1 & -0.02 \\ 0.05 & 0 & -0.05 \\ -0.02 & 0 & 0.17 \end{vmatrix}}{5.57 \times 10^{-3}}$$

$$= -0.1(0.05 \times 0.17 - 0.05 \times 0.02)/(5.57 \times 10^{-3})$$
$$= (-8.5 \times 10^{-4})/(5.7 \times 10^{-3}) = -0.15 \text{ V}$$

$$V_C = \frac{\begin{vmatrix} 0.17 & -0.05 & 0.1 \\ 0.05 & 0.2 & 0 \\ -0.02 & -0.05 & 0 \end{vmatrix}}{5.7 \times 10^{-3}}$$

$$= 0.1(0.05 \times -0.05 + 0.2 \times 0.02)/(5.7 \times 10^{-3}) = 0.26 \text{ V}$$

Therefore the currents in the circuit are:

$I_{10L} = 0.55/10 = 55$ mA (*Ans.*)

$I_{50} = (0.55 - 0.26)/50 = 5.8$ mA (*Ans.*)

$I_{20L} = (0.55 + 0.15)/20 = 35$ mA (*Ans.*)

$I_{10M} = -0.15/10 = -15$ mA (*Ans.*)

$I_{10R} = 0.26/10 = 26$ mA (*Ans.*)

$I_{20R} = (-0.15 - 0.26)/20 = -20.5$ mA (*Ans.*)

Example 5.10

Calculate the current in the 2 Ω resistor in the circuit given in Fig. 5.9 using (*a*) the current dividing rule and (*b*) nodal analysis.

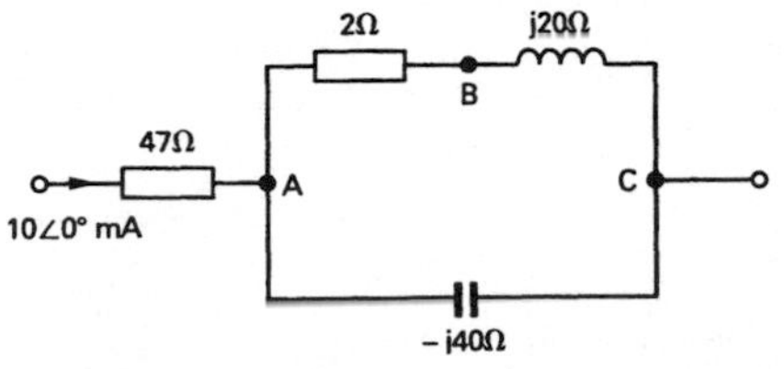

Fig. 5.9

Solution

(*a*) $i_2 = (10 \times -j40)/(2 + j20 - j40) = 19.8 - j1.98$ mA
$= 19.9\angle{-5.7°}$ mA (*Ans.*)

(*b*) Take node C as the reference node. Then at node A

$$10 \times 10^{-3} = V_A/-j40 + V_A/(2 + j20)$$
$$V_A = 16.23 + j392.2 = 399\angle 79° \text{ mV}$$

Current in 2 Ω resistor is

$0.399\angle 79°/(2+j20) = 19.9\angle -5.3°$ mA (*Ans.*)

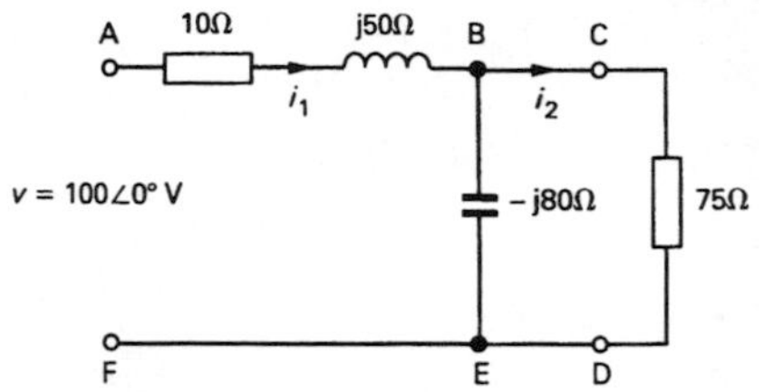

Fig. 5.10

Example 5.11

For the circuit shown in Fig. 5.10 calculate the current in the 75 Ω resistor using (*a*) loop analysis and (*b*) nodal analysis.

Solution

(*a*) Loop ABEF:

$$100 = (10+j50)i_1 - j80(i_1 - i_2) = (10 - j30)i_1 + j80i_2 \quad (5.19)$$

Loop BCDE:

$$0 = 75i_2 - [-j80(i_1 - i_2)] = j80i_1 + (75 - j80)i_2 \quad (5.20)$$

From equation (5.20),

$$i_1 = [(75 - j80)i_2]/-j80 = (1 + j0.9375)i_2 \quad (5.21)$$

Substituting equation (5.21) into (5.19) gives $100 = i_2(38.125 + j59.375)$ or

$$i_2 = 0.766 - j1.193 = 1.42\angle -57.3° \text{ A} \quad (\textit{Ans.})$$

(*b*) $(100 - V_B)/(10 + j50) = V_B/-j80 + V_B/75$

$$0.385 - j1.923 = V_B(j0.0125 + 0.0133 + 3.85 \times 10^{-3} - j0.0192)$$
$$= V_B(0.01715 - j6.7 \times 10^{-3})$$

$V_B = 57.481 - j89.672 = 106.5\angle -57.3°$ V and

$I_{75} = V_B/75 = 1.42\angle -57.3°$ A (*Ans.*)

With a complex circuit it is often easier if each voltage source is converted into a current source and all impedances are converted into the corresponding admittances.

Example 5.12

For the circuit shown in Fig. 5.11 determine, using nodal analysis, the voltage across the 75 Ω resistor.

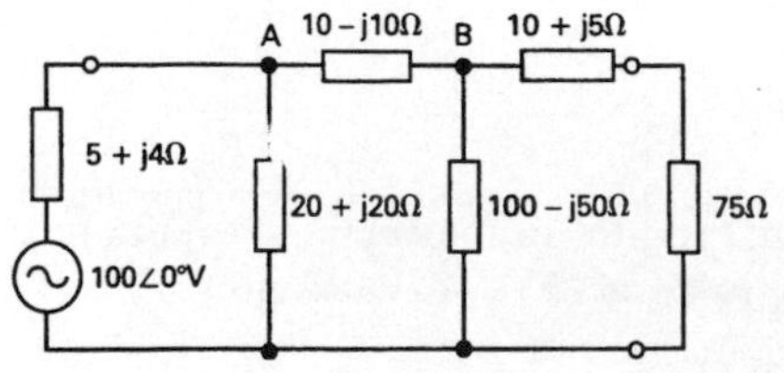

Fig. 5.11

Solution

The voltage source is converted into a current source of $100/(5+j4) = 12.2 - j9.76$ A. The impedances in the circuit are converted into the admittances:

$5 + j4\ \Omega = 0.122 - j0.098$ S

$20 + 20\ \Omega = 0.025 - j0.025$ S

$10 - j10\ \Omega = 0.05 + 0.05$ S

$100 - j50\ \Omega = (8 + j4) \times 10^{-3}$ S

and $10 + j5 + 75\ \Omega = 0.012 - j6.9 \times 10^{-4}$ S.

Node A:

$$V_A[(0.122 - j0.098) + (0.025 - j0.025) + (0.05 + j0.05)] - V_B(0.05 + j0.05) = 12.2 - j9.76$$

$$V_A(0.197 - j0.073) - V_B(0.05 + j0.05) = 12.2 - 9.76 \quad (5.22)$$

Node B:

$$V_B[(0.05 + j0.05) + (8 + 4) \times 10^{-3} + (0.012 - j6.9 \times 10^{-4})] - V_A(0.05 + j0.05) = 0$$

$$V_B(0.07 + j0.053) - V_A(0.05 + j0.05) = 0 \quad \text{or}$$

$$V_A = (1.23 - j0.17)V_B \quad (5.23)$$

Substituting this value for V_A into equation (5.22) gives

$$(1.23 - j0.17)(0.197 - j0.073)V_B - V_B(0.05 + j0.05) = 12.2 - j9.76$$

or $V_B(0.18 - j0.475) = 12.2 - j9.76$ and $V_B = 26.48 + j15.65$ V

Therefore,

$$V_{75} = (26.48 + j15.65)75/(85 + j5) = 24.1 + j12.4$$
$$= 27.1\angle 27.2° \text{ V} \quad (\textit{Ans.})$$

Mutual Inductance

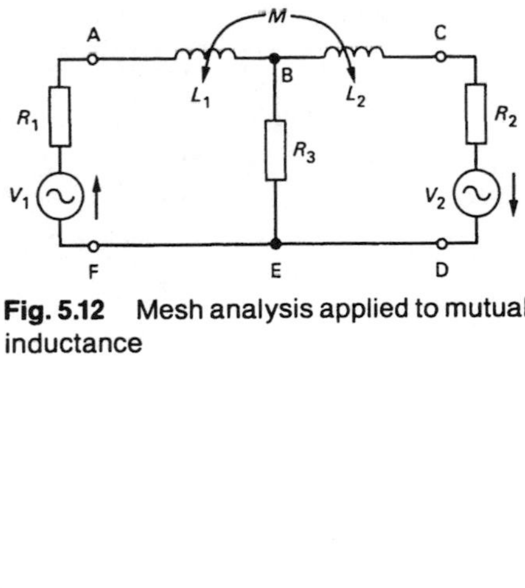

Fig. 5.12 Mesh analysis applied to mutual inductance

Mesh analysis may also be employed to solve circuits that contain mutual inductance (mutual inductance is discussed in Chapter 10). Figure 5.12 shows a T-network in which two inductors L_1 and L_2 are coupled together by mutual inductance M. The inductors are wound so that the mutual inductance is series aiding. If loop currents i_1 and i_2 are assumed to flow around the loops ABEF and BCDE respectively, then

$$V_1 = (R_1 + j\omega L_1)i_1 + j\omega M i_2 + R_3(i_1 - i_2)$$
$$= (R_1 + R_3 + j\omega L_1)i_1 - (R_3 - j\omega M)i_2 \quad (5.24)$$

$$V_2 = (R_2 + j\omega L_2)i_2 + j\omega M i_1 + R_3(i_2 - i_1)$$
$$= (R_2 + R_3 + j\omega L_2)i_2 - (R_3 - j\omega M)i_1 \quad (5.25)$$

$$i_1 = [(R_2 + R_3 + j\omega L_2)i_2 - V_2]/(R_3 - j\omega M) \quad (5.26)$$

Example 5.13

The circuit shown in Fig. 5.12 has $R_1 = R_2 = R_3 = 100\ \Omega$, $L_1 = L_2 = 31.8$ mH and $M = 15.9$ mH. If $V_1 = 100\angle 0°$ V and $V_2 = 100\angle 90°$ V at 1 kHz, calculate the current i_2.

Solution

Substitute equation (5.26) into (5.24) to obtain,

$$(R_3 - j\omega M)V_1 = (R_1 + R_3 + j\omega L_1)(R_2 + R_3 + j\omega L_2)i_2 - (R_1 + R_3 + j\omega L_1)V_2 - (R_3 - j\omega M)^2 i_2$$

therefore,

$$i_2 = \frac{(R_3 - j\omega M)V_1 + (R_1 + R_3 + j\omega L_1)V_2}{(R_1 + R_3 + j\omega L_1)(R_2 + R_3 + j\omega L_2) - (R_3 - j\omega M)^2}$$

Hence,

$$i_2 = \frac{(100 - j100)100 - (200 + j200)j100}{(200 + j200)^2 - (100 - j100)^2}$$

$$= 0.424\angle -135° \text{ A} \quad (\textit{Ans.})$$

Superposition theorem

The *superposition theorem* states that

> The current at any point in a *linear* network containing two, or more, voltage or current sources is the sum of the currents that would flow at that point if only one source is considered at a time, all other sources being replaced by impedances equal in value to their internal impedances.

This means that a voltage source is short circuited and a current source is open circuited.

Example 5.14

Calculate the current flowing in the 20 Ω resistor of Fig. 5.13(a) if the two generators operate at the same frequency and their e.m.f.s are (a) in phase and (b) in antiphase with one another.

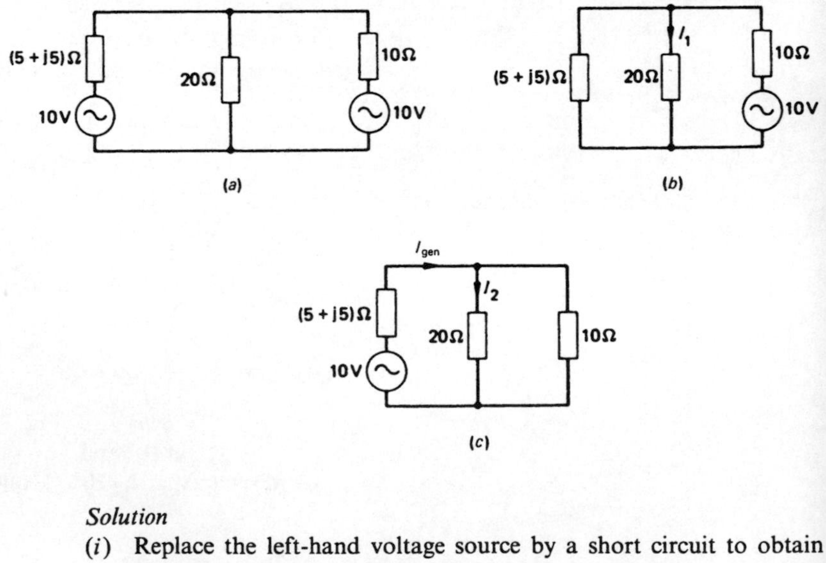

Fig. 5.13

Solution

(i) Replace the left-hand voltage source by a short circuit to obtain the circuit shown in Fig 5.13(b). The impedance seen by the right-hand voltage source is $Z_1 = 10 + [20(5 + j5)]/(25 + j5) = 14.62 + j3.08$ Ω. The current taken from this source is $I = 10/(14.62 + j3.08) = 0.655 - j0.13 = 0.67\angle -11.9°$ A. Hence, $I_1 = [(0.67\angle -11.9°) \times (7.07\angle 45°)]/(25.5\angle 11.3°) = 0.186\angle 21.8°$ A.

(*ii*) Now replace the right-hand source by a short circuit to obtain the circuit shown in Fig 5.13(*c*). The impedance seen by the source is $Z_2 = 5 + j5 + 6.67 = 11.67 + j5\ \Omega$. Hence the current supplied by the source is

$$I = 10/(11/67 + j5) = 0.724 - j0.31 = 0.788\angle{-23.2^\circ}\ \text{A}$$

$$\text{Current } I_2 = (0.788\angle{-23.2^\circ}) \times 10/30 = 0.263\angle{-23.2^\circ}\ \text{A}$$

(*a*) $I_{20} = I_1 + I_2 = 0.186\angle 21.8^\circ + 0.263\angle{-23.2^\circ}$
$= 0.416\angle{-4.8^\circ}$ A (*Ans.*)

(*b*) $I_{20} = I_1 - I_2 = 0.186\angle 111.7^\circ$ A (*Ans.*)

Example 5.15

Figure 5.14(*a*) shows a T network that has two voltage sources applied to it. Use the Superposition theorem to find the voltage across the capacitor.

Fig. 5.14

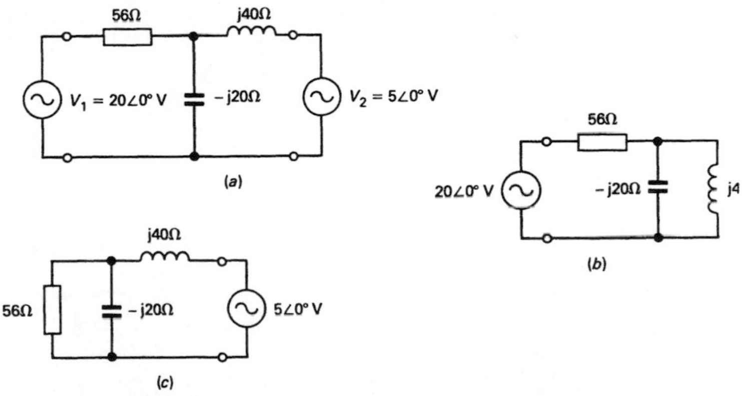

Solution

(*a*) Replace the voltage source V_2 by a short circuit to obtain the circuit given in Fig. 5.14(*b*). The impedance of the two components in parallel is $Z = (j40 \times -j20)/(j40 - j20) = -j40\ \Omega$. The voltage across the capacitor is

$$V_{C1} = (20 \times -j40)/(56 - j40) = 6.76 - j9.46\ \text{V}$$

(*b*) Replace voltage source V_1 by a short circuit, see Fig. 5.14(*c*). The impedance of the two components in parallel is

$$(-j20 \times 56)/(56 - j20) = 6.34 - j17.74\ \Omega$$

The voltage across the capacitor is

$$V_{C2} = 5(6.34 - j17.74)/(6.34 - j17.64 + j40) = 4.06 - j0.27\ \text{V}$$

(*c*) The capacitor voltage $= V_{C1} - V_{C2}$
$= (6.76 - j9.46) - (4.06 - j0.27) = 10.82 - j9.73$
$= 14.55\angle{-42^\circ}$ V (*Ans.*)

Example 5.16

Use the Superposition theorem to calculate the current in the 100 Ω resistor in Fig. 5.15.

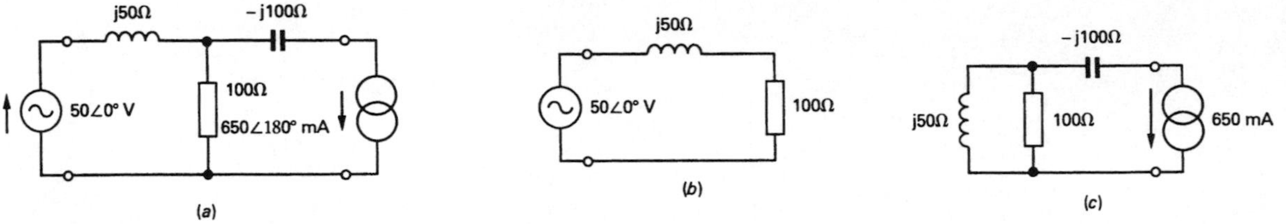

Fig. 5.15

Solution

(*a*) Open circuit the current source to give Fig 5.15(*b*). The current i_1 taken from the 50 V voltage source is, $i_1 = 50/(100 + j50) = 0.4 - j0.2$ A, downwards through the resistor

(*b*) Short circuit the voltage source to give the circuit shown in Fig. 5.15(*c*). Using the current dividing rule, $i_2 = (650 \times j50)/(100 + j50) = 130 + j260$ mA, upwards through the resistor

Therefore,

$$i_{100} = i_1 - i_2 = (0.4 - j0.2) - (0.13 + j0.26) = 0.27 - j0.46$$
$$= 0.533 \angle -59.6^\circ \text{ A} \quad (Ans.)$$

Example 5.17

Use the Superposition theorem to calculate the power dissipated in the circuit of Fig. 5.16(*a*).

Solution

Power is only dissipated in the 100 Ω resistor so it is necessary to find the current in that component.

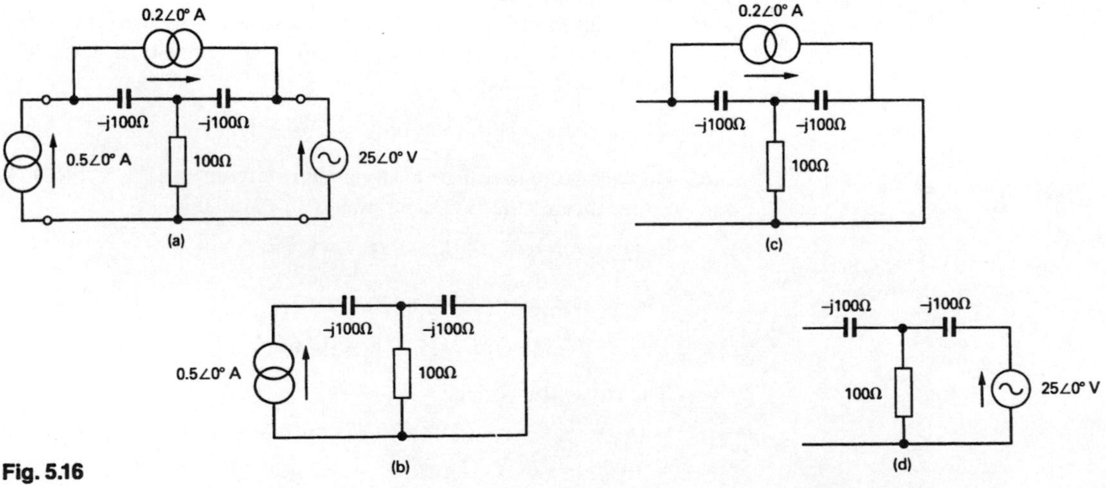

Fig. 5.16

(*a*) Short circuit the 25 V voltage source and open circuit the 0.2 A current source to obtain Fig. 5.16(*b*). Using the current dividing rule,

$$i_1 = (-j100 \times 0.5)/(100 - j100) = 0.25 - j0.25 \text{ A, downwards}$$

(*b*) Short circuit the voltage source and open circuit the 0.5 A current source. Now the circuit is shown by Fig. 5.16(*c*).

$$i_2 = (-j100 \times 0.2)/(100 - j100) = 0.1 - j0.1 \text{ A, upwards}$$

(*c*) Open circuit both of the current sources. The circuit is now shown in Fig. 5.16(*d*). From this,

$$i_3 = 25/(100 - j100) = 0.125 + j0.125 \text{ A, downwards}$$

(*d*) $i_{100} = (0.25 - j0.25) - (0.1 - j0.1) + (0.125 + j0.125)$
$= 0.275 - j0.025 = 0.276\angle -5.2° \text{ A}$

Power dissipated $= 0.276^2 \times 100 = 7.62$ W (*Ans.*)

Very often in electronic circuits both a.c. and d.c. current and/or voltage sources are present and then the Superposition theorem can be used to analyse the circuit.

Example 5.18

An a.c. voltage source of e.m.f. 6 V r.m.s. and internal resistance 600 Ω is connected in series with a 10 V d.c. source. The two sources are used to supply current to a load that consists of a 200 Ω resistor connected in series with a 20 mH inductance. If the frequency of the a.c. supply is 796 Hz calculate the total power dissipated in the load.

Solution

(*a*) Short circuit the a.c. source.

$I_{DC} = 10/800 = 12.5$ mA $\quad P_{DC} = (12.5 \times 10^{-3})^2 \times 200$
$= 31.25$ mW

(*b*) Short circuit the d.c. source.

$\omega L = 2\pi \times 796 \times 20 \times 10^{-3} = 100\ \Omega$

$I_{AC} = 6/(800 + j100) = 7.385 - j9.231 \text{ mA} = 11.82\angle 51.3° \text{ mA}$

$P_{AC} = (11.82 \times 10^{-3})^2 \times 200 = 27.94$ mW

Total power $= 31.25 + 27.94 = 59.19$ mW (*Ans.*)

Exercises 5

5.1 Use nodal analysis to determine the voltage between the nodes B and C in the circuit shown in Fig. 5.17.

5.2 For the circuit of Fig. 5.18 calculate the current in each impedance.

5.3 Use mesh analysis to find the current in the inductor in Fig 5.19.

5.4 Use loop analysis to show that when the bridge in Fig 5.20 is balanced, that is $V_{out}/V_{in} = 0$, $r = 1/\omega^2 C^2 R$ and $L = 2/\omega^2 C$.

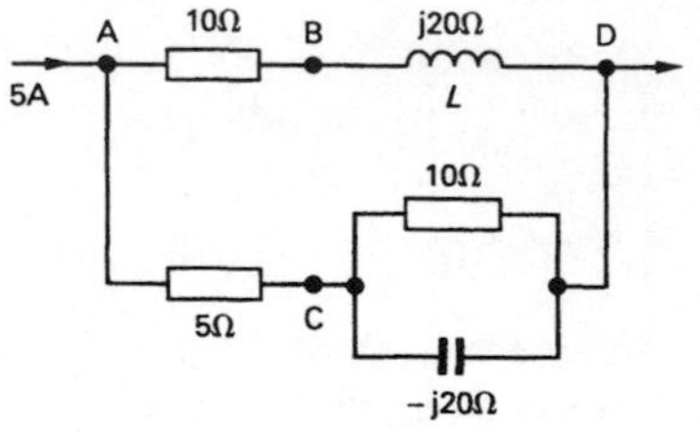

Fig. 5.17

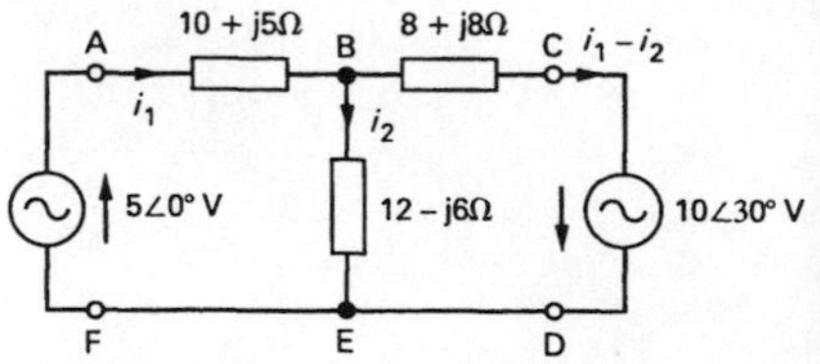

Fig. 5.18

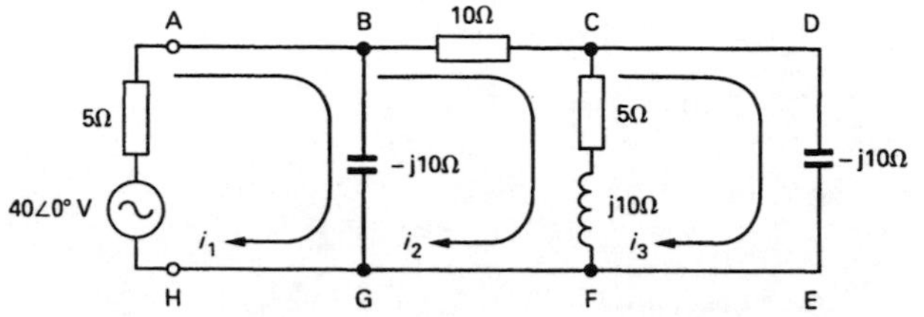

Fig. 5.19

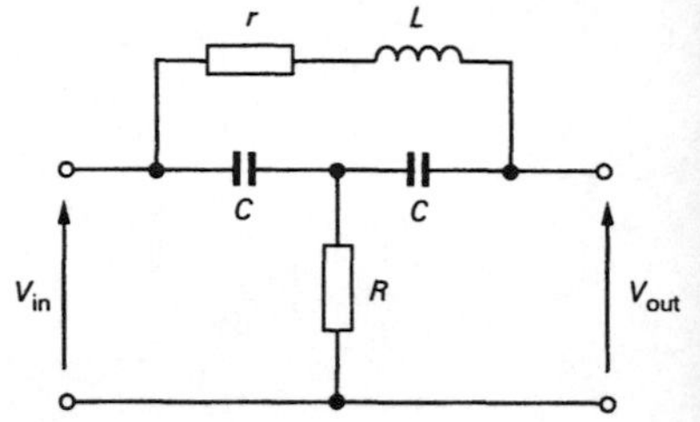

Fig. 5.20

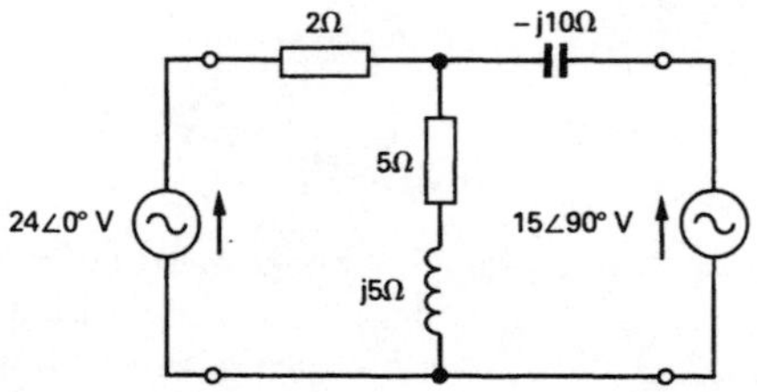

Fig. 5.21

5.5 For the circuit given in Fig. 5.21 calculate using nodal analysis the current flowing in the inductor.

5.6 Two voltages $V_1 = (100 - j100)$ volts and $V_2 = (100 + j100)$ volts having internal impedances of $Z_1 = (30 + j30)$ ohms and $Z_2 = (50 + j0)$ ohms respectively are connected in parallel across an impedance $Z_3 = (20 + j100)$ ohms. Calculate the value of the current flowing in the impedance Z_3 and its phase angle relative to V_2.

5.7 Calculate the current flowing in each of the impedances shown in Fig. 5.22

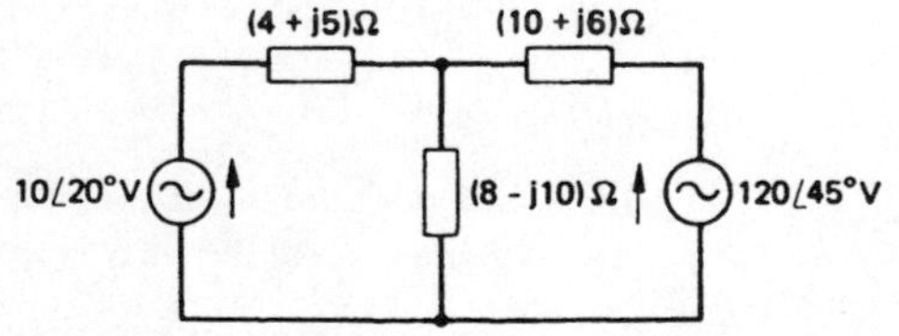

Fig. 5.22

5.8 In the circuit of Fig. 5.23 a current of 3 A in phase with V_1 flows in the unknown impedance Z_3. Use Kirchhoff's laws to find the value of Z_3 and also the currents flowing in the other impedances.

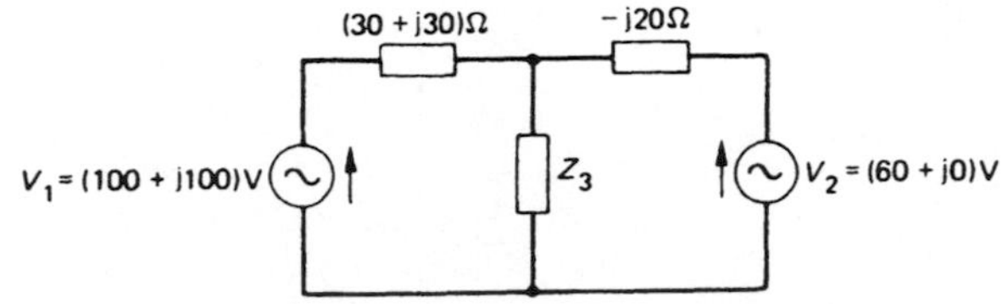

Fig. 5.23

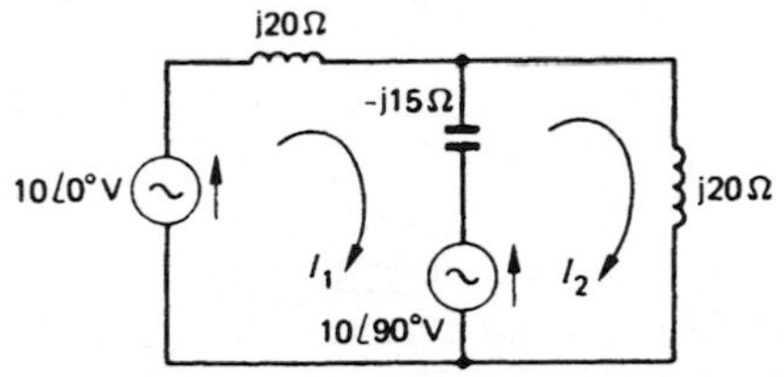

Fig. 5.24

5.9 Use Maxwell's circulating currents to calculate the current I_1 in the circuit given in Fig. 5.24.

5.10 Repeat exercise 5.9 using nodal analysis.

5.11 A T network has series arms of j50 Ω and −j200 Ω and a shunt arm of 120 Ω. An a.c. voltage source is applied between the j50 Ω arm and common, and an a.c. current source of 650 ∠0° mA is applied between the −200 Ω arm and common. Use the Superposition theorem to calculate the voltage across the capacitor.

5.12 The voltage $v = 20\ \sin(3000t)$ V is applied across the series connection of a 500 Ω resistor, a 300 mH inductance and a d.c. source of e.m.f. E. Determine the minimum value for E if the voltage developed across the 500 Ω resistor is not to go negative.

5.13 A T network has a series resistor of 10 Ω and a 10 Ω shunt resistor. A voltage source of e.m.f. 10 V and internal resistance 10 Ω is connected to one of the series resistors and a 1 A current source is connected to the other series resistor. If both sources are in phase with one another use the Superposition theorem to calculate the current in the shunt resistor.

5.14 A T network has series arms of impedance −j10 Ω and −j80 Ω respectively and a shunt arm of impedance j50 Ω. A 50 ∠0° V source is connected between the −j10 Ω impedance and the common line, and a 20 ∠0° V source is connected to the −j80 Ω impedance. Calculate the current in the j50 Ω impedance using mesh analysis.

5.15 Repeat exercise 5.14 using nodal analysis.

5.16 Repeat exercise 5.15 using the Superposition theorem.

5.17 A 10 V d.c. voltage source is applied to a circuit that consists of a 2 Ω resistor in series with a 6 Ω resistor that is in parallel with a 12 Ω resistor. Use (*a*) mesh and (*b*) nodal analysis to determine the current flowing in the 12 Ω resistor.

5.18 A T network has series arms of 5 − j5 Ω and 4 + j2 Ω and a shunt arm of 2 Ω. A voltage source of 10 ∠0° V is applied to the capacitive arm and a voltage source of $2\angle 0°$ V is applied to the inductive arm. Use nodal analysis to find the voltage across the shunt arm.

5.19 A π network has a series arm of 10 Ω resistance and shunt arms of 6 Ω and 20 Ω resistance. A 10 V source is applied across the 6 Ω resistor and a 20 Ω load resistor is connected in parallel with the 20 Ω shunt resistor. Calculate using (*a*) mesh and (*b*) nodal analysis, the current in the 20 Ω load resistor.

5.20 Determine the currents supplied by the two voltage sources in the circuit given in Fig 5.25.

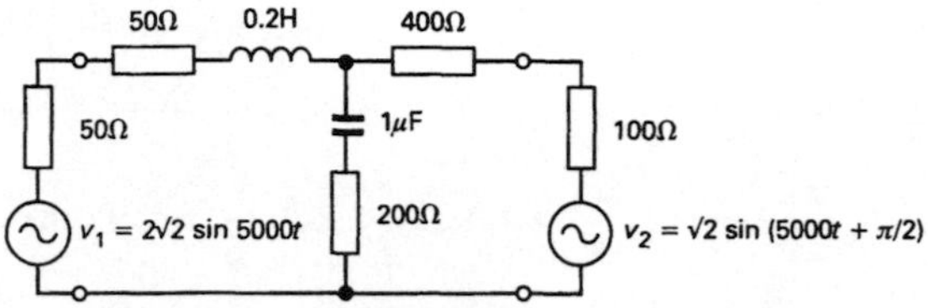

Fig. 5.25

5.21 A T network has series arms of 20 Ω resistance and $-j8$ Ω reactance, and a shunt arm of $j16$ Ω reactance. A $24\angle 0°$ V voltage source is applied to the 20 Ω arm and a voltage source of $12\angle 0°$ V is applied to the $-j8$ Ω arm. Calculate the current flowing in the capacitor.

5.22 A π network has a series arm of $-j1$ kΩ reactance and shunt arms of 1 kΩ resistance and $-j1$ kΩ reactance. A current source of $10\angle 0°$ mA is connected across the 1 kΩ shunt resistance and a voltage source of internal impedance $j2$ kΩ is connected in parallel with the $-j1$ kΩ arm. Calculate the voltage across the $j2$ kΩ reactance.

5.23 A π network has a series arm of 5 Ω resistance, and shunt arms of $-j10$ Ω and $-j20$ Ω reactance. A current source of $0.2\angle 0°$ A is applied across the $-j10$ Ω reactance. Determine the magnitude and phase of the current that must be supplied by another current source connected in parallel with the $-j20$ Ω reactance for the voltage across the $-j20$ Ω reactance to be $2\angle 0°$ V.

5.24 Calculate the current in the 100 Ω resistor in Fig. 5.26. Use the Superposition theorem.

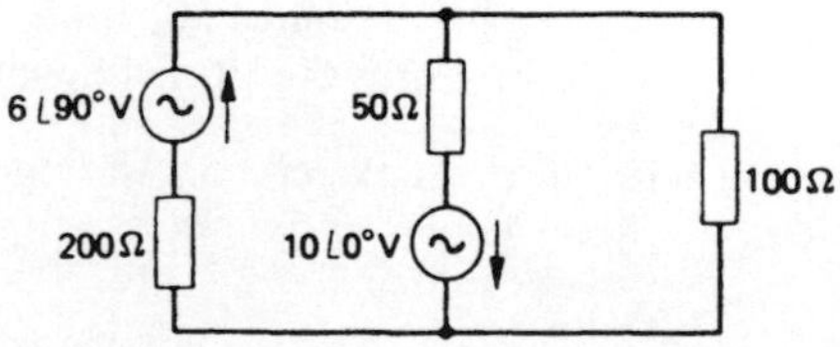

Fig. 5.26

5.25 Repeat 5.24 using mesh analysis.

5.26 Repeat 5.24 using nodal analysis.

6 Circuit theorems

The solution of an electric circuit using mesh or nodal analysis, or the Superposition theorem, can be a lengthy and error-prone process. Simplification of the work involved, with a consequent increase in both speed and accuracy can result if the circuit is first reduced to a simpler form. Three network theorems are in common use for the simplification of circuits before they are analysed and two of these, namely *Thevenin's theorem* and *Norton's theorem*, are considered in this chapter, while a third, known as the *star-to-delta transform*, is the subject of Chapter 7. Another very useful network theorem that allows the conditions for the maximum possible power to be transferred from a source to a load to be investigated is also covered in this chapter, this is the *maximum power transfer theorem.*

Thevenin's theorem

Thevenin's theorem states that

> The current flowing in an impedance Z_L connected across the output terminals of a linear network will be the same as the current that would flow if Z_L were connected to a voltage source of e.m.f. V_{oc} and internal impedance Z_{oc}.
>
> V_{oc} is the voltage that appears across the open-circuited output terminals of the network and Z_{oc} is the output impedance of the network with all internal sources replaced by their internal impedances.

The statement of the theorem is rather lengthy but it can be summarized by means of the circuits shown in Fig. 6.1.

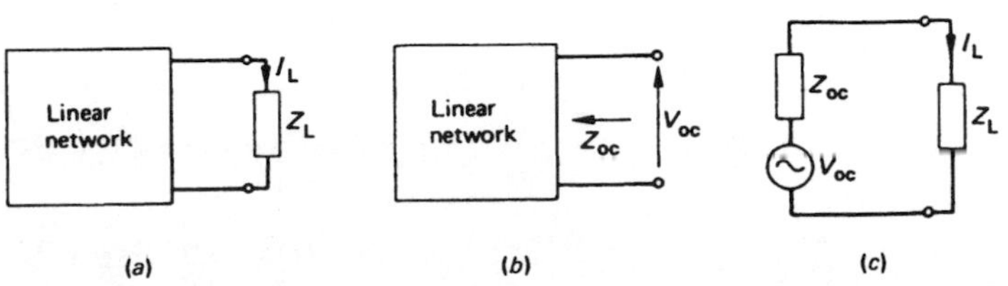

Fig. 6.1 Thevenin's theorem

The term *linear* means that each impedance in the network obeys Ohm' law, that is, the current in each impedance is directly proportional to the voltage across it. This means that any linear circuit can always be reduced to a single voltage source in series with a single impedance. No matter how complex the circuit is, or how many voltage and/or current sources it might contain. In the case of the more complex circuits it may well prove necessary to apply Thevenin's theorem more than once. The procedure to be followed for finding the Thevenin equivalent of a circuit is as follows:

- Open-circuit the terminals across which the Thevenin's equivalent circuit is wanted.
- Calculate the impedance Z_{oc} seen looking into the open-circuited terminals, with each voltage source replaced by a short circuit and each current source replaced by an open circuit.
- Calculate the voltage V_{oc} which will appear across the open-circuited terminals.
- Replace the original circuit by its Thevenin equivalent circuit and then replace the load impedance across the terminals.

Example 6.1

Use Thevenin's theorem to determine the current flowing in an impedance of (250 – j500) Ω when it is connected across the terminals AB of the network in Fig. 6.2(*a*)

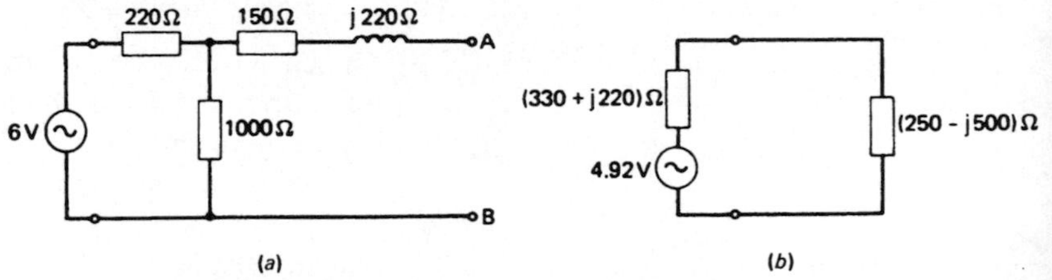

Fig. 6.2

Solution
The voltage V_{oc} appearing across the open-circuited terminals AB is $V_{oc} = 6 \times 1000/(220 + 1000) = 4.92$ V. The output impedance of the network is

$$Z_{oc} = 150 + j220 + (1000 \times 220)/(1000 + 220) = (330 + j220)\ \Omega$$

The Thevenin equivalent circuit is shown by Fig. 6.2(*b*) and from this

$$I = 4.9/(580 - j280) = 7.64\angle 25.8^\circ \text{ mA} \quad (\textit{Ans.})$$

Example 6.2

For the circuit shown in Fig. 6.3, use Thevenin's theorem to calculate the power dissipated in the 100 Ω load impedance.

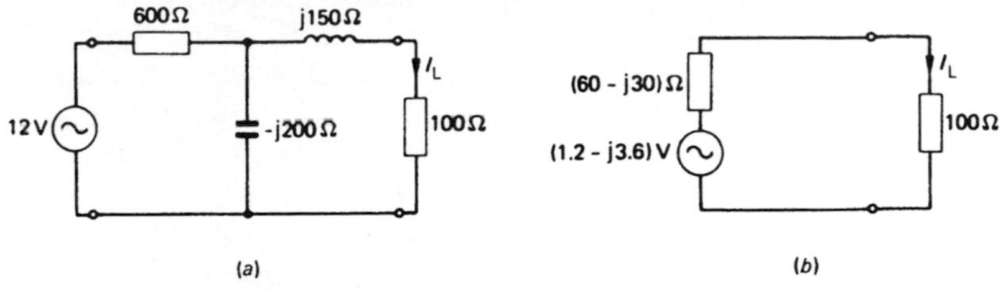

Fig. 6.3

Solution

$$V_{oc} = (12 \times -j200)/(600 - j200) = (48 - j144)/40$$
$$= (1.2 - j3.6)\ \text{V}$$
$$Z_{oc} = j150 + (600 \times -j200)/(600 - j200) = j150 + (2400 - j7200)/40$$
$$= (60 - j30)\ \Omega$$

The Thevenin equivalent circuit is shown by Fig. 6.3(*b*) and from this figure the current flowing in the 100 Ω load is

$$I = (1.2 - j3.6)/(160 - j30) = (3.795\angle{-71.6^\circ})/(162.79\angle{-10.6^\circ})$$
$$= 23.3\angle{-61^\circ}\ \text{mA}$$

The power dissipated in the 100 Ω load is

$$(23.3 \times 10^{-3})^2 \times 100 = 54.3\ \text{mW} \quad (\textit{Ans.})$$

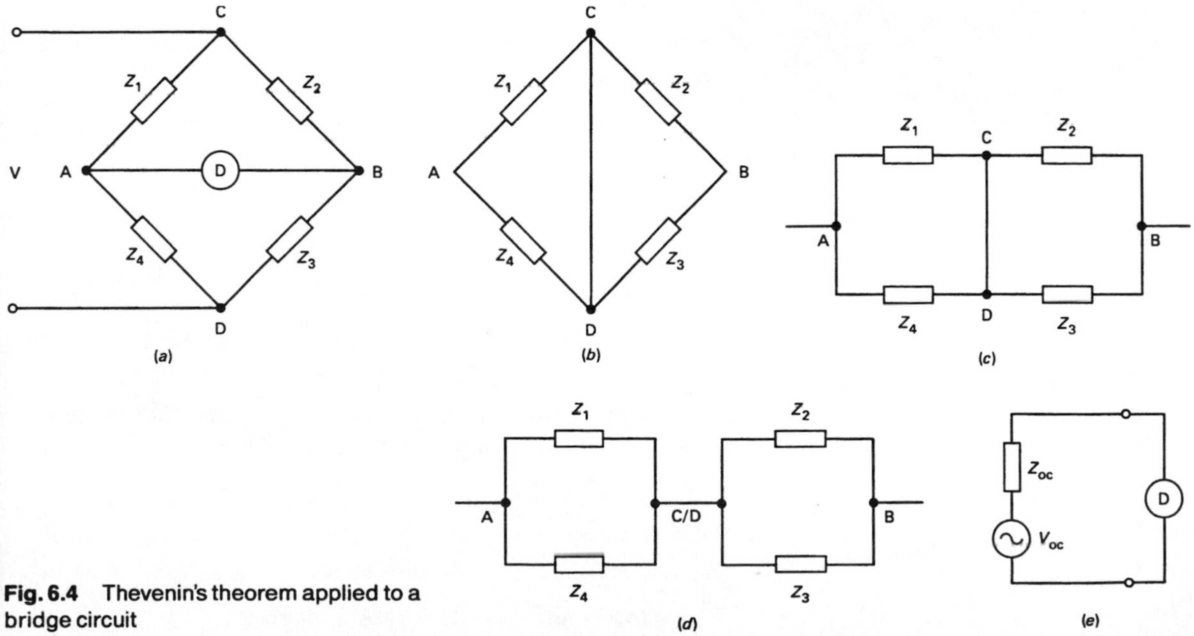

Fig. 6.4 Thevenin's theorem applied to a bridge circuit

Bridge circuit

Figure 6.4(a) shows a bridge circuit with a voltage source V applied between the points marked as C and D and the detector connected between points A and B. With the detector removed from the circuit, the voltage V_{CA} between points A and C is $V_{CA} = VZ_4/(Z_1 + Z_4)$, and the voltage between C and B is $V_{CB} = VZ_3/(Z_2 + Z_3)$. The voltage between A and B is $V_{oc} = V_{CA} - V_{CB} = V[Z_4/(Z_1 + Z_4) - Z_3/(Z_2 + Z_3)]$.

Replacing the voltage source by a short circuit gives the circuit shown in Fig. 6.4(b), and this is re-drawn in Fig. 6.4(c), and further simplified in Fig. 6.4(d). From this last figure, $Z_{oc} = Z_1Z_4/(Z_1 + Z_4) + Z_2Z_3/(Z_2 + Z_3)$. The Thevenin equivalent circuit is shown in Fig. 6.4(e).

Example 6.3

Use Thevenin's theorem to determine the current in the 10 Ω resistor in the bridge circuit of Fig. 6.5(a).

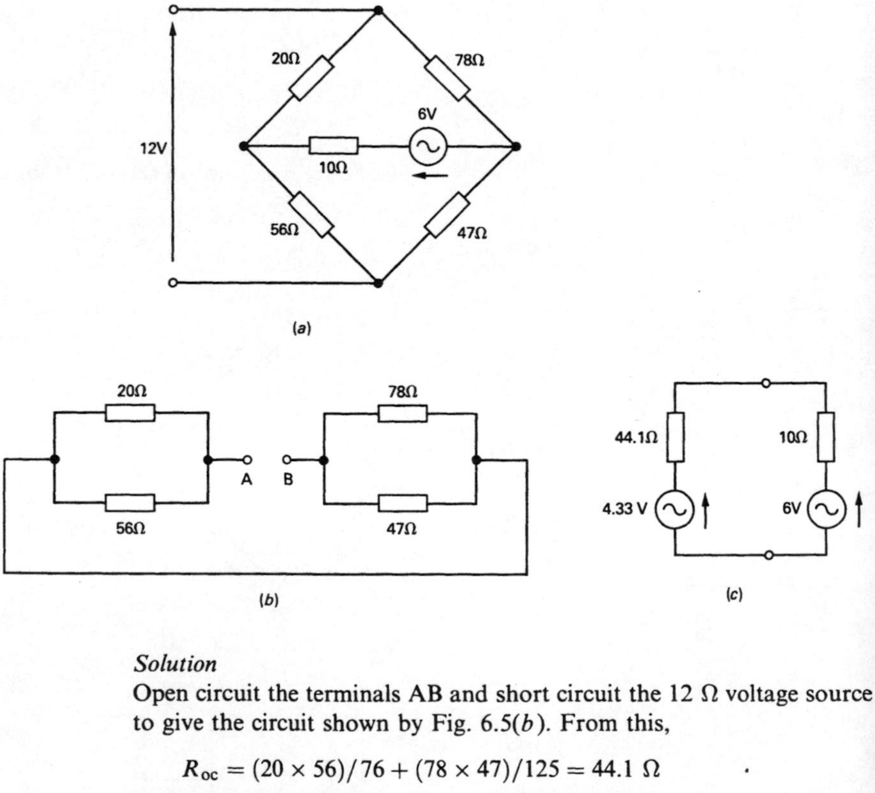

Fig. 6.5

Solution

Open circuit the terminals AB and short circuit the 12 Ω voltage source to give the circuit shown by Fig. 6.5(b). From this,

$$R_{oc} = (20 \times 56)/76 + (78 \times 47)/125 = 44.1\ \Omega$$

and $V_{oc} = (12 \times 56)/76 - (12 \times 47)/125 = 4.33$ V

The Thevenin equivalent circuit is shown in Fig. 6.5(c), from which

$$i = (4.33 - 6)/54.1 = 30.87\ \text{mA} \quad (\textit{Ans.})$$

More complex circuits

In some more complex circuits with two, or more, voltage and/or current sources, it may be necessary to employ the Superposition theorem in the calculation of the open-circuit voltage V_{oc}. This is illustrated by the next example.

Example 6.4

Use Thevenin's theorem to calculate the voltage across the inductance in the circuit given in Fig. 6.6(*a*).

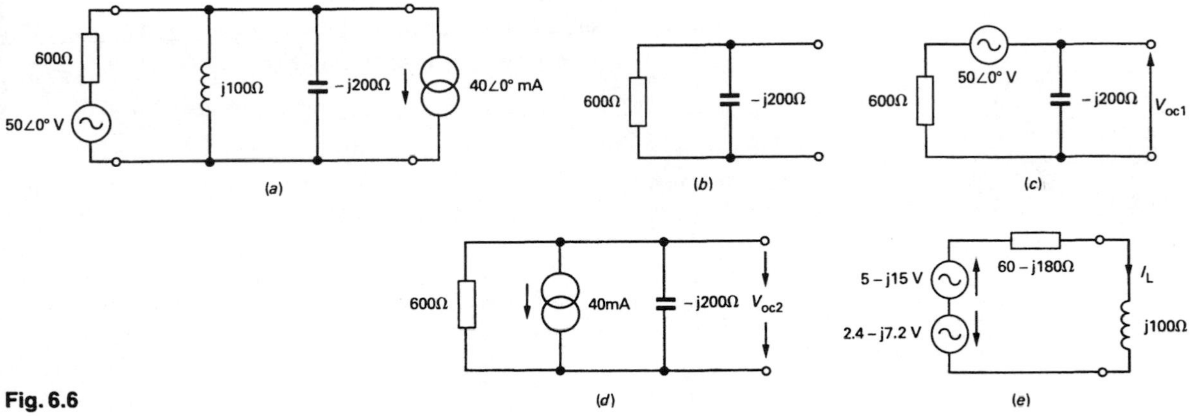

Fig. 6.6

Solution

(*a*) Short circuit the voltage source and open circuit the current source. Remove the inductance from the circuit. The circuit is then as shown in Fig. 6.6(*b*). From this,

$$Z_{oc} = (600 \times -j200)/(600 - j200) = 60 - j180\ \Omega$$

(*b*) Open circuit the current source, Fig 6.6(*c*).

$$V_{oc1} = (50 \times -j200)/(600 - j200) = 5 - j15\ \text{V}$$

(*c*) Short circuit the voltage source to obtain Fig. 6.6(*d*).

$$V_{oc2} = (40 \times 10^{-3}) \times (60 - j180) = 2.4 - j7.2\ \text{V}$$

V_{oc2} has the opposite polarity to V_{oc1}. The Thevenin equivalent circuit is given in Fig 6.6(*e*). From this,

$$i_L = [(5 - j15) - (2.4 - j7.2)]/(60 - j180 + j100)$$
$$= 0.078 - j0.026 = 0.082\angle{-18.4°}\ \text{A}.$$

Voltage across inductance is

$$0.082\angle{-18.4°} \times 100\angle{90°} = 8.22\angle{71.6°}\ \text{V} \quad (\textit{Ans.})$$

Example 6.5

Calculate, using Thevenin's theorem, the current flowing in the capacitor in the circuit given in example 5.2 (Fig. 6.7).

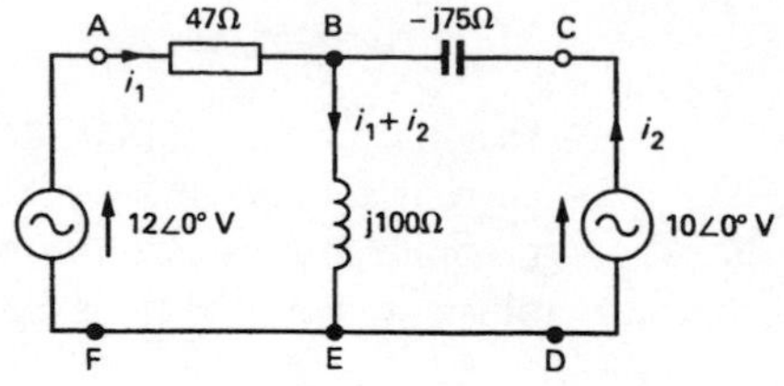

Fig. 6.7

Solution

(*a*) $V_{oc} = (12 \times j100)/(47 + j100) = 9.83 + j4.62$ V

(*b*) $Z_{oc} = -j75 + (47 \times j100)/(47 + j100) = 38.5 - j56.9\ \Omega$. The Thevenin equivalent circuit is shown by Fig 6.8. From this;

$$I_C = [(9.831 + j4.62) - 10]/(38.5 - j56.9) = -0.057 + j0.036$$
$$= 0.068\angle 147.8° \text{ A, downwards}$$

or $0.068\angle -32.3°$ A, upwards (*Ans.*)

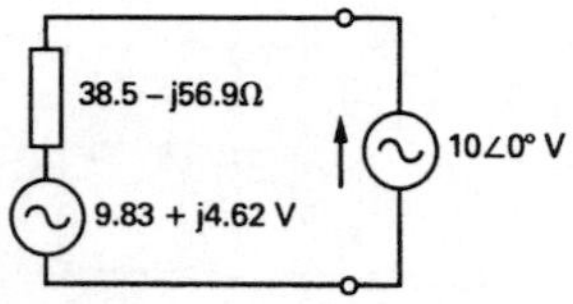

Fig. 6.8

Repeated use of Thevenin's theorem

In a more complex network it may prove necessary to apply Thevenin's theorem more than once. Figure 6.9 shows a resistive ladder network. Suppose it is required to find the current which flows in the 600 Ω load resistor.

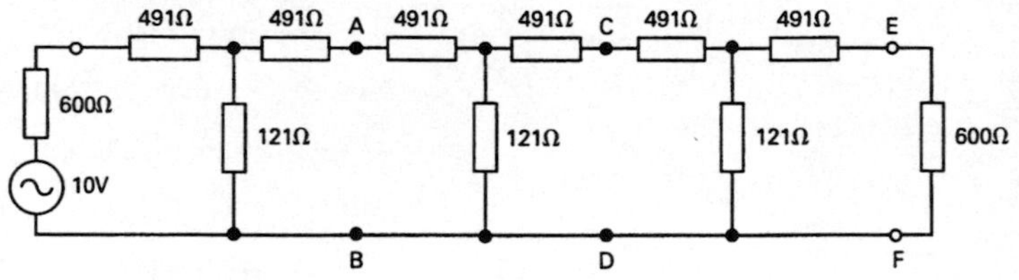

Fig. 6.9

- Apply Thevenin's theorem to the left of the terminals AB.

  $R_{oc} = [(600 + 491)(121)]/(600 + 491 + 121) = 600\ \Omega$

  $V_{oc} = (10 \times 121)/(600 + 491 + 121) = 1$ V

- If, now, this Thevenin circuit is connected across terminals AB the theorem can next be applied to terminals CD. The circuit is shown by Fig. 6.10(*a*). Clearly, R_{oc} is again equal to 600 Ω and V_{oc} is 0.1 V giving Fig. 6.10(*b*).
- Now connecting Fig. 6.10(*b*) across terminals EF gives the simple arrangement shown in Fig. 6.10(*c*) from which

  $i = 0.01/1200 = 8.33\ \mu$A (*Ans.*)

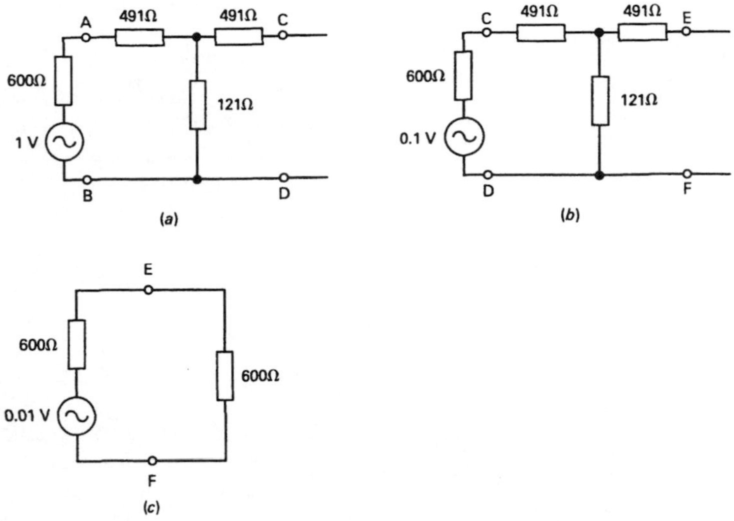

Fig. 6.10

Example 6.6

Apply Thevenin's theorem repeatedly to the ladder network shown in Fig. 6.11 to obtain the current which flows in the 5 Ω load resistor.

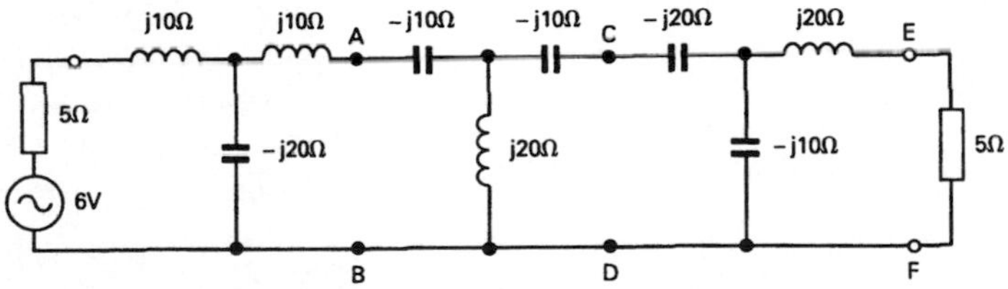

Fig. 6.11

Solution

Apply Thevenin's theorem to terminals AB.

$$Z_{oc} = (5 + j10)(-j20)/(5 - j10) + j10 = 16 + j22\ \Omega$$
$$V_{oc} = (6 \times -j20)/(5 - j10) = 9.6 - j4.8\ \text{V}$$

The circuit is now shown by Fig 6.12(*a*). Apply Thevenin's theorem to terminals CD:

$$Z_{oc} = [(16 + j22 - j10)j20]/(16 + j22 - j10 + j20) - j10 = 5\ \Omega$$
$$V_{oc} = [(9.6 - j4.8)(j20)]/(16 + j22 - j10 + j20) = 6\ \text{V}$$

The circuit is now shown by Fig 6.12(b). From this:

$$Z_{oc} = [(5 - j20)(-j10)]/(5 - j30) + j20 = 0.54 + j13.24\ \Omega \approx j13.24\ \Omega$$
$$V_{oc} = (6 \times -j10)/(5 - j30) = 1.95 - j0.32\ \text{V}$$

The ladder network is finally reduced to the circuit given by Fig. 6.12(*c*); hence

$$i = (1.95 - j0.32)/(5 + j13.24) = 139.6\angle -78.6°\ \text{mA} \quad (\textit{Ans.})$$

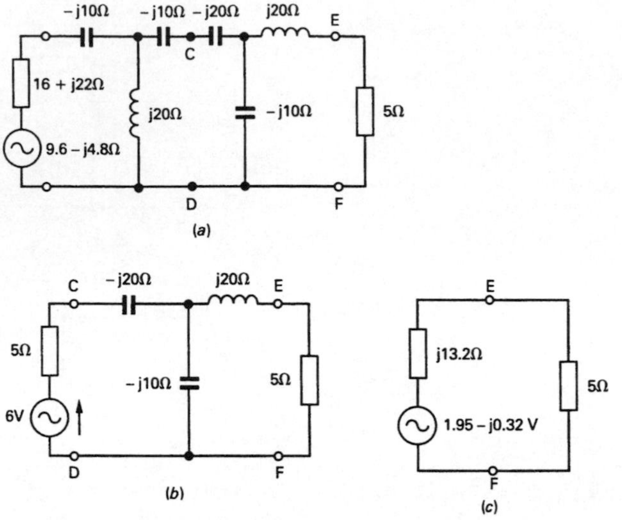

Fig. 6.12

Norton's theorem

Norton's theorem states that

> The current flowing in an impedance Z_L connected across the output terminals of a linear network is the same as the current that would flow if Z_L were to be connected across a current generator in parallel with an impedance. The generated current is the current I_{sc} that would flow in the short-circuited output terminals of the network and the impedance is the output impedance Z_{oc} of the network with all internal sources replaced by their internal impedances.

Norton's theorem is summarized by Fig. 6.13. The procedure to be followed is:

- Short circuit the terminals across which the Norton equivalent circuit is wanted.
- Calculate the current that would flow in the short circuit.
- Determine the impedance looking into the open-circuited terminals.
- Replace the circuit by its Norton equivalent circuit and then replace the load across the terminals.

The Norton equivalent of a circuit can always be found by first finding its Thevenin equivalent circuit and then converting the voltage source into the corresponding current source. Conversely,

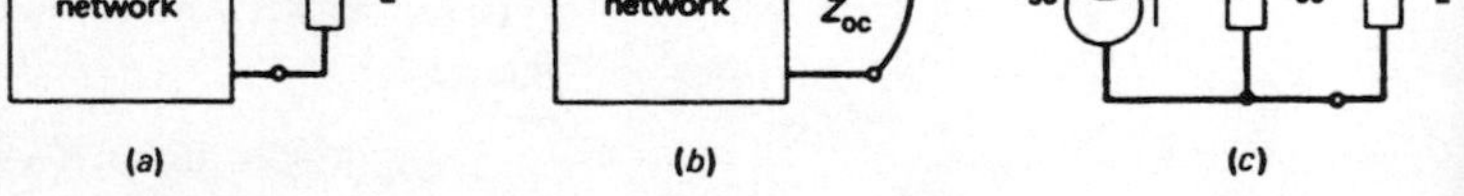

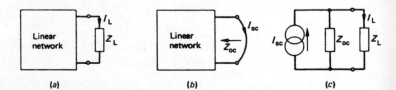

Fig. 6.13 Norton's theorem

a Norton equivalent circuit may always be converted into the equivalent Thevenin circuit by merely converting the current source into a voltage source. The Thevenin equivalent circuit in Fig. 6.14(*a*) can be converted into the corresponding Norton circuit by short circuiting its output terminals. Then $I_{sc} = 2/50 = 0.04$ A and so the Norton equivalent circuit is given in Fig. 6.14(*b*). If, now, the terminals of the Norton circuit are open circuited $V_{oc} = 0.04 \times 50 = 2$ V and the Thevenin circuit can be obtained.

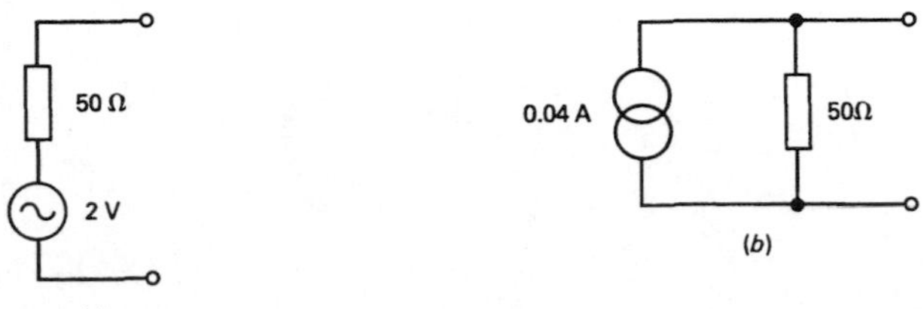

Fig. 6.14

Example 6.7

For the network shown in Fig. 6.15(*a*) calculate the current flowing in a 1000 Ω resistor connected across terminals AB using (*a*) Thevenin's theorem and (*b*) Norton's theorem.

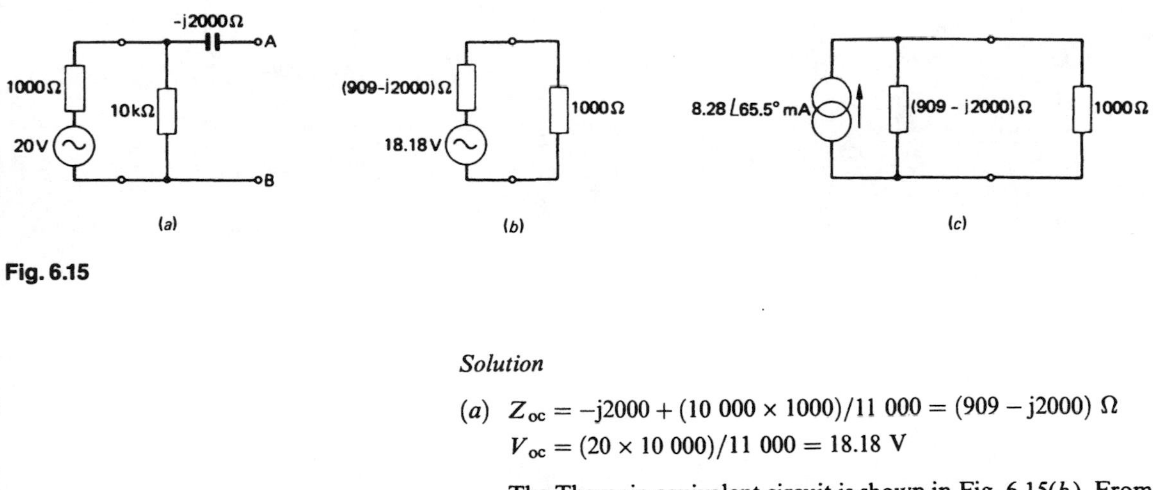

Fig. 6.15

Solution

(*a*) $Z_{oc} = -j2000 + (10\,000 \times 1000)/11\,000 = (909 - j2000)\ \Omega$

$V_{oc} = (20 \times 10\,000)/11\,000 = 18.18$ V

The Thevenin equivalent circuit is shown in Fig. 6.15(*b*). From this figure,

$$I_L = 18.18/(1909 - j2000) = 18.18/(2765\angle -46.3°)$$
$$= 6.58\angle 46.3° \text{mA} \quad (\textit{Ans.})$$

(*b*) As in (*a*), $Z_{oc} = (909 - j2000)\ \Omega$. The impedance 'seen' by the voltage source when the load is short circuited is

$$1000 + [10\,000 \times (-j2000)]/(10\,000 - j2000) = 1385 - j1923\ \Omega$$

The input current to the network is 20/(1385 − j1923) and so the short-circuit current I_{sc} is equal to

$$20/(1385 - j1923) \times 10000/(10000 - j2000) = 8.28\angle 65.5^\circ \text{ mA}$$

The Norton equivalent circuit is shown in Fig. 6.15(*c*), from which

$$I_L = 8.28\angle 65.5^\circ \times (909 - j2000)/(1909 - j2000)$$
$$= 6.58\angle 46.3^\circ \text{ mA} \quad (Ans.)$$

If Norton's theorem is applied to Fig. 6.15(*b*),

$$I_{sc} = 18.18/(909 - j2000) = 18.18\angle 0^\circ/2196\angle -65.5^\circ$$
$$= 8.28\angle 65.5^\circ \text{ mA}$$

as in Fig. 6.15(*c*). Alternatively, if Thevenin's theorem is applied to Fig 6.15(*c*),

$$V_{oc} = 8.28\angle 65.5^\circ \times 10^{-3} \times (909 - j2000) = 18.18 \text{ V}$$

as in Fig. 8.15(*b*). This means, of course, that either of the two equivalent circuits can be quickly converted into the other.

Example 6.8

Use (*a*) Thevenin's theorem and (*b*) Norton's theorem to determine the current flowing in the 12 Ω resistor in the circuit shown by Fig. 6.16(*a*).

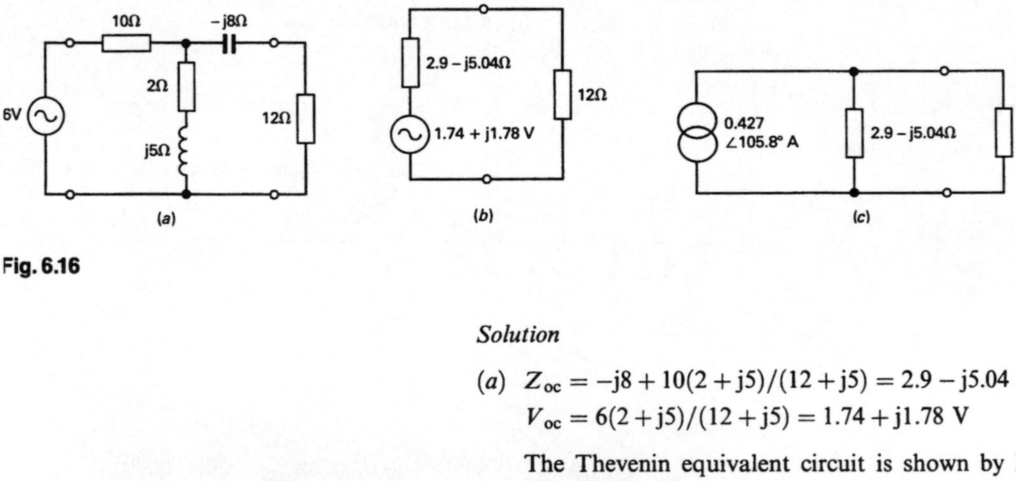

Fig. 6.16

Solution

(*a*) $Z_{oc} = -j8 + 10(2 + j5)/(12 + j5) = 2.9 - j5.04\ \Omega$

$V_{oc} = 6(2 + j5)/(12 + j5) = 1.74 + j1.78$ V

The Thevenin equivalent circuit is shown by Fig. 6.16(*b*). From this,

$$i_L = (1.74 + j1.78)/(14.9 - j5.04)$$
$$= 0.07 + j0.14 = 0.157\angle 64.4^\circ \text{ A} \quad (Ans.)$$

(*b*) $Z_{oc} = 2.9 - j5.04\ \Omega$ as before. With the output terminals short circuited the impedance seen by the 6 V voltage source is

$$10 + [(-j8)(2 + j5)]/(2 - j3) = 19.85 + j6.77\ \Omega$$

The current supplied by the voltage source is

$$6/(19.85 + j6.77) = 0.27 - j0.09 \text{ A}$$

Therefore, the short-circuit current is

$$I_{sc} = [(0.27 - j0.09)(2 + j5)]/(2 - j3) = -0.12 + j0.41$$
$$= 0.427\angle 106.3^\circ \text{ A}$$

The Norton equivalent circuit is shown in Fig 6.16(c). From this circuit,

$$i = [(0.427\angle 106.3^\circ) \times (5.815\angle -60.1^\circ)]/(15.73\angle -18.7^\circ)$$
$$= 0.158\angle 67^\circ \text{ A} \quad (\textit{Ans.})$$

Example 6.9

For the circuit shown in Fig. 6.17(a) use (a) Norton's theorem and (b) Thevenin's theorem to find the current flowing in the 5 kΩ resistor.

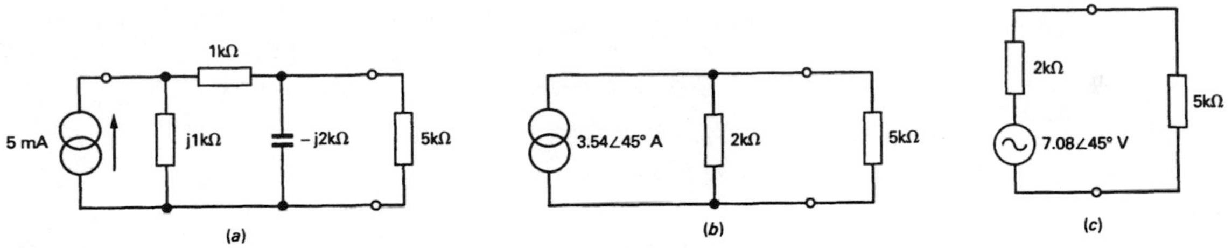

Fig. 6.17

Solution
Referring to Fig 6.17(a),

(a) With the output terminals short circuited

$$I_{sc} = j5/(1 + j1) = 2.5 + j2.5 = 3.54\angle 45^\circ \text{ A}$$
$$Z_{oc} = [-j2(1 + j1)]/(1 + j1 - j2) = 2 \text{ k}\Omega$$

The Norton equivalent circuit is shown in Fig. 6.17(b). From the figure,

$$I_L = [(3.54\angle 45^\circ \times 2)]/(2 + j5) \approx 1\angle 45^\circ \text{ mA} \quad (\textit{Ans.})$$

(b) $Z_{oc} = 2$ kΩ. Convert the Norton current source into a voltage source of $3.54\angle 45^\circ \times 2 = 7.08\angle 45^\circ$ V. The Thevenin equivalent circuit is shown in Fig. 6.16(c), from which,

$$I_L = (7.072\angle 45^\circ)/7000 \approx 1\angle 45^\circ \text{ mA} \quad (\textit{Ans.})$$

Example 6.10

For the circuit given in Fig. 6.18 calculate the current flowing in the 1 kΩ resistor.

Solution
(a) Remove the 1 kΩ resistor from the circuit, short circuit the voltage source and open circuit the current source to give the circuit shown in Fig. 6.19(a). From this,

$$Z_{oc} = 5000 - j1000 = 5099\angle -11.3^\circ \ \Omega$$

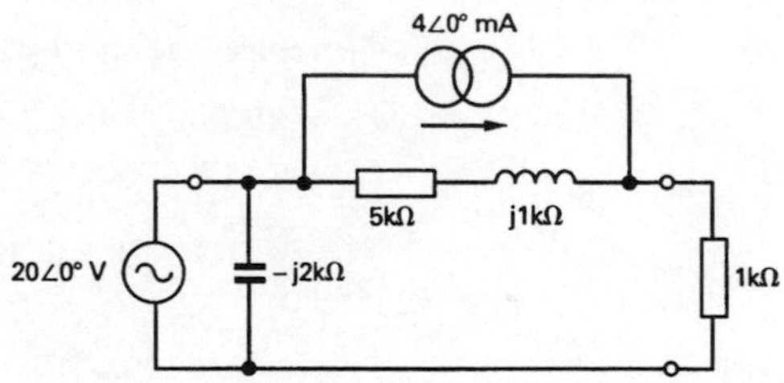

Fig. 6.18

(*b*) Fig. 6.19(*b*) shows the circuit with the output terminals short circuited and the current source replaced by an open circuit. From this,

$$i_1 = 20/(5 + j1) = 3.85 - j0.77 \text{ mA}$$

(*c*) Short circuit both the output terminals and the voltage source to get the circuit shown in Fig 6.19(*c*). Clearly, the $5000 + j1000\ \Omega$ impedance is short circuited and so $i_2 = 4$ mA.

(*d*) The short-circuit current is

$$i_{sc} = i_1 + i_2 = 7.85 - j0.77 \text{ mA} = 7.89\angle{-5.6^\circ} \text{ mA}.$$

The Norton equivalent circuit is shown in Fig 6.19(*d*). Hence,

$$i = (7.89\angle{-5.6^\circ}) \times [1/(5.1\angle{-11.3^\circ} + 1)]$$
$$= 1.297\angle 3.9^\circ \text{ mA} \quad (\textit{Ans.})$$

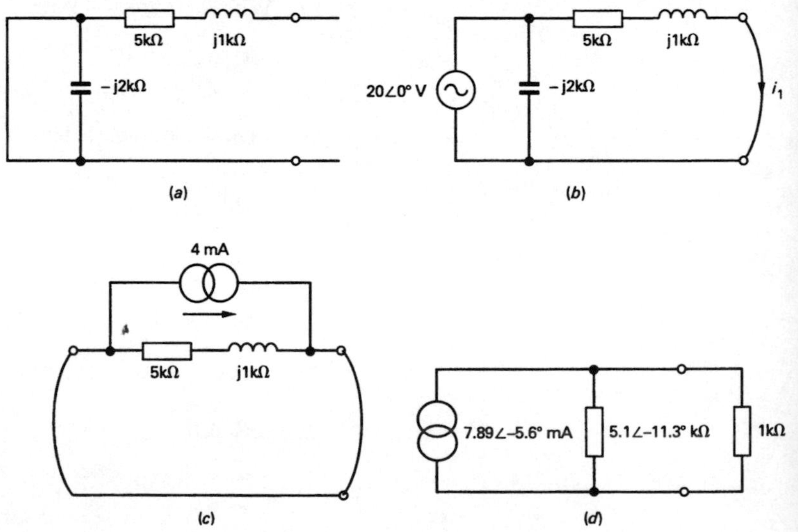

Fig. 6.19

Maximum power transfer theorems The *maximum power transfer theorems* state the conditions for the maximum power to be transferred from a source to a load.

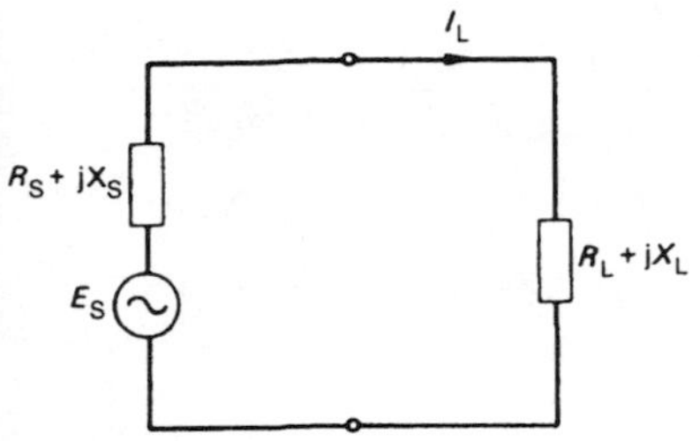

Fig. 6.20

There are four possible cases:

- The maximum power is transferred from a source to a load when the load impedance is the conjugate of the source impedance.

  Figure 6.20 shows a voltage source of e.m.f. E_S and internal impedance $R_S + jX_S$ connected to a load of impedance $R_L + jX_L$. The current I_L flowing in the load is

  $$I_L = E_S/[R_S + R_L + j(X_S + X_L)] \quad (6.1)$$

  and the load power P_L is $|I_L|^2 R_L$ or

  $$P_L = E_S^2 R_L/[(R_S + R_L)^2 + (X_S + X_L)^2] \quad (6.2)$$

  For the load power to be a maximum it is clear that the total reactance of the circuit must be zero, so that $X_L = -X_S$. For this condition

  $$P_L = E_S^2 R_L/(R_S + R_L)^2$$

  Differentiating with respect to R_L and equating the result to zero gives

  $$dP_L/dR_L = \frac{(R_S + R_L)^2 E_S^2 - 2E_S^2 R_L(R_S + R_L)}{(R_S + R_L)^4} = 0$$

  or $R_S^2 = R_L^2$

  The condition for maximum power transfer is therefore

  $$R_S = R_L \quad \text{and} \quad X_C = -X_L \quad (6.3)$$

  i.e. the load impedance is the conjugate of the source impedance

- If a *purely resistive source* and a *purely resistive load* are considered,

  $$i = E_S/(R_S + R_L), \; P_L = E_S^2/R_L(R_S + R_L)^2$$

  The maximum power is transferred to the load when $dP_L/R_L = 0$.

  $$dP_L/R_L = E_S^2[(R_S + R_L)^2 - 2R_L(R_S + R_L)]/(R_S + R_L)^2$$

  or $(R_S + R_L)^2 = R_S^2 + 2R_S R_L + R_L^2 = 2R_L^2 + 2R_L R_S$

  and hence

  $$R_S = R_L \quad (6.4)$$

- Very often it is only possible to control the effective magnitude of the load impedance, its angle remaining unchanged. This is the case, for example, when a transformer is used for impedance transformation. Re-writing equation (6.2),

$$P_L = \frac{E_S^2|Z_L|\cos\varphi_L}{(R_S+|Z_L|\cos\varphi_L)^2+(X_S+|Z_L|\sin\varphi_L)^2}$$

For P_L to be a maximum

$$\begin{aligned}\mathrm{d}P_L/\mathrm{d}|Z_L| = {} & E_S^2\cos\varphi_L[(R_S+|Z_L|\cos\varphi_L)^2 \\ & +(X_S+|Z_L|\sin\varphi_L)^2] \\ & - E_S^2|Z_L|\cos\varphi_L[2(R_S+|Z_L|\cos\varphi_L)\cos\varphi_L \\ & +2(X_S+|Z_L|\sin\varphi_L)\sin\varphi_L]=0\end{aligned}$$

(numerator only). Hence

$$\begin{aligned}& R_S^2+2R_S|Z_L|\cos\varphi_L+|Z_L|^2\cos^2\varphi_L \\ & \quad +X_S^2+2X_S|Z_L|\sin\varphi_L+|Z_L|\sin^2\varphi_L \\ & =2R_S|Z_L|\cos\varphi_L+2|Z_L|^2\cos^2\varphi_L \\ & \quad +2X_S|Z_L|^2\sin\varphi_L+2|Z_L|\sin^2\varphi_L\end{aligned}$$

or $R_S^2+X_S^2=|Z_L|^2(\cos^2\varphi_L+\sin^2\varphi_L)$ and so

$$|Z_S|=|Z_L| \qquad (6.5)$$

Thus, if the angle of the load impedance cannot be altered, the maximum possible transfer of power from a source to a load occurs when the source and load impedances are of the same magnitude.

- The current I_L supplied to a purely resistive load R_L by a source of complex impedance R_S+jX_S is

$$I_L=E_S/(R_S+R_L+jX_S) \qquad (6.6)$$

and the load power is

$$P_L=|I_L|^2R_L=E_S^2R_L/[(R_S+R_L)^2+X_S^2] \qquad (6.7)$$

For maximum load power $\mathrm{d}P_L/\mathrm{d}R_L=0$. Hence,

$$(R_S+R_L)^2+X_S^2=2R_L(R_S+R_L)$$

$$R_S^2+X_S^2=R_L^2 \quad \text{and} \quad R_L=\sqrt{(R_S^2+X_S^2)} \qquad (6.8)$$

This means that for the maximum power to be transferred from a source with a complex internal impedance to a purely resistive load, the load resistance should be equal to the magnitude of the source impedance.

- The current I_L in the load and the load power P_L are given by equations (6.1) and (6.2) respectively. For a load of variable resistance and constant reactance

$$\begin{aligned}\mathrm{d}P_L/R_L = {} & E_S^2\{[(R_S+R_L)^2+(X_S+X_L)^2-2R_L(R_S+R_L)]/ \\ & [(R_S+R_L)^2+(X_S+X_L)^2]^2\}=0\end{aligned}$$

Therefore,

$$(R_S + R_L)^2 + (X_S + X_L)^2 = 2R_L^2 + 2R_SR_L \quad \text{or}$$
$$R_L = \sqrt{[(R_S^2 + (X_S + X_L)^2]} \qquad (6.9)$$

Example 6.11

Determine the Thevenin equivalent circuit of the network given in Fig. 6.21. Use it to find the load impedance that will dissipate the maximum power and the value of this power.

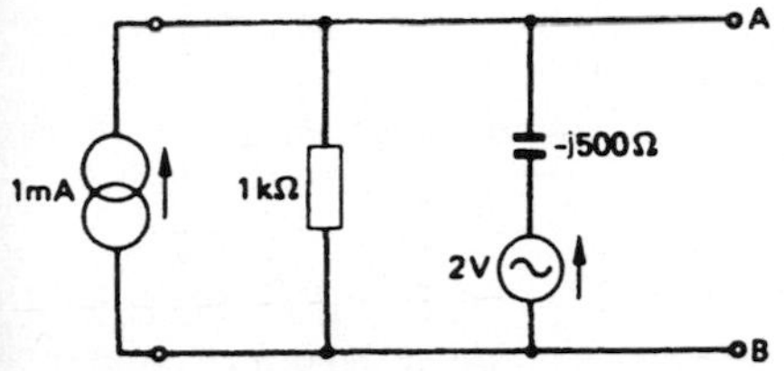

Fig. 6.21

Solution

The constant-current source can be converted into a voltage source by the use of Thevenin's theorem. The voltage source has an e.m.f. of $1 \times 10^{-3} \times 1000 = 1$ V and an impedance of 1000 Ω. The equivalent circuit is shown in Fig. 6.22(*a*).

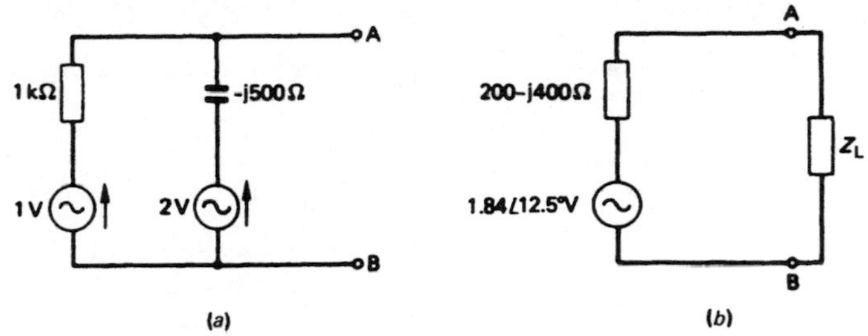

Fig. 6.22

Using the superposition theorem

$$V_1 = (1 \times -j500)/(1000 - j500) = -j/(2 - j) \text{ V}$$

and

$$V_2 = 2000/(1000 - j500) = 4/(2 - j) \text{ V}$$

The voltage V_{AB} across the terminals AB is

$$V_{AB} = (4 - j)/(2 - j) = 1.84\angle 12.5° \text{ V}$$

and the open-circuit impedance is

$$Z_{oc} = [1000 \times (-j500)]/(1000 - j500) = -j1000/(2 - j)$$
$$= 200 - j400 \ \Omega$$

The Thevenin equivalent circuit is given in Fig. 6.22(*b*). The load impedance for maximum power is,

(200 + j400) Ω (*Ans.*)

With this value of load impedance connected across terminals A and B

$$I_L = 1.84\angle 12.5°/400 = 4.6\angle 12.5° \text{ mA}$$

The load power is

$$(4.6 \times 10^{-3})^2 \times 200 = 4.23 \text{ mW} \quad (\textit{Ans.})$$

Example 6.12

A voltage source has an e.m.f. of 10 V and an internal impedance of (600 + j100) ohms. The source is to supply power to a 20 Ω load resistor. Determine the turns ratio of the transformer needed to connect load to source in order to obtain the maximum load power. Calculate the value of this maximum load power, assuming the transformer losses are negligibly small.

Solution
The transformer will only match the magnitudes of the source and load impedances. Since, for a transformer,

$$\text{Impedance ratio} = (\text{turns ratio})^2$$
$$n = \sqrt{[\sqrt{(600^2 + 100^2)}/20]} = 5.51 : 1 \quad (\textit{Ans.})$$

The current supplied by the source is

$$10/[600 + \text{j}100 + (5.51^2 \times 20)] = 10/(1207 + \text{j}100)$$

Therefore, $|I| = 8.23$ mA and so the maximum load power P_L is

$$P_L = (8.26 \times 10^{-3})^2 \times 607 = 41.1 \text{ mW} \quad (\textit{Ans.})$$

Example 6.13

A voltage source of e.m.f. 5 V and internal impedance 100 + j100 Ω is connected in series with a load that consists of a capacitive reactance of −j10 Ω and a variable resistor R. Calculate (*a*) the value of R for maximum power transfer and (*b*) the value of this maximum power.

Solution
(*a*) For maximum power transfer,

$$R = \sqrt{(100^2 + 90^2)} = 134.5 \ \Omega \quad (\textit{Ans.})$$

(*b*) $I = 5/(100 + \text{j}100 + 134.5 - j10) = 19.94\angle{-21^\circ}$ mA

Load power $P_L = (19.94 \times 10^{-3})^2 \times 134.5 = 53.48$ mW (*Ans.*)

Example 6.14

A voltage source of e.m.f. 2 V and internal impedance 600 + j150 Ω is to supply power to a purely resistive load. Calculate (*a*) the required resistance of the load for the maximum power to be dissipated, and (*b*) the value of the maximum load power. (*c*) If a variable capacitor is connected in series with the load what should be its reactance for maximum load power? What should then be the resistance of the load and what is the power in that load?

Solution

(*a*) $R_L = \sqrt{(600^2 + 150^2)} = 618.5\ \Omega$ (*Ans.*)

(*b*) $I = 2/(600 + 618.5 + j150) = 1.61 - j0.2 = 1.63\angle{-7°}$ mA

$P_L = (1.63 \times 10^{-3})^2 \times 618.5 = 1.64$ mW (*Ans.*)

(*c*) $X_C = -j150\ \Omega$ and $R_L = 600\ \Omega$ (*Ans.*)

Maximum power $= (2/1200)^2 \times 600 = 1.67$ mW (Ans.)

Exercises 6

6.1 State Thevenin's theorem. Determine the Thevenin equivalent circuit for Fig. 6.23. Thence find the load impedance that will dissipate the maximum power and the value of that power.

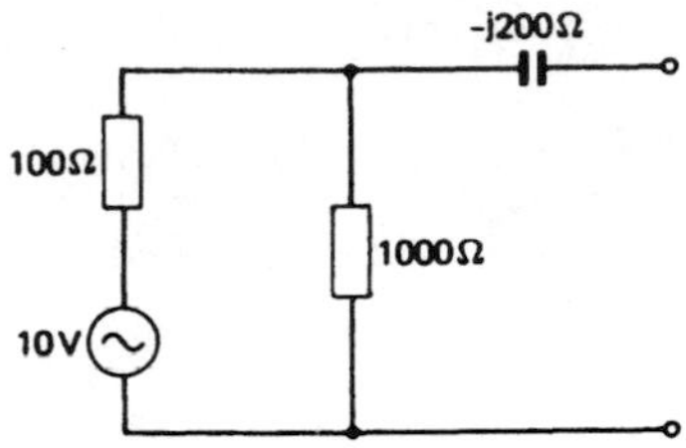

Fig. 6.23

6.2 A voltage source has an e.m.f. of 15 V and an internal impedance of (500 + j30) Ω. The source is connected to an 8 Ω load by a transformer whose losses may be neglected. Calculate the transformer turns ratio for the maximum load power to be dissipated and the value of this power.

6.3 A voltage source has an e.m.f. of 20 sin 10^3t volts and an internal impedance of 600 Ω. A 1 μF capacitance is effectively connected across the terminals of the source. The voltage source is connected across an inductor of 60 mH inductance and 50 Ω resistance. Calculate the magnitudes of the current in, and the voltage across, the inductor.

6.4 For the circuit in Fig. 6.24 calculate (*a*) the two components which when connected in series and (*b*) the two components which, when connected in parallel, across the terminals AB, will dissipate the maximum power.

6.5 Calculate the power dissipated in the 100 Ω load resistor in Fig. 6.25.

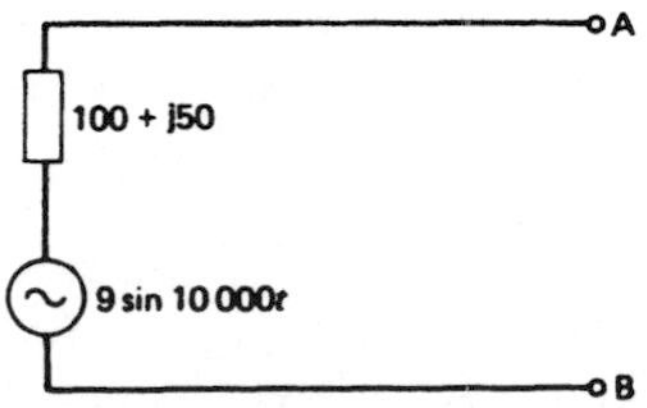

Fig. 6.24

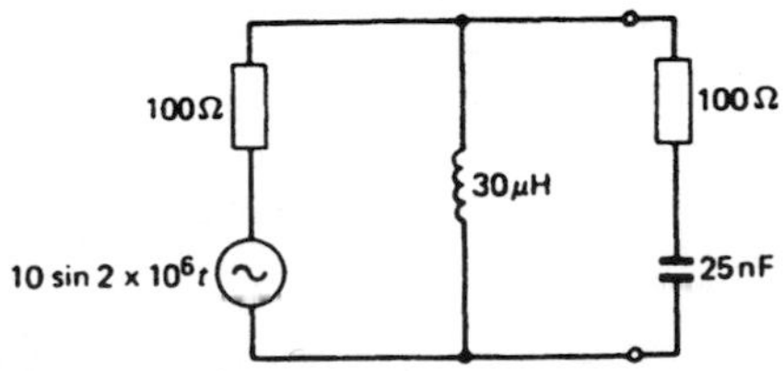

Fig. 6.25

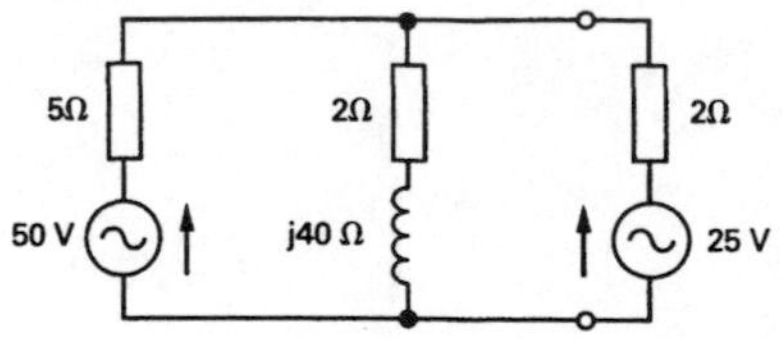

Fig. 6.26

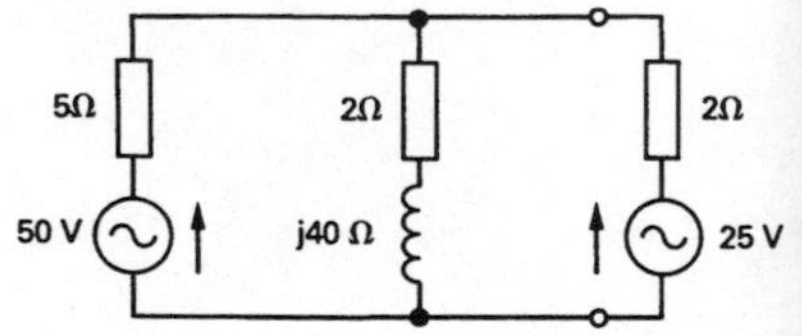

Fig. 6.27

6.6 Use Thevenin's theorem to calculate the power dissipated in the 20 kΩ resistor in Fig 6.26.

6.7 Calculate the current flowing in the inductance in Fig 6.27 using (*a*) Thevenin's theorem, (*b*) the Superposition theorem and (*c*) Nodal analysis.

6.8 Use Thevenin's theorem to calculate the power dissipated in the 30 Ω resistor in Fig. 6.28.

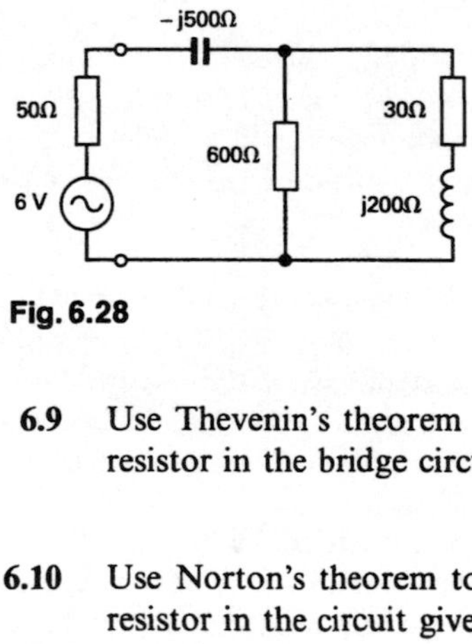

Fig. 6.28

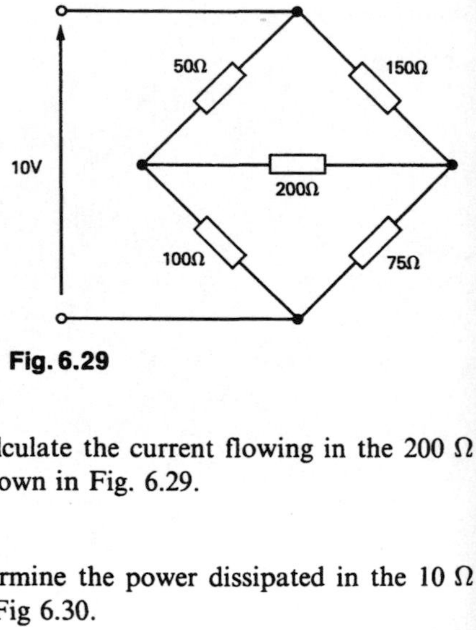

Fig. 6.29

6.9 Use Thevenin's theorem to calculate the current flowing in the 200 Ω resistor in the bridge circuit shown in Fig. 6.29.

6.10 Use Norton's theorem to determine the power dissipated in the 10 Ω resistor in the circuit given in Fig 6.30.

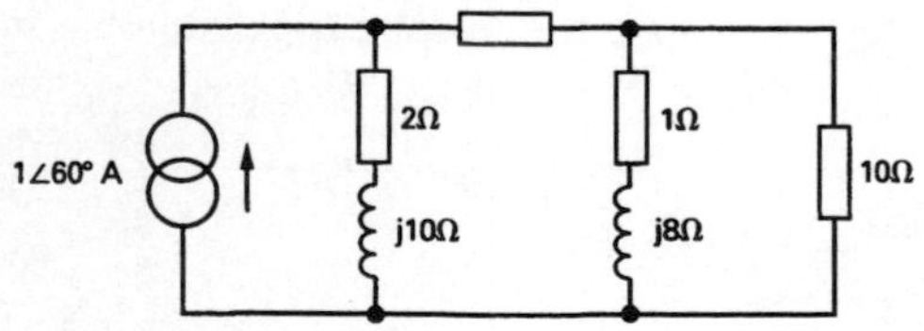

Fig. 6.30

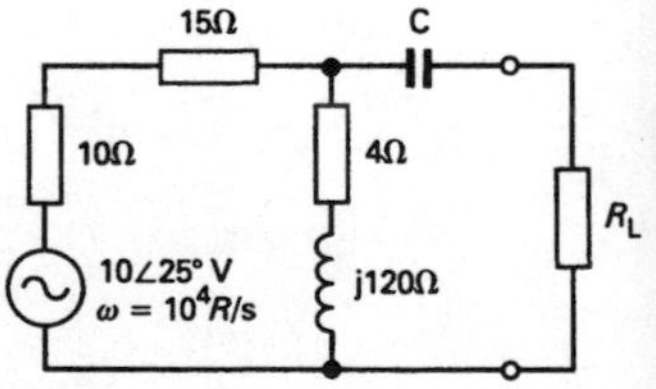

Fig. 6.31

6.11 For the network shown in Fig 6.31, calculate the values of C and R_L for maximum power dissipation in R_L. Determine the maximum power.

6.12 (*a*) A Thevenin equivalent circuit has $V_{oc} = 250\angle 0°$ V and $Z_{oc} =$ j2.7 kΩ. Calculate the equivalent Norton circuit. (*b*) A Norton equivalent circuit has $I_{sc} = 0.67\angle 180°$ mA and $Z_{oc} = 3$ kΩ. Determine the equivalent Thevenin circuit.

6.13 A 5 kHz voltage source has an internal impedance of $1\angle 60°$ kΩ. Calculate the components of the load that will absorb the maximum power from the source.

6.14 Repeat example 5.3 using Thevenin's theorem.

6.15 A 4.5 V voltage source has an internal impedance consisting of a 500 Ω resistance in parallel with a 500 Ω inductive reactance. The source is connected to a purely resistive 600 Ω load. Use Thevenin's theorem to determine the load voltage.

6.16 A voltage source of e.m.f. 12 V and internal impedance 4 + j5 Ω is connected to a load via a capacitive reactance of −j2 Ω. (*a*) Calculate the load impedance that will give maximum power transfer. (*b*) Calculate the load power.

6.17 A 6 V voltage source has an internal impedance of 600 + j400 Ω. It is connected to a load of 50 + j30 Ω by a transformer of turns ratio *n* : 1. Determine the value of *n* for maximum power to be dissipated in the load and the value of this power.

6.18 Obtain the Thevenin and Norton equivalent circuits of the network shown in Fig. 6.32.

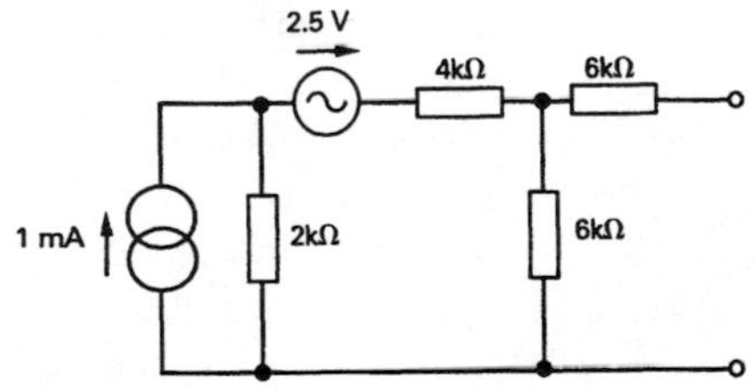

Fig. 6.32

6.19 A voltage source is connected to a variable resistor. When the resistance is varied the voltage across it varies in the manner shown by the table.

R (Ω)	100	200	400
V (V)	7	10	12.73

Calculate the maximum power the source is able to deliver to a resistive load.

6.20 A source of impedance (6000 − j125) Ω at a frequency of 1000/2π Hz is to be connected to a resistive load of 6000 Ω. Calculate the value of the component that should be connected in series with the load in order that the load should dissipate the maximum possible power.

7 Star–delta Transform

The terms *star* and *delta* refer to two types of network which are commonly employed in electrical engineering. Star and delta networks are shown in Figs 7.1(*a*) and (*b*) respectively. In light current engineering the two networks are usually drawn upside down when they are known as the T and π networks, see Figs 7.1(*c*) and (*d*) respectively. Each network has three terminals, one of which is common to both its input and output. Some circuits have their components so connected that they are neither in parallel nor in series and such circuits are difficult to analyse using the methods considered in earlier chapters. Often, however, it is possible to transform a part of a circuit so that the modified circuit is easier to analyse. This transformation may be carried out using either the *star-to-delta transform*, or the *delta-to-star transform*. Also many circuits, particularly in three-phase systems, are connected in either the star or the delta configuration. The voltage at the star point in an unbalanced star network without a neutral line may be difficult to determine. This difficulty may be overcome if the star network is first transformed into the equivalent delta network which, of course, does not have a star point.

Very often the solution of a circuit can be simplified if a star (T) network is converted into the equivalent delta (π) network or vice versa.

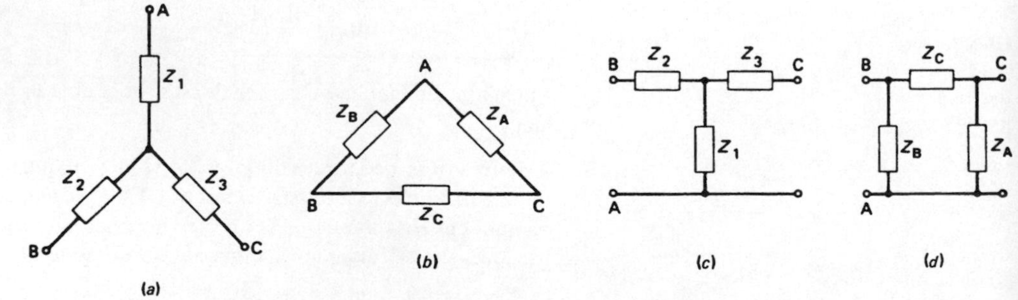

Fig. 7.1 Networks: (a) star network; (b) delta network; (c) T network; (d) π network

Delta–star transform

For two star (T) and delta (π) networks to be equivalent, the total impedance across any pair of terminals, with the third terminal open circuited, must be equal to the impedance across the corresponding terminals in the other network.

- With terminal C open circuited the impedances across terminals A and B are:

$$Z_1 + Z_2 = [Z_B(Z_A + Z_C)]/(Z_A + Z_B + Z_C) \qquad (7.1)$$

- With terminal B open circuited,

$$Z_1 + Z_3 = [Z_A(Z_B + Z_C)]/(Z_A + Z_B + Z_C) \qquad (7.2)$$

- With terminal A open circuited terminals B and C have impedances

$$Z_2 + Z_3 = [Z_C(Z_A + Z_B)]/(Z_A + Z_B + Z_C) \qquad (7.3)$$

Add equations (7.1) and (7.2),

$$2Z_1 + Z_2 + Z_3 = [2Z_AZ_B + Z_C(Z_A + Z_B)]/(Z_A + Z_B + Z_C) \qquad (7.4)$$

Subtract equation (7.3) from equation (7.4),

$$Z_1 = (Z_AZ_B)/(Z_A + Z_B + Z_C) \qquad (7.5)$$

Substitute equation (7.5) into equation (7.1),

$$Z_2 = (Z_BZ_C)/(Z_A + Z_B + Z_C) \qquad (7.6)$$

Substitute equation (7.6) into equation (7.3),

$$Z_3 = (Z_AZ_C)/(Z_A + Z_B + Z_C) \qquad (7.7)$$

It is difficult to remember these three equations correctly and so it is better to note that each impedance in the star (T) network is equal to the product of the delta (π) network's impedances which are connected to the same labelled terminal divided by the sum of the three impedances.

Example 7.1

Convert the delta network shown in Fig. 7.2(*a*) into the equivalent star network.

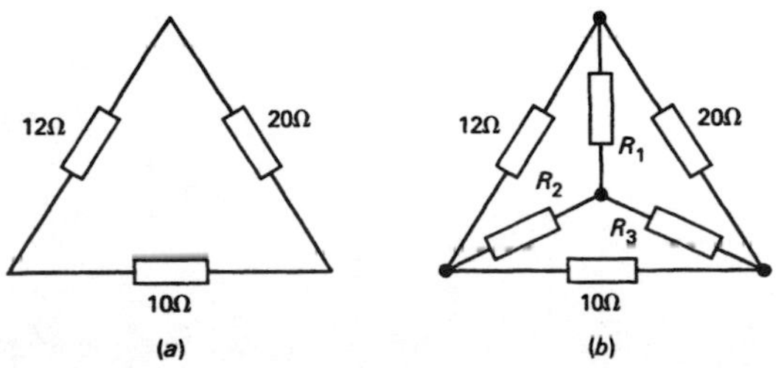

Fig. 7.2

Solution
It may assist if the star network in drawn inside the delta network as in Fig 7.2(*b*).

R_1 = (Product of resistances connected to the same terminal)/(Sum of resistances)

$= (12 \times 20)/(12 + 20 + 10) = 5.71\,\Omega$

$R_2 = (12 \times 10)/42 = 2.86\,\Omega$

$R_3 = (20 \times 10)/42 = 4.76\,\Omega$ (*Ans.*)

Star–delta transform

The corresponding equations for transforming a star network into the equivalent delta network *can* be obtained by re-arranging equations (7.5), (7.6) and (7.7). It is easier, however, to derive the necessary equations by considering the input admittances of each pair of terminals. Thus,

- With terminals B and C short circuited

$$Y_A + Y_B = [Y_1(Y_2 + Y_3)]/(Y_1 + Y_2 + Y_3) \qquad (7.8)$$

- With terminals A and C short circuited,

$$Y_B + Y_C = [Y_2(Y_1 + Y_3)]/(Y_1 + Y_2 + Y_3) \qquad (7.9)$$

- With terminals A and B short circuited,

$$Y_A + Y_C = [Y_3(Y_1 + Y_2)]/(Y_1 + Y_2 + Y_3) \qquad (7.10)$$

Clearly these three equations are of the same form as those derived for the delta–star transformation and they will hence lead to similar results.

Adding equations (7.8) and (7.9) and then subtracting equation (7.10) from the sum will give

$$Y_B = (Y_1 Y_2)/(Y_1 + Y_2 + Y_3) \qquad (7.11)$$

or

$$\begin{aligned} Z_B &= (Y_1 + Y_2 + Y_3)/(Y_1 Y_2) \\ &= (1/Y_2) + (1/Y_1) + (Y_3/Y_1 Y_2) \\ &= Z_1 + Z_2 + (Z_1 Z_2)/Z_3 \end{aligned}$$

Therefore,

$$Z_B = (Z_1 Z_3 + Z_2 Z_3 + Z_1 Z_2)/Z_3 \qquad (7.12)$$

Substitute equation (7.11) into equation (7.9),

$$Y_C = (Y_2 Y_3)/(Y_1 + Y_2 + Y_3) \qquad (7.13)$$

or

$$Z_C = (Z_1 Z_3 + Z_2 Z_3 + Z_1 Z_2)/Z_1 \qquad (7.14)$$

Also

$$Y_A = (Y_1 Y_3)/(Y_1 + Y_2 + Y_3) \quad (7.15)$$

and $Z_A = (Z_1 Z_3 + Z_1 Z_2 + Z_2 Z_3)/Z_2$ (7.16)

As a memory aid, a delta impedance is equal to the sum of the products of pairs of star impedances divided by the star impedance on the opposite side of the network.

Example 7.2

Convert the star network shown in Fig. 7.2(*b*) into the equivalent delta network.

Solution

$$Z_A = \frac{\text{Product of pairs of star impedances}}{\text{Opposite impedance}}$$

$$= [(5.71 \times 2.86) + (5.71 \times 4.76) + (2.86 \times 4.76)]/2.86 \approx 20\,\Omega$$

$Z_B = 57.12/4.76 = 12\,\Omega$ and $Z_C = 57.12/5.71 = 10\,\Omega$ (*Ans.*)

Example 7.3

Reduce the circuit given in Fig. 7.3 to a single equivalent resistance in series with a voltage source. Hence calculate the power dissipated in the network.

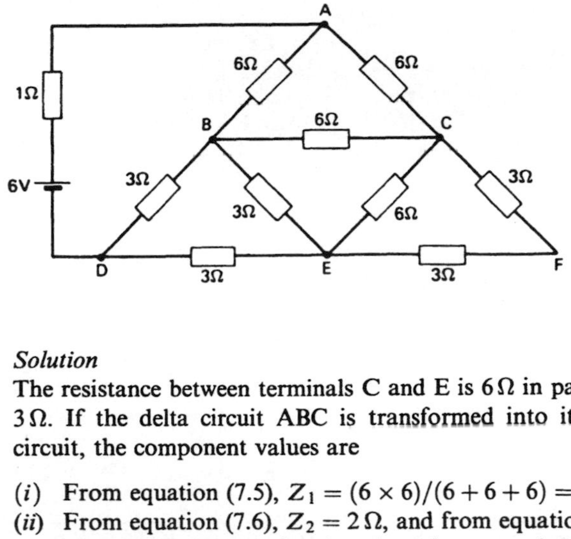

Fig. 7.3

Solution

The resistance between terminals C and E is 6 Ω in parallel with 6 Ω or 3 Ω. If the delta circuit ABC is transformed into its equivalent star circuit, the component values are

(*i*) From equation (7.5), $Z_1 = (6 \times 6)/(6 + 6 + 6) = 2\,\Omega$

(*ii*) From equation (7.6), $Z_2 = 2\,\Omega$, and from equation (7.7), $Z_3 = 2\,\Omega$ also. Similarly, transforming the delta network BDE gives

$Z_1' = (3 \times 3)/(3 + 3 + 3) = 1\,\Omega$ $\quad Z_2' = 1\,\Omega$

$Z_3' = 1\,\Omega$

Figure 7.3 can therefore be re-drawn as shown in Fig. 7.4(*a*). This circuit can be still further reduced, since the paths XBY and XCEY are in parallel, to the circuit of Fig. 7.4(*b*). Thus, there is a single equivalent resistance of 5 Ω across the terminals AD (*Ans.*)

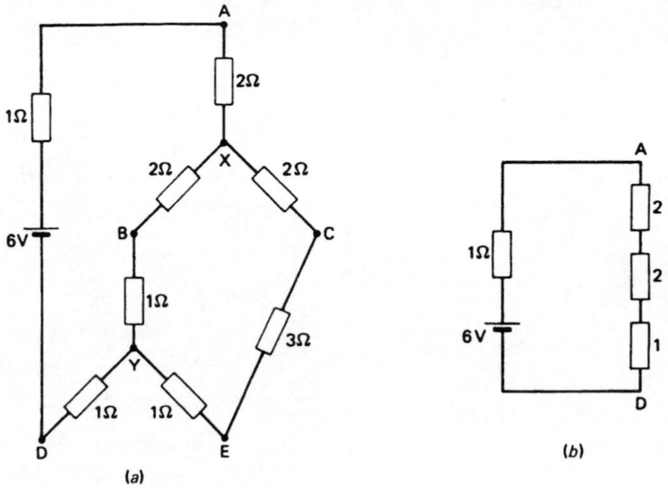

Fig. 7.4

Therefore, the current flowing is 6/6 = 1 A and the power dissipated is

$1^2 \times 5 = 5\,\text{W}$ (*Ans.*)

Example 7.4

Use the delta–star transform to find the current flowing in the 40 Ω load in the circuit given in Fig. 7.5.

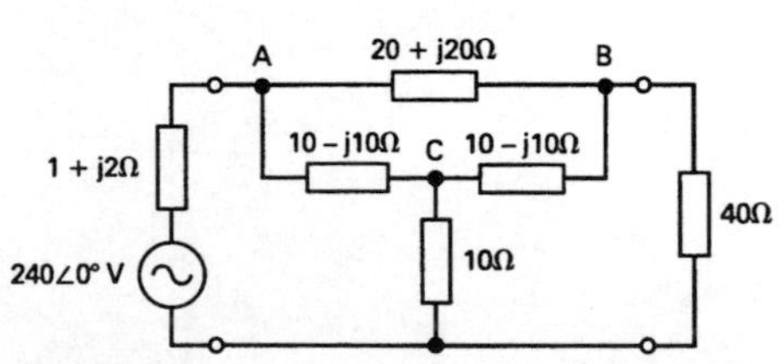

Fig. 7.5

Solution

For the delta ABC the equivalent star circuit has:

$$Z_1 = Z_A Z_B / \sum Z = \frac{(10 - j10)(10 - j10)}{20 + j20 + 10 - j10 + 10 - j10} = -j5\,\Omega$$

$$Z_2 = Z_B Z_C / \sum Z = [(10 - j10)(20 + j20)]/40 = 10\,\Omega$$

$$Z_3 = Z_A Z_C / \sum Z = [(10 - j10)(20 + j20)]/40 = 10\,\Omega$$

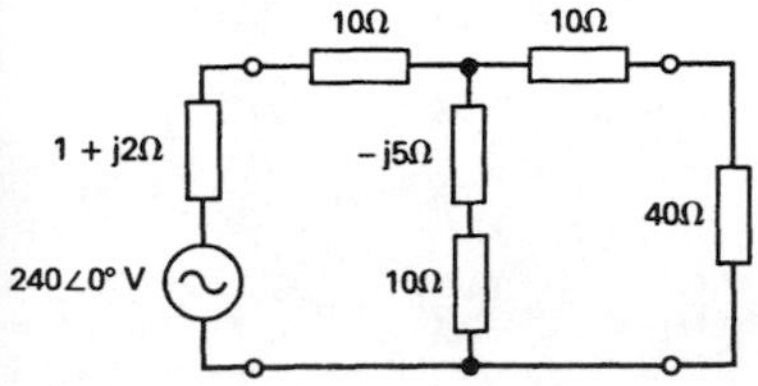

Fig. 7.6

The circuit may be re-drawn as shown by Fig. 7.6 from which its input impedance is

$$Z = 10 + [50(10 - j5)]/(50 + 10 - j5) = 18.62 - j3.45\ \Omega$$

Input current $= 240/(19.62 - j1.45) = 12.17 + j0.9 = 12.2\angle 4.2°$ A. Therefore the current flowing in the 40 Ω resistor is

$$I_{40} = [(12.2\angle 4.2°) \times (11.18\angle -26.6°)]/(60.2\angle -4.8°)$$
$$= 2.27\angle -17.6° \quad (\textit{Ans.})$$

Example 7.5

Determine the equivalent delta network to the star circuit shown by Fig. 7.7(*a*).

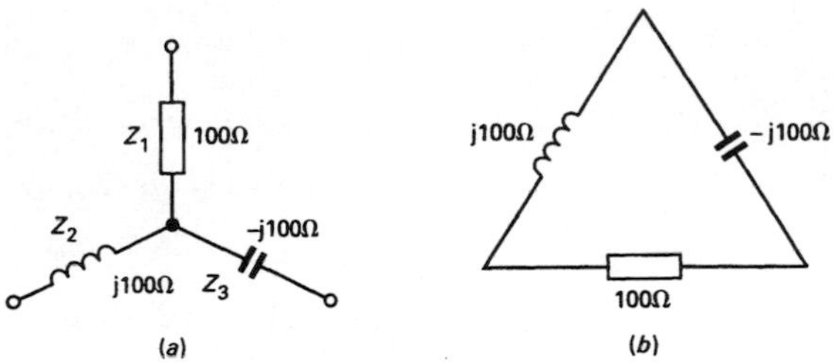

Fig. 7.7

Solution

$$Z_A = [(100 \times j100) + (100 \times -j100) + (j100 \times -j100)]/j100$$
$$= -j100\ \Omega$$
$$Z_B = 10^4/-j100 = j100\ \Omega \quad \text{and} \quad Z_C = 10^4/100 = 100\ \Omega.$$

The equivalent delta circuit is shown by Fig. 7.7(*b*).

Example 7.6

For the circuit shown in Fig. 7.8(*a*) calculate the r.m.s. currents supplied by the two voltage sources.

Solution

Applying the delta–star transform to the middle part of the network gives

$$Z_1 = 100(100 - j100)/300 = (100/3)(1 - j)$$
$$Z_2 = 100(100 + j100)/300 = (100/3)(1 + j) \quad \text{and} \quad Z_3 = 200/3\ \Omega$$

The re-drawn circuit is given in Fig. 7.8(*b*). From this circuit:

$$1 = (400/3 - j100/3)i_1 + (200/3)(i_1 - i_2) \quad \text{and}$$
$$j1 = (400/3 + j100/3)i_2 + (200/3)(i_2 - i_1)$$

Hence,

$$3/100 = (6 - j1)i_1 - 2i_2 \qquad (7.17)$$

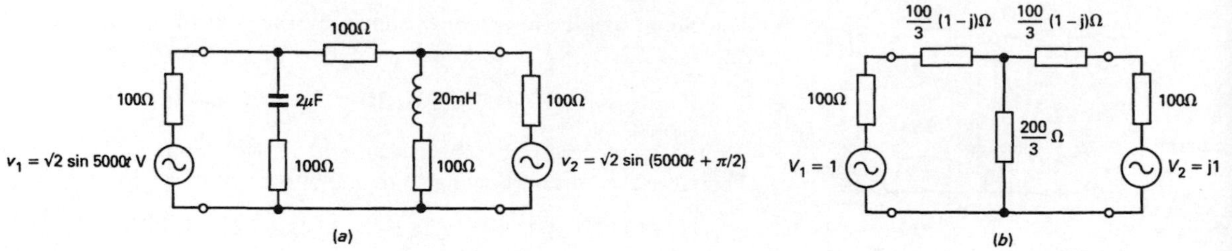

Fig. 7.8

and

$$j3/100 = (6 + j1)i_2 - 2i_1 \quad (7.18)$$

From equation (7.17)

$$i_2 = 0.5[(6 - j1)i_1 - 3/100] \quad (7.19)$$

Substituting into equation (7.18) gives

$$j3/100 = \{[(6 + j1)(6 - j1)]/2\}i_1 - 2i_1 - (3/100)(6 + j1)/2 \quad \text{or}$$

$$i_1 = 6.1\angle 26.6° \text{ mA} \quad (Ans.)$$

Substituting this value into equation (7.19),

$$i_2 = 6.1\angle 63.2° \text{ mA} \quad (Ans.)$$

Bridge circuits

The star–delta transform can also be employed for the simplification of a bridge circuit. Figure 7.9(*a*) shows an unbalanced bridge that consists of four impedances Z_1, Z_2, Z_3 and Z_4 with a fifth impedance Z_5 connected between the junctions of the two opposite arms. To reduce the bridge to a simpler form convert the delta (π) circuit ACD into the

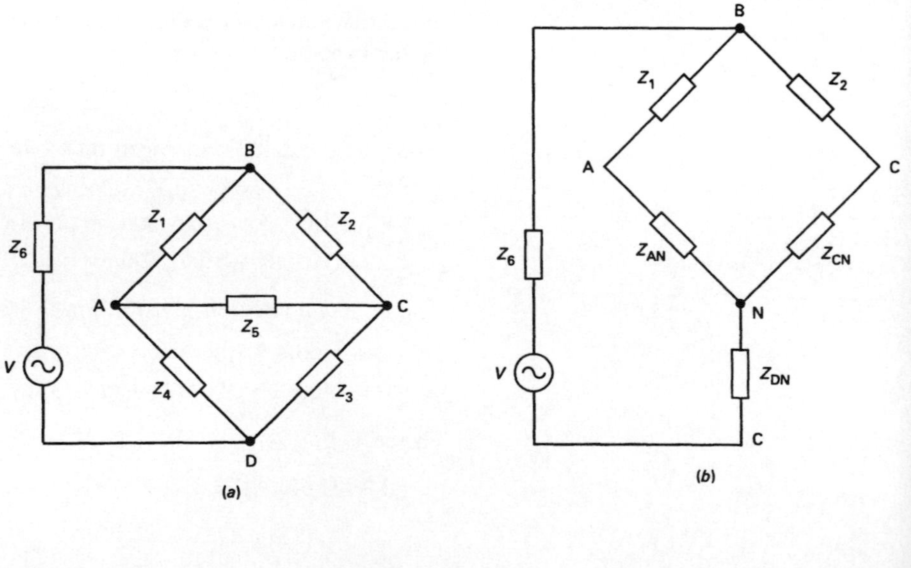

Fig. 7.9

equivalent star (T) network. Then, Fig. 7.9(*b*),

$$Z_{CN} = Z_3Z_4/(Z_3 + Z_4 + Z_5)$$
$$Z_{DN} = Z_3Z_5/(Z_3 + Z_4 + Z_5) \quad \text{and}$$
$$Z_{AN} = Z_4Z_5/(Z_3 + Z_4 + Z_5)$$

The loop ABCN is now just a parallel circuit.

Example 7.7

The bridge circuit of Fig. 7.9(*a*) has $Z_6 = 0.5\ \Omega$, $Z_1 = 5\ \Omega$, $Z_2 = 20\ \Omega$, $Z_3 = 15\ \Omega$, $Z_4 = 4\ \Omega$ and $Z_5 = 15\ \Omega$. Calculate the current supplied by the voltage source of 6 V.

Solution
$Z_{CN} = 1.77\ \Omega$, $Z_{DN} = 6.62\ \Omega$ and $Z_{AN} = 1.77\ \Omega$. Hence the resistance of the circuit is

$$0.5 + 6.62 + (21.77 \times 6.77)/28.54 = 5.16\ \Omega.$$

Therefore,

$$\text{Current supplied} = 6/5.16 = 1.163\ \text{A} \quad (\textit{Ans.})$$

Example 7.8

A bridge circuit has $Z_1 = -\text{j}10\ \Omega$, $Z_2 = 40 + \text{j}20\ \Omega$, $Z_3 = -\text{j}20\ \Omega$ and $Z_4 = 20 + \text{j}40\ \Omega$. A 10 Ω resistor is connected between the junction of Z_1/Z_4 and the junction of Z_2/Z_3 and a voltage source of $12\angle 0°$ V is applied across the junctions of Z_1/Z_2 and Z_3/Z_4. Apply the delta–star transform to the Z_3Z_4 10 Ω delta to determine the current taken from the source.

Solution

$$Z_{AN} = [10(20 + \text{j}40)]/(10 + 20 + \text{j}40 - \text{j}20) = 10.77 + \text{j}6.15\ \Omega$$
$$Z_{DN} = [(20 + \text{j}40))(-\text{j}20)]/(30 + \text{j}20) = 12.31 - \text{j}21.54\ \Omega$$
$$Z_{CN} = (10 \times -\text{j}20)/(30 + \text{j}20) = -3.08 - \text{j}4.62\ \Omega$$

There are now two impedances in parallel:

$$Z_A = -\text{j}10 + 10.77 + \text{j}6.15 = 10.77 - \text{j}3.85\ \Omega \quad \text{and}$$
$$Z_B = 40 + \text{j}20 - 3.08 - \text{j}4.62 = 36.92 + \text{j}15.38\ \Omega$$

The total impedance is then

$$Z = (10.77 - \text{j}3.85)(36.92 + \text{j}15.38)/(47.69 + \text{j}11.53)$$
$$= 9.17 - \text{j}1.73 = 9.33\angle -10.7°\ \Omega$$

This impedance is in series with Z_{DN} so the total impedance is

$$9.17 - \text{j}1.73 + 12.31 - \text{j}21.54 = 31.67\angle -47.3°\ \Omega$$

Current taken from supply is

$$12/(31.67\angle -47.3°) = 379\angle 47.3°\ \text{mA} \quad (\textit{Ans.})$$

Bridged-T networks

Bridged-T networks are often used for the measurement of component values at both audio and radio frequencies. When a bridged-T network is balanced its output voltage is zero. The bridged-T arrangement offers the advantage over the conventional bridge circuit (Chapter 8) in that it allows one side of both the a.c. voltage source and the detector to be connected to earth. This is a desirable practice since it reduces the adverse effects of stray capacitances and hence increases the accuracy of measurement. Even better accuracy can be achieved if a *twin-T network* is employed. A number of bridged-T networks exist and two examples follow.

- The bridged-T network shown in Fig 7.10(*a*) may be used to measure the inductance and resistance of an inductor. Alternatively, it is sometimes used to measure frequency.

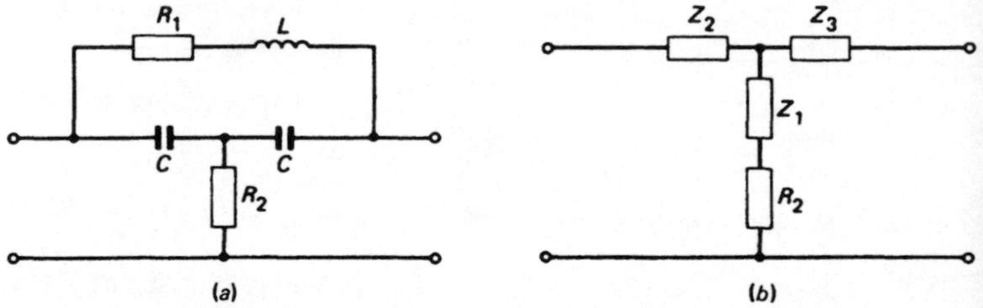

Fig. 7.10 Bridged-T network

Let $Z_C = R_1 + j\omega L$ and $Z_A = Z_B = 1/j\omega C$ and convert the bridge into its equivalent T network.

(*i*) From equation (7.7).

$$Z_3 = [(R + j\omega L)/j\omega C]/[R_1 + j(\omega L - 2/\omega C)]$$
$$= [(L/C) - (jR_1/\omega C)]/[R_1 + j(\omega L - 2/\omega C)]$$

This is also Z_2, since $Z_A = Z_B$.

(*ii*) From equation (7.5)

$$Z_1 = \frac{(1/j\omega C) \times (1/j\omega C)}{R_1 + j[\omega L - (2/\omega C)]} = \frac{-1/\omega^2 C^2}{R_1 + j[\omega L - (2/\omega C)]}$$

The network will have zero output voltage at the frequency at which the shunt path has zero impedance. (The values of Z_2 and Z_3 are unimportant.) Hence, from Fig. 7.10(*b*)

$$R_2 - [1/(\omega_0^2 C^2)]/[R_1 + j(\omega_0 L - 2/\omega_0 C)] = 0$$
$$R_1 R_2 + jR_2[\omega_0 L - (2/\omega_0 C)] = 1/\omega_0^2 C^2$$

Equating the real terms,

$$R_1 R_2 = 1/\omega_0^2 C^2 \qquad \omega_0^2 = 1/R_1 R_2 C^2$$

or $f_0 = 1/[2\pi C\sqrt{(R_1 R_2)}]$ Hz (7.20)

Equating the imaginary parts,

$$R_2[\omega_0 L - (2/\omega_0 C)] = 0 \qquad \omega_0^2 = 2/LC$$

or $f_0 = (1/2\pi)\sqrt{(2/LC)}$ Hz (7.21)

Both equation (7.20) and equation (7.21) must be satisfied for the network to have a null in its output voltage–frequency characteristic at frequency f_0 (in practice, a fairly sharp minimum is obtained).

Alternatively, from the real terms,

$$R_1 = 1/(\omega_0^2 C^2 R_2) \tag{7.22}$$

and from the imaginary terms,

$$L = 2/(\omega_0^2 C) = 1/(2\pi^2 f_0^2 C) \tag{7.23}$$

Example 7.9

The bridged-T network of Fig. 7.11(*a*) is to be used to measure an inductor at a frequency of 2 kHz. The output voltage of the circuit is set to zero when $C = 0.1\ \mu$F and $R_2 = 21$ kΩ. Calculate the inductance and self-resistance of the inductor.

Solution

$$L = 1/(2\pi^2 \times 4 \times 10^6 \times 0.1 \times 10^{-6}) = 126.7 \text{ mH} \quad (\textit{Ans.})$$

$$R_1 = 1/(4\pi^2 \times 4 \times 10^6 \times 0.1^3 \times 10^{-12} \times 21 \times 10^3)$$
$$= 30.2\ \Omega \quad (\textit{Ans.})$$

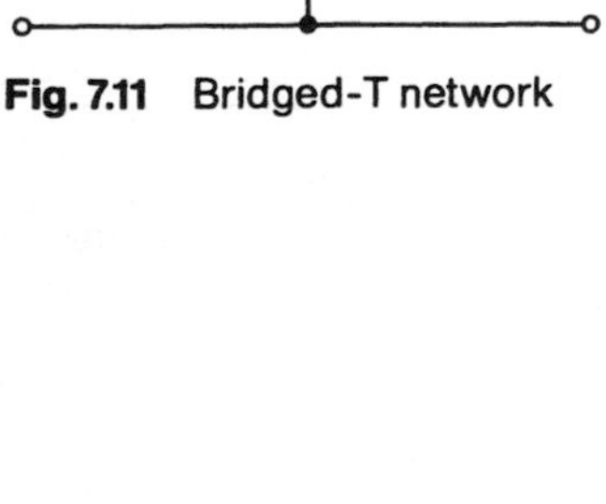

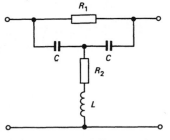

Fig. 7.11 Bridged-T network

- Figure 7.11 shows a bridged-T network that is employed for the measurement of the inductance and resistance of a resistor. The π circuit, R_1, C and C can be converted into the equivalent T network with $Z_1 = Z_A Z_B/\sum Z = (1/j\omega C)^2/(R_1 - 2j/\omega C)$. The values of Z_2 and Z_3 are not needed. The shunt arm of the transformed circuit consists of Z_1 in series with the resistor being measured. At balance the output voltage of the network is zero because then

$$R_2 + j\omega L = (1/\omega C)^2/(R_1 - j2/\omega C)$$
$$= (1/\omega^2 C^2)(R_1 + j2/\omega C)/(R_1^2 + 4/\omega^2 C^2)$$
$$= (R_1 + j2/\omega C)/[\omega^2 C^2 (R_1^2 + 4/\omega^2 C^2)]$$

Equating real terms:

$$R_2 = R_1/(\omega^2 C^2 R_1^2 + 4)$$

Equating imaginary terms:

$$L = 2/[\omega C(\omega^2 C^2 R_1^2 + 4)]$$

Twin-T networks

A twin-T network employs two different T networks that are connected in parallel and a typical circuit is shown by Fig. 7.12(*a*). The network is balanced when its output voltage is zero and this occurs when the total admittance of the series path is zero. Each of the T networks must be transformed into its equivalent π network to give Fig. 7.12(*b*).

$$Y_C = (j\omega C_1 \times j\omega C_1)/(1/R_1 + j2\omega C_1)$$
$$= -\omega^2 C_1^2 R_1/(1 + j2\omega C_1 R_1)$$
$$Y'_C = (1/R_2 \times 1/R_2)/(2/R_2 + j\omega C_2)$$
$$= (1/R_2)/(2 + j\omega C_2 R_2)$$

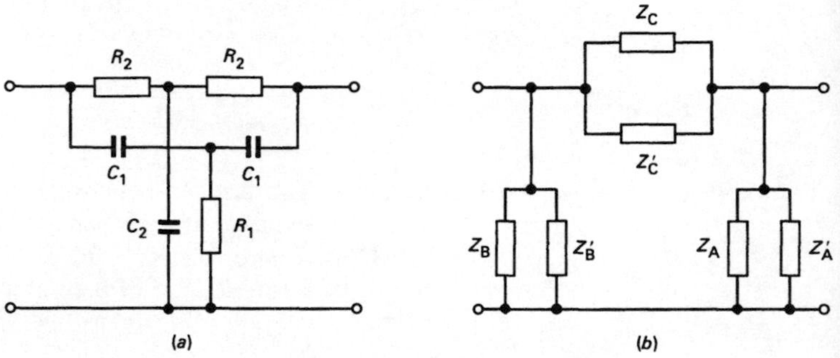

Fig. 7.12 Twin-T network

For zero output voltage,

$$(1/R_2)/(2 + j\omega C_2 R_2) = (\omega^2 C_1^2 R_1)/(1 + j2\omega C_1 R_1)$$
$$1 + j2\omega C_1 R_1 = 2\omega^2 C_1^2 R_1 R_2 + j\omega^3 C_2 C_1^2 R_1 R_1^2$$

Equating the real parts:

$$1 = 2\omega^2 C_1^2 R_1 R_2. \text{ Hence, } \omega = 1/[C_1\sqrt{(2R_1R_2)}] \quad \text{and}$$
$$f = 1/[2\pi C_1 \sqrt{(2R_1R_2)}] \quad (7.24)$$

Equating the imaginary parts:

$$2\omega C_1 R_1 = \omega^3 C_2 C_1^2 R_1 R_2^2. \text{ Hence, } \omega = \sqrt{2}/[R_2\sqrt{(C_1C_2)}]$$
and $f = 1/[\pi R_2 \sqrt{(2C_1C_2)}]$ (7.25)

Example 7.10

The twin-T network of Fig. 7.12 has $R_2 = 47$ kΩ and $C_1 = 4.7$ nF and has zero output voltage at a frequency of 796 Hz. Calculate the values of R_1 and C_2.

Solution
From equation (7.24),

$$R_1 = 1/[8\pi^2 \times 796^2 \times (4.7 \times 10^{-9})^2 \times 47 \times 10^3]$$
$$= 19.25 \text{ k}\Omega \quad (\textit{Ans.})$$

From equation (7.25),

$$C_2 = 1/[2\pi^2 \times 796^2 \times (47 \times 10^3)^2 \times 4.7 \times 10^{-9}]$$
$$= 7.7 \text{ nF} \quad (\textit{Ans.})$$

Three-phase cables

The star–delta transform can also be employed to determine the effective total capacitance of a three-conductor three-phase cable. Such a cable has inter-conductor capacitances C_1 and a capacitance C_2 between each conductor and the normally earthed sheath, see Fig. 7.13. Effectively, the three conductor-earth capacitances form a star network with earth as the star point. If this star network is transformed into the equivalent delta network the transformed capacitances will appear in parallel with the inter-conductor capacitances:

$$Z_A = Z_B = Z_C = (X_{C2}^2 + X_{C2}^2 + X_{C2}^2)/X_{C2} = 3X_{C2}$$

Thus, the value of each transformed capacitance is one-third of the original capacitance between conductor and earth.

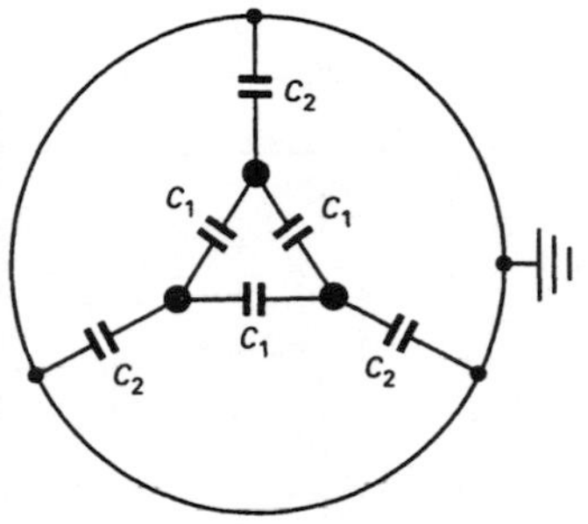

Fig. 7.13 Capacitances in a three-phase cable

Example 7.11

A three-conductor three-phase line has inter-conductor capacitances of 0.25 μF and conductor-to-earth capacitances of 0.3 μF. Determine the effective inter-conductor capacitance of the cable.

Solution
Inter-conductor capacitance $= 0.25 + 0.1 = 0.35$ μF (*Ans.*)

Exercises 7

7.1 Use both the star–delta transformation and Thevenin's theorem to reduce the circuit shown in Fig. 7.14 to simpler form. Then determine the value of R_L for it to dissipate the maximum possible power and the value of this power.

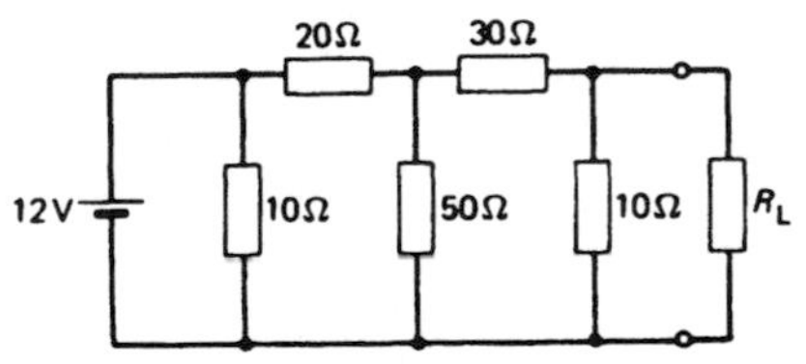

Fig. 7.14

7.2 Three 10 Ω resistors are connected in delta. Calculate the components of the equivalent star network.

7.3 Three 10 Ω resistors are star connected. Calculate the components of the equivalent delta network.

7.4 A star network has $Z_1 = 20$ Ω, $Z_2 = (10 + j10)$ Ω and $Z_3 = (20 - j5)$ Ω. Calculate the values of the components of the equivalent delta circuit.

7.5 For the circuit given in Fig. 7.15 calculate (*a*) the current taken from the 9 V voltage source and (*b*) the output voltage of the circuit.

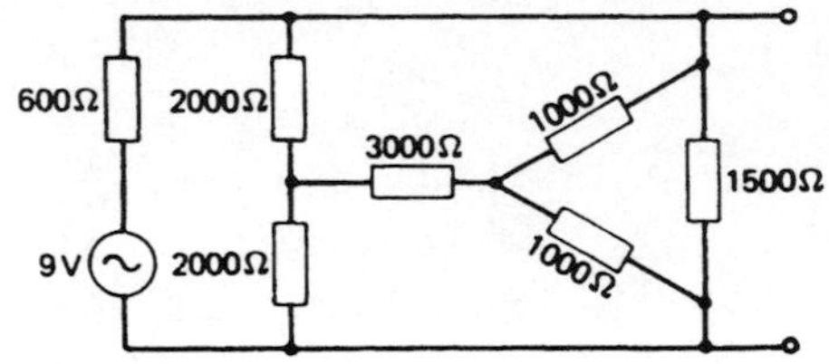

Fig. 7.15

7.6 Calculate the current supplied to the bridge circuit given in Fig. 7.16.

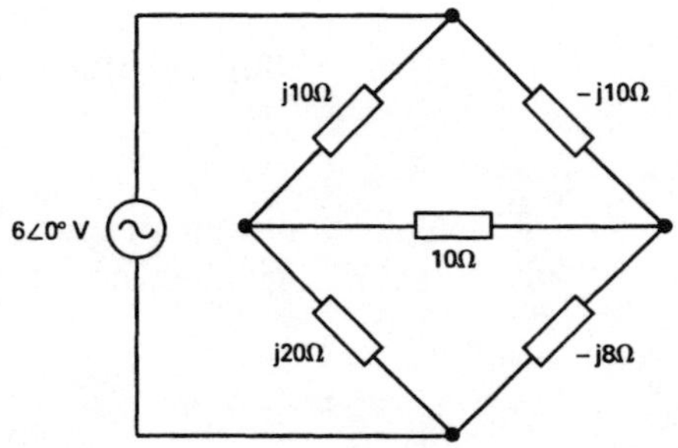

Fig. 7.16

7.7 A delta network has $Z_A = j200\ \Omega$, $Z_B = -j600\ \Omega$ and $Z_C = 300\ \Omega$. Calculate the impedances of the equivalent star network.

7.8 A bridge circuit has $Z_1 = -j40\ \Omega$, $Z_2 = j20\ \Omega$, $Z_3 = 40\ \Omega$ and $Z_4 = 50\ \Omega$. An impedance of j40 Ω is connected between the junction of Z_1/Z_4 and Z_2/Z_3 and a voltage source of $15\angle 0°$ V applied across the junctions of Z_1/Z_2 and Z_3/Z_4. Calculate the current taken from the supply.

7.9 A T network has series impedances of 20 + j10 Ω and 10 − j20 Ω, and a shunt impedance of 20 Ω. A 6 V voltage source is applied to the 20 + j10 Ω impedance. Calculate the current flowing in the 50 Ω load resistor by first transforming the T network into the equivalent π circuit and then applying Thevenin's theorem.

7.10 The bridged-T circuit in Fig. 7.10 has $L = 10$ mH and $R_1 = 5\ \Omega$ and has zero output voltage at a frequency of 15.9 kHz. Calculate the values of C and R_2.

7.11 A resistive bridge has opposite arms of 40 Ω and 45 Ω, and 120 Ω and 90 Ω resistance. A 270 Ω resistor is connected between the junctions of the 40 Ω and 90 Ω, and the 120 Ω/45 Ω resistors. A 6 V source is applied between the junctions of the 40 Ω and 120 Ω, and the 90 Ω and 45 Ω resistors. Calculate the voltage across the 90 Ω resistor.

7.12 For the bridged-T network in Fig. 7.10 calculate the input impedance when the network is connected to a load of 5 Ω. $R_1 = 10\ \Omega$, $R_2 = 50\ \Omega$, $X_L = j10\ \Omega$ and $X_C = -j20\ \Omega$.

7.13 Derive the balance equations for the Anderson bridge shown on page 147 by applying the star–delta transform to the loop DEF.

7.14 A bridged-T network has a series impedance Z. To each side of Z are connected two capacitors C_1, and C_2. The other terminals of the capacitors are connected together and their junction is connected to a resistor R. The other side of R is connected to the common line. Find the admittance of Z which would result in zero output voltage at an angular frequency of 1×10^6 rad/s, if $C_1 = C_2 = 200$ pF and $R = 125$ kΩ.

7.15 A twin-T network has one T with series resistors of 10 kΩ and a shunt capacitor of 0.2 μF, and the other T with series capacitors of 0.1 μF and a shunt resistor of 5 kΩ. Calculate the frequency at which the output voltage of the circuit is zero.

7.16 A bridged-T network has a series resistor R shunted by two 0.1 μF capacitors in series with the junction of the capacitors connected to one side of a 10 Ω resistor. The other side of the 10 Ω resistor is connected to the common line. The output voltage of the circuit is zero at 1592 Hz. Calculate the value of R.

7.17 For the bridged-T network shown in Fig. 7.11 show that when the circuit is balanced the current in the inductor is $I_L = V_{in}[2/R + j\omega C]$.

7.18 A twin-T network has series components $R = 10$ kΩ, $C = 0.05$ μF, and parallel components $R = 5$ kΩ and $C = 0.1$ μF. Calculate the frequency of infinite attenuation.

8 A.C. bridges

An a.c. bridge is used for the measurement of an unknown impedance. A comparison technique is employed in which the impedance to be measured is compared with a known impedance. Most a.c. bridges are based upon the same principle as the Wheatstone bridge and contain an impedance in each of their four arms. The values of three of the impedances are known while the value of the fourth is the unknown which is to be measured. For a measurement it is necessary for both the magnitude and the phase of the unknown impedance to be balanced. Usually, the impedance to be measured is in one arm of the bridge. The other three arms contain either fixed or variable components. To balance a bridge the voltage across the detector is reduced to a minimum – in theory to zero – by successive adjustments to the two variable components. Once the bridge has been balanced, or the nearest possible approach to it has been obtained, the values of the fixed and variable components are inserted into the balance equation and the value of the unknown calculated. Several different forms of a.c. bridge exist; they each have their relative merits and usually one particular bridge is preferred for a particular measurement. A substitution method is often employed in which the bridge is first balanced with the unknown impedance inserted; the unknown impedance is then replaced by a variable standard component and the bridge is re-balanced. A substitution method may be employed because it is (*a*) cheaper since accurate standard (expensive) components are not necessary, and (*b*) a Wagner earth is not required.

Four-arm A.C. bridges

The basic arrangement of a four-arm bridge is shown in Fig. 8.1. It consists of four impedances connected in a Wheatstone bridge arrangement. Any one of the four impedances may be the unknown, and any one, or more, of the other impedances may

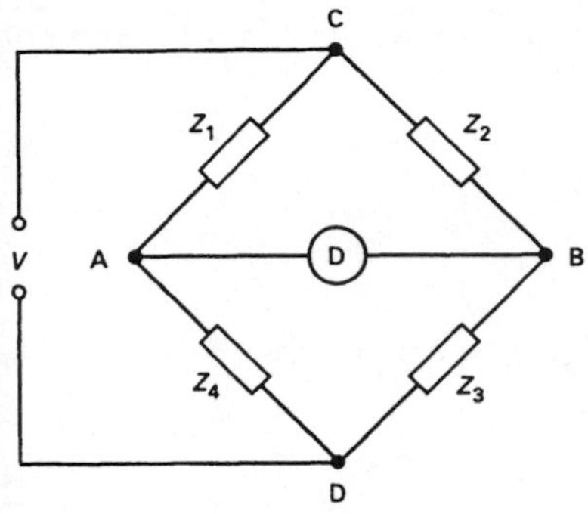

Fig. 8.1 A.C. bridge

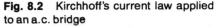

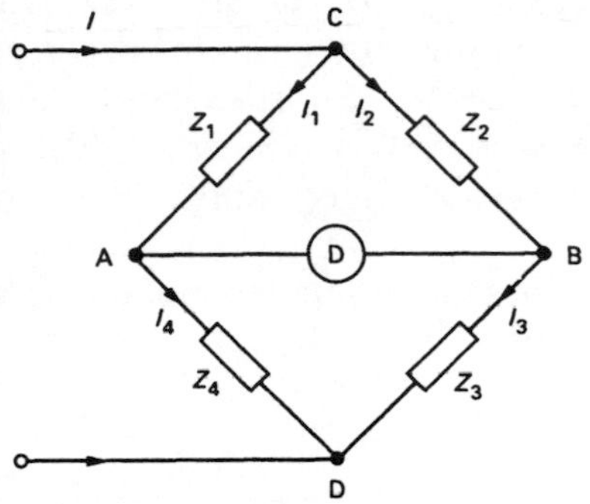

Fig. 8.2 Kirchhoff's current law applied to an a.c. bridge

be a variable component. When an a.c. bridge is used for a measurement, the variable impedance is adjusted until the indication of the detector D is zero, (in practice, usually until the minimum achievable indication is obtained). Once *balance* has been obtained the potentials at points A and B are equal to one another and zero current flows through the detector. Then, referring to Fig. 8.2, the input current I to the bridge divides into two parts I_1 and I_2. Since there is zero detector current all of the current I_1 must flow in both Z_1 and Z_4 and all of the current I_2 must flow in impedances Z_2 and Z_3. This means that $I_1 = I_4$ and $I_2 = I_3$. Also,

$$I_1 Z_1 = I_2 Z_2 \qquad (8.1)$$

and $$I_4 Z_4 = I_3 Z_3 \qquad (8.2)$$

Dividing equation (8.1) by (8.2) gives $Z_1/Z_4 = Z_2/Z_3$, or

$$Z_1 Z_3 = Z_2 Z_4 \qquad (8.3)$$

Thus, at balance the products of the impedances of diagonally opposite arms are equal to one another.

Example 8.1

In an a.c. bridge with fixed impedances $Z_1 = 1000\angle 45°\ \Omega$ and $Z_3 = 780\angle 0°\ \Omega$ balance is obtained when the variable impedance $Z_2 = 884\angle 32°\ \Omega$. Determine the value of the unknown impedance.

Solution

From equation (8.3),

$$(1000\angle 45°) \times (780\angle 0°) = 874\angle 32° Z_4$$

Therefore,

$$Z_4 = (1000\angle 45° \times 780\angle 0°)/874\angle 32° = 892.5\angle 13°\ \Omega$$
$$= 869.6 + j200.8\ \Omega \quad (\textit{Ans.})$$

The unknown impedance is a 869.6 Ω resistance in series with a 200.8 Ω inductive reactance.

Detectors

A number of different kinds of detector may be used with an a.c. bridge. Commonly employed detectors are (*a*) the cathode-ray oscilloscope (CRO), (*b*) earphones, (*c*) an electronic detector and (*d*) a vibration galvanometer. A CRO can be employed over a wide range of frequencies but the use of earphones must be restricted to audio frequencies. Vibration galvanometers are only suitable for use as detectors at low frequencies, most commonly at the mains frequency of 50 Hz.

Ratio bridge

If Z_1 is the unknown impedance then $Z_1 = Z_2Z_4/Z_3$. If the ratio Z_2/Z_3 is constant the bridge can be balanced by suitable adjustment of Z_4 only. Obtaining balance is much easier if it is possible to balance the bridge for R_1 and $\pm jX_1$ by adjustment of R_4 and $\pm jX_4$ alone. For this to be possible the ratio Z_2/Z_3 must either be wholly real or it must be wholly imaginary. This means that the fixed components must be pure; for example, resistors with no self-inductance or self-capacitance, and capacitors with no resistance or inductance. Such a bridge is known as a *ratio bridge*. Four examples of ratio bridges are given in Fig. 8.3.

A ratio bridge is only able to measure components of the same type.

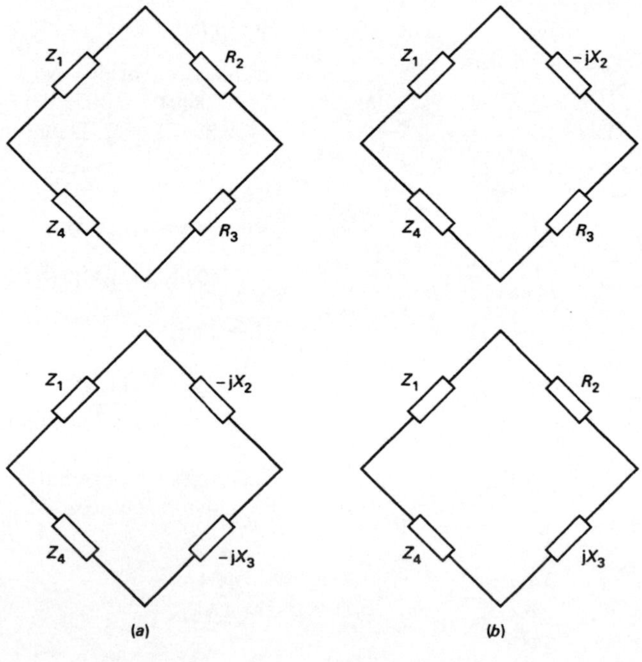

Fig. 8.3 Ratio bridges

Product bridge

If, again, Z_1 is the unknown impedance, $Z_1 = Z_2Z_4/Z_3$, but now the product Z_2Z_4 is made a constant quantity and the bridge is balanced by adjustment of impedance Z_3. The product Z_2Z_4 must either be wholly real or wholly imaginary. This type of bridge is called a *product bridge*. Two examples of product bridges are shown in Fig. 8.4. The variable reactance part of the variable impedance Z_3 must be of the opposite sign to the reactance of the unknown impedance or the bridge will not balance.

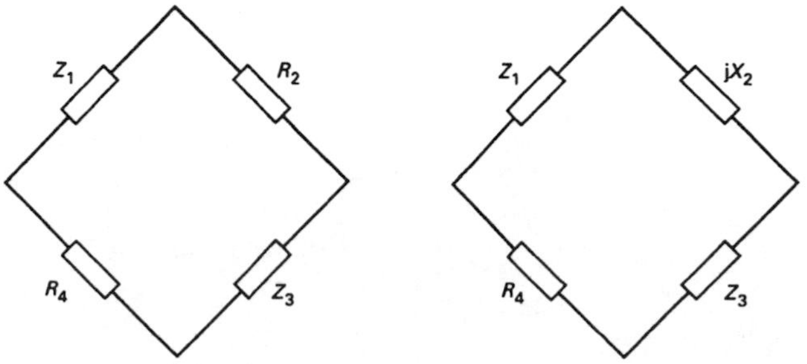

Fig. 8.4 Product bridges

A.C. bridge circuits

A number of different a.c. bridge circuits are in existence and several of them offer a particular advantage for the measurement of a certain type of component in a given frequency range. A few bridge circuits may also be employed to measure frequency.

Maxwell bridge

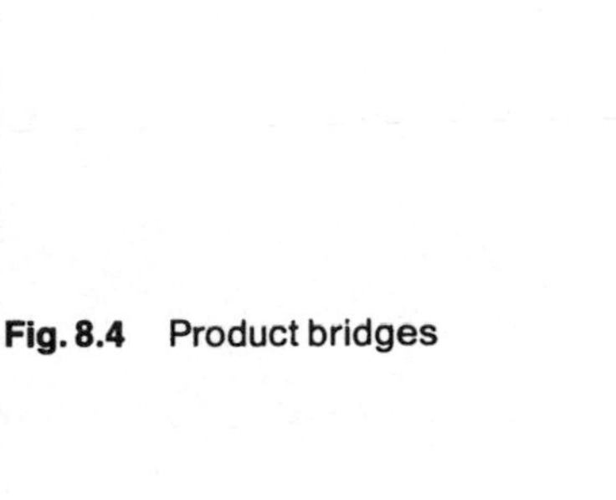

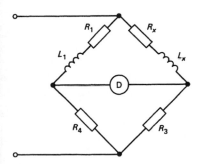

Fig. 8.5 Maxwell bridge

The Maxwell bridge is used to measure the resistance and inductance of a high Q factor inductor and it is an example of a ratio bridge. The inductance is measured against a standard inductance. The circuit of a Maxwell bridge is given in Fig. 8.5. At balance,

$$R_3(R_1 + j\omega L_1) = R_4(R_x + j\omega L_x)$$

$$R_x + j\omega L_x = (R_3/R_4)(R_1 + j\omega L_1)$$

Equating real parts:

$$R_x = R_3R_1/R_4 \tag{8.4}$$

Equating imaginary parts:

$$L_x = R_3L_1/R_4 \tag{8.5}$$

The bridge has the disadvantages that (a) the balance conditions cannot be satisfied independently since both equations contain the ratio R_3/R_4, and (b) standard inductors are expensive components. On the other hand the balance condition is independent of frequency; this is always an advantage since it means that balance will not be upset by the presence of harmonics in the source voltage waveform.

Maxwell–Wien bridge

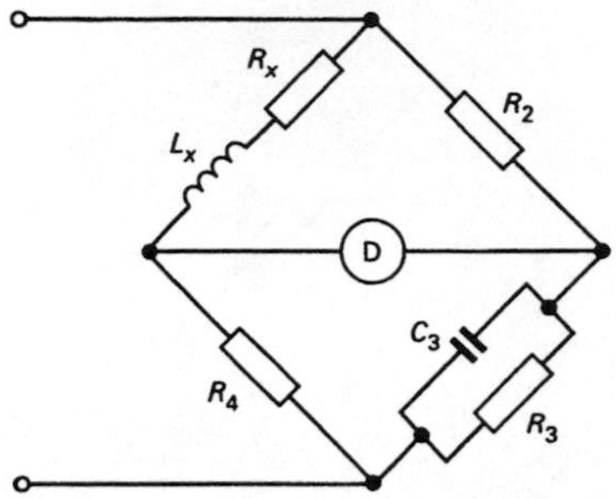

Fig. 8.6 Maxwell–Wien bridge

The Maxwell–Wien bridge is an example of a product bridge that is employed to measure the resistance and inductance of a low Q factor inductor. The circuit of the bridge is shown by Fig 8.6. At balance,

$$[(R_x + j\omega L_x)R_3]/(1 + j\omega C_3 R_3) = R_2 R_4$$
$$R_x + j\omega L_x = (R_2 R_4 / R_3)(1 + j\omega C_3 R_3)$$

Equating real parts:

$$R_x = R_2 R_4 / R_3 \tag{8.6}$$

Equating imaginary parts;

$$L_x = C_3 R_2 R_4 \tag{8.7}$$

Once again, the balance condition is independent of frequency. The unknown inductance always contains some self-capacitance and this may make it difficult to obtain a good balance. The two balance equations are only independent of one another if the capacitance C_3 is the variable component. In a measurement, a preliminary balance is obtained for R_x by varying R_3 alone and then, keeping R_3 constant, C_3 is varied to get the best possible balance.

Owen bridge

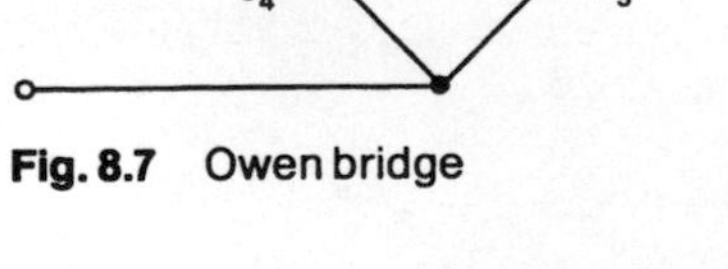

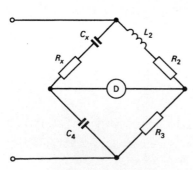

Fig. 8.7 Owen bridge

The Owen bridge is shown in Fig 8.7. It is a ratio bridge that is used to measure either capacitance or large values of inductance, with the latter the more common. At balance,

$$R_3(R_x + 1/j\omega C_x) = (R_2 + j\omega L_2)/(j\omega C_4)$$
$$[R_3 R_x + R_3/(j\omega C_x)]j\omega C_4 = R_3 C_4/C_x + j\omega R_3 R_x C_4$$
$$= R_2 + j\omega L_2$$

Equating real parts:

$$C_x = R_3 C_4 / R_2 \tag{8.8}$$

Equating imaginary parts:

$$R_x = L_2 / R_3 C_4 \tag{8.9}$$

Alternatively, when inductance is measured,

$$R_2 = R_3 C_4 / C_x \quad \text{and} \quad L_2 = R_x R_3 C_4 \tag{8.10}$$

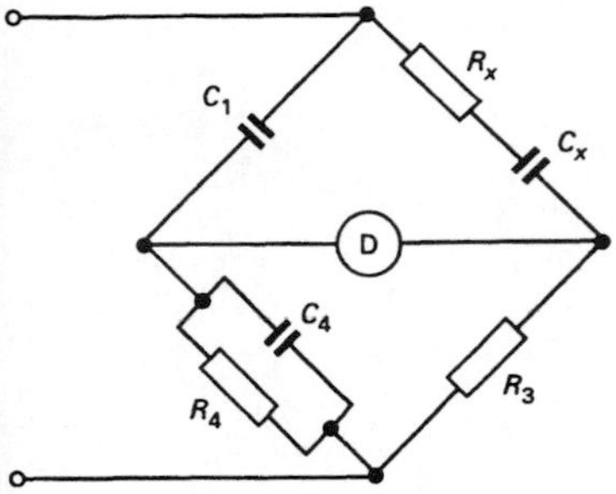

Fig. 8.8 Schering bridge

Schering bridge

Figure 8.8 shows the Schering bridge. It is commonly employed for high-voltage (typically 100 kV or more) dielectric measurements. At balance,

$$(R_x + 1/j\omega C_x)[R_4/(1 + j\omega C_4R_4)] = R_3/j\omega C_1$$
$$R_xR_4 + R_4/j\omega C_x = [R_3(1 + j\omega C_4R_4)]/j\omega C_1$$
$$R_xR_4 - jR_4/\omega C_x = R_3R_4C_4/C_1 - jR_3/(\omega C_1)$$

Equating real parts:

$$R_x = R_3C_4/C_1 \qquad (8.11)$$

Equating imaginary parts:

$$C_x = R_4C_1/R_3 \qquad (8.12)$$

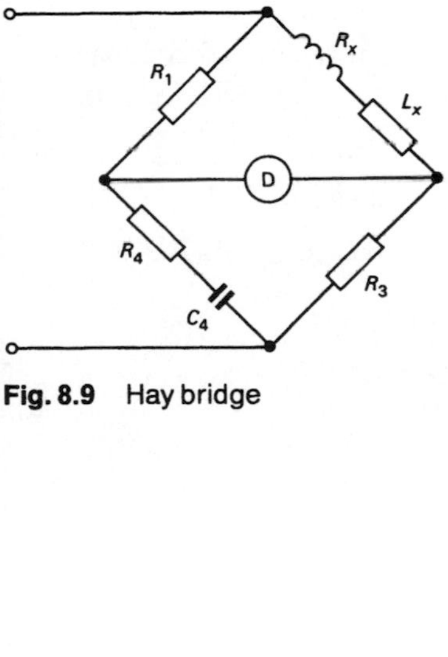

Fig. 8.9 Hay bridge

Hay bridge

The measurement of a high-Q factor inductor is often carried out using the Hay bridge whose circuit is shown in Fig. 8.9. At balance,

$$R_1R_3 = (R_x + j\omega L_x)(R_4 + 1/j\omega C_4)$$
$$R_x + j\omega L_x = R_1R_3/(R_4 - j/\omega C_4)$$
$$= R_1R_3(R_4 + j/\omega C_4)/(R_4^2 + 1/\omega^2C_4^2)$$

Equating real parts:

$$R_x = R_1R_3R_4/(R_4^2 + 1/\omega^2C_4^2)$$
$$= \omega^2C_4^2R_1R_3R_4/(1 + \omega^2C_4^2R_4^2) \qquad (8.13)$$

Equating imaginary parts:

$$\omega L_x = (R_1R_3/\omega C_4)/(R_4^2 + 1/\omega^2C_4^2)$$
$$L_x = R_1R_3C_4/(1 + \omega^2C_4^2R_4^2) \qquad (8.14)$$

Clearly, the balance equations are frequency dependent and this means that the source must be purely sinusoidal or balance will prove difficult to obtain.

Example 8.1

A Hay bridge has the following component values: $R_1 = R_3 = 1500\ \Omega$, $R_4 = 3404\ \Omega$ and $C_4 = 12$ nF. If the frequency of the measurement is 2 kHz determine the values of the unknown inductor.

Solution

$$L_x = (1500^2 \times 12 \times 10^{-9})/(1 + 4\pi^2 \times 4 \times 10^6 \times 144 \times 10^{-18} \times 3404^2) = 21.38 \text{ mH} \quad (\textit{Ans.})$$

$$R_x = (4\pi^2 \times 4 \times 10^6 \times 144 \times 10^{-18} \times 1500^2 \times 3404)/(1 + 4\pi^2 \times 4 \times 10^6 \times 144 \times 10^{-18} \times 3404^2) = 138.8\ \Omega \quad (\textit{Ans.})$$

De Sauty bridge

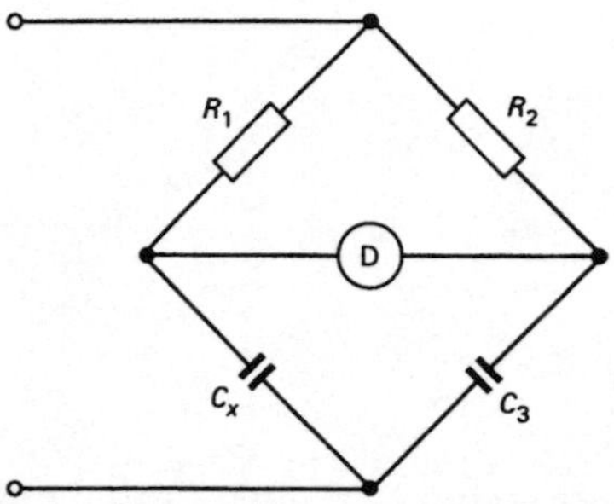

Fig. 8.10 De Sauty bridge

The De Sauty bridge may be used to measure a pure capacitance only, since it is unable to measure the self-resistance of a capacitance which may often be of some importance. The circuit of the De Sauty bridge is given by Fig. 8.10. At balance,

$$R_2/(j\omega C_x) = R_1/(j\omega C_3) \quad \text{or} \quad C_x = R_2C_3/R_1 \qquad (8.15)$$

Wien bridge

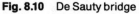

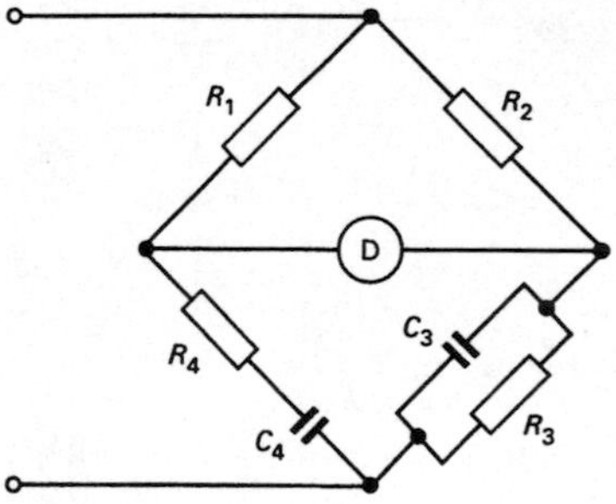

Fig. 8.11 Wien bridge

The Wien bridge may be employed to measure either capacitance or frequency and its circuit is shown by Fig. 8.11. At balance,

$$R_2(R_4 + 1/j\omega C_4) = R_1R_3/(1 + j\omega C_3R_3)$$
$$(R_4 + 1/j\omega C_4)(1 + j\omega C_3R_3) = R_1R_3/R_2$$
$$R_4 + j\omega R_3R_4C_3 - j/(\omega C_4) + C_3R_3/C_4 = R_1R_3/R_2$$

Equating real parts:

$$R_4 + C_3R_3/C_4 = R_1R_3/R_2$$

Equating imaginary parts:

$$\omega C_3R_3R_4 - 1/(\omega C_4) = 0, \quad \omega C_3R_3R_4 = 1/\omega C_4,$$
$$\omega^2 = 1/(R_3R_4C_3C_4) \quad \text{and}$$
$$f = 1/[2\pi\sqrt{(R_3R_4C_3C_4)}] \qquad (8.16)$$

More complex bridges

The balance equations of some bridges cannot be obtained by equating the products of the impedances of the opposite arms and some examples of such bridges follow.

Anderson bridge

The Anderson bridge is used to measure the inductance and resistance of an inductor with increased sensitivity. Figure 8.12

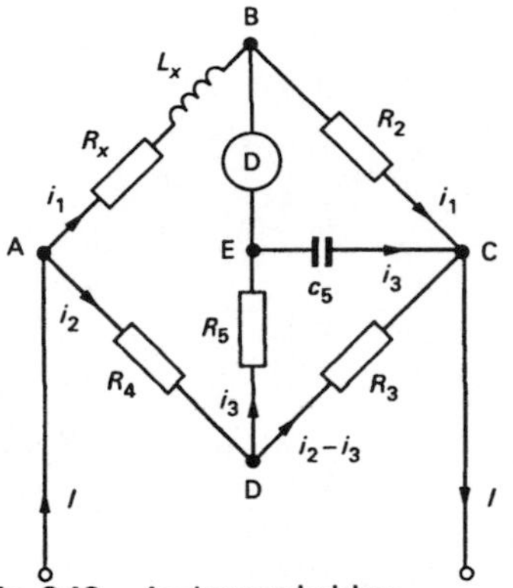

Fig. 8.12 Anderson bridge

shows the circuit of an Anderson bridge. A known value of resistance may sometimes be inserted in series with the inductor. The derivation of the balance equations for the bridge requires the use of Kirchhoff's laws. At balance:

Mesh ABEDA:

$$i_1(R_x + j\omega L_x) - i_3R_5 - i_2R_4 = 0 \tag{8.17}$$

Mesh BCDEB:

$$i_1R_2 - R_3(i_2 - I_3) + i_3R_5 = 0$$
$$i_1R_2 - i_2R_3 + i_3(R_3 + R_5) = 0 \tag{8.18}$$

Mesh ECDE:

$$i_3/j\omega C_5 - R_3(i_2 - i_3) + i_3R_5 = 0$$
$$i_3(R_3 + R_5 + 1/j\omega C_5) - i_2R_3 = 0$$

or $$i_2 = i_3(R_3 + R_5 + 1/j\omega C_5)/R_3 \tag{8.19}$$

From equation (8.18),

$$i_1R_2 + i_3(R_3 + R_5) = i_2R_3 = i_3(R_3 + R_5 + 1/j\omega C_5)$$
$$i_1 = i_3(R_3 + R_5 + 1/j\omega C_5 - R_3 - R_5)/R_2$$
$$= i_3/j\omega C_5R_2 \tag{8.20}$$

Substituting equations (8.19) and (8.20) into equation (8.17) gives,

$$i_3\left[\frac{(R_1 + j\omega L_x)}{j\omega C_5R_2} - R_5 - \frac{R_4}{R_3}\left(\frac{R_3 + R_5 + 1}{j\omega C_5}\right)\right] = 0$$

Equating imaginary parts:

$$R_x = R_2R_4/R_3 = R_2 \quad \text{if} \quad R_3 = R_4 \tag{8.21}$$

Equating real parts:

$$L_x = C_5R_2(R_5R_4/R_3 + R_4 + R_5) \tag{8.22}$$

Often, $R_3 = R_4$ and then

$$L_x = C_5R_2(2R_5 + R_4) \tag{8.23}$$

Example 8.2

An Anderson bridge consists of the following components:

$R_2 = 1\ \text{k}\Omega$, $R_3 = 1\ \text{k}\Omega$, $R_4 = 560\ \Omega$, $R_5 = 220\ \Omega$ and $C_5 = 2\ \mu\text{F}$

It is used to measure an inductance. Calculate the inductance and resistance of the measured component.

Solution

$$L_x = (2 \times 10^{-6} \times 1000)[(220 \times 560/1000) + 780] = 1.81 \text{ H} \quad (Ans.)$$
$$R_x = 1000 \times 560/1000 = 560\ \Omega \quad (Ans.)$$

Mutual inductance bridge

The circuit of the mutual inductance bridge is shown by Fig. 8.13. As with the Anderson bridge Kirchhoff's laws must be used to obtain the balance equations.

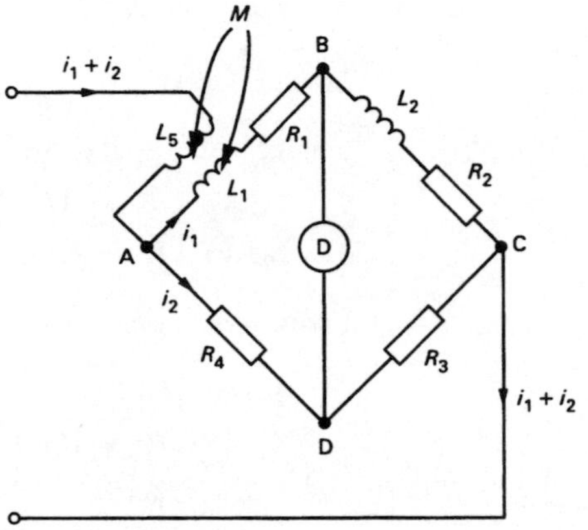

Fig. 8.13 Mutual inductance bridge

Mesh ABDA:

$$i_1(R_1 + j\omega L_1) + j\omega M(i_1 + i_2) = i_2 R_4 \qquad (8.24)$$

Mesh BCDB:

$$i_2(R_2 + j\omega L_2) = i_2 R_3 \qquad (8.25)$$

From equation (8.25),

$$i_2 = i_1(R_2 + j\omega L_2)/R_3$$

This value of i_2 is substituted into equation (8.24) to give,

$$i_1(R_1 + j\omega L_1) + j\omega M i_1 + i_1 j\omega M(R_2 + j\omega L_2)/R_3$$
$$= i_1 R_4(R_2 + j\omega L_2)/R_3$$

Equating real parts:

$$R_1 - (\omega^2 L_2 M/R_3) = R_2 R_4/R_3 \quad \text{or}$$
$$M = (R_1 R_3 - R_2 R_4)/\omega^2 L_2 \qquad (8.26)$$

Equating imaginary parts:

$$\omega L_1 + \omega M + (\omega M R_2/R_3) = \omega L_2 R_4/R_3$$
$$M = (L_2 R_4 - L_1 R_3)/(R_2 + R_3) \qquad (8.27)$$

Carey-Foster bridge

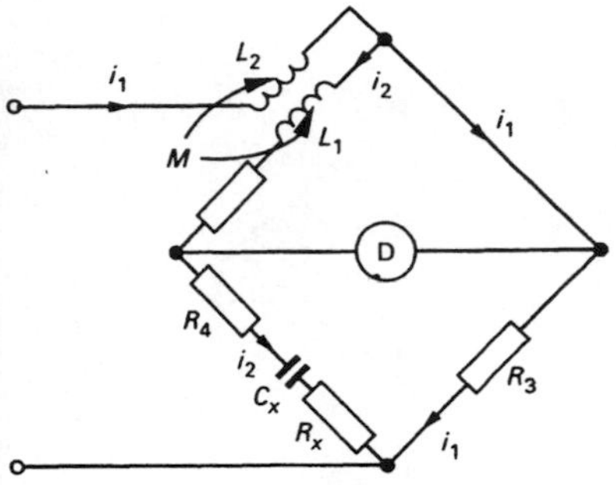

Fig. 8.14 Carey-Foster bridge

Figure 8.14 shows the circuit of a Carey-Foster bridge which may either be used to measure capacitance in terms of mutual inductance, or mutual inductance in terms of capacitance. The voltage dropped across L_1 and R_1 must be equal to zero and hence,

$$i_2(R_1 + j\omega L_1) - j\omega M(i_1 + i_2) = i_1 j\omega M$$
$$= i_2(R_1 + j\omega L_1 - j\omega M) \qquad (8.28)$$

Also, the voltage dropped across R_3 must be equal to the voltage dropped across the arm that contains the unknown impedance. Therefore,

$$i_1 R_3 = i_2(R_4 + R_x - j/\omega C_x) \qquad (8.29)$$

Dividing equation (8.28) by (8.29) gives

$$i_1 j\omega M / i_1 R_3 = \frac{i_2(R_1 + j\omega L_1 - j\omega M)}{i_2(R_4 + R_x - j/\omega C_x)}$$
$$j\omega M(R_4 + R_x - j/\omega C_x) = R_3(R_1 + j\omega L_1 - j\omega M),$$
$$M/C_x + j\omega M(R_4 + R_x) = R_1 R_3 + j\omega(L_1 - M)$$

Equating real terms:

$$M/C_x = R_1 R_3, \quad C_x = M/R_1 R_3 \qquad (8.30)$$

Equating imaginary terms:

$$M(R_4 + R_x) = L_1 - M, \; R_x = (L_1 - M)/M - R_4 \qquad (8.31)$$

Alternatively, if the value of C_x is known the bridge may be employed to measure the mutual inductance between two coils.

Z Short-circuit bridges

In some a.c. bridges the unknown impedance Z is connected in series with some known values of components. The bridge is then balanced first with Z in circuit and then with Z short circuited. This technique removes the potential inaccuracies which may be caused by unwanted capacitances in the bridge circuit (e.g. capacitances between components and earth and between the detector and earth). Two examples of such bridges are shown in Figs 8.15(*a*) and (*b*). When the bridge in figure (*a*) is at balance,

$$R_4(1/R_1 + j\omega C_1) = [Z + R_3 + (1/j\omega C_3)]/(1/j\omega C_2)$$
$$Z = (R_4/j\omega C_2)[(1/R_1) + j\omega C_1] - R_3 - 1/(j\omega C_3)$$
$$= [(R_4 C_1/C_2) - R_3] + (1/j\omega)[(R_4/R_1 C_2) - 1/C_3]$$
$$= \left(\frac{R_4 C_1}{C_2} - R_3\right) + \left(\frac{1}{j\omega C_1}\right)\left(\frac{R_4 C_1}{R_1 C_2} - \frac{C_1}{C_3}\right)$$

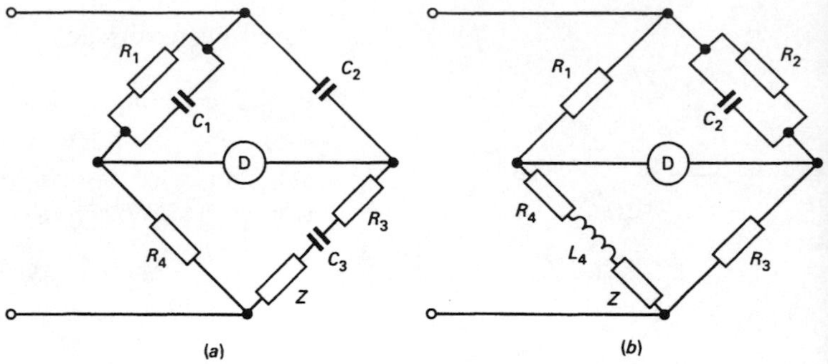

Fig. 8.15 Two bridges in which the unknown impedance *Z* is short circuited in the balance procedure

Example 8.3

In the bridge circuit of Fig. 8.15(*a*) balance is obtained at 1 kHz with Z short circuited when $R_1 = R_3 = 1$ kΩ. When the short circuit is removed, balance is obtained with $R_1 = 1250$ Ω and $R_3 = 900$ Ω. $C_2 = 159$ nF and $R_4 = 1$ kΩ. Calculate the value of Z.

Solution
When $Z = 0$, $R_4C_1/C_2 = R_3 = 1000$ Ω and $C_1/C_3 = 1$. Hence for any other value of Z,

$$Z = (1000 - R_3) + (1/j\omega C_1)[(1000/R_1) - 1)$$

When $R_3 = 900$ Ω and $R_1 = 1250$ Ω

$$Z = 100 + (1/j\omega C_1)[1000/1250 - 1]$$
$$= 100 + j10^9/(5 \times 2\pi \times 1000 \times 159) = 100 + j200\ \Omega \quad (\textit{Ans.})$$

Example 8.4

The bridge circuit shown in Fig. 8.15(*b*) balances at $\omega = 500$ rad/s when $R_2 = 8500$ Ω and $C_2 = 110$ nF. R_4 is constant at 1.7 Ω. When Z is short circuited the bridge balances with $R_2 = 10$ kΩ and $C_2 = 100$ nF. Calculate the values of L and of Z.

Solution
At balance,

$$R_1R_3 = [R_2/(1 + j\omega C_2R_2)](R_4 + j\omega L_4 + Z)$$
$$R_1R_3 + j\omega C_2R_1R_2R_3 = R_2R_4 + j\omega L_4R_2 + R_2Z$$

Hence,

$$Z = [R_1R_3 + j\omega C_2R_1R_2R_3 - R_2R_4 - j\omega L_4R_2]/R_2$$
$$= (R_1R_3 - R_2R_4)/R_2 + j\omega(C_2R_1R_3 - L_4)$$

When $Z = 0$, $R_2 = 10$ kΩ, $C_2 = 100$ nF and $R_4 = 1.7$ Ω. Then,

$$R_1R_3 = 17 \times 10^3 \quad \text{and} \quad L = 17 \times 10^{-4}\omega = 850\ \text{mH} \quad (\textit{Ans.})$$

When $Z \neq 0$, $R_2 = 8500$ Ω and $C_2 = 110$ nF, so that

$$Z = (17 \times 10^3 - 8500 \times 1.7)/10^4 + j(0.935 - 0.85)$$
$$= 0.3 + j0.09\ \Omega$$

Therefore, $Z = 0.3$ Ω in series with $L = 180$ μH (*Ans.*)

Exercises 8

8.1 An a.c. bridge used to measure inductance has opposite arms containing resistors R_2 and R_4, and the arm opposite the unknown inductor containing a resistor R_3 and a capacitor C_3 connected in series. (*a*) Determine the balance equations. (*b*) If $R_2 = R_4 = 2000\ \Omega$, $R_3 = 2980\ \Omega$ and $C_3 = 20$ nF calculate (*i*) r_x, (*ii*) L_x, and (*iii*) the Q factor. The frequency of the measurement is 2 kHz.

8.2 At balance the component values of an Anderson bridge are: $R_1 = 1$ kΩ, $R_3 = R_4 = 1.2$ kΩ, $R_5 = 110\ \Omega$ and $C_5 = 2.5\ \mu$F. Calculate the resistance and inductance of the inductor being measured.

8.3 An a.c. bridge circuit has 600 Ω resistors in three of its arms and the impedance $600 + j\omega L - j/\omega C\ \Omega$ in the fourth arm. A 600 Ω resistor is connected between the junction A of two 600 Ω resistors and the junction B of the impedance and the third 600 Ω resistor. A voltage source at a frequency of 5000 rad/s is applied across the bridge. The current in the 600 Ω resistor connected between A and B is 10 mA when $L = 200$ mH and $C = 200$ nF, and 20 mA when $L = 100$ mH and $C = 100$ nF. Calculate the percentage second harmonic in the voltage of the supply.

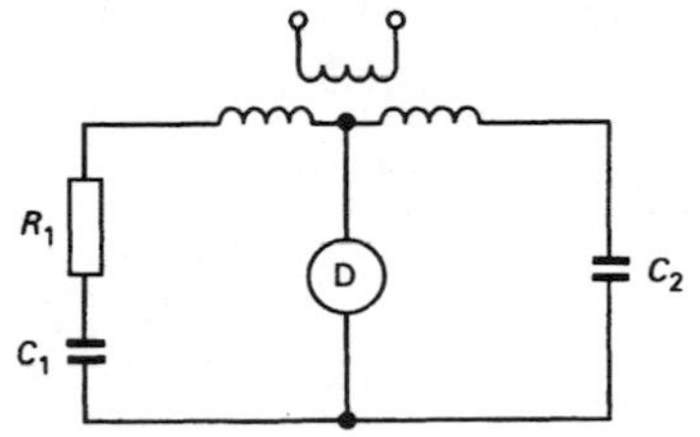

Fig. 8.16

8.4 Figure 8.16 shows an a.c. bridge in which all the transformer windings have an equal number of turns. Balance is obtained at 1000 Hz with $C_1 = 10$ nF and $R_1 = 50\ \Omega$. Calculate the capacitance and power factor of C_2. The voltage of the source is 1 V r.m.s. Calculate also (*a*) the voltage across C_2, (*b*) the voltage across the detector when the difference between the two capacitor values is 1 per cent.

8.5 A Carey–Foster bridge is used to measure the coefficient of coupling between two coils of equal inductance L. If R_3 and R_4 can each be varied over the range 120 Ω to 1200 Ω calculate the range of coupling coefficients which can be measured.

8.6 For the a.c. bridge given in Fig. 8.15(*a*) derive the balance equations. At a frequency of 2 kHz balance is obtained with $R_1 = 3500\ \Omega$ and $R_3 = 2800\ \Omega$. When Z is short circuited balance occurs when $R_1 = R_3 = 3$ kΩ. If $C_1 = 0.2\ \mu$F calculate the components of impedance Z.

8.7 Apply the star–delta transform to the Anderson bridge of Fig. 8.12 to obtain the equivalent Maxwell–Wien bridge circuit shown in Fig. 8.6.

8.8 The four arms af an a.c. bridge contain: arm A, the unknown capacitor and its loss resistance; arm B, resistor R_2; arm C, resistor R_3; and arm D, resistor R_4 in series with capacitor C_4. Show that at balance $C_x = C_4R_3/R_2$ and $r_x = R_2R_4/R_3$.

8.9 A Schering bridge is balanced at a frequency of 1200 Hz when $C_1 = 0.25\ \mu$F, $C_4 = 3$ nF, $R_4 = 560\ \Omega$ and $R_3 = 180\ \Omega$. Calculate (*a*) the resistance and capacitance of the unknown capacitor and (*b*) the power factor of the unknown capacitor.

8.10 A four-arm bridge has: arm A, 1000 Ω in series with a 0.62 μF capacitor, arm B, the unknown impedance Z; arm C, a 0.2 μF capacitor, arm D, a 2000 Ω resistor in parallel with a 0.7 μF capacitor. Balance is obtained at a frequency of 1 kHz. Calculate the components of Z if they are (*a*) series connected and (*b*) parallel connected.

8.11 A Hay bridge has $R_4 = 14.8\ \text{k}\Omega$, $C_4 = 450$ pF, $R_1 = 10\ \text{k}\Omega$ and $R_3 = 10\ \text{k}\Omega$. The bridge is balanced at 1 kHz. Calculate the inductance and self-resistance of the unknown inductor.

8.12 A Schering bridge has $R_4 = 2300\ \Omega$, $C_4 = 800$ pF, $C_1 = 5$ nF and $R_3 = 1\ \text{k}\Omega$. Calculate the resistance and capacitance of the unknown capacitor.

8.13 A Maxwell–Wien bridge has the following component values when it is balanced: $R_3 = 8300\ \Omega$, $C_3 = 0.1\ \mu\text{F}$, $R_2 = 100\ \Omega$, $R_4 = 2420\ \Omega$. Calculate the resistance and inductance of the inductor measured.

8.14 For the bridge circuit given in Fig. 8.15(a) calculate Z if, with Z short circuited, balance is obtained at 1 kHz when $R_1 = R_3 = 1\ \text{k}\Omega$. With the short circuit removed balance is obtained with $R_1 = 1250\ \Omega$ and $R_3 = 900\ \Omega$. Calculate Z.

8.15 A Hay bridge is balanced at 1 kHz when $R_4 = 100\ \text{k}\Omega$, $C_4 = 200$ pF, $R_3 = 1\ \text{k}\Omega$ and $R_1 = 10\ \text{k}\Omega$. Calculate r_x and L_x.

8.16 Two series LC circuits, L_1C_1 and L_2C_2, are coupled together by mutual inductance M and the two capacitors are coupled by a third capacitor C_3. A voltage source is applied across the L_1C_1 circuit and a detector is connected across the L_2C_2 circuit. If $L_1 = L_2 = 100$ mH and $C_1 = C_2 = 1\ \mu\text{F}$ determine the range of frequencies over which the mutual inductance can be measured from 10 to 95 mH.

9 Magnetic and dielectric materials

The passive components which are widely used in electrical and electronic circuits are the resistor, the inductor and the capacitor. Although the first-named component, the resistor, is used to provide a predetermined amount of resistance into a circuit with consequent power dissipation, the other two components should, ideally, dissipate zero energy.

Unfortunately, all the dielectric and magnetic materials available for use possess inherent losses and so their practical performance must inevitably fall short of the ideal. Dielectric materials are also used to insulate conductors in cables and these should also have the minimum possible loss.

Most engineering materials are *diamagnetic*; this means that they show little, if any, reaction to an applied magnetic field. A *paramagnetic* material can be magnetized by an external magnetic field but it is unable to retain that magnetism once the field has been removed. A few materials are *ferromagnetic*; these materials are easily magnetized and have the ability to retain the magnetism for a long period of time after the magnetizing field had been removed. Ferromagnetic materials are employed in the manufacture of permanent magnets.

Dielectric materials and capacitors

The insulating material between the plates of a capacitor or around the conductors of a cable is known as the *dielectric*. A wide variety of different materials have been used as dielectrics but all of them exhibit various imperfections. The various parameters of a dielectric material will be discussed and then some typical figures for various materials will be given.

Permittivity

Capacitance is a measure of how much electric energy can be stored by an insulating medium that is bounded by two conductors. Capacitance always exists between two non-touching conductors. The amount of stored energy depends upon both the distance separating the two conductors and upon the nature of the insulating medium. The reference medium is, strictly speaking, a vacuum but practically it is normally taken as being air. Air is said to have a *permittivity* ε_0 equal to 8.854×10^{-12} F/m [$1/(36\pi \times 10^9)$ F/m]. Any other medium will increase the capacitance between the two conductors because its permittivity will be greater than ε_0. Thus the capacitance of two parallel plates of area A m^2 separated by a distance d metres will possess a capacitance of $\varepsilon_0 A/d$ farads when air is the dielectric material. When some other dielectric material is employed the capacitance will be increased to $\varepsilon_0 \varepsilon_r A/d$ farads where ε_r is a dimensionless quantity known as the *relative permittivity* of the medium.

The dielectric material of a capacitor must have a high relative permittivity in order to obtain a large capacitance in a small volume. Conversely, the dielectric used to insulate the conductors in a cable must have a low relative permittivity since a low inter-conductor capacitance is required.

Some typical figures for the relative permittivity of some commonly employed dielectrics are given in Table 9.1.

Table 9.1 Relative permittivities of common dielectrics

Air	1.0006	Polyethylene	2.3
Aluminium oxide	7	Polypropylene	2.5
Bakelite	4–8	Polystyrene	2.55
Ceramic	20–100	Polythene	1.5
Glass	4–10	PFTE	2
Mica	2.5–8	Rubber	2
Mylar	3	Tantalum oxide	10–20
Oil	2.5	Teflon	2
Paper	2–4	Water	80
Paraffin wax	2.3		

Electric field strength

The *electric field strength*, often known as the dielectric strength, of a dielectric is the electric field in volts/metre at which the dielectric breaks down.

The narrower the dielectric the lower will be the electric field that causes breakdown. All components with a dielectric, such as capacitors and conductors in cables, have a stated maximum safe working voltage. The maximum voltage is quoted by the manufacturer. Dielectrics vary considerably in their ability to resist breakdown and some typical figures are given in Table 9.2.

Table 9.2 Dielectric strength

Dielectric material	*E* (V/m)	*Dielectric material*	*E* (V/m)
Air	2×10^6	Paraffin wax	7.5×10^6
Glass	$8–20 \times 10^6$	Polythene	42×10^6
Mica	65×10^6	PTFE	16×10^6
Oil	6×10^6	Rubber	30×10^6

Leakage current

The leakage current of a dielectric is determined by its insulation resistance. When a d.c. voltage is applied across a dielectric, a current will flow which will reach a steady value after a time determined by a time constant *CR*. Obviously, the insulation resistance should be as high as possible in order to minimize the leakage current.

Dielectric absorption

When a capacitor is charged up from a constant voltage source and is then discharged, and then the capacitor is left on open-circuit for some time, it is found that a new, but smaller, charge accumulates. This effect arises because some of the original charge was absorbed by the dielectric. The practical result of this is that there must always be a difference between the rates at which a capacitor can be charged and discharged. This means that the effective value of the capacitance falls with increase in the frequency of an applied voltage. *Dielectric absorption* is normally quoted as a percentage and Table 9.3 gives a few typical values.

Table 9.3 Dielectric absorption values

Mica	3%	Polypropylene	0.1%
Paper	2%	Polystyrene	0.1%

The mechanical and thermal characteristics of dielectric materials

The choice of a dielectric material for a particular application will be decided by the mechanical and thermal characteristics of the various alternatives. Depending upon the intended use the main requirement may be for mechanical strength, such as compression, shearing, or tensile, for resistance to attack by acids, or for high thermal resistance. The insulator used for the conductors in a telephone cable must be able to withstand repeated bending but its resistance to a direct pressure need not be high.

Once the main dielectric used in both capacitors and cables was a combination of air and paper. It is obvious that the mechanical strength of such a dielectric is very small and, of course, the resistance to water is very limited. Nowadays the most commonly used dielectric for cables is some kind of plastic and this is used for many types of capacitor also.

The plastic materials are of low density, have a high resistance to attack by chemicals, are of high thermal resistivity, and are relatively easy to fabricate into any required shape. The main disadvantage of plastics is their low mechanical strength and elasticity and their relatively high thermal expansion coefficient. Polyethylene has a high resistance to most chemicals and solvents and it remains both tough and flexible over quite a wide range of temperatures. There are two types: one of which, known as low-density, will soften if immersed in boiling water, while the other, known as high density, does not soften in boiling water. Polypropylene becomes softer at a higher temperature than polyethylene and it is particularly good for use in any environment where it will be subjected to repeated bending, since it is highly resistant to cracking. Polystyrene is of high resistance to most acids but is very vulnerable to some others. It is easily moulded into a required shape and is very light in weight.

The particular characteristic of rubber, real or synthetic, is its ability to show considerable deformation when subject to stress and yet return to its original shape and dimensions immediately the stress is removed. Rubber is not rigid and its mechanical strength is not high.

Bakelite is a very hard and rigid material that has a high thermal resistivity and good resistance to acids, oils, etc. Ceramics are also of high strength and rigidity and also exhibit a high resistance to abrasion and to wear. These attributes also apply to aluminium oxide.

The thermal characteristics of dielectrics are many and any one or more of them may be of importance in a particular application. The characteristics include (*a*) the temperature at which the material will melt, (*b*) the temperature at which it may catch fire, (*c*) its ageing-temperature characteristic, (*d*) its thermal resistance and (*e*) its expansion with increase in temperature.

Loss angle and power factor

When a capacitance is first charged up and then is discharged it is always found that the energy that can be taken out of the capacitance is less than the energy that was originally put into it. The missing energy is lost because of one, or more, of the following reasons:

- Surface leakage.
- Leakage paths within the dielectric itself because of various imperfections.

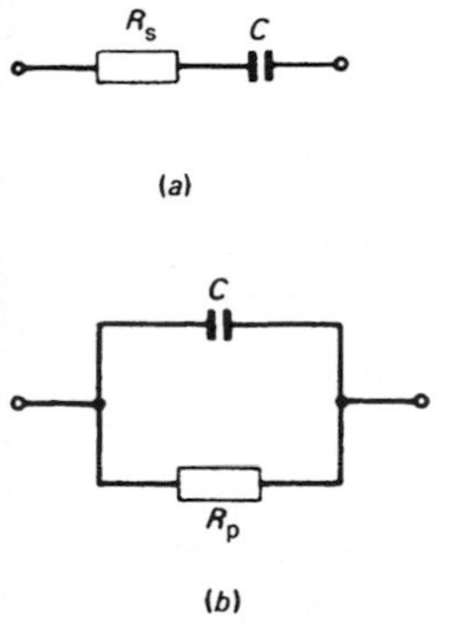

Fig. 9.1 Representation of capacitor losses: (a) by series resistance; (b) by parallel resistance

- The finite insulation resistance of the dielectric.
- The work done in storing and then releasing charge.

It is neither possible nor particularly desirable to distinguish between these various effects and so their combined effect is usually taken into account by assuming that all the energy loss is caused by I^2R or V^2/R power dissipation in a resistance. This *equivalent loss resistance* may be considered to be connected *either* in series *or* in parallel with the (assumed) perfect capacitance. This is shown by Fig. 9.1(*a*) and (*b*). The series loss resistance has only a low resistance value, typically less than 0.1 Ω, while the parallel loss resistance is of very much higher value, usually 0.5 MΩ or more.

Series loss resistance

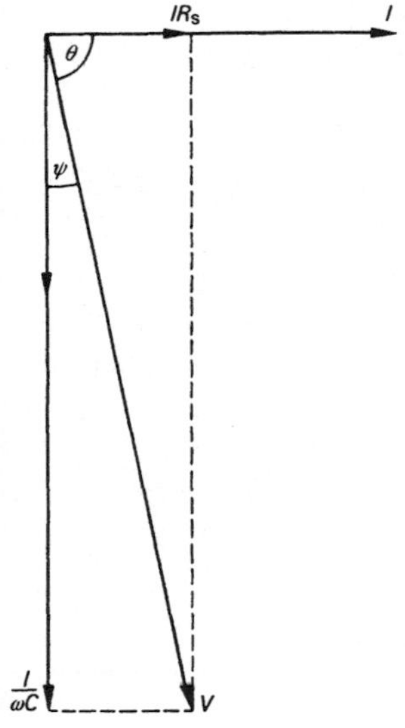

Fig. 9.2 Phasor diagram of a capacitor with series loss resistance

The phasor diagram representing the current and voltages in the case of the *series loss resistance* is shown by Fig. 9.2. The phase difference between the sinusoidal applied voltage V and the current I that flows is the angle θ. Since the applied voltage is sinusoidal, the power factor of the dielectric is equal to cos θ.

The *loss angle* ψ of the dielectric is the angle by which θ falls short of 90° (the zero loss case), i.e.

$$\psi = (90 - \theta)^\circ \tag{9.1}$$

The loss angle is always a very small angle, certainly less than 1°, and this means that

$$\text{Power factor} = \cos\theta = \sin\psi \approx \psi \tag{9.2}$$

Thus, the power factor of a dielectric is very nearly equal to its loss angle.

Referring to Fig. 9.2; since ψ is small $\sin\psi \approx \tan\psi \approx \psi$ and hence

$$\sin\psi = IR_s/(I/\omega C) = \omega CR_s \quad \text{or} \quad R_s = \psi/\omega C \tag{9.3}$$

Further, from equation (4.14) the Q-factor of R_s in series with C is $Q = 1/\omega CR_s$, hence

$$Q = 1/\psi = 1/(\text{Power factor}) \tag{9.4}$$

provided the loss angle is small the dissipation and power factors are very nearly equal to one another.

Example 9.1

At a frequency of 5000 rad/s the losses of a 0.1 μF capacitor can be represented by a 0.1 Ω resistor. Calculate (*a*) the power factor, (*b*) the Q

factor, and (*c*) the power dissipated in the capacitor when a 2 V r.m.s. sinusoidal voltage at a frequency of 5000 rad/s is applied across the capacitor.

Solution

(*a*) Power factor $= \omega CR_s = 5000 \times 0.1 \times 10^{-6} \times 0.1 = 50 \times 10^{-6}$ (*Ans.*)

(*b*) $Q = 1/(\text{Power factor}) = 20000$ (*Ans.*)

(*c*) The current flowing in the capacitor is

$$I = 2/\{\sqrt{[0.1^2 + 1/(5000 \times 0.1 \times 10^{-6})^2]}\} \approx 2/2000 = 1 \text{ mA}$$

and the power dissipated is

$$I^2R_s = (1 \times 10^{-3})^2 \times 0.1 = 0.1\ \mu\text{W} \quad (Ans.)$$

Parallel loss resistance

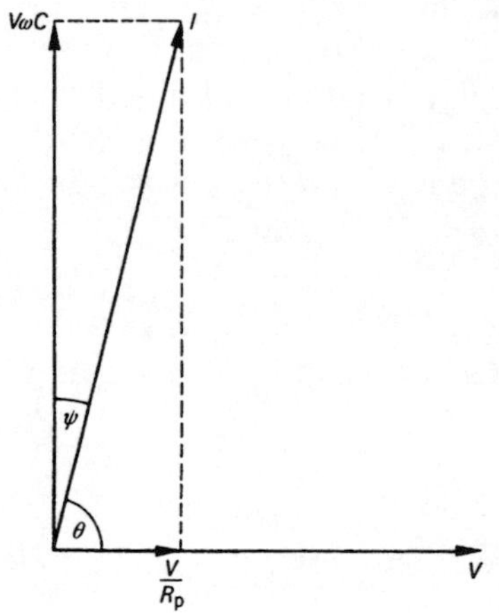

Fig. 9.3 Phasor diagram of a capacitor with parallel loss resistance

The phasor diagram of the currents and voltage in the *parallel loss resistance* circuit of Fig. 9.1(*b*) is given in Fig. 9.3. Once again the loss angle is the angle by which θ falls short of 90°. From the phasor diagram, $\sin\psi \approx \tan\psi = (V/R_p)/V\omega C = 1/\omega CR_p$, or since $\sin\psi = \psi$,

$$R_p = 1/\psi\omega C \tag{9.5}$$

In general, the series resistance representation of capacitor losses should be used for small capacitance values and the parallel resistance representation for large capacitance values.

Some typical loss angles for some commonly used dielectrics are listed in Table 9.4.

Table 9.4 Loss angles for common dielectrics

Aluminium oxide	0.06	Ceramic	0.005
Mica	0.01	Paper	0.01
Polypropylene	0.0003	Polystyrene	0.003

Example 9.2

A sinusoidal voltage of 10 V at a frequency of $3000/2\pi$ Hz is applied across a 1 μF capacitor of loss angle 2.5×10^{-4} rad. Calculate the equivalent series and parallel loss resistances of the capacitor. For each resistance calculate the power dissipated in the capacitor.

Solution

From equation (9.3)

$$R_s = (2.5 \times 10^{-4})/(3000 \times 1 \times 10^{-6}) = 0.083\ \Omega \quad (Ans.)$$

From equation (9.5)

$$R_p = 1/(2.5 \times 10^{-4} \times 3000 \times 1 \times 10^{-6}) = 1.33\ \text{M}\Omega \quad (\textit{Ans.})$$

For the series loss resistance

$$I = V\omega C = 10 \times 3000 \times 1 \times 10^{-6} = 30\ \text{mA}$$

and the power dissipated is

$$(30 \times 10^{-3})^2 \times 0.083 = 75\ \mu\text{W} \quad (\textit{Ans.})$$

For the parallel loss resistance the power dissipated is

$$P = 10^2/(1.33 \times 10^6) = 75\ \mu\text{W} \quad (\textit{Ans.})$$

Example 9.3

When the voltage $v = 20\sqrt{2}\sin(100\pi t)$ volts is applied across a 1 μF capacitor 200 mW power is dissipated in the dielectric. Calculate (*a*) the parallel loss resistance, (*b*) the loss angle, (*c*) the Q factor, (*d*) the power factor and (*e*) the series loss resistance.

Solution

(*a*) $200 \times 10^{-6} = 20^2/R_p$, $R_p = 20^2/(200 \times 10^{-6}) = 2\ \text{M}\Omega$ (*Ans.*)

(*b*) $2 \times 10^6 = 1/(100\pi\Psi \times 1 \times 10^{-6})$, or

$$\Psi = 1/(2 \times 10^6 \times 100\pi \times 1 \times 10^{-6}) = 1.59 \times 10^{-3}\ \text{rad} \quad (\textit{Ans.})$$

(*c*) $Q = 1/\Psi = 628$ (*Ans.*)

(*d*) $\Psi = 0.091°$, power factor $= \cos(90 - 0.091°) = \sin 0.091°$

$$= 1.59 \times 10^{-3} = 1/Q \quad (\textit{Ans.})$$

(*e*) $1.59 \times 10^{-3} = \omega C R_s$, $R_s = (1.59 \times 10^{-3})/(100\pi \times 1 \times 10^{-6})$

$$= 5.07\ \Omega \quad (\textit{Ans.})$$

Example 9.4

A capacitor has an equivalent parallel resistance of 1 MΩ at 1 kHz. Calculate its equivalent series resistance.

Solution

$$\Psi = R_s\omega C = 1/(R_p\omega C)$$

$$R_s = 1/(1 \times 10^6 \times 4\pi^2 \times 10^6 \times 0.1^2 \times 10^{-12}) = 2.53\ \Omega \quad (\textit{Ans.})$$

Inductance of a capacitor

All capacitors possess some inductance and a leakage current and hence their equivalent circuit should really be somewhat more complex (see Fig. 9.4). In this circuit L is the inductance of the capacitor, R'_s a resistor that represents all the real losses of the capacitor (i.e. the lead, plate and contact resistances), R_L

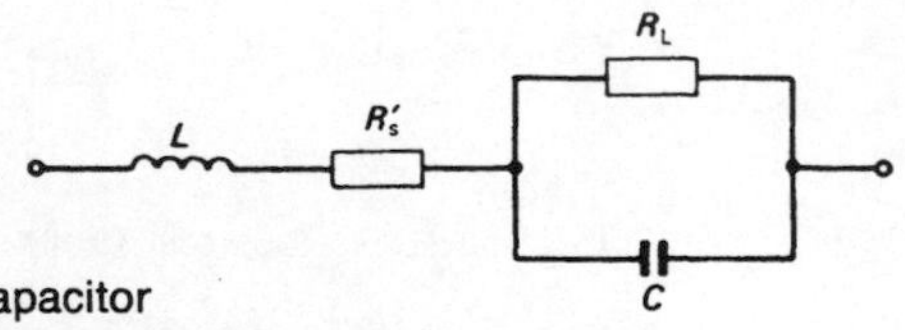

Fig. 9.4 Complete equivalent circuit of a capacitor

represents the dielectric losses, and C is the actual capacitance. Much of the self-inductance arises from the connecting leads and the manner in which they are connected to the component.

The impedance of Fig. 9.4 is

$$Z = R_s' + j\omega L + (R_L/j\omega C)/[R_L + (1/j\omega C)]$$
$$= R_s' + R_L/[1 + (\omega^2 C^2 R_L^2)]$$
$$+ j[\omega L - \omega R_L^2 C + \omega^3 R_L^2 L C^2]/(1 + \omega^2 C^2 R_L^2)$$

The real part of this expression is the equivalent series loss resistance R_s of the capacitor. Thus

$$R_s \approx R_s' + (1/\omega^2 C^2 R_L)$$

Magnetic materials

The materials which exhibit the phenomenon of magnetism are chiefly iron and steel. A variety of alloys of iron, such as nickel and cobalt, are also widely used as magnetic materials. When a piece of iron or steel is placed in a magnetic field it will become magnetized by a process known as *induction*. The iron provides a path of smaller reluctance for the lines of magnetic flux and so they tend to concentrate within the iron.

The degree to which the magnetic field is concentrated in the iron is expressed by the *relative permeability* μ_r, of the iron. Relative permeability is defined as being the ratio of the magnetic flux density produced in the iron (or other material) to the flux density produced in air (strictly speaking a vacuum) by the same magnetic field. The permeability of air is known as the *absolute permeability* of free space μ_0 and it is equal to $4\pi \times 10^{-7}$ H/m.

The permeability μ of a magnetic material is then equal to the product $\mu_r \mu_0$. The relative permeability μ_r of a magnetic material may have a value of several thousand but it is not a constant quantity. μ_r is equal to the ratio of the magnetizing force to the flux density and the relationship between these two parameters is not a linear one. Permeability curves for various magnetic materials are available from their manufacturers.

A diamagnetic material has a straight-line B-H plot and the value of its μ_r is less than unity.

Ferrites

Ferrites are a class of non-metallic material which have a very high resistivity and a high permeability. Because of their high resistivity the problems associated with eddy currents are largely overcome. The ferrite material is a compound of iron oxide, zinc and nickel particles held together by a binding agent and formed by a high temperature process. Typically, the permeability may be somewhere in the range 200–10 000.

Ferrite-based magnetic cores allow components to be much smaller for the same power output, are easier to manufacture in smaller dimensions and operate at higher frequencies with much smaller eddy current losses.

Hysteresis

If a magnetic material is taken through a complete cycle of magnetization and the corresponding values of magnetizing force and flux density are plotted, the *hysteresis loop* of the material will be obtained.

Consider Fig. 9.5(*a*). If the material is initially demagnetized and a positive magnetizing force is applied, in a number of discrete steps, the flux density follows the line marked as OA. If, then, the magnetizing force is removed step-by-step it is found that the fall in the flux density does *not* follow the path OA but instead it follows the path AB. It can be seen that, when the magnetizing force has been reduced to zero, some *residual magnetism* remains, represented by the distance OB. This residual magnetism is known as the *remanant flux density*.

To reduce the flux density to zero a negative magnetizing force OC must be applied to the material. This force is known as the *coercive force*.

Further increase in the negative magnetizing force will increase the flux density in the opposite direction to before, see path CD. If, now, the magnetizing force is removed the material will again exhibit some residual magnetism since the flux density will only reduce to the value OE when the magnetizing force has reduced to zero.

Eventually, the point is reached at which it becomes difficult to obtain any further increase in the flux density. The material is then said to be *magnetically saturated*. The distance from the magnetizing force axis and the point marked as A is the *saturation flux density*.

If the material is taken into *magnetic saturation* by either, or both, the positive and negative magnetizing forces, then (*a*) the remnant flux density is known as the *remanence* and (*b*) the coercive force is called the *coercivity*. These terms are illustrated by Fig. 9.5(*b*).

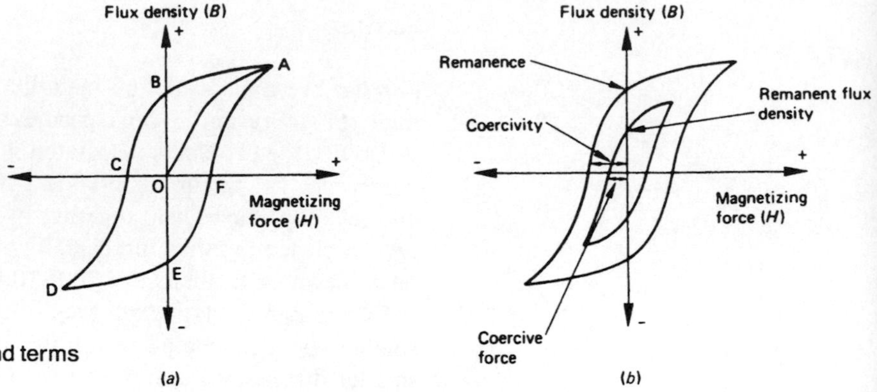

Fig. 9.5 Hysteresis loop (a) and terms employed (b)

Typical hysteresis loops for a permanent magnet material and for an electromagnet material are shown in Fig. 9.6(a) and (b). It should be noted that a material which is suitable for use as a permanent magnet has a wide hysteresis loop but a material that is best suited for use as an electromagnet has a narrow hysteresis loop.

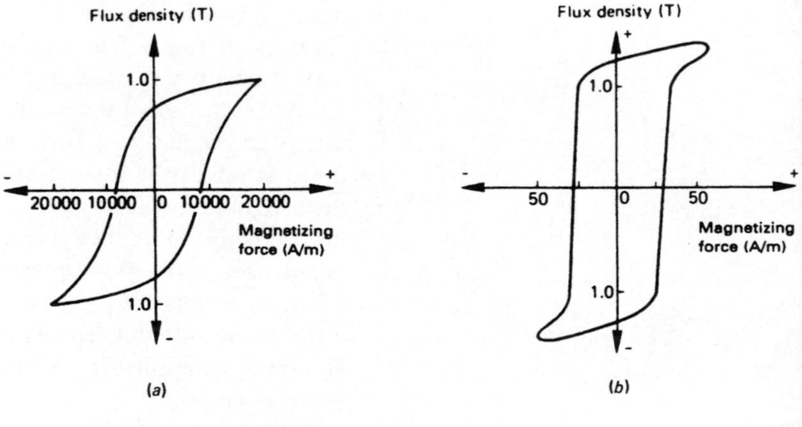

Fig. 9.6 Hysteresis loops of: (a) a permanent magnet material; (b) an electromagnet material

Some materials have a very nearly square hysteresis loop and need only a small change in the magnetizing force to change from one saturation state to the other. Such materials are therefore eminently suitable for use in magnetic digital devices. Work must be done to take a magnetic material through a cycle of magnetization and so energy will be dissipated. This dissipated energy appears in the form of heat. The amount of energy dissipated is proportional to the *area* of the hysteresis loop. The area increases with increase in the magnetizing force and the maximum area is obtained when the material is taken into magnetic saturation. Based upon this observed fact an empirical formula has been obtained by Steinmetz for the hysteresis power loss:

$$P_{\mathrm{H}} = \eta f B_{\mathrm{max}}^{x} \qquad (9.6)$$

where η is the hysteresis coefficient of the material,

f is the frequency,
B_{max} is the maximum flux density,
x is the Steinmetz index and this is often taken as being equal to 1.6.

In the case of a transformer core, B_{max} is proportional to voltage/frequency (provided the change in frequency is fairly small) and then

$$P_H = k_1 f^{(1-x)} V^x \text{ watts} \quad \text{where } k_1 \text{ is a constant} \tag{9.7}$$

Example 9.5

A magnetic material has a Steinmetz index of 1.6 and a hysteresis loss of 100 W at 50 Hz when the maximum flux density is 1.25 T. Determine the hysteresis loss when the maximum flux density is (*a*) 1 T and (*b*) 1.4 T.

Solution

$$100 = 50\eta \times 1.25^{1.6}, \quad \eta = 100/(50 \times 1.429) = 1.4$$

(*a*) $P_H = 1.4 \times 50 \times 1^{1.6} = 70$ W (*Ans.*)

(*b*) $P_H = 1.4 \times 50 \times 1.4^{1.6} = 120$ W (*Ans.*)

Eddy current loss

The alternating flux set up in a core of magnetic material not only cuts any wires wound around it but also cuts the core itself. This means that the flux induces voltages into the core and these, in turn, cause currents – known as *eddy currents* – to flow within the core. The eddy currents produce an I^2R power loss that is given by

$$P_E = k_2 f^2 B_{max}^2 \tag{9.8}$$

where f is the frequency, B_{max} is the maximum flux density in the core and k_2 is a constant determined by the resistivity of the core material.

To reduce the magnitude of the eddy current loss the core of an inductor or a transformer is made up of a number of thin strips, known as laminations, which are insulated from one another to provide poor electrical conduction in the direction perpendicular to the laminations. The resistance of the core is increased and consequently the eddy currents are reduced in magnitude. This technique was developed for devices operating at frequencies of 50/60 Hz and it is not effective for higher frequency applications, such as switched-mode power supplies.

At higher frequencies, therefore, eddy currents are minimized by the use of ferrite cores.

The total core loss is the sum of the hysteresis and the eddy current losses.

Example 9.6

A transformer core has an eddy current loss of 4 W at a frequency of 50 Hz. With the maximum flux density remaining constant the frequency is increased to 400 Hz. Calculate the new value of the eddy current loss.

Solution

$$4 = 50^2 \times k_2 B^2, \quad \text{or} \quad k_2 B^2 = 4/50^2 = 1.6 \times 10^{-3}$$

At 400 Hz:

$$P_E = 1.6 \times 10^{-3} \times 400^2 = 256 \text{ W} \quad (Ans.)$$

Alternatively,

$$P_E = 4 \times (400/50)^2 = 256 \text{ W} \quad (Ans.)$$

Example 9.7

A transformer had a hysteresis loss of 300 W and an eddy current loss of 120 W when the applied voltage was 2000 V at 60 Hz. Calculate the total loss if this voltage is changed to 1000 V at 50 Hz. Take the Steinmetz index as 1.6.

Solution

From equations (9.7) and (9.8)

$$300 = k_1 \times 60^{-0.6} \times 2000^{1.6} \quad \text{or} \quad k_1 = 300 \times 60^{0.6} \times 2000^{-1.6}$$

and $120 = k_2 \times 2000^2 \times 60^2$ or $k_2 = 120/(2000^2 \times 60^2)$

Therefore, when the applied voltage is 1000 V at 50 Hz, hysteresis loss is

$$P_H = (300 \times 60^{0.6} \times 2000^{-1.6})50^{-0.6} \times 1000^{1.6} = 110.4 \text{ W}$$

The, eddy current loss is

$$P_E = \left[\frac{120}{(60^2 \times 2000^2)}\right] \times 1000^2 \times 50^2 = 20.83 \text{ W}$$

The total iron loss is

$$P = 110.4 + 20.83 = 131.23 \text{ W} \quad (Ans.)$$

Example 9.8

A transformer core has a hysteresis loss of 100 W and an eddy current loss of 75 W at 50 Hz. If the maximum flux density remains constant calculate the two losses when the frequency is increased to 100 Hz.

Solution

$$P_H = (100 \times 60)/50 = 120 \text{ W} \quad (\textit{Ans.})$$

$$P_E = 75 = k_2 50^2 B^2 \quad \text{or} \quad k_2 = 75/(50B)^2$$

At 60 Hz,

$$P_E = [75/(50B)^2] \times 60^2 B^2 = 108 \text{ W} \quad (\textit{Ans.})$$

Separation of losses

The total loss in a magnetic core is the sum of the hysteresis and eddy current losses, i.e. $P_T = P_H + P_E$.

Hysteresis loss is proportional to frequency and eddy current loss is proportional to frequency squared. Hence, $P_T = mf + nf^2$, where m and n are constants. Dividing by f, $P_T/f = m + nf$, which is the equation of a straight line. If P_T/f is plotted against frequency the slope of the plotted line is equal to the constant n, while the intersection of the line with the vertical axis gives the constant m.

Example 9.9

The total power loss in a magnetic core is 120 W at 50 Hz and 162 W at 60 Hz. Calculate (a) the hysteresis loss and (b) the eddy current loss at 50 Hz.

Solution
At 50 Hz,

$$P_T/f = 120/50 = 2.4 = m + 50n \tag{9.9}$$

At 60 Hz,

$$P_T/f = 162/60 = 2.7 = m + 60n \tag{9.10}$$

Subtracting equation (9.9) from (9.10) gives, $0.3 = 10n$, and $n = 0.03$. Hence, $2.4 = m + 1.5$ and $m = 0.9$. Therefore,

$$P_T = 0.9 \times 50 + 0.03 \times 50^2 = 45 + 75 = 120 \text{ W}$$

Hysteresis loss = 45 W; eddy current loss = 75 W (*Ans.*)

Properties of magnetic materials

Materials that possess the ability to be magnetized are known as *ferromagnetic* materials and these, in turn, are either suited to use as permanent magnets or as electromagnets.

- Permanent magnet materials are those which have high values of both remanence and coercivity and (unfortunately) a high hysteresis loss. Because of these features, once such a material has become magnetized, it will retain its magnetism for a long while after the magnetizing force has been removed.

Most permanent magnet materials consist of some kind of iron or steel alloy. In the past, alloys containing cobalt, tungsten and chromium have been used but most modern materials consist of some particular alloy of iron, and one or more of aluminium, nickel, cobalt and copper such as Alnico and Alcom.

- Materials suited for use in an *electromagnet* should possess the following properties: (*a*) high electrical resistivity, to reduce eddy currents, and (*b*) high permeability to give a high flux density and so limit the size of a core and hence its weight and cost, (*c*) low remanence, (*d*) low coercivity and (*e*) minimum hysteresis loss. Also of importance is the flux density at which the material saturates and the constancy of the magnetic properties with the passage of time.

Suitable materials are often nickel–iron alloys such as Permalloy or Mumetal, sometimes with the addition of either some copper or some cobalt. Other electromagnetic materials are various alloys of cobalt and iron. Table 9.5 compares some of the more popular alloys.

Table 9.5 Electromagnetic materials

Magnetic material	*Typical permeability*	*Resistivity*
Silicon – iron	3 500	high
Nickel – iron	100 000	low
Nickel – iron – copper	80 000	medium
Nickel – iron – copper – chromium	75 000	high
Nickel – iron cobalt	10 000	high

The magnetic materials used in transformers and in electrical generators and motors are generally of high permeability so that a high flux density can be obtained with minimum current. This reduces the size and weight of the equipment. Whenever an alternating flux is employed both hysteresis and eddy current losses should be minimized. It is necessary to ensure that the material is not taken into magnetic saturation and this requirement often means that pure iron cannot be employed. Often iron with a small percentage, typically 3 per cent, of silicon is employed since it is cheaper, but it has a lower saturation flux density and increased resistivity. More than about 4 per cent silicon content is not used because it makes the alloy become mechanically brittle.

A reduction in the eddy current loss is obtained by constructing the core from thin laminations of nickel–iron alloy but at higher frequencies it becomes necessary to employ a ferrite material for the core.

Exercises 9

9.1 Figure 9.7 shows the equivalent circuit of an iron-cored inductor. Derive an expression for the equivalent series resistance and reactance of the inductor. At a particular frequency the reactance of the inductor is 800 Ω, $R_s = 50$ Ω and $R_p = 30$ kΩ. Calculate the equivalent resistance and reactance of the inductor.

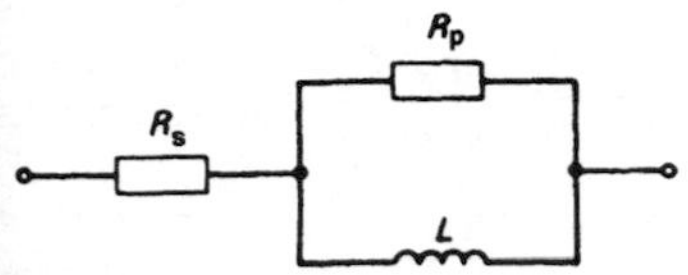

Fig. 9.7

9.2 Two capacitors of 0.4 μF each and power factor 3×10^{-4} and 4.5×10^{-4} are connected first in series and then in parallel with one another. For each connection calculate the total capacitance and power factor.

9.3 A 0.1 μF capacitor has a loss angle of 2.5×10^{-4} rad. If the frequency of the sinusoidal supply voltage is $8000/2\pi$ Hz calculate (*a*) the reactance of the capacitor, (*b*) its series loss resistance and (*c*) its parallel loss resistance.

9.4 A capacitance of 0.159 μF is connected to a 20 V supply at 2 kHz. The power dissipated in the capacitor is then 200 μW. Calculate (*a*) the parallel loss resistance, (*b*) the power factor and (*c*) the Q factor.

9.5 Explain with the aid of a phasor diagram the meaning of the term loss angle when applied to a dielectric. A 12 V signal at 4 kHz is applied to a capacitor. If the reactance of the capacitor is 1200 Ω calculate (*a*) the capacitance, (*b*) the capacitor current and (*c*) if the power factor is 4×10^{-4} calculate the power dissipated and the equivalent loss resistance.

9.6 A capacitor may be represented at 1 MHz by a capacitance of 110 pF in series with a resistance of 0.05 Ω. Calculate (*a*) the power factor, (*b*) the Q factor and (*c*) the parallel loss resistance of the capacitor.

9.7 A 0.5 H inductor has a resistance of 10 Ω. Its losses at a frequency of $5000/2\pi$ Hz may be represented by a parallel resistor of 20 kΩ. Determine the effective series resistance and series inductance of the inductor. A current of 100 mA flows in this component at the same frequency. Calculate the power dissipated in the inductor.

9.8 Figure 9.8 shows a parallel plate capacitor with two dielectrics arranged as shown. Prove that the effective capacitance is given by

$$C = \varepsilon_0 \varepsilon_1 \varepsilon_2 dB / [\varepsilon_1 (l_2 - l_1) + \varepsilon_2 l_1]$$

Derive an expression for the potential across one of the dielectrics when a voltage V is applied between the plates. If $V = 3500$ V, $\varepsilon_1 = 4$, $\varepsilon_2 = 6$, $l_2 = 10$ mm and $l_1 = 4$ mm calculate the voltage gradient in each dielectric.

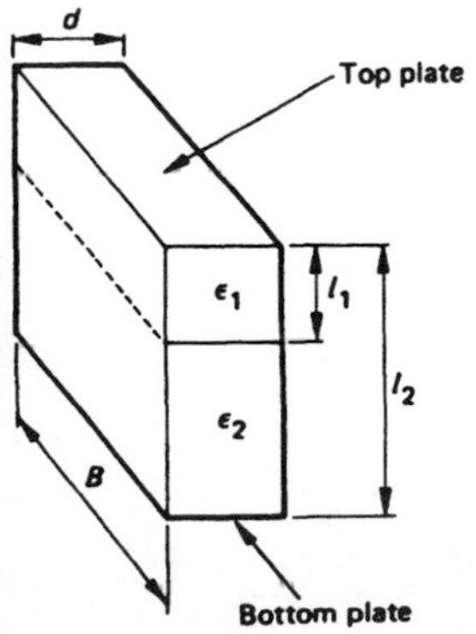

Fig. 9.8

9.9 A 100 kVA 6600/330 V single-phase transformer has a hysteresis loss of 400 W and an eddy current loss of 180 W when a voltage of 6600 V at 60 Hz is applied to the primary winding. Calculate the total iron loss when the primary voltage is 6600 V at 50 Hz. Take the Steinmetz index to be 1.6.

9.10 Explain how the iron losses of a transformer may be separated from one another. A transformer has a hysteresis loss of 220 W and an eddy current loss of 160 W at 50 Hz when a particular voltage is applied to the primary. When the voltage is reduced to 75 per cent of its initial value with the frequency unchanged at 50 Hz the total iron loss is 220 W. Calculate the Steinmetz index for the core material. Also find the total iron loss if the frequency is increased to 60 Hz and the voltage remains at 75 per cent of its initial value.

9.11 A magnetic core with a maximum flux density of 1.5 T and a Steinmetz index of 1.6 at 50 Hz has a hysteresis loss of 125 W. Calculate its hysteresis loss when the maximum flux density is 1.1 T at a frequency of 25 Hz.

9.12 The maximum flux density in a magnetic circuit is 12 mwb at a frequency of 50 Hz. The cross-sectional area of the circuit is 100 cm^2 and the Steinmetz index is 1.6. If the hysteresis loss of the circuit is 120 W determine the loss at 100 Hz when the maximum flux is reduced to 10 mwb.

9.13 An inductor core has a hysteresis loss of 28 W and an eddy current loss of 18 W at 50 Hz. Calculate the total core loss at 100 W.

9.14 The total loss in a magnetic core is 40 W at 40 Hz and 60 W at 50 Hz. Calculate (*a*) the hysteresis loss and (*b*) the eddy current loss at 50 Hz.

10 Coupled circuits

Two coupled circuits are magnetically linked to one another by mutual inductance. When the current flowing in one of the circuits changes it sets up a changing magnetic field. Some of this changing magnetic field links with the other circuit and induces an e.m.f. into it. If the other circuit is closed in an impedance a current will flow in that circuit, this will set up another magnetic field that will link with the first circuit and induce an e.m.f. into it.

Mutual Inductance

Fig. 10.1 Magnetic flux linking two coils of wire

Consider Fig. 10.1 which shows a coil of wire connected across an alternating voltage source having an e.m.f. of E_p volts. A current I_p flows in the *primary winding* an this sets up a magnetic flux around the winding. As shown by the dotted lines, *some* of the magnetic flux links with the turns of the other (secondary) coil. As the current in the primary changes, the number of flux linkages changes also and consequently an e.m.f. is induced into the secondary winding.

The magnitude of the secondary e.m.f. is given by

$$e = M(\mathrm{d}I_p/\mathrm{d}t) \tag{10.1}$$

where M is the *mutual inductance* between the two windings.

The two coils are said to possess a *mutual inductance* of 1 henry (H) when the primary current changing at the rate of 1 A/s induces a voltage of 1 V into the secondary winding.

When the magnetic flux set up by the primary current changes by amount $\mathrm{d}\varphi$ in a time $\mathrm{d}t$ it induces an e.m.f. in each turn of the secondary winding equal to $-\mathrm{d}\varphi/dt$ volts. If the secondary winding has N_s turns the induced voltage is

$$E_s = -N_s\, \mathrm{d}\varphi/\mathrm{d}t \text{ V} \tag{10.2}$$

$$E_s = -(N_p\, \mathrm{d}\varphi/\mathrm{d}I_p)(\mathrm{d}I_p/\mathrm{d}t)$$

$$= -M\, \mathrm{d}I_p/\mathrm{d}t, \text{ where } M = N_s\, \mathrm{d}\varphi/\mathrm{d}I_p \tag{10.3}$$

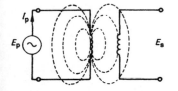

The mutual inductance between two coils is the same whether the flux is set up by the primary current or by the secondary current. Hence,

$$M = N_p \, d\varphi/dI_s \tag{10.4}$$

Now, $N \, d\varphi/dt = L \, dI/dt$, or $L = N \, d\varphi/dI$ and

$$L_p = N_p\varphi_p/I_p \quad \text{or} \quad \varphi_p = L_pI_p/N_p \tag{10.5}$$

When the primary current is of sinusoidal waveform, i.e. $i = I_p \sin \omega t$, the rate of change of the primary current is

$$di_p/dt = \omega I_p \cos \omega t = \omega I_p \sin(\omega t + 90°) = j\omega I_p \tag{10.6}$$

Then, from equation (10.1)

$$E_s = j\omega M I_p \tag{10.7}$$

Dot notation

The polarity of the voltage induced into another coil depends upon the way in which the two coils are wound and mounted. Dots may be drawn alongside the windings to identify the terminals of equivalent polarity for two mutual inductance coupled coils. The dots shown in Fig 10.2(*a*) and (*b*) indicate the terminals that have the same polarity. Thus, the coil terminals marked with a dot are both positive, and then both negative, at the same time.

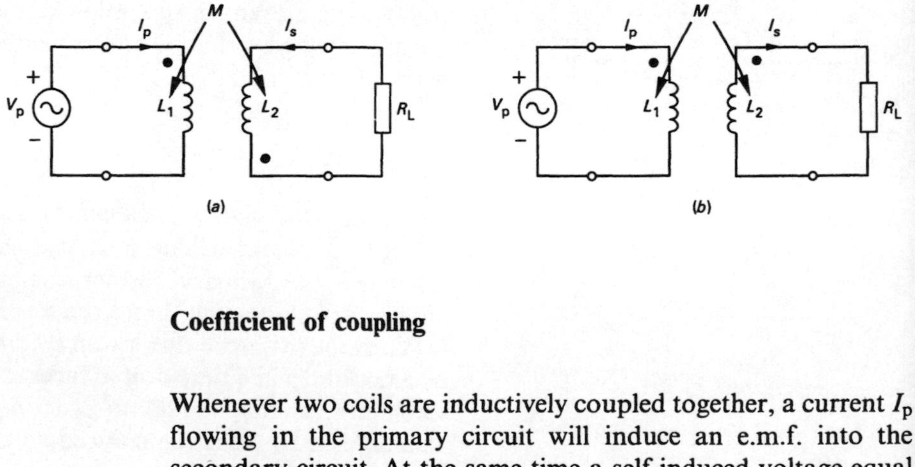

Fig. 10.2 Dot notation

Coefficient of coupling

Whenever two coils are inductively coupled together, a current I_p flowing in the primary circuit will induce an e.m.f. into the secondary circuit. At the same time a self-induced voltage equal to $L \, dI_p/dt$ will also be induced into the primary circuit. The ratio of the two induced voltages is

$$V_s/V_p = M/L_p \tag{10.8}$$

If the coupling between the coils is perfect then

$$V_s = N_s \, d\Phi/dt \quad \text{and} \quad V_p = N_p \, d\Phi/dt$$

where N_p and N_s are the number of primary and secondary turns respectively, and Φ is the common flux set up by the primary current. The inductance of a winding is proportional to the *square* of the number of turns so that $L_p/L_s = (N_p/N_s)^2$. Hence,

$$V_s/V_p = N_s/N_p = \sqrt{(L_s/L_p)}$$

and combining with equation (10.8) gives

$$M/L_p = \sqrt{(L_s/L_p)} \quad \text{or} \quad M = \sqrt{(L_p/L_s)}$$

For all practical cases, however, 100 per cent coupling between the windings is never achieved and this fact is taken into account by the introduction of a multiplying factor k known as the *coefficient of coupling*. Therefore,

$$M = k\sqrt{(L_p L_s)} \tag{10.9}$$

The value of k may be anywhere in between 0 (for zero coupling) and 1 (for perfect coupling).

When the coupling is *tight*, i.e. k is near unity, a magnetic core has been employed to increase the mutual inductance between two coils. Iron-cored transformers employ tight coupling and are discussed in the following chapter. When the coupling between two coils is *loose*, i.e. k is very small, the two coils are generally known as coupled circuits.

Coupled circuits

Open-circuit secondary

The voltage induced into the secondary winding effectively appears in *series* with the secondary winding. Because the flux set up by the primary current also links with the primary winding, there will also be an e.m.f. equal to $j\omega L_p I_p$ induced into the primary winding, where L_p is the inductance of the primary winding.

Figure 10.3 shows two coils, having self-inductances L_p and L_s, which are inductively coupled together by a mutual inductance M. The primary winding has a voltage source of e.m.f. E_p volts connected across its terminals. The internal resistance of the source has been included in the primary resistance R_p. The secondary winding is left open circuited and has a resistance of R_s.

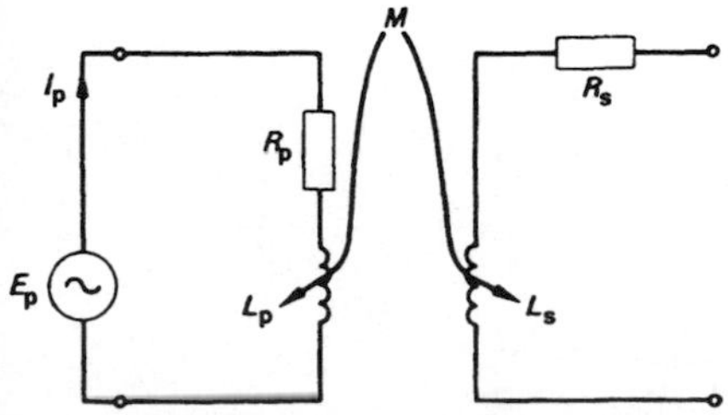

Fig. 10.3 Two coils coupled together by mutual inductance

Applying Kirchhoff's voltage law to the primary circuit gives

$$E_p = I_p R_p + L_p \, dI_p/dt$$

If E_p and I_p are *both* of sinusoidal waveshape this expression can be written as $E_p = I_p(R_p + j\omega L_p)$. Therefore, $I_p = E_p/(R_p + j\omega L_p)$ and, from equation (10.3),

$$E_s = \pm j\omega M E_p/(R_p + j\omega L_p) \tag{10.10}$$

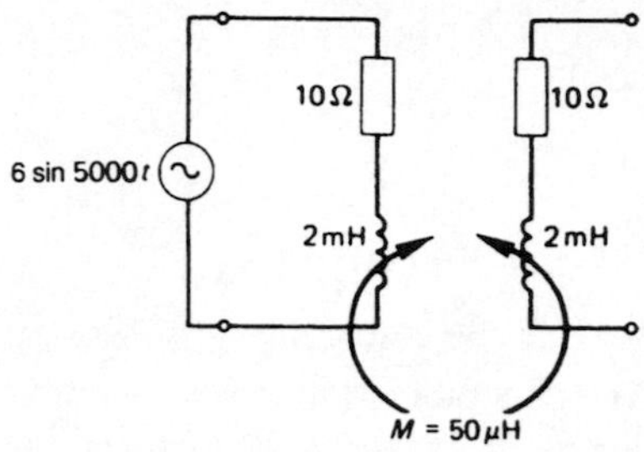

Fig.10.4

Example 10.1

Calculate the voltage that appears across the open-circuited secondary winding of the circuit given in Fig. 10.4.

Solution

The impedance of the primary circuit is

$$Z_p = 10 + j5000 \times 2 \times 10^{-3} = (10 + j10)\ \Omega$$

Hence, $I_p = E_p/Z_p = 6/(10 + j10)$ and so the secondary voltage is

$$E_s = \pm j\omega M I_p = [\pm j5000 \times 50 \times 10^{-6} \times 6]/(10 + j10)$$
$$= (1.5\angle \pm 90°)/(14.14\angle 45°)$$
$$E_s = 106\angle 45°\ \text{mV} \quad \text{or} \quad E_s = 106\angle -135°\ \text{mV} \quad (\textit{Ans.})$$

Untuned secondary circuit

When the secondary terminals are closed in an impedance so that a secondary current flows, the situation is rather more complex. Consider Fig. 10.5 which is similar to Fig. 10.3 except that the secondary terminals are terminated by a load resistance R_L. The secondary circuit resistances R_s' and R_L can be combined to form a total resistance R_s. When now an e.m.f. is induced into the secondary winding, a current I_s will flow and this will *induce an e.m.f. into the primary winding*. This e.m.f. must be taken into account when determining the primary current.

Applying Kirchhoff's voltage law to both the primary and the secondary circuits of Fig. 10.5. gives

$$E_p = I_p(R_p + j\omega L_p) \pm j\omega M I_s \tag{10.11}$$

and $$0 = I_s(R_s + j\omega L_s) \pm j\omega M I_p \tag{10.12}$$

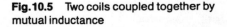

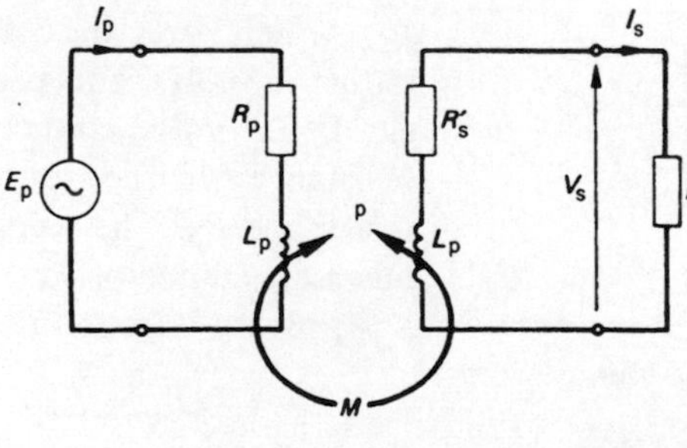

Fig.10.5 Two coils coupled together by mutual inductance

From equation (10.12)

$$I_s = \mp j\omega M I_p/(R_s + j\omega L_s)$$

and substituting for I_s into equation (10.11) gives

$$E_p = I_p[R_p + j\omega L_p + (\omega M)^2/(R_s + j\omega L_s)]$$

Hence,

$$E_p = I_p\{R_p + j\omega L_p + [\omega^2 M^2(R_s - j\omega L_s)/(R_s^2 + \omega^2 L_s^2)]\}$$

The *effective* primary impedance $Z_{p(eff)}$ of the circuit is

$$Z_{p(eff)} = \frac{E_p}{I_p} = R_p + \frac{\omega^2 M^2 R_s}{R_s^2 + \omega^2 L_s^2} + j\left(\omega L_p - \frac{\omega^3 M^2 L_s}{R_s^2 + \omega^2 L_s^2}\right) \quad (10.13)$$

The effective primary resistance $R_{p(eff)}$, of the primary winding is the real part of equation (10.13) and it is always greater than the resistance R_p of the primary on its own. The effective reactance $X_{p(eff)}$ of the primary circuit is the imaginary part of equation (10.13) and it is smaller than the reactance of the primary on its own.

The effect of a coupled secondary circuit is to add an impedance $(\omega M)^2/Z_s$ in series with the primary impedance. Inductive reactance in the secondary circuit appears as capacitive reactance in the primary circuit and vice versa. Once the effective primary impedance is known the primary current I_p is given by $I_p = E_p/Z_{p(eff)}$. Lastly, the voltage induced into the secondary winding is $E_s = j\omega M I_p$ volts. Clearly, the coupled impedance is very small whenever the mutual inductance is small and this means the primary current is almost the same as it would be with the secondary winding open circuited.

The effective Q factor of the primary circuit is the ratio $X_{p(eff)}/R_{p(eff)}$ and it will always be smaller than the Q factor of the uncoupled primary circuit.

Example 10.2

For the circuit shown in Fig. 10.6 calculate the current in the secondary winding.

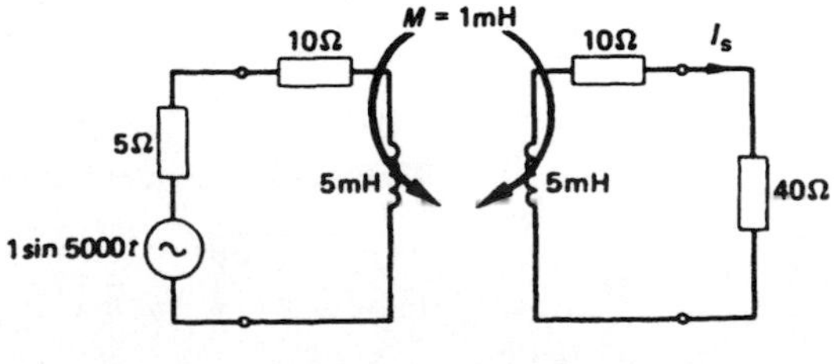

Fig. 10.6

Solution

$$\omega L_p = \omega L_s = 5000 \times 5 \times 10^{-3} = 25\ \Omega \quad \text{and} \quad \omega M = 5\ \Omega$$

From equation (10.13)

$$R_{p(eff)} = 15 + (25 \times 50)/(50^2 + 25^2) = 15.4\ \Omega$$
$$X_{p(eff)} = 25 - (25 \times 25)/(50^2 + 25^2) = 24.8\ \Omega$$

Hence

$$I_p = 1/(15.4 + j24.8) \quad \text{and} \quad E_s = j5/(15.4 + j24.8)$$
$$I_s = E_s/Z_s = j5/[(15.4 + j24.8)(50 + j25)]$$
$$= (5\angle 90°)/[(29.19\angle 58.2°) \times (55.9\angle 26.6°)]$$

or $I = 3.06\angle 5.2°$ mA (*Ans.*)

The expression for the output impedance of a coupled circuit is of the same form as that for the input impedance. Thus:

$$\text{Output impedance} = \text{Secondary winding impedance} + \frac{(\omega M)^2}{\text{Total primary impedance}} \qquad (10.14)$$

Example 10.3

A sinusoidal voltage source of e.m.f. 10 V and internal resistance 100 Ω is connected across the primary winding of an air-cored transformer. The primary winding has $L_p = 0.1$ H and $R_p = 100\ \Omega$ and the secondary winding has $L_s = 0.05$ H and $R_s = 200\ \Omega$. If the mutual inductance M between the windings is 0.06 H calculate (*a*) the voltage across the open-circuited secondary terminals, (*b*) the output impedance of the circuit and (*c*) the current that flows in the secondary circuit when its terminals are closed by a 100 Ω resistor. The frequency is $5000/2\pi$ Hz.

Solution

$$X_{Lp} = 5000 \times 0.1 = 500\ \Omega,\ X_{Ls} = 5000 \times 0.05 = 250\ \Omega$$
and $X_M = 5000 \times 0.06 = 300\ \Omega$

(*a*) $I_p = 10/(100 + 100 + j500) = 1/(20 + j50) = (6.9 - j17.2) \times 10^{-3}$
$= 18.53\angle -68.1°$ mA

$V_s = 300\angle 90° \times 18.53 \times 10^{-3}\angle -68.1° = 5.56\angle 21.9°$ V (*Ans.*)

(*b*) Output impedance $= 200 + j250 + 300^2/(200 + j500) = 262 + j95\ \Omega$
$= 279\angle 19.9°\ \Omega$ (*Ans.*)

(*c*) $I_s = (5.56\angle 21.9°)/(100 + 262 + j95)$
$= (5.56\angle 21.9°)/(374\angle -14.7°) = 14.9\angle 36.6°$ mA (*Ans.*)

Resonant secondary circuit

The secondary circuit of a pair of coupled coils may be closed by a capacitor when the total secondary reactance is

$jX_s = j(\omega L_s - 1/\omega C_s)$. At some particular frequency the secondary circuit will be resonant. When the secondary circuit is resonant its total reactance is zero and hence the coupled impedance (see equation (10.13)) is a pure resistance equal to $\omega^2 M^2 R_s / R_s^2 = (\omega M)^2 / R_s$. For the maximum transfer of power to the secondary circuit the magnitude of the primary resistance should be equal to the coupled resistance. Thus:

$$\sqrt{(R_P^2 + \omega^2 L_p^2)} = \omega^2 M^2 / R_s \qquad (10.15)$$

Example 10.4

Two coils have a mutual inductance of 10 μH. One coil has a resistance of 10 Ω and an inductance of 15.9 μH and it has the voltage 10 mV at 5/2π MHz applied to it. The other coil has a resistance of 10 Ω and an inductance of 159 μH and connected across it is a variable capacitor *C*. The capacitor is adjusted to resonate the secondary circuit. Calculate (*a*) the input impedance of the primary circuit, (*b*) the current flowing in the secondary winding, (c) the voltage across the capacitor and (*d*) the value of *C*.

Solution

$X_{Lp} = 5 \times 15.9 = 79.5\ \Omega$. $X_{Ls} = 5 \times 159 = 795\ \Omega$, so $X_c = 795\ \Omega$ also. $X_M = 5 \times 10 = 50\ \Omega$.

(*a*) $Z_p = 10 + j79.5 + 50^2/10 = 260 + j79.5 = 271.9\angle 17°\ \Omega$ (*Ans.*)

(*b*) $I_p = (10 \times 10^{-3})/271.9\angle 17°) = 36.8\angle -17°\ \mu A$

$E_s = j50 \times 36.8 \times 10^{-6}\angle -17° = 1.84\angle 73°\ mV$

$I_s = (1.84 \times 10^{-3}\angle 73°)/10 = 184\angle 73°\ \mu A$ (*Ans.*)

(*c*) $V_C = I_s X_C = (184 \times 10^{-6}\angle 73°) \times (795\angle -90°)$
$= 0.15\angle -17°\ V$ (*Ans.*)

(*d*) $C = 1/(795 \times 5 \times 10^6) \approx 252\ pF$ (*Ans.*)

Primary and secondary circuits resonant

Very often both the primary and the secondary circuit may be caused to resonate at particular frequencies – not necessarily equal to one another – by means of tuning capacitors. The tuning capacitors can be connected either in series with, or in parallel with, the two windings. Figure 10.7 shows two windings each tuned by series-connected capacitors C_p and C_s. The analysis of the circuit is very similar to that already given for the circuit of Fig. 10.6 except that now the total reactances of the two windings are

$$jX_p = [j(\omega L_p - (1/\omega C_p)] \quad \text{and} \quad jX_s = j[\omega L_s - (1/\omega C_s)]$$

respectively.

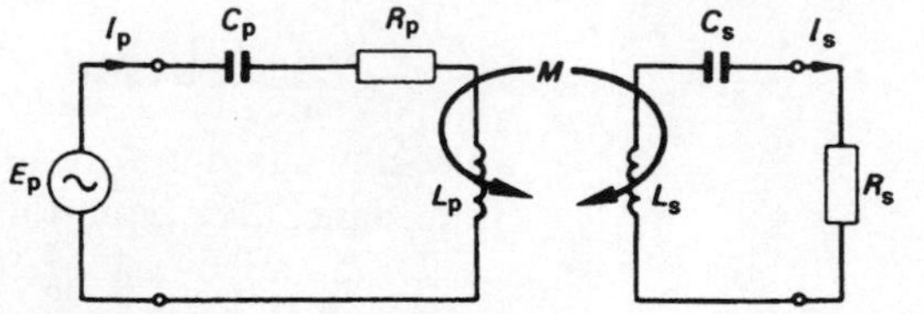

Fig. 10.7 Two series tuned circuits coupled together by mutual inductance

When both the primary and the secondary circuits are at resonance their total reactances cancel out and the coupled impedance is purely resistive. For maximum power transfer to the secondary circuit

$$R_p = \omega^2 M^2/R_s, \quad \text{or} \quad M = \sqrt{(R_p R_s)}/\omega \tag{10.16}$$

The coupling coefficient may be expressed in terms of the Q factors of the two windings. Thus,

$$k = M/\sqrt{(L_p L_s)} = M/\sqrt{[(Q_p R_p/\omega_p)(Q_s R_s/\omega_s)]}$$

If $\omega_p = \omega_s = \omega$, then

$$k = \omega M/\sqrt{(Q_p Q_s R_p R_s)} \tag{10.17}$$

Example 10.5

In the circuit of Fig. 10.8 the two windings are tuned to resonate at the same frequency. Calculate (a) the resonant frequency, (b) the value of C_s, (c) the voltage appearing across C_s and (d) the coefficient of coupling.

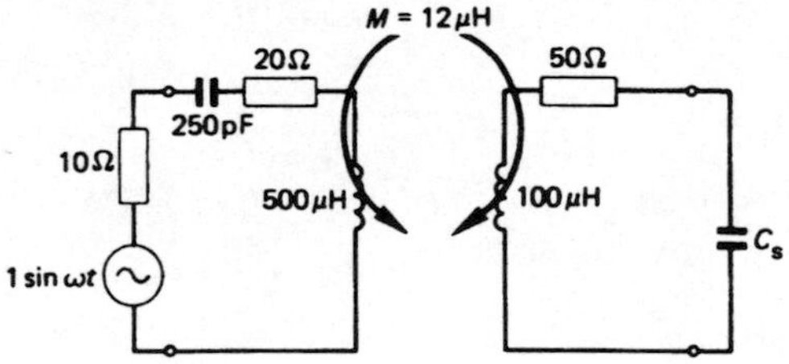

Fig. 10.8

Solution

(a) $f_0 = 1[2\pi\sqrt{(500 \times 10^{-6} \times 250 \times 10^{-12})}] = 450.16$ kHz (*Ans.*)

(b) $f_0 = 1/[2\pi\sqrt{(C_p L_p)}] = 1/[2\pi\sqrt{(L_s C_s)}]$ so that $C_p L_p = C_s L_s$

$$C_s = C_p L_p/L_s = 250 \times 500/100 = 1.25 \text{ nF} \quad (\textit{Ans.})$$

(c) Since both the primary and the secondary circuits are resonant, the effective primary impedance of the circuit is

$$Z_{p(eff)} = R_{p(eff)} = 30 + [(2\pi \times 450.16 \times 10^3)^2 \times (12 \times 10^{-6})^2]/50$$
$$= 53\ \Omega$$

$$I_p = E_p/Z_{p(eff)} = 10/53 = 0.189 \text{ A}$$

$$E_s = \omega M I_p = 2\pi \times 450.16 \times 10^3 \times 12 \times 10^{-6} \times 0.189 = 6.4 \text{ V}$$

$$I_s = E_s/Z_s = 6.4/50 = 0.128 \text{ A}$$

Therefore,

$$V_{Cs} = I_s/\omega C_s = 0.128/[2\pi \times 450.16 \times 10^3 \times 1.25 \times 10^{-9}]$$
$$= 36.2 \text{ V} \quad (Ans.)$$

Alternatively,

$$Q_s = \omega L_s/R_s = [2\pi \times 450.16 \times 10^3 \times 100 \times 10^{-6}]/50 = 5.66$$

and $V_{Cs} = QE_s = 5.66 \times 6.4 = 36.2$ V (*Ans.*)

(*d*) $k = M/[\surd(L_pL_s)] = 12/\surd(500 \times 100) = 0.054$ (*Ans.*)

Secondary current–frequency characteristic

If the frequency of the voltage applied to the primary of two coupled circuits is varied either side of the resonant frequency, the shape of the *secondary current–frequency characteristic* will depend upon the value of the coupling coefficient.

When the coupling coefficient is very small, the secondary current will also be small and the voltage induced into the primary winding by the secondary current is so minute that it has little effect upon the primary current. The primary current–frequency characteristic is then determined solely by the selectivity of the primary circuit. This is shown by Figs 10.9(*a*) and (*b*) by the curves marked *loose coupling*.

When the coupling between the windings is increased, the e.m.f. induced into the secondary winding by the primary current will also increase. This produces an increase in the secondary current, particularly in the neighbourhood of the resonant frequency. The e.m.f. induced into the primary circuit by the secondary current will be larger and so tends to oppose the flow of the primary current. The primary current at, and near, resonance is therefore reduced and the primary current–frequency characteristic develops two peaks as shown by Fig. 10.9(*a*).

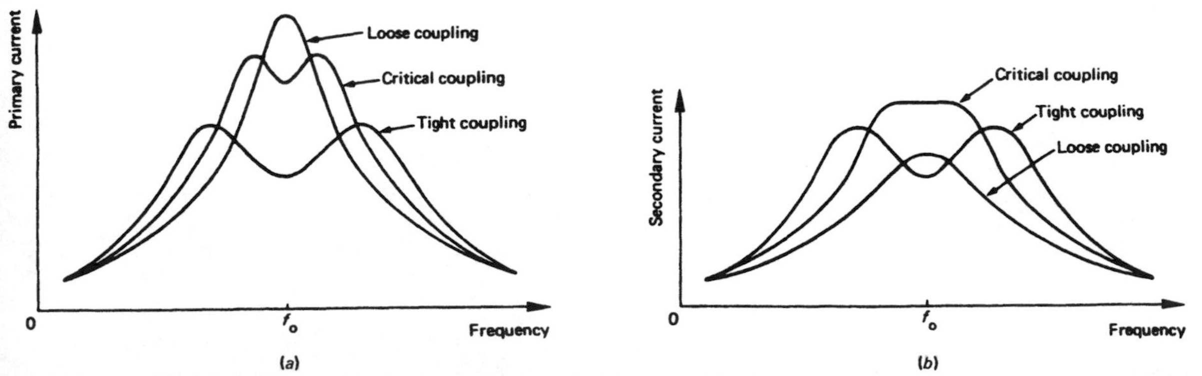

Fig. 10.9 Current–frequency curves for two mutual-inductance coupled coils: (a) primary current; (b) secondary current

If the coupling between the two windings is still further increased, the secondary current will rise with the result that the trough in the primary current response deepens. When the coupling coefficient has been increased to its *critical value* k_{crit} the secondary current has reached its maximum possible value and at this point its frequency characteristic exhibits a more or less flat top. This is shown in Fig. 10.9(*b*) by the curve marked as *critical coupling*.

If, now, the coupling coefficient is still further increased to become greater than the critical value k_{crit} the secondary current–frequency characteristic will also develop two peaks as shown by Fig. 10.9(*b*). Any further increase in the coupling coefficient will not produce a corresponding increase in the secondary current but will only cause the twin peaks to move even further apart from one another. The two peaks occur at frequencies symmetrically spaced either side of the resonant frequency f_0 with a spacing equal to kf_0 Hertz.

When both the primary and the secondary circuits are resonant, the effective primary resistance $R_{p(eff)}$ is equal to $R_p + (\omega_0^2 M^2/R_s)$. If the coupling is critical, the coupled resistance is equal to the primary resistance proper. Thus

$$R_p = \omega_0^2 M^2/R_s \tag{10.18}$$

Substituting equation (10.18) into (10.17) gives

$$k_{crit} = (\omega_0 M)/\sqrt{(Q_p Q_s \omega_0^2 M^2)} = 1/\sqrt{(Q_p Q_s)} \tag{10.19}$$

If $Q_p = Q_s$, then

$$k_{crit} = 1/Q \tag{10.20}$$

The 3 dB bandwidth of the secondary current of two *identical critically coupled* resonant circuits is given by

$$B_{3dB} = (f_0\sqrt{2})/Q \tag{10.21}$$

Example 10.6

Two 300 μH coils are inductively coupled together and are each tuned to resonate at 300 kHz by series-connected capacitors. The 3 dB bandwidth of the secondary current is 10 kHz. Calculate (*a*) the critical coefficient of coupling, (*b*) the critical value of the mutual inductance, (*c*) the capacitor value, (*d*) the separation between the secondary current peaks when the coupling coefficient is increased to 125 per cent of its critical value.

Solution

(*a*) From equation (10.21), $Q = \sqrt{2} \times 300/10 = 42.4$ and therefore

$$k_{crit} = 1/Q = 0.024 \quad (\textit{Ans.})$$

(*b*) $M = 0.024 \times 300 = 7.2\ \mu H$ (*Ans.*)
(*c*) $C_p = C_s = 1/(4\pi^2 \times 300^2 \times 10^6 \times 300 \times 10^{-6}) = 938$ pF (*Ans.*)
(*d*) The coupling coefficient is $1.25 \times 0.024 = 0.03$ and therefore

$$\text{Peak separation} = 0.03 \times 300 \times 10^3 = 9 \text{ kHz} \quad (Ans.)$$

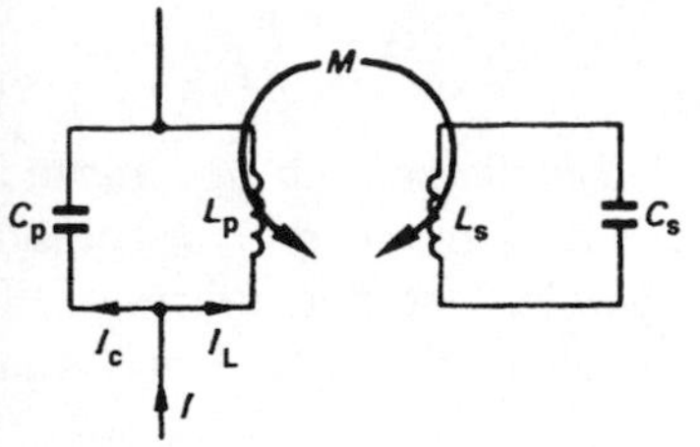

Fig.10.10. Air-cored transformer with primary tuned by a parallel-connected capacitor

Very often the primary winding has its tuning capacitor connected in parallel as shown by Fig. 10.10. The current I_L flowing in the primary inductance will be equal to Q_pI. Alternatively, Thevenin's theorem can be used to convert the primary circuit into the equivalent series circuit. If $1/\omega C_p \ll$ (the source resistance), the series circuit previously discussed is obtained.

Probably the main application of air-cored transformers lies in radio receivers where they are often used in r.f. amplifiers.

Equivalent T circuit

The T equivalent circuit of two mutual inductance coupled circuits can be obtained by the use of *z* parameters (p. 245). Figure 10.11(*a*) shows two circuits coupled by mutual inductance *M* and Fig. 12.11(*b*) shows a T network with impedances Z_1, Z_2 and Z_3. The z parameters of the two circuits are:

	Transformer	T network
$Z_i = V_p/I_p$ with $I_s = 0$	$= R_p + j\omega L_p$	$= Z_1 + Z_3$
$Z_f = V_s/I_p$ with $I_s = 0$	$= j\omega M$	$= Z_3$
$Z_r = V_p/I_s$ with $I_p = 0$	$= j\omega M$	$= Z_3$
$Z_o = V_s/I_s$ with $I_p = 0$	$= R_s + j\omega L_s$	$= Z_2 + Z_3$

Equating the transformer and T parameters: $Z_1 = Z_i - Z_r = R_p + j\omega(L_p - M)$, $Z_2 = Z_o - Z_r = R_s + j\omega(L_s - M)$ and $Z_3 = Z_f = Z_r = j\omega M$. The equivalent T network is shown in Fig 10.11(*c*).

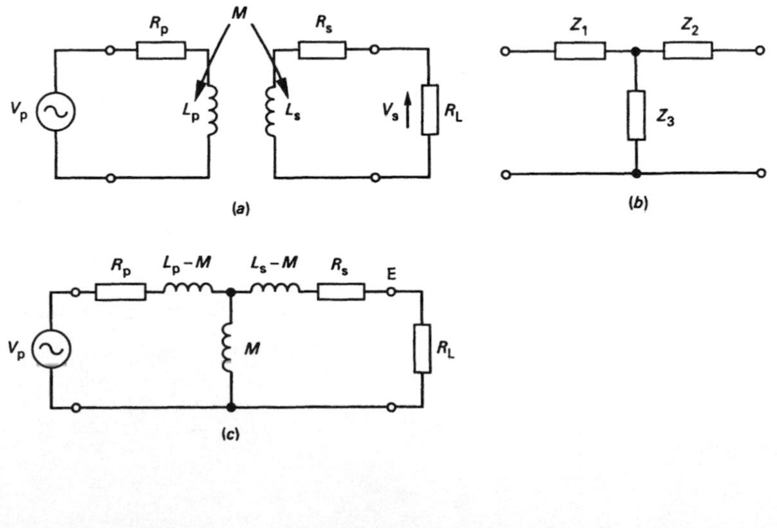

Fig.10.11 Equivalent T circuit of two mutual inductances

Circuits employing mutual inductance

The use of mutual inductance is not restricted to the coupling together of two electrically isolated circuits.

Inductors in series

Two inductors L_1 and L_2 connected in series with one another will have some mutual inductance if they are sited close enough together and this inductive coupling will affect their effective self-inductance. If a voltage V is applied across the inductors so that a current i flows then,

$$V = L_1 \, di/dt \pm M \, di/dt + L_2 \, di/dt \pm M \, di/dt$$
$$= (L_1 + L_2 + 2M) \, di/dt$$

The equivalent inductance is

$$L_s = L_1 + L_2 \pm 2M \qquad (10.22)$$

The term $2M$ takes into account the flux linkages in each inductor due to the current flowing in the other inductor. The mutual inductance term may add to, or subtract from, the individual self-inductances depending upon the sense of the windings.

Inductors in parallel

When two inductors of self inductances L_1 and L_2 are connected in parallel their effective self-inductance depends upon both their individual self-inductances and their mutual inductance. When a voltage V is applied across the paralleled inductors a current i_1 flows into L_1 and a current i_2 flows into L_2. Since $i = i_1 + i_2$ then

$$V = L_1 \, di_1/dt \pm M \, di_2/dt = L_2 \, di_2/dt \pm M \, di_1/dt$$
$$(L_1 \mp M) di_1/dt = (L_2 \mp M) di_2/dt,$$
$$di_2/dt = [(L_1 \mp M)/(L_2 \mp M)](di_1/dt)$$

di_2/dt is also equal to $di/dt - di_1/dt$ and hence

$$di/dt = [(L_1 \mp M + L_2 \mp M)/(L_2 \mp M](di_1/dt),$$
$$di_1/dt = [(L_2 \mp M)/(L_1 + L_2 \mp 2M)](di/dt)$$

Also,

$$di_2/dt = [(L_1 \mp M)/(L_1 + L_2 \mp 2M)](di/dt)$$
$$V = \frac{L_1(L_2 \mp M)}{L_1 + L_2 \mp 2M}\left(\frac{di}{dt}\right) \pm \frac{M(L_1 \mp M)}{L_1 + L_2 \mp 2M}\left(\frac{di}{dt}\right)$$
$$= [(L_1 L_2 - M^2)/(L_1 + L_2 \mp 2M)](di/dt)$$

Therefore, the equivalent inductance is

$$L_p = (L_1 L_2 - M^2)/(L_1 + L_2 \mp 2M) \qquad (10.23)$$

Example 10.7

Two inductors, of 1 H and 0.64 H respectively, are mutually coupled with a coupling coefficient of 0.5. Calculate their effective inductance when connected (*a*) in series assisting and (*b*) in parallel assisting.

Solution

(*a*) $L_s = 1 + 0.64 + (2 \times 0.5)\sqrt{(1 \times 0.64)} = 2.44$ H (*Ans.*)

(*b*) $L_p = \dfrac{1 \times 0.64 - (0.5\sqrt{(1 \times 0.64)})^2}{1 + 0.64 - (2 \times 0.5)\sqrt{(1 \times 0.64)}} = 0.57$ H (*Ans.*)

If the resistances of the two inductors are not negligibly small the impedance of the circuit can be determined in a similar manner.

$$V = (R_1 + j\omega L_1)i_1 \pm j\omega M i_2 = (R_2 + j\omega L_2)i_2 \pm j\omega M i_1$$

$$[R_1 + j\omega(L_1 \mp M)]i_1 = [R_2 + j\omega(L_2 \mp M)]i_2, \text{ or}$$

$$i_2 = [R_1 + j\omega(L_1 \mp M)]/[R_2 + j\omega(L_2 \mp M)]i_1.$$

$$\begin{aligned} i_1 + i_2 &= i_1\left[1 + \frac{R_1 + j\omega(L_1 \mp M)}{R_2 + j\omega(L_2 \mp M)}\right] \\ &= i_1\left[\frac{R_1 + R_2 + j\omega(L_2 \mp M) + j\omega(L_1 \mp M)}{R_2 + j\omega(L_2 \mp M)}\right] \\ &= i_1\left[\frac{R_1 + R_2 + j\omega(L_1 + L_2 \mp 2M)}{R_2 + j\omega(L_2 \mp M)}\right] \end{aligned}$$

$$Z = \frac{V}{i} = \frac{R_1 + j\omega L_1 \pm j\omega M\left[\dfrac{R_1 + j\omega(L_1 \mp M)}{R_2 + j\omega(L_2 \mp M)}\right]}{\dfrac{R_1 + R_2 + j\omega(L_1 + L_2 \mp 2M)}{R_2 + j\omega(L_2 \mp M)}}$$

$$= \frac{R_1R_2 - \omega^2(L_1L_2 - M^2) + j\omega(L_1R_2 + L_2R_1)}{R_1 + R_2 + j\omega(L_1 + L_2 \mp 2M)} \qquad (10.24)$$

Note that if $R_1 = R_2 = 0$, equation (10.24) reduces to the expression given in equation (10.23).

Example 10.8

For the circuit given in Fig. 10.12 calculate the voltage across the 510 Ω resistor.

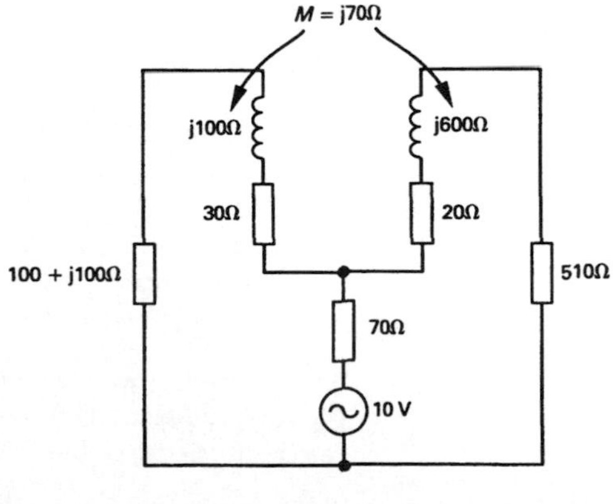

Fig. 10.12

Solution

The circuit is re-drawn in Fig. 10.13. From this:

$$10 = (70 + j70)(i_1 + i_2) + (30 + j30)i_1 + (100 + j100)i_1$$
$$= i_1(200 + j200) + i_2(70 + j70) \qquad (10.25)$$
$$10 = (70 + j70)(i_1 + i_2) + (20 + j530)i_2 + 510i_2$$
$$= i_1(70 + j70) + i_2(600 + j600) \qquad (10.26)$$

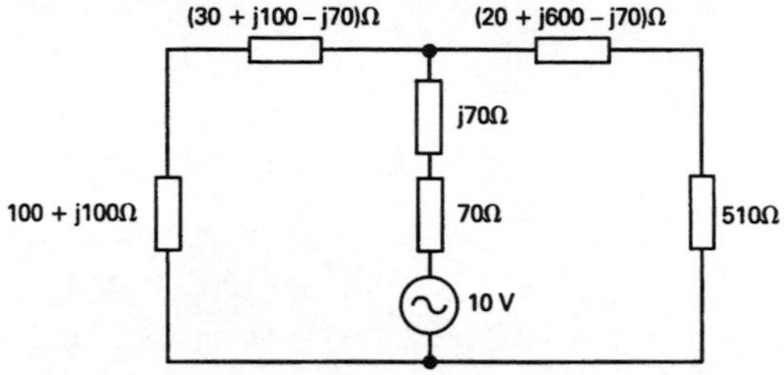

Fig. 10.13

From equation (10.25), $i_1 = [10 - (70 + j70)i_2]/(200 + j200)$. Substitute into equation (10.26) to get:

$$10 = (70 + j70)[10 - (70 + j70)i_2]/(200 + j200) + i_2(600 + j600)$$
$$= 0.35[10 - (70 + j70)i_2] + i_2(600 + j600)$$
$$6.5 = i_2(575.5 + j575.5) \quad \text{or} \quad i_2 = 5.647 - j5.647 \text{ mA}$$

$$V_{510} = 510i_2 = 2.88 - j2.88 = 4\angle{-45°} \text{ V} \quad (Ans.)$$

Exercises 10

10.1 The secondary and the primary circuits of Fig. 10.14 have equal inductances and are tuned to resonance at a frequency of $500\,000/2\pi$ Hz. When an e.m.f. E_p at this frequency is applied to the primary, currents of 100 mA and 50 mA flow in the primary and secondary circuits respectively. Calculate (*a*) the inductances L_p and L_s, (*b*) the capacitances C_p and C_s, (*c*) the mutual inductance M, (*d*) the coupling coefficient and (*e*) the critical value of the coupling coefficient.

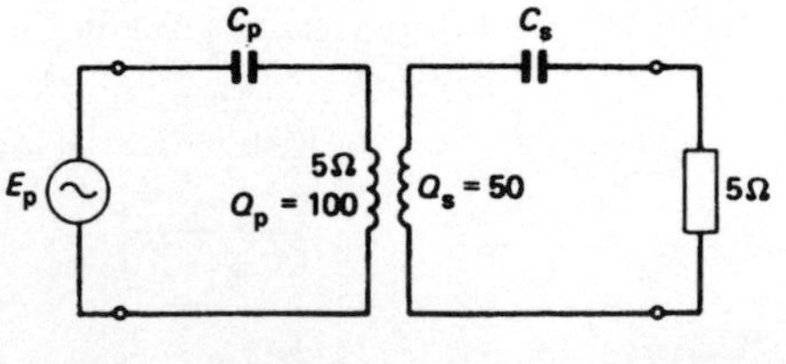

Fig. 10.14

10.2 The circuit of Fig. 10.7 has the following values: $R_p = R_s = 10\ \Omega$, $M = 10\ \mu$H and $k = 0.01$. Also $L_p = L_s$ and $C_p = C_s = 2.533$ nF. Calculate the resonant frequency of each winding, the effective primary impedance and the secondary current.

10.3 An air-cored transformer has primary circuit values $L_p = 160\ \mu$H, $Q_p = 80$ and secondary values $L_s = 125\ \mu$H and $Q_s = 40$. The coupling coefficient between the windings is 0.20. A 1 V signal at the resonant frequency of 85 kHz is applied to the primary winding. Calculate the primary current if the secondary is short-circuit. Calculate also the effective Q factor of the primary winding.

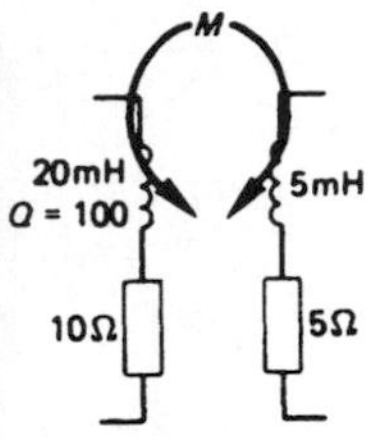

Fig. 10.15

10.4 For the circuit given in Fig. 10.15 calculate the values of the series-connected capacitors required to resonate each winding at the same frequency. If the coupling between the windings is critical calculate the effective primary resistance at the resonant frequency.

10.5 For the circuit of Fig. 10.7, $Q_p = 100$, $Q_s = 50$, $M = 10\ \mu$H and $C_p = C_s = 1$ nF. If a voltage of 1 V at the resonant frequency of $10^6/2\pi$ Hz is applied to the primary calculate the voltage that appears across the secondary capacitor C_s.

10.6 Two identical tuned circuits each having $L = 300\ \mu$H, $R = 100\ \Omega$ and $C = 1.5$ nF are coupled together by a mutual inductance of 50 μH. An e.m.f. of 12 V at the resonant frequency is applied to the primary winding. If the secondary is short circuited calculate (*a*) the primary current, (*b*) the coupling coefficient and (*c*) the secondary current.

10.7 Two coils of 0.4 H and 0.1 H inductance are connected in series on a common core. When the current flowing is 5 A and the stored energy is 2.25 J. Calculate the mutual inductance between the coils.

10.8 Two coils, of 9 mH and 4 mH inductance respectively, have a mutual inductance of 4.5 mH. The 9 mH coil has a resistance of 4 Ω and the 9 mH coil has 3 Ω resistance. A voltage source of e.m.f. 5 mV at $500/\pi$ radians is applied to the 4 mH coil. The terminal of the 9 mH coil are closed in a resistance of 5 Ω. Calculate (*a*) the secondary current, (*b*) the primary current and (*c*) the power dissipated in the 5 Ω resistor.

10.9 Two inductors of 100 mH and 50 mH are coupled together by a mutual inductance of 60 mH. The total primary resistance is 20 Ω and the total secondary resistance is 10 Ω. The secondary terminals are closed by a 100 nF capacitor. Calculate the frequency at which the ratio (capacitor voltage)/(applied primary voltage) has its maximum value.

10.10 An air-cored transformer has identical primary and secondary circuits inductively coupled together. The inductance of each winding is 458 μH and the 3 dB bandwidth of the transformer is 20 kHz. If the resonant frequency of the circuit is 1.2 MHz calculate, for critical coupling (*a*) k, (*b*) the mutual inductance M, (*c*) the Q factor of each winding and the overall Q factor and (*d*) the tuning capacitances.

10.11 A parallel-tuned circuit consists of a capacitor of negligible loss connected in parallel with an inductor of $Q = 80$ and $L = 50\ \mu$H. The circuit is at resonance at 1.2 MHz and takes a current of 6 mA from the source. Calculate the current flowing in the inductor. Also find the e.m.f. that is induced into an open-circuited coil that is inductively coupled by a mutual inductance of 2 μH.

10.12 Two mutual-inductance coupled coils have $L_p = L_s = 20$ mH, $k = 0.01$, and $Q_p = Q_s = 30$. Both the primary and secondary windings are tuned to resonance by series capacitors and 1 V at the resonant frequency is applied to the primary circuit. Calculate the voltage across the secondary capacitor.

10.13 A 5 mH inductor is coupled by a mutual inductance of 5 mH to a 9 mH inductor. A 4 kΩ resistor is connected across the terminals of the 9 mH inductor. A voltage at a frequency of 10 kiloradians is applied to the primary. Calculate the phase angle between the applied voltage and the voltage across the 4 kΩ resistor. The resistance of the inductors is negligibly small.

10.14 A coil of inductance 7 μH and resistance 8 Ω is coupled to another coil of inductance 2 μH. If the total secondary resistance is 75 Ω and the mutual inductance is 1.5 μH calculate the effective Q factor of the primary circuit at 7.9578 μH.

10.15 A 100 μH coil is connected in parallel with a capacitor and a voltage of 1 kV is applied across the circuit. A 50 μH coil is coupled by a mutual inductance of 20 μH. Calculate the value of the capacitance required to resonate the primary circuit and the power delivered to the 500 Ω resistor that terminates the secondary circuit.

10.16 A current of 1 mA at a frequency of 100 kHz flows in an inductance of 1 mH. If the inductance is coupled to another 0.25 mH coil with a mutual inductance of 100 μH calculate the voltage induced into the 0.25 mH coil.

10.17 A coil of 100 μH inductance and Q factor 156 is coupled to a coil of 10 μH inductance and Q factor 50. If the mutual inductance is 1.6 μH calculate the coefficient of coupling. Also calculate the effective primary impedance at 2.5 MHz if the secondary load impedance is $200\angle 90°$ Ω.

11 Iron-cored transformers

Essentially, a transformer consists of two or more coils coupled together by mutual inductance. The behaviour of a transformer is determined by the core upon which the coils are wound. If the core is made from some magnetic material the coefficient of coupling k between the primary and the secondary windings is very nearly equal to unity and then the component is known as an *iron-cored transformer*.

The basic arrangement of an iron-cored transformer is shown in Fig. 11.1. Because of the eddy current losses discussed in Chapter 9 the iron core is constructed from a number of thin insulated laminations. Since the two windings are wound around a core made from a high-permeability material, the flux produced by a given current is much larger than that which would be produced using an air core.

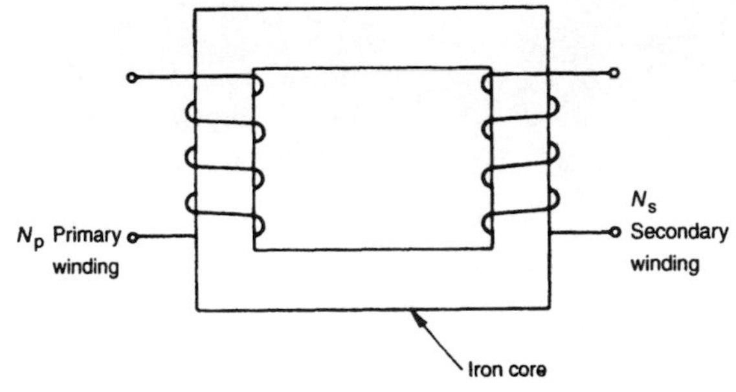

Fig.11.1 Basic iron-cored transformer

An iron-cored transformer therefore possesses large values of self- and mutual inductance but these are obtained at the cost of introducing both eddy current and hysteresis losses. Another disadvantage, that will be discussed in Chapter 19, is that the relationship between the primary current and the flux it produces may be a non-linear one, particularly if the core material is at, or near, its saturated condition.

The main use of the iron-cored transformer is the stepping-up, or the stepping-down, of an a.c. current or voltage. The voltages involved may be very high, as in electricity supply distribution, or may be from the UK standard a.c. mains supply voltage of 240 V to some much lower voltage needed to operate an electronic equipment such as, for example, a radio receiver. In such cases the transformer is operated as a constant-voltage and constant-frequency device.

The *rating* of a transformer specifies the maximum current it can handle safely when subjected to its maximum working primary voltage. The rating is quoted in VA. A 6600/330 V transformer rated at 20 kVA will have a secondary voltage of 330 V when the applied primary voltage is 6600 V and with these voltages the apparent power will be 20 kVA.

High-frequency, high-power transformers used in switched-mode power supplies may have to handle powers in the kilowatt range at frequencies of some tens of kilohertz, (typically, 10 kW at 20 kHz). At such frequencies silicon-steel laminated cores would have excessive eddy current and hysteresis losses and hence such transformers always employ a ferrite core.

Example 11.1

The primary winding of an ideal transformer has 100 turns and a self-inductance of 0.25 H. It is connected to a 240 V, 50 Hz sinusoidal voltage supply. If the secondary winding has 400 turns calculate (*a*) the mutual inductance and (*b*) the self-inductance of the secondary winding.

Solution

(*a*) $V = L_p \,\mathrm{d}i/\mathrm{d}t$, $\mathrm{d}i/\mathrm{d}t = V/L_p = 240/0.25 = 960$ A/s.
$V_s = 240 \times 400/100 = 960\ V = M\,\mathrm{d}i/\mathrm{d}t$
Therefore, $M = 960/960 = 1$ H (*Ans.*)

(*b*) $L_s = L_p(N_s/N_p)^2 = 0.25 \times (400/100)^2 = 4$ H (*Ans.*)

The e.m.f. equation

Suppose that the primary current in the transformer is of sinusoidal waveform with a peak value of I_p. Then the instantaneous value of the current is given by $i = I_p \sin \omega t$. Assuming the operation of the transformer to be linear the flux Φ set up in the core is

$$\Phi = \Phi_{max} \sin \omega t \text{ webers} \tag{11.1}$$

Since the coefficient of coupling k is approximately equal to unity, the e.m.f. induced into the N_s secondary turns is

$$e_s = N_s \mathrm{d}\Phi/\mathrm{dt} \quad \text{or} \quad e_s = N_s \omega \Phi \cos \omega t \text{ volts}$$

The r.m.s. value of this voltage is

$$E_s = (N_s\omega\Phi_{max})/\sqrt{2} = (2\pi f N_s \Phi_{max})/\sqrt{2}$$
$$= 4.44\, f N_s \Phi_{max} \qquad (11.2)$$

This equation is often written in the form

$$E = 4.44 f N_s B_{max} A \qquad (11.3)$$

where B_{max} is the maximum flux density set up in the core of effective cross-sectional area A m^2. Similarly the r.m.s. value of the voltage induced into the primary winding is

$$E_p = 4.44 f N_p B_{max} A \qquad (11.4)$$

The magnetic area of an iron core is smaller than its physical area because of the laminations making up the core. The thickness of the insulation of each lamination is approximately one-tenth of the lamination thickness. Because of this it is customary to assume that the area of the magnetic core is about 0.9 times the physical area. This reduction is known as the *stacking factor*. The maximum flux density in a core is normally restricted to about 1.4 T.

Example 11.2

The core of a 1200/240 V, 50 Hz transformer has an effective area of 0.0324 m^2. The maximum permissible flux density in the core is 1.35 T. Determine the number of (*a*) primary turns and (*b*) secondary turns.

Solution

(*a*) Area of core = 0.0324 m^2
Magnetic area ≈ 0.9 × 0.0324 = 0.0292 m^2

$1200 = 4.44 \times 50 \times N_p \times 1.35 \times 0.0292$, or $N_p \approx 137$ (*Ans.*)

(*b*) $N_s = 137 \times 240/1200 \approx 27$ (*Ans.*)

The ideal transformer

The ideal iron-cored transformer would have each of the following:

(*a*) A coupling coefficient equal to unity.
(*b*) Zero losses.
(*c*) Infinite primary inductance.
(*d*) Zero winding capacitances.

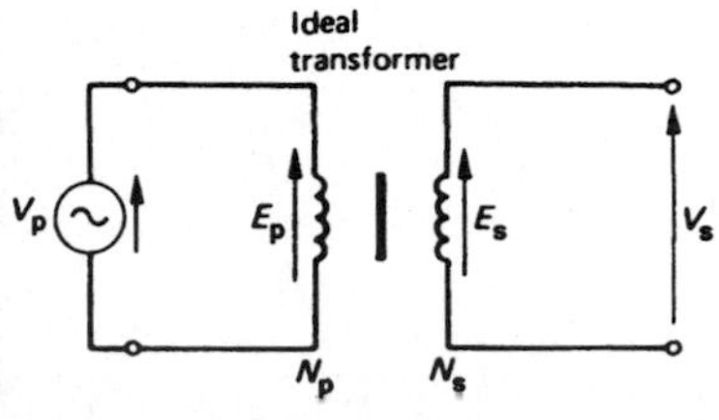

Fig. 11.2 Ideal transformer secondary – open-circuit

Secondary open circuited

Figure 11.2 shows an ideal transformer with its primary circuit supplied by a sinusoidal voltage source V_p and its secondary terminals open circuited.

When a sinusoidal current I_p flows in the primary winding, the flux set up in the core links with both the windings and induces voltages

$$E_p = 4.44 f N_p \Phi_{max} \quad \text{and} \quad E_s = 4.44 f N_s \Phi_{max}$$

into them. The voltage ratio is therefore $E_p/E_s = N_p/N_s$. Further, since the windings are assumed to have zero resistance, the turns ratio is equal to the voltage ratio, i.e.

$$V_p/V_s = N_p/N_s \qquad (11.5)$$

Secondary terminated by a resistance

If the secondary terminals are closed in a resistance R_L, a secondary current I_s will flow and the transformer then will possess a current ratio. Fig. 11.3(*a*) shows the circuit and Fig. 11.3(*b*) gives the phasor diagram representing its currents and voltages. The flux Φ is taken as the reference phasor since it is common to both of the windings.

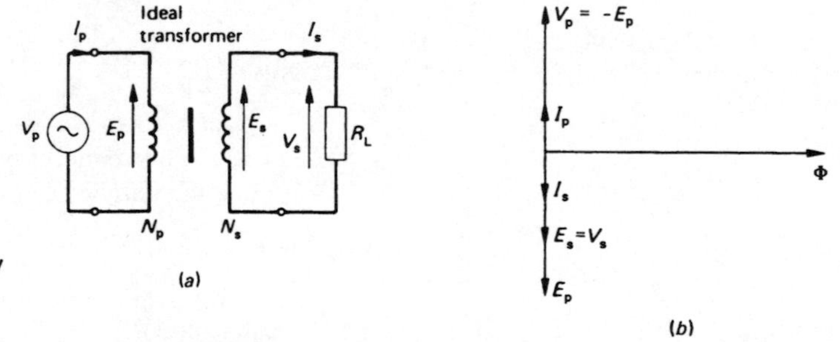

Fig. 11.3 Ideal transformer: (a) secondary terminated by a resistance; (b) phasor diagram

Since the primary inductance is assumed to be infinitely large, zero current is needed to magnetize the core. The applied voltage V_s leads the core flux Φ by 90° and the primary e.m.f. E_p is equal in magnitude to V_p but is of the opposite phase. The e.m.f. E_s induced into the secondary winding is in phase with E_p and its magnitude is equal to $N_s E_p/N_p$.

The secondary current I_s is equal to $E_s/R_L = V_s/R_L$ and it is in phase with E_s. In order to maintain the core flux Φ at a constant value, a primary current I_p must flow, 180° out of phase with I_s. The magnitude of I_p must be such that the m.m.f. it produces is equal to the m.m.f. produced by the secondary current. Therefore,

$$I_p N_p = I_s N_s \quad \text{or} \quad I_p/I_s = N_s/N_p$$

This means that the current ratio of an ideal transformer is the inverse of its turns and voltage ratios.

Further, the input resistance R_{in} of the transformer is

$$R_{in} = V_p/I_p = [(N_p V_s)/N_s]/[(N_s I_s)/N_p]$$
$$= (N_p^2/N_s^2)R_L \quad \text{or} \quad R_{in}/R_L = (N_p/N_s)^2 \qquad (11.6)$$

Thus the impedance ratio of an ideal transformer is the square of the turns ratio.

When the load connected across the secondary terminals is either an inductive or a capacitive impedance, the primary and secondary currents will be out of phase with the primary and the secondary voltages respectively by an angle φ. This is shown by Figs 11.4(*a*) and (*b*).

Now $|Z_{in}| = |V_p|/|I_p|$ and $|Z_L| = |V_s|/|I_s|$ so that

$$|Z_{in}|/|Z_L| = (N_p/N_s)^2$$

Thus, the transformer action will only alter the *magnitude* of the load impedance but its phase angle will remain unchanged.

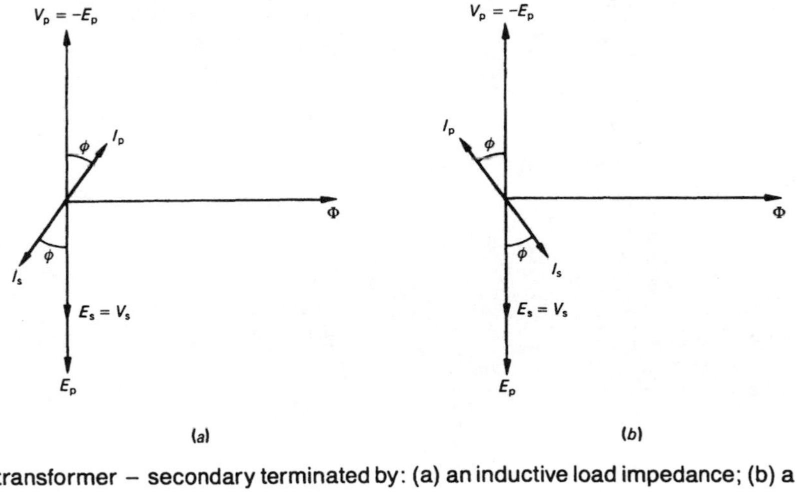

Fig. 11.4 Phasor diagram for an ideal transformer – secondary terminated by: (a) an inductive load impedance; (b) a capacitive load impedance

Example 11.3

An ideal transformer has a turns ratio of 5 : 1. Calculate its input impedance when the load connected across the secondary terminals is (*a*) a 1000 Ω resistor; (*b*) a 1 μF capacitor and (*c*) a 1000 Ω resistor in series with a 1 μF capacitor. The frequency is 5000/2π Hz.

Solution

(*a*) $R_{in} = 25 \times 1000 = 25$ kΩ (*Ans.*)

(*b*) $X_C = 1/(j5000 \times 10^{-6}) = -j200$ Ω. Therefore
$Z_{in} = -j200 \times 25 = -j5$ kΩ (*Ans.*)

(c) $Z_L = 1000 - j200\ \Omega = 1020\angle{-11.3°}\ \Omega$. Therefore
$Z_{in} = 1020 \times 25\angle{-11.3°}\ \Omega = 25.5\angle{-11.3°}\ k\Omega$ (*Ans.*)

Practical transformers

Any practical transformer will not satisfy the listed requirements for an ideal transformer.

No-load losses

Magnetizing current

The *magnetizing current* I_m is the current that must flow in the primary winding in order to set up the required flux in the core because of the finite value of the primary inductance L_m. The magnetizing current I_m is purely inductive ($I_m = V_p/j\omega L_m$) and it therefore lags the primary voltage by 90°. Since I_m does not contribute to the secondary current, the primary inductance is drawn in the equivalent circuit of the transformer in parallel with the primary winding (Fig. 11.5(*a*)). I_m is considered to flow only in L_m. The windings are wound upon a core made from a high permeability material so that L_m is of high value and the required magnetizing current is small.

Iron losses

The practical transformer is subject to both *eddy current* and *hysteresis losses* in its core. The total loss is constant at a particular frequency and it is therefore represented in the equivalent circuit of Fig. 11.5(*a*) by a resistor R_c connected in parallel with the primary inductance. The total iron power losses are then equal to

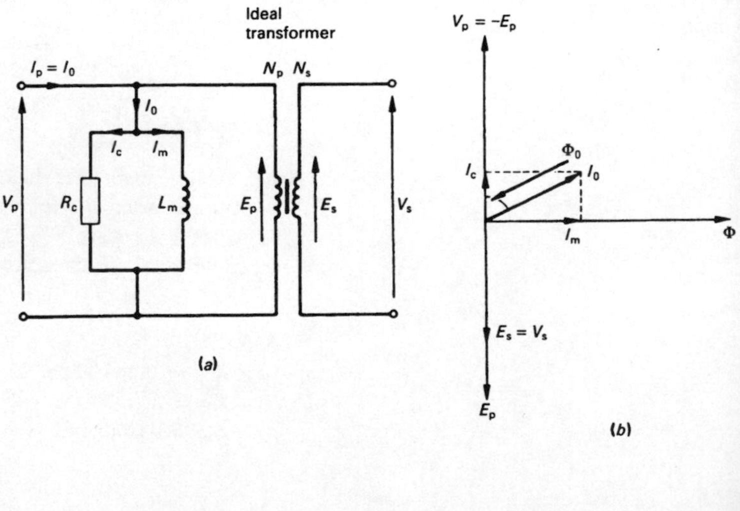

Fig. 11.5 Magnetizing current and iron losses: (a) equivalent circuit; (b) phasor diagram of an iron-cored transformer secondary open-circuited

$I_c^2 R_c$. The total no-load primary current I_0 is the phasor sum of the magnetizing current and the core-loss current.

$$I_0 = \sqrt{(I_c^2 + I_m^2)} \qquad (11.7)$$

This is shown by the phasor diagram given in Fig. 11.5(*b*). Usually I_c is much smaller than I_m.

Example 11.4

A transformer has a no-load primary current of 0.5 A at a lagging power factor of 0.3 when the primary voltage is 240 V at 50 Hz. Calculate the values of (*a*) the magnetizing current and (*b*) the no-load power loss.

Solution

(*a*) All of the primary current supplies the transformer losses. Hence, $I_p = I_0$, and

Magnetizing current $I_m = I_0 \sin\varphi_0 = 0.5 \times 0.954 = 0.477$ A (*Ans.*)

Loss component of $I_0 = I_c = I_0 \cos\varphi_0 = 0.5 \times 0.3 = 0.15$ A, and

The no-load power loss $= VI_0 \cos\varphi_0 = 240 \times 0.15 = 36$ W (*Ans.*)

Example 11.5

A 440/240 V 50 Hz transformer supplies a no-load primary current of 0.6 A at a lagging power factor of 0.28. Calculate the values of the components of the equivalent circuit.

Solution

$I_c = I_0 \cos\varphi_0 = 0.6 \times 0.28 = 0.168$ A

$R_c = 440/0.168 = 2619\ \Omega$ (*Ans.*)

$I_m = I_0 \sin\varphi_0 = 0.6 \times 0.96 = 0.576$ A

$X_{Lm} = 440/0.576 = 763.9\ \Omega$

so $L_m = 763.9/(100\pi) = 2.43$ H (*Ans.*)

Example 11.6

When a 240 V, 50 Hz voltage is applied to the 400 turn primary winding of a transformer a no-load current of 0.4 A flows. The power loss in the transformer is 52 W. If the length of the transformer core is 0.38 m and the magnetic material has a permeability of 1500 calculate the flux density in the core.

Solution

Power factor $= \cos\varphi_0 = P/VI = 52/(240 \times 0.4) = 0.542.$

Therefore, $\varphi_0 = 57.2°$
Magnetizing current $I_m = 0.4 \sin 57.2° = 0.336$ A
Flux density B

$$= \mu_0\mu_r H = (4\pi \times 10^{-7} \times 1500 \times 0.336 \times 400)/0.38$$
$$= 0.67 \text{ T} \quad (Ans.)$$

Loaded losses

When a load is connected across the secondary terminals of a transformer, a secondary current I_s must flow (Fig. 11.6(*a*)). As before, a primary current $I'_p = N_s I_s / N_p$ must flow to maintain the m.m.f. in the core. The total primary current I_p is then the phasor sum of I_0 and $N_s I_s / N_p$ so that the phasor diagram must be modified to that shown by Fig. 11.6(*b*). In most cases $I'_p \gg I_0$ and then $I'_p \approx I_p$.

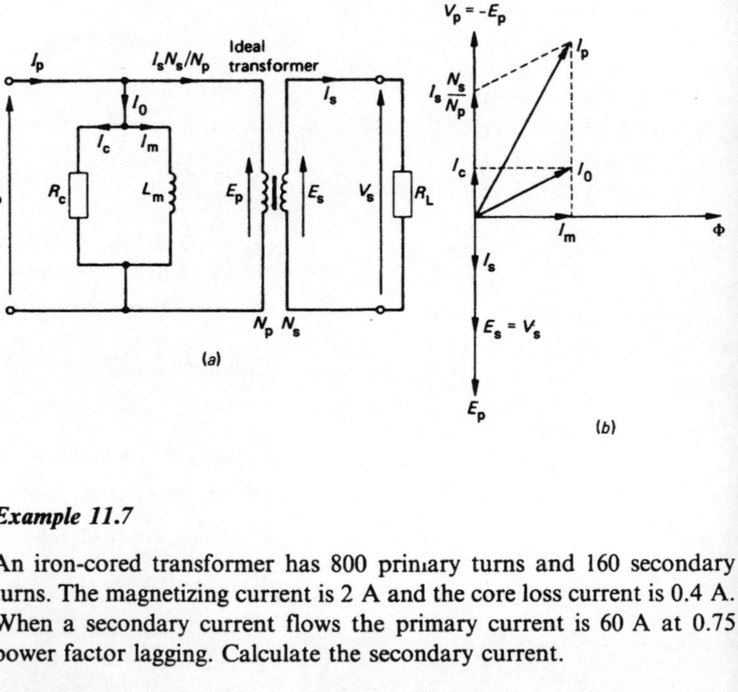

Fig. 11.6 Transformer losses with the secondary terminals closed by a resistance

Example 11.7

An iron-cored transformer has 800 primary turns and 160 secondary turns. The magnetizing current is 2 A and the core loss current is 0.4 A. When a secondary current flows the primary current is 60 A at 0.75 power factor lagging. Calculate the secondary current.

Solution

The primary current is the phasor sum of the current I_0 and the current $N_s I_s / N_p$. I_0 is the phasor sum of the magnetizing current I_m and the core-loss current I_c. Therefore, from Fig. 11.6(*b*), taking V_p as the reference,

$$I_0 = (0.4 - j2) \text{ A}$$

and $I_p = (60 \times 0.75) - (j60 \times 0.66) = (45 - j39.7)$ A

Hence,

$$I_p = 45 - j39.7 = (0.4 - j2) + 800I_s/160, \quad 5I_s = 44.6 - j37.7$$

$$I_s = 8.92 - j7.54 = 11.68\angle{-40.2°}\ \text{A} \quad (\textit{Ans.})$$

$$= 11.68 \text{ at } 0.76 \text{ lagging power factor} \quad (\textit{Ans.})$$

Copper losses

Both the primary and the secondary windings inevitably possess resistance and this is represented in the equivalent circuit by resistors R_p and R_s connected in series with the respective windings as shown by Fig. 11.7(*a*). A voltage will be dropped across each resistor by the current flowing through it and hence Fig. 11.7(*b*) gives the phasor diagram. The applied voltage to the primary terminals V_p is the phasor sum of $-E_p$ and the voltage dropped across the primary resistance R_p, i.e. I_pR_p. The voltage V_s appearing across the load resistor R_L is smaller than the secondary e.m.f. E_s because of the voltage drop I_sR_s across the secondary resistance R_s. Because $I_0 \ll I_sN_s/N_p$, little error is introduced if the components R_c and L_m are drawn across the input terminals instead (Fig. 11.7(*c*)). Often R_c and L_m can be omitted altogether.

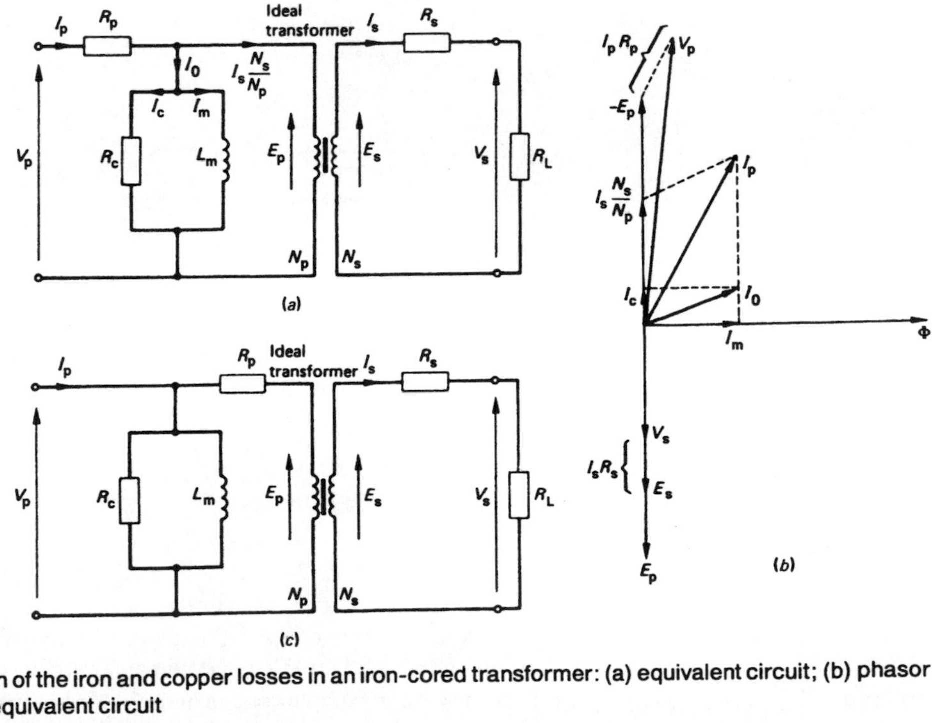

Fig. 11.7 Representation of the iron and copper losses in an iron-cored transformer: (a) equivalent circuit; (b) phasor diagram; (c) alternative equivalent circuit

Example 11.8

A transformer has a primary winding resistance of 0.5 Ω and a secondary winding resistance of 0.2 Ω. The transformer supplies 20 W power to its 12 Ω resistive load. If the primary current is 0.4 A and the total core losses are 1 W, calculate the efficiency of the transformer.

Solution

$$P_L = 20 = 12I_s^2, \text{ so } I_s = \sqrt{(20/12)} = 1.29 \text{ A}$$

Copper loss in secondary winding $= 1.29^2 \times 0.2 = 0.333$ W, and copper loss in primary winding $= 0.4^2 \times 0.5 = 0.08$ W. Therefore,

$$\text{The total loss} = 1 + 0.333 + 0.08 = 1.413 \text{ W}$$

$$\text{Input power} = 20 + 1.413 = 21.413 \text{ W}$$

Hence,

$$\text{Efficiency} = 100P_{out}/P_{in}\% = 2000/21.413 = 93.4\%$$

Leakage inductance

In a practical transformer the coefficient of coupling between the primary and the secondary windings is always less than unity. The fraction of the flux set up by the current in one winding that does not link with the other winding is known as the *leakage flux*. Leakage flux is represented in the equivalent circuit by an inductance connected in series with the winding resistance. Thus, in the equivalent circuit given in Fig. 11.8 L_p and L_s represent, respectively, the primary and secondary leakage inductances.

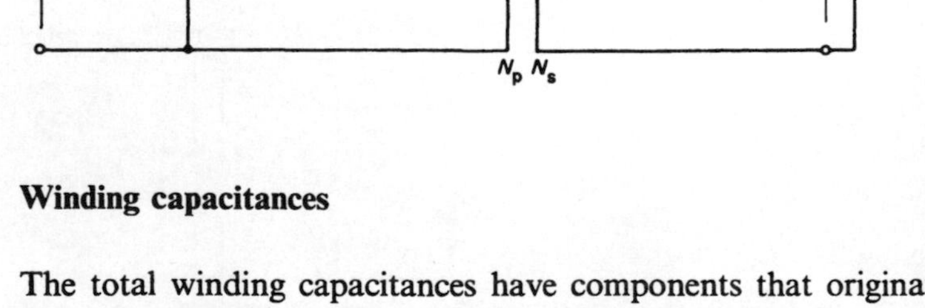

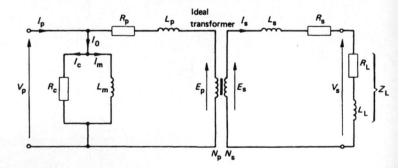

Fig. 11.8 Representation of leakage inductance in an iron-cored transformer

Winding capacitances

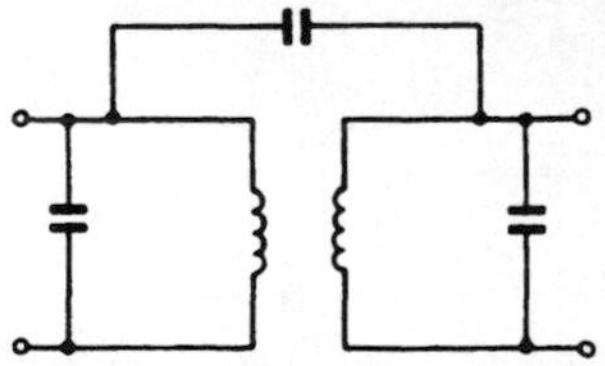

Fig. 11.9 Transformer capacitances

The total winding capacitances have components that originate from a number of sources such as the inter-turn, inter-layer, and inter-winding capacitances. These capacitances are usually represented in the equivalent circuit by capacitors drawn across the input and output terminals of the transformer (see Fig. 11.9).

Equivalent circuit of a transformer

The winding capacitances of a transformer are normally only of importance at the higher audio-frequencies and may be neglected at all lower frequencies. The no-load primary current I_0 of a transformer is usually only a small fraction of the total primary current and hence the components R_c and L_m can often be omitted without the introduction of undue error. Further, the remaining components of the circuit may be referred to *either* the primary *or* the secondary circuit.

- If the secondary components R_s and X_s $(= j\omega L_s)$ are referred to the *primary circuit* they become

  $R_s' = R_s(N_p/N_s)^2$ and $X_s' = X_s(N_p/N_s)^2$ respectively

  The total primary resistance is then equal to

  $R_{p(eff)} = R_p + R_s(N_p/N_s)^2$

  Similarly, the total primary reactance is then

  $X_{p(eff)} = X_p + X_s(N_p/N_s)^2$

  The simplified equivalent circuit of the transformer is then given by Fig 11.10. V_s' is the secondary load voltage referred to the primary, i.e. $V_s' = N_p V_s/N_s$ and $I_p' = I_s N_s/N_p$. The phasor diagram is given in Fig. 11.10(b).

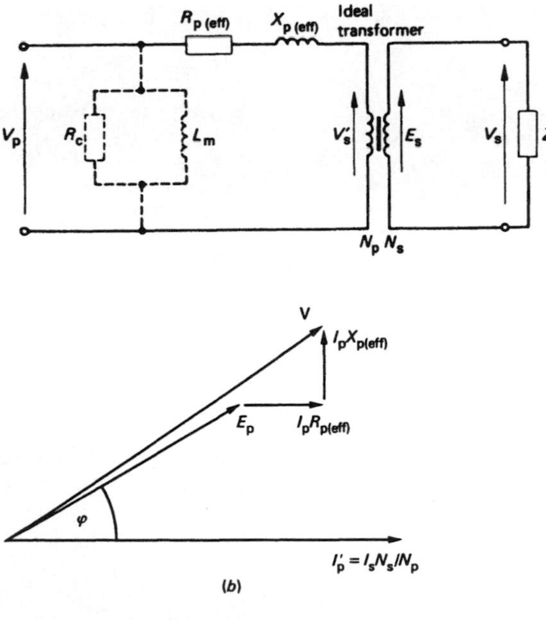

Fig. 11.10 Simplified transformer equivalent circuit (a) with all components referred to the primary circuit; (b) phasor diagram

- Referring the primary components R_p and $X_p(= j\omega L_p)$ to the secondary circuit gives total secondary component values of

$$R_{s(eff)} = R_s + R_p(N_s/N_p)^2 \quad \text{and}$$
$$X_{s(eff)} = X_s + X_p(N_s/N_p)^2$$

The simplified equivalent circuit with all components referred to the secondary is shown in Fig. 11.11 in which $V'_p = V_p N_s/N_p$ and is the primary voltage referred to the secondary.

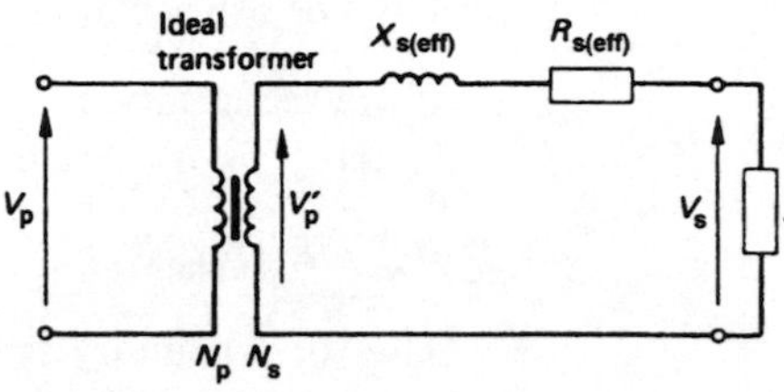

Fig. 11.11 Simplified transformer equivalent circuit: all components referred to the secondary circuit

Example 11.9

A 40 kVA 4000/230 V transformer has the following data: primary resistance = 4 Ω, primary reactance = 15 Ω, secondary resistance = 0.05 Ω, secondary reactance = 0.1 Ω. Calculate the primary voltage needed to obtain the full-load current when the secondary terminals are short circuited.

Solution
Effective primary resistance

$$R_{p(eff)} = 4 + (4000/230)^2 \times 0.05 = 19.12 \ \Omega$$

Effective primary reactance

$$X_{p(eff)} = 15 + (4000/230)^2 \times 0.1 = 45.25 \ \Omega$$

Full-load primary current = (40000/4000) = 10 A. Therefore,

$$V_p = 10(19.12 + j45.25) = 191.2 + j452.5 \text{ V} \quad \text{and}$$
$$|V_p| = 491.2 \text{ V} \quad (\textit{Ans.})$$

Example 11.10

The equivalent circuit of a 6 kVA, 240/600 V transformer has R_c = 650 Ω, ωL_m = 300 Ω, $R_{p(eff)}$ = 0.5 Ω and $X_{p(eff)}$ = 0.8 Ω. When 240 V is applied to the primary circuit the full-load current flows with a power factor of 0.8. Calculate (*a*) the secondary voltage, (*b*) the primary current and (*c*) the efficiency of the transformer.

Solution
(*a*) From the phasor diagram shown in Fig. 11.10(*b*),

$$V - E_p = I'_p(R_{p(eff)} \cos\varphi + X_{p(eff)} \sin\varphi)$$

The full-load secondary current is 6000/600 = 10 A and hence $I'_p = 10 \times 600/240 = 25$ A. Thus;

$$V - E_p = 25(0.5 \times 0.8 + 0.8 \times 0.6) = 22 \text{ V} \quad \text{and}$$
$$E_p = 240 - 22 = 218 \text{ V}$$

Therefore,

$$V_s = 218 \times 600/240 = 545 \text{ V} \quad (Ans.)$$

(*b*) The component of I_0 in phase with the applied voltage = $240/650 = 0.369$ A. The component of I_c in phase with $I'_p = 0.369 \times 0.8 = 0.295$ A. The component of I_c in quadrature with $I'_p = 0.369 \times 0.6 = 0.221$ A. The magnetizing current $I_m = 240/300 = 0.8$ A. The component of I_m in phase with $I'_p = 0.8 \times 0.6 = 0.48$ A. The component of I_m in quadrature with $I'_p = -0.8 \times 0.8 = -0.64$ A. Therefore, the total primary current is

$$I_p = \sqrt{[(25 + 0.295 + 0.48)^2 + (0.221 - 0.64)^2]} = 25.78 \text{ A} \quad (Ans.)$$

(*c*) Input power $= VI\cos\varphi = 240 \times 25.78 \times 0.8 \approx 4950$ W (ignoring the fact that I_p is not quite in phase with I'_p). Copper loss = $25^2 \times 0.5 = 312.5$ W; iron loss $= 240 \times 0.369 = 88.6$ W, so that the total loss $= 312.5 + 88.6 = 401.1$ W.

Output power $= 4950 - 401.1 = 4548.9$ W

Efficiency $= (4548.9/4950) \times 100 = 91.9$ % (*Ans.*)

Voltage regulation

The *voltage regulation* of a transformer is the change in the secondary voltage as the current taken by a load connected across the secondary terminals varies from zero to the maximum permitted current.

The voltage regulation of a transformer is defined by

$$\text{Regulation} = \frac{\text{No-load secondary voltage} - \text{Full-load secondary voltage}}{\text{No-load secondary voltage}} \qquad (11.8)$$

The voltage regulation can be determined using equation (11.8) as a *per-unit value* or it may be multiplied by 100 and thereby expressed as a percentage.

If the primary components R_p and X_p of the equivalent circuit are referred to the secondary (Fig. 11.12(*a*)), the phasor diagram of the transformer, assuming a lagging power factor, is given by Fig. 11.12(*b*). The no-load current I_0 has been assumed to be negligibly small with respect to $N_s I_s / N_p$.

The voltage $V'_s = N_s V_p / N_p$ applied to the equivalent secondary circuit is the phasor sum of the secondary voltage V_s and the voltage dropped across $R_{s(\text{eff})}$ and $X_{s(\text{eff})}$. From Fig. 11.12(*b*),

$$V'_s = \sqrt{[(V_s + I_s R_{s(\text{eff})}\cos\varphi + I_s X_{s(\text{eff})}\sin\varphi)^2 + (I_s X_{s(\text{eff})}\cos\varphi - I_s R_{s(\text{eff})}\sin\varphi)]^2} \qquad (11.9)$$

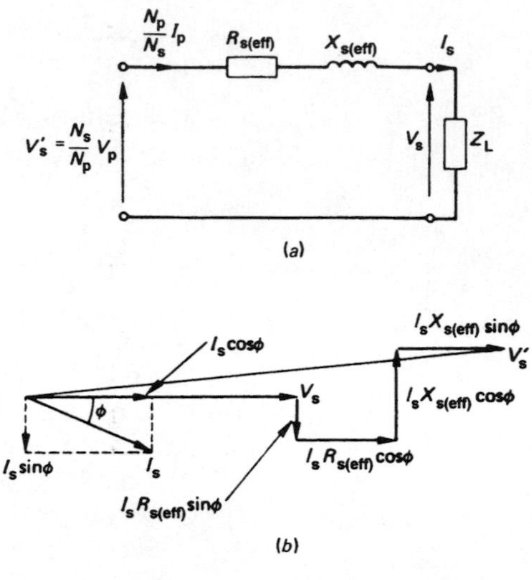

Fig. 11.12 Voltage regulation of a transformer

It can be seen from the phasor diagram that $I_s X_{s(eff)} \cos\varphi$ and $I_s R_{s(eff)} \sin\varphi$ tend to cancel out and generally little error is introduced if these terms are neglected. If this is done then

$$V_s' = V_s + I_s R_{s(eff)} \cos\varphi + I_s X_{s(eff)} \sin\varphi \qquad (11.10)$$

The per unit *voltage regulation* of the transformer is then given by

$$\text{Regulation} = (V_s' - V_s)/V_s'$$
$$= [I_s R_{s(eff)} \cos\varphi + I_s X_{s(eff)} \sin\varphi]/V_s' \qquad (11.11)$$

If the load has a leading power factor then equation (11.11) must be altered as follows:

$$\text{Regulation} = [I_s R_{s(eff)} \cos\varphi - I_s X_{s(eff)} \sin\varphi]/V_s' \qquad (11.12)$$

Example 11.11

Calculate the percentage regulation of the transformer of Example 11.9 if the load power factor is (*a*) unity, (*b*) 0.85 lagging and (*c*) 0.85 leading.

Solution

Referring all values to the secondary circuit, $I_s = 400.00/230 = 173.91$ A

$$R_{s(eff)} = 0.05 + (230/4000)^2 \times 4 = 0.063\ \Omega$$
$$X_{s(eff)} = 0.1 + (230/4000)^2 \times 15 = 0.15\ \Omega$$

(*a*) % Regulation = {[(173.91 × 0.063 × 1) − 0]/230} × 100
= 4.76% (*Ans.*)

(*b*) % Regulation = {[(173.91 × 0.063 × 0.85)
+ (173.91 × 0.15 × 0.53)]/230} × 100
= 10.06% (*Ans.*)

(*c*) % Regulation = {[(173.91 × 0.063 × 0.85)
− (173.91 × 0.15 × 0.53)]/230} × 100
= −4.51% (*Ans.*)

The negative sign means that the terminal voltage increases with increase in the load current.

Efficiency of a transformer

The efficiency of a transformer is always less than 100 per cent because of the inevitable copper and iron losses. The core flux is, more or less, constant regardless of the load and so it is usual to assume that the iron losses are constant for all values of load current. Copper losses, on the other hand, are the result of I^2R dissipation in both of the windings and therefore increase in proportion to the squares of the primary and secondary currents.

The *efficiency* η of a transformer is given by

$$\eta = \frac{\text{Output power}}{\text{Input power}} \times 100\% \qquad (11.13)$$

The input power to a transformer is the sum of the output power and the total losses within the transformer and so equation (11.13) can be written as

$$\eta = \frac{I_s V_s \times \text{Power factor}}{(I_s V_s \times \text{Power factor}) + P_c + I_p^2 R_p + I_s^2 R_s} \times 100\% \qquad (11.14)$$

$$= \frac{I_s V_s \times \text{Power factor}}{(I_s V_s \times \text{Power factor}) + P_c + I_p^2 R_{s(\text{eff})}} \times 100\% \qquad (11.15)$$

$$= \frac{I_s V_s \times \text{Power factor}}{(I_s V_s \times \text{Power factor}) + \text{Iron loss} + \text{Copper loss}} \times 100\% \qquad (11.16)$$

Example 11.12

A 1 kVA transformer has an iron loss of 25 W and a full load copper loss of 30 W. Calculate the efficiency of the transformer (*a*) on full load and (*b*) on half load if the load power factor is 0.85.

Solution
(*a*) $\eta = [(1000 \times 0.85)/(850 + 25 + 30)] \times 100 = 93.92\%$ (*Ans.*)
(*b*) On half load the copper losses will be reduced to

$$(\tfrac{1}{2})^2 \times 30 = 7.5 \text{ W}$$

Therefore,

$$\eta = [(500 \times 0.85)/(425 + 25 + 7.5)] \times 100 = 92.9\% \quad (Ans.)$$

Maximum efficiency

To determine the necessary conditions for the maximum efficiency of a transformer, equation (11.15) must be differentiated with respect to I_s and the result equated to zero. Thus

$$\frac{d\eta}{dI_s} = \frac{V_s \times PF(I_s V_s \times PF + P_c + I_s^2 R_{s(eff)}) - (I_s V_s \times PF)(V_s \times PF + 2I_s R_{s(eff)})}{(I_s V_s \times PF) + P_c + I_s^2 R_{s(eff)}} = 0$$

Hence,

$$(I_s V_s \times PF) + 2I_s^2 R_{s(eff)} = (I_s V_s \times PF) + P_c + I_s^2 R_{s(eff)}$$

or $I_s^2 R_{s(eff)} = P_c$

This result means that a transformer will operate with its maximum possible efficiency when the load current is such that the copper losses are equal to the iron losses.

Example 11.13

A 22 kVA, 1100/240 V transformer has an iron loss of 520 W and, at full load, a copper loss of 580 W. Calculate the efficiency of the transformer when (*a*) it is delivering its full load at a lagging power factor of 0.8 and (*b*) it is delivering 60 per cent of its full load at unity power factor. (*c*) Calculate the load at which the transformer operates with its maximum possible efficiency.

Solution
(*a*) Full-load power = 22 × 0.8 = 17.6 kW. Full-load power losses = 520 + 580 = 1100 W.
Efficiency = (17.6/18.7) × 100 = 94.1% (*Ans.*)
(*b*) 60% of full load = 13.2 kVA. Output power = 13.2 × 1 = 13.2 kW.
Copper loss = 580 × $(0.6)^2$ = 208.8 W.
Total loss = 520 + 208.8 = 728.8 W.
Efficiency = (13 200/13 928.8) × 100 = 94.8% (*Ans.*)
(*c*) Maximum efficiency when copper loss = 520 W. This loss occurs when load = *y* times full load. Hence,

$$520 = 580y^2 \quad \text{or} \quad y = \sqrt{(520/580)} = 0.95$$

Load for maximum efficiency = 0.95 × 22 = 20.9 kVA (*Ans.*)

Transformer tests

The testing of a transformer to ascertain its important parameters such as its voltage regulation and its efficiency is normally carried out by means of two tests. The first of these tests, known as the *open-circuit test*, measures the iron losses of the transformer. The other test, known as the *short-circuit test*, obtains the copper losses of the transformer.

Open-circuit test

The circuit employed to carry out the open-circuit test is shown in Fig. 11.13. In order to keep the copper losses to the minimum possible value, the secondary terminals of the transformer are open-circuited. For this condition there are no copper losses in the secondary circuit and only a very small copper loss in the primary (which is due to I_0).

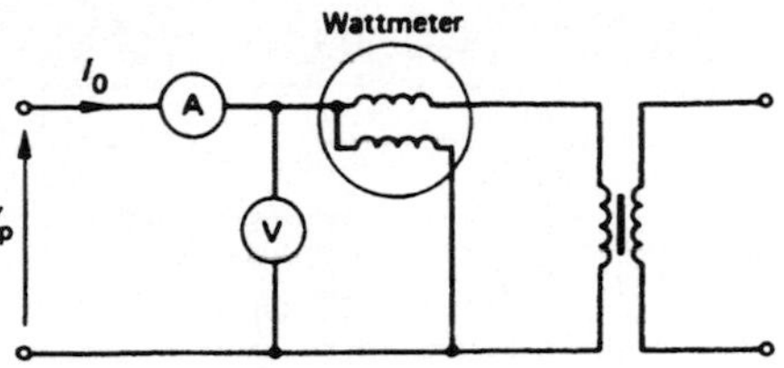

Fig. 11.13 Open-circuit test of a transformer

With the rated input voltage applied to the primary winding the indication of the wattmeter W_c will be very nearly equal to the iron losses. Further,

$$W_c = I_0 V_p \cos\varphi \quad \text{or} \quad \cos\varphi = W_c / V_p I_0$$

and finally

$$I_c = I_0 \cos\varphi \quad \text{and} \quad I_m = I_0 \sin\varphi$$

Short-circuit test

The circuit used for the short-circuit test is shown by Fig. 11.14. Since the secondary winding is short circuited it is possible to obtain the full-load current with only a low voltage applied across

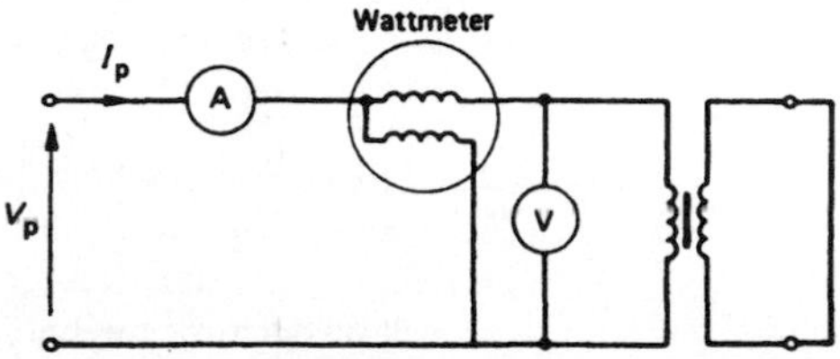

Fig. 11.14 Short-circuit test of a transformer

the primary winding. This means that the iron losses are negligibly small and so the wattmeter indicates the full-load copper loss of the transformer.

Also the magnitude $|Z_{p(eff)}|$ of the effective primary impedance is equal to V_p/I_p, and the phase angle between V_p and I_p is obtained from $\cos\beta = W_c/V_pI_p$. Then,

$$\text{Effective primary resistance } R_{p(eff)} = |Z_{p(eff)}|\cos\beta$$

$$\text{and Effective primary reactance } X_{p(eff)} = |Z_{p(eff)}|\sin\beta$$

Example 11.14

The results given in Table 11.1 were obtained from open- and short-circuit tests on a 8 kVA, 400/240 V transformer at a frequency of 50 Hz. Use the results to obtain the values of the components in the transformer's equivalent circuit.

Table 11.1

	Primary voltage V_p (V)	*Primary current* I_p (A)	*Input power* (W)
Open-circuit test	400	0.48	65
Short-circuit test	20	20	108

Solution

(*a*) Open circuit: $R_c = 400^2/65 = 2461.5\ \Omega$ (*Ans.*)

$\cos\varphi_o = 65/(400 \times 0.48) = 0.339$ or $\varphi_o = 70.2°$

$I_m = 0.48 \sin 70.2° = 0.452$ A

$X_m = 400/0.452$ and $L_m = X_m/100\pi = 400/(0.452 \times 100\pi)$
$= 2.82$ H (*Ans.*)

(*b*) Short circuit: $Z_{p(eff)} = 20/20 = 1\ \Omega$

$R_{p(eff)} = 108/20^2 = 0.27$ V (*Ans.*)

$X_{p(eff)} = \sqrt{(1^2 - 0.27^2)} = 0.963\ \Omega$ and

$L_{p(eff)} = 0.963/(100\pi) = 3.07$ mH (*Ans.*)

Exercises 11

11.1 A 10 kVA, 3200/400 V transformer has primary and secondary resistances of 2.5 Ω and 0.2 Ω respectively and a total reactance referred to the secondary of 0.3 Ω. Calculate the primary voltage needed to obtain the full-load secondary current when the secondary terminals are short circuited.

11.2 A 100 kVA transformer has a nominal secondary voltage of 230 V. Its iron and full-load copper losses are 1000 W and 1500 W respectively. Calculate (*a*) its full-load efficiency (assume unity power factor), (*b*) its efficiency on half-load, (*c*) the load at which the maximum efficiency occurs and (*d*) the maximum efficiency (assume unity power factor).

11.3 The primary and secondary windings of a 20 kVA, 3200/230 V transformer have resistances of 5 Ω and 0.01 Ω respectively. The total reactance referred to the primary is 25 Ω. Calculate the voltage regulation of the transformer, (*a*) for unity load power factor and (*b*) for a load power factor of 0.82 lagging.

11.4 A 6600/400 V, 10 kVA single-phase transformer has a load voltage of 385 V when supplying its full-load current at a power factor of 0.8 lagging. When the full-load current is supplied at a power factor of 0.62 leading, the load voltage is 401.6 V. Calculate the total resistance and reactance of the transformer when referred to the secondary.

11.5 A 6600/1100 V, 10 kVA single-phase transformer has the following data:

primary resistance = 3.8 Ω, secondary resistance = 0.12 Ω
primary reactance = 8.8 Ω, secondary reactance = 0.26 Ω

Use the approximate regulation expression to calculate the percentage regulation at the low-voltage terminals when the transformer supplies the full load current at (*a*) 0.86 lagging power factor and (*b*) unity power factor.

11.6 A 6600/330 V, 100 kVA single-phase transformer has its maximum efficiency at a 0.9 full load. When the secondary terminals are short circuited 100 V are necessary to produce the full-load current in the short circuit. Calculate (*a*) the total resistance and reactance when referred to the secondary winding and (*b*) the voltage across a load that takes the full-load current at a power factor of 0.8 lagging. Iron losses are 3000 W.

11.7 A 100 kVA, 6600/330 V, 50 Hz single-phase transformer is supplied at 6600 volts and 60 Hz and then has a hysteresis loss of 350 W and an eddy current loss of 150 W. (*a*) Calculate the total iron loss when the transformer is supplied at 6600 V at 50 Hz. Assume the Steinmetz index to be 1.6. (*b*) If the maximum efficiency of the transformer occurs at 0.9 full load, and 100 V is needed to circulate the full-load current when the secondary terminals are short circuited, calculate the values of the resistance and reactance referred to the primary circuit.

11.8 A 2000/400 V, 10 kVA single-phase transformer has an effective primary impedance of $(5+j10)$ Ω and a secondary winding impedance of $(0.25+j1.2)$ Ω. Use the approximate regulation expression to calculate the terminal voltage on the low voltage side when supplying the full-load current at a power factor of (*a*) 0.8 lagging and (*b*) 0.6 leading.

11.9 The circuit given in Fig. 11.10 could represent a 4 kVA 50 Hz single-phase transformer with a voltage ratio of 2000/400. When this transformer was tested using the open- and short-circuit tests, with the measurements taken on the low-voltage side, the data obtained is given in Table 11.2. Determine the component values for Fig. 11.10.

Table 11.2

Secondary	V_p(V)	I_p(A)	*Input power* (W)
Open circuit	220	0.7	72
Short circuit	8.2	2	8.0

11.10 A 10 kVA, 6600/330 V, 50 Hz single-phase transformer has a primary resistance of 9.2 Ω and a secondary resistance of 0.1 Ω. The primary and secondary reactances are, respectively, 30 Ω and 0.3 Ω. Calculate (*a*) the total equivalent resistance and reactance referred to the primary, (*b*) the voltage regulation at a power factor of (*i*) 0.8 lagging and (*ii*) 0.8 leading.

11.11 A 20 kVA transformer has an iron loss of 1000 W. What value of copper loss will produce the maximum efficiency? If the power factor of the load is then 0.85 calculate the power dissipated in the load.

11.12 A 3200/400 V transformer has the following data: primary resistance = 5 Ω, primary reactance = 0.15 Ω, secondary resistance = 0.03 Ω, secondary reactance = 0.1 Ω. Determine the effective primary impedance.

11.13 In a short-circuit test carried out on a 6 kVA transformer an applied voltage of 24 V caused the full-load current of 14.5 A to flow in the primary circuit. The power then indicated by the wattmeter was 80 W. Calculate the per unit resistance and reactance of the transformer.

11.14 A 3300/240 V, 25 kVA single-phase transformer has a maximum efficiency of 95 per cent at 0.9 power factor. Calculate its iron losses.

11.15 A small transformer has $R_p = 0.5\ \Omega$, $R_s = 0.1\ \Omega$ and a copper loss of 0.24 W when the power delivered to the load is 24 W. If the primary current is 0.4 A and the efficiency of the transformer is 92 per cent determine the iron loss.

11.16 A 4 kV/400 V, 20 kVA transformer has $R_p = 10\ \Omega$, $R_s = 0.2\ \Omega$, $X_{p(\text{eff})} = 40\ \Omega$, $X_m = 5\ \Omega$ and $R_c = 10\ \text{k}\Omega$. Calculate the input current (*a*) with the secondary terminals open circuit and (*b*) with a secondary current of 20 A at a lagging power factor of 0.8.

11.17 A 40 kVA, 6.6 kV/240 V transformer has $R_p = 12\ \Omega$, $R_s = 0.05\ \Omega$ and $X_{p(\text{eff})} = 30\ \Omega$. Calculate (*a*) the voltage which must be applied to get the full-load current when the secondary terminals are short circuited and (*b*) the full-load regulation if the power factor is 0.8 lagging.

11.18 A 20 kVA transformer has a iron loss of 240 W and a full-load copper loss of 280 W. (*a*) Determine the load for maximum efficiency. (*b*) Determine the maximum efficiency when the load power factor is (*i*) unity and (*ii*) 0.8 lagging.

11.19 A 440/240 V, 50 Hz transformer has a primary inductance L_m of 1.5 H, an iron loss of 20 W and a loss current of 0.4 A. The secondary terminals are connected to a load of $120\angle 26°\ \Omega$. Calculate the primary current.

11.20 A 1600/240 V transformer has $R_p = 0.9\ \Omega$, $R_s = 0.1\ \Omega$, $X_p = 5\ \Omega$ and $X_s = 0.2\ \Omega$. Calculate the primary voltage which will cause a short-circuit current of 100 A to flow in the secondary winding.

11.21 A 240/20 V transformer has a primary winding resistance of 1 Ω and a secondary winding resistance of 0.04 Ω. The leakage reactances are $X_p = 5\ \Omega$ and $X_s = 0.4\ \Omega$. The transformer is connected to a 100 V supply. Calculate the voltage across the 12 Ω load.

12 Electric and magnetic fields

A field is defined as a region of space in which a certain physical state prevails and in which certain actions take place. Thus an *electric field* is a region in the vicinity of an electric charge in which a force will be exerted upon any other electric charge. The electric charge is supposed, for convenience, to consist of a number of lines of electric flux that can be drawn or *mapped*.

Whenever a current flows in a conductor, a *magnetic field* will be set up around that conductor. The magnetic field is the region in which a force may be exerted upon any other conductor and it is also supposed to consist of a number of lines of flux.

Both electric and magnetic fields can be represented by an equation of the form

$$\text{Flux density} = \text{Constant} \times \text{Potential gradient} \qquad (12.1)$$

Several other fields can also be accurately represented by an equation of the same form; among these are included current conduction, fluidic dynamic and thermal conduction fields. Because of the analogy between the various fields it is possible to simulate the behaviour of one field by a consideration of the analogous electric field.

Many of the problems encountered in electrical engineering involve the determination of the potentials existing at various points in a field. The field in question might be an electric field or a magnetic field but, equally likely, the problem may be concerned with the flow of current in a resistive field or the determination of the temperature at various points in a thermal field, or perhaps the flow of a liquid in a cooling system.

Several different kinds of field can be described mathematically by the use of Laplace's equation, i.e.

$$\partial^2\theta/\partial x^2 + \partial^2\theta/\partial y^2 + \partial^2\theta/\partial z^2 = 0 \qquad (12.2)$$

where θ is the scalar value of a variable in the field.

Each such field is essentially of the form

Flow density = Constant × Potential gradient

For example, a current can be written as

Current density = Conductivity × Voltage gradient

and an electric field can be written as

Flux density = Permittivity × Voltage gradient

The relationships between electric, magnetic, current and thermal fields are given in Table 12.1.

Table 12.1

Field	*Potential difference*	*Potential gradient*	*Flow density*
Current	Voltage V	$E = V/d$	$J = \sigma E$
Electric	Voltage V	$E = V/d$	$D = \varepsilon E$
Magnetic	Ampere-turns NI	$H = NI/l$	$B = \mu H$
Thermal	Degrees θ	degrees/m θ/d	$\theta/R_{th} \cdot A$

When a field has simple boundaries, such as two parallel planes, it is not difficult to solve problems analytically.

In most other cases, however, the problem is usually too difficult to be tackled in this manner and an approximate method of solution must be adopted. For a paper solution, a *mapping* method can be adopted which is capable of producing reasonably simple and accurate solutions although often at the expense of considerable effort.

Experimental methods are also available in which an electric current is set up to simulate the field to be investigated. Such simulation is possible because of the similarities between the mathematical equations describing the electric field and the other fields mentioned earlier. Two experimental methods are often used: one, known as the electrolytic tank, uses a conducting liquid medium to simulate the field under investigation; the other method uses a special resistance paper.

The electric field

When an electric charge is brought into the vicinity of another electric charge, a force of either attraction or repulsion (depending on the signs of the charges) is experienced. The space throughout which the force exists is said to be an *electric field.*

Figure 12.1 shows the electric fields in (*a*) a parallel-plate capacitor, (*b*) two concentric cylinders, and a twin line carrying current in (*c*) opposite directions, and (*d*) in the same direction.

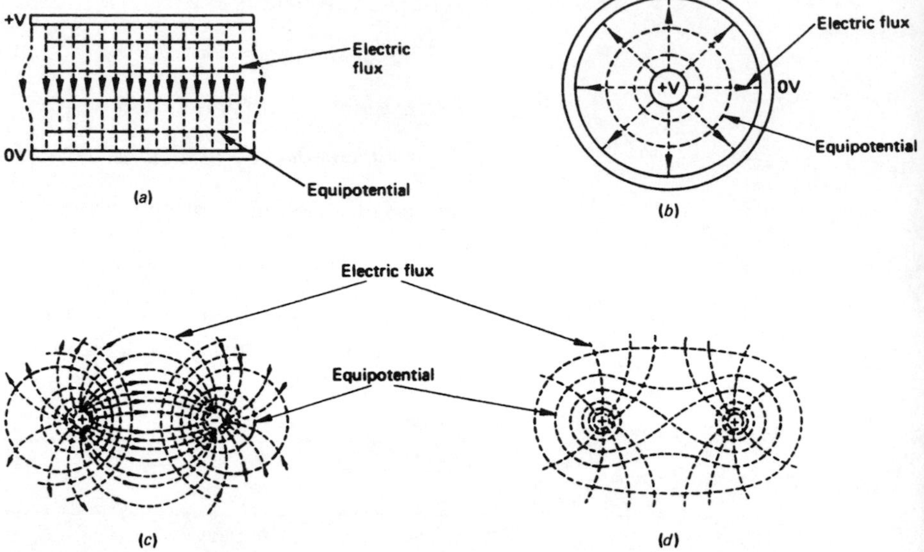

Fig. 12.1 Electric fields of: (a) two parallel metal plates; (b) two concentric metal cylinders; (c) & (d) two parallel conductors

In each case the lines of electric flux are really three-dimensional and leave a positive charge and terminate on a less positive or negative charge. In Fig. 12.1(*d*) it is assumed that both positive charges are of equal magnitude and so the flux lines are only shown leaving a charge.

Equi-potential lines may be drawn to connect all points within a field that have the same potential. Four equi-potential lines are shown in Fig. 12.1(*a*); their potentials are $V/5$, $2V/5$, $3V/5$ and $4V/5$. The upper and lower plates are two further equi-potentials at potentials of V and 0 respectively. The equi-potential lines are everywhere at right angles to the lines of flux; in the case of the parallel-plate capacitor this means that the equi-potentials are parallel to the plates. The flux lines are sometimes known as *streamlines* or *flowlines*.

Electric field strength

The strength of the force exerted is measured in terms of the *electric field strength* E of the field. The field strength at any point is the force exerted on a unit charge at that point and it is measured in volts/metre.

The *electric potential* V at any point in the field is the work done, in joules, in moving unit charge from a point of zero potential to that point.

$$V = W/Q \text{ N/C (or J)} \qquad (12.3)$$

Within an electric field there will be always a number of points at the same potential and these points are said to be *equi-potentials*. If a line is drawn to join up a number of equi-potentials, an equi-potential line or surface is formed.

There can be no current flowing along an equi-potential surface and hence lines of force, known as *streamlines*, must intersect such a surface at right angles. Then there are zero components of the current flowing along the equi-potential surface.

The *potential difference* between two points d metres apart is the work done in moving unit charge between these two points. If the potential difference between two parallel metal plates is v volts then the electric field strength between the two points is

$$E = -V/d \text{ V/m}, \quad \text{and} \quad V = -Ed. \tag{12.4}$$

Consider two points A and B in an electric field which are distance dx apart and the line between them is at an angle θ to the electric field then $\mathrm{d}V = -E\cos\theta\,\mathrm{d}x$, $\mathrm{d}V/\mathrm{d}x = -E\cos\theta = -E_x$ and $V = -E\,\mathrm{d}x$.

Electric flux

The *electric flux* Ψ is the quantity of charge that is moved across a given area in a dielectric. The flux leaving a closed surface which encloses a charge of Q coulombs is equal to that charge, i.e. $\Psi = Q$. This is known as *Gauss's theorem*.

Electric flux density

The *electric flux density* D is the flux per metre, i.e. $D = \Psi/A$ C/m^2. At any point in the field the flux density is equal to the product of the electric field strength and the permittivity of the dielectric, i.e.

$$D = \varepsilon E \text{ C/m}^2 \tag{12.5}$$

- For a point charge Q the lines of force leave in all directions over the area of a sphere, $4\pi r^2$ m^2, where r is the radius of the sphere. Hence,

  $$D = Q/(4\pi r^2) \text{ C/m}^2 \tag{12.6}$$

- For a straight length of current-carrying conductor,

  $$D = Q/(2\pi x) \text{ C/m}^2 \tag{12.7}$$

  where x is the distance from the conductor.

- For two parallel metal plates of area A m^2,

$$D = Q/A \text{ C/m}^2 \tag{12.8}$$

Example 12.1

Two charged metal plates of area 4 cm by 5 cm are mounted parallel to one another and 1 cm apart. The flux lines between the two plates are both straight and parallel. An area measuring 6 mm × 8 mm in between the two plates and parallel with them has a flux of 100 μC passing through it. Calculate (*a*) the flux density in the region between the plates and (*b*) the total flux between the plates.

(*a*) Flux density $D = (100 \times 10^{-6})/(48 \times 10^{-6}) = 2.083$ C/m^2 (*Ans.*)
(*b*) Flux $= DA = 2.083 \times 20 \times 10^{-4} = 4.17$ mC (*Ans.*)

Capacitance

The *capacitance* of a system enclosing an electric field relates the charge stored to the potential difference: $C = Q/V$.

For a parallel-plate capacitor the electric flux is equal to the charge Q stored and the flux density D is equal to Q/A C/m, where A is the area of the plates. In turn, the electric field strength is $E = D/\varepsilon = Q/\varepsilon A$ and the voltage across the plates is $V = Ed = Qd/\varepsilon A$, where d is the plate separation. Therefore,

$$C = Q/V = \varepsilon A/d \text{ F} \tag{12.9}$$

The magnetic field

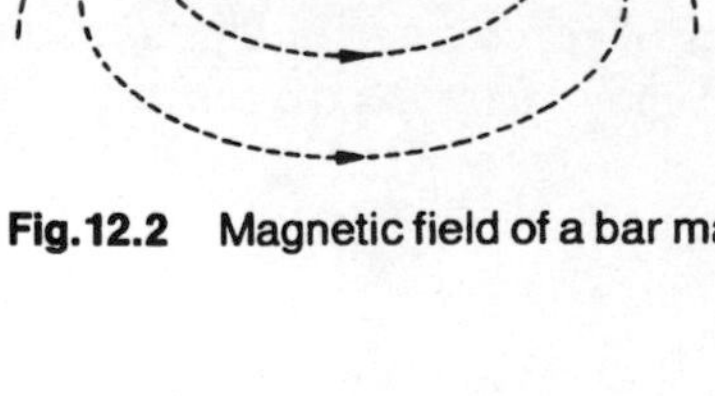

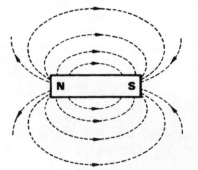

Fig. 12.2 Magnetic field of a bar magnet

It is well known that a magnet is able to exert an attractive force upon bodies made of certain materials. The region of space within which this force is experienced is said to be a region in which the magnetic field of the magnet exists. The field is supposed to consist of a number of lines of magnetic flux. The direction of the flux lines at any point in the field is chosen to indicate the direction in which the force exerted acts at that point. Fig. 12.2, for example, shows the magnetic field of a bar magnet; the direction of each of the flux lines is from North to South.

In a magnetic field streamlines are parallel to the lines of magnetic flux.

Electromagnetism

When a current I flows in a conductor a magnetic field is set up around that conductor. The direction of the field is determined by the direction in which the current flows in the conductor.

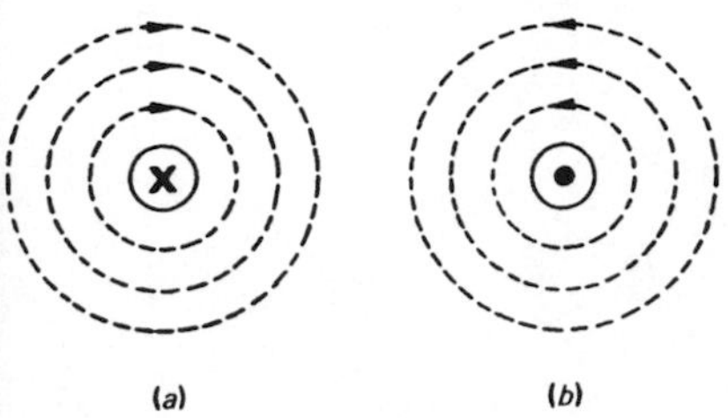

Fig. 12.3 Magnetic field set up around a current-carrying conductor

When the current flows into the paper (Fig. 12.3(a)), the magnetic field is in the clockwise direction. Conversely (Fig. 12.3(b)), when the current flows out of the paper an anti-clockwise field is set up.

When the current first starts to flow in the conductor, a back e.m.f. is produced. Work must be done to overcome this e.m.f. so that the current can increase and the magnetic field can build up. Some of the energy supplied is used to provide the inevitable I^2Rt heat energy developed in the resistance of the conductor. The remainder of the energy is stored in the magnetic field. When the current ceases to flow, the magnetic field collapses back into the conductor and the energy stored in the field is returned to the source.

The magnetic field strength H at a distance r from the conductor is given by

$$H = I/2\pi r \qquad (12.10)$$

acting at right angles to the conductor.

Field plotting

Mapping method

For a field to be represented by a mapping method a *flat* plane must be used and so the term $\partial^2\theta/\partial z^2$ in Laplace's equation becomes equal to zero. The plane of the field is divided into a number of 'squares' formed by equi-potentials and streamlines. The two sets of lines must be drawn at right angles to one another at each point of intersection. Unless the field is uniform the two sets of lines will *not* be straight and then only approximate squares can be drawn. For this reason the 'squares' are known as *curvilinear squares*. The smaller the curvilinear squares are drawn the nearer they will become to true squares.

An example of the method is shown by Fig. 12.4 in which the area enclosed by the equi-potentials AA and BB and the lines CC and DD has been divided into a number of actual squares in Fig. 12.4(a) and curvilinear squares in Fig. 12.4(b). No problem exists for the linear field of Fig. 12.4(a) but greater care is needed for the non-linear field. Once the pattern of equi-potentials has been drawn, e.g. lines AA, BB and all those in between, the conjugate pattern remains to be drawn. This must be done by (a) making sure that all lines cross the equi-potentials at right angles and (b) forming approximate (curvilinear) squares such as AEFG and FEIH for which AE/AG $\approx$ EI/EF. Several attempts may be needed to obtain a reasonable map in which the majority of the 'squares' are very nearly actual squares.

Greater accuracy can always be obtained by drawing smaller squares but obviously this will entail more work. Thus a

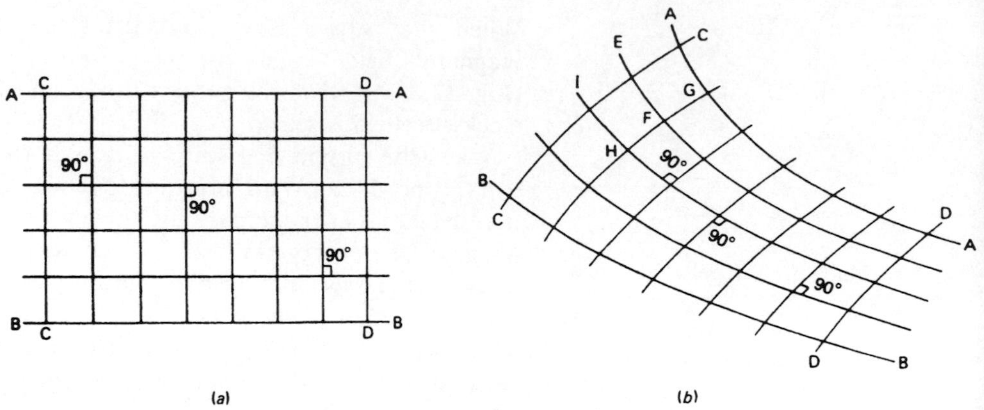

Fig.12.4 Mapping a field: (a) linear; (b) non-linear

reasonable compromise between accuracy and time must be settled upon.

Example 12.2

Figure 12.5 shows two metal plates with the upper plate at a potential 100 V more positive than the lower plate. Map the electric field between the plates.

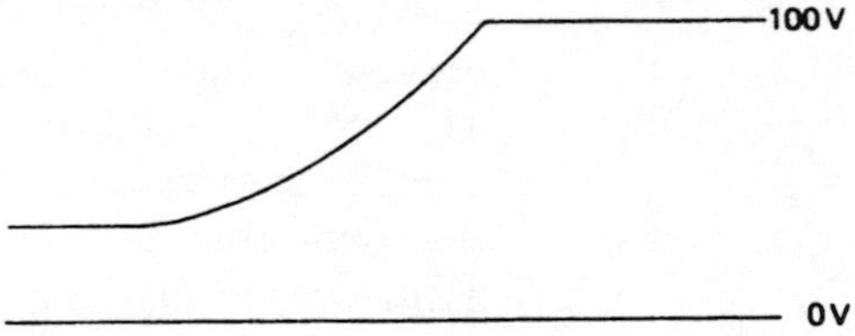

Fig.12.5

Solution

The first step is to assume that the electric field is uniform on both the left-hand and the right-hand sides of the system where the two plates are parallel to one another. Other equi-potentials can then be drawn at regular intervals between the plates; this has been done in Fig. 12.6(*a*) for the equi-potentials at 75 V, 50 V and 25 V.

The next step is to draw smooth curves, following the curvature of the upper plate, to join up the equi-potentials already drawn. This step has been carried out in Fig.12.6(*b*).

The conjugate pattern of streamlines must now be drawn. The attempt must be made to form curvilinear squares that are as near actual squares as possible. It should be remembered that all intersections must be made at right angles. Fig. 12.6(*c*) shows the map after this step has been taken.

The final step is to increase the accuracy of the mapping by reducing the size of the squares by inserting extra equi-potentials and streamlines and adjusting the pattern if and where necessary. The final mapping is shown in Fig. 12.6(*d*).

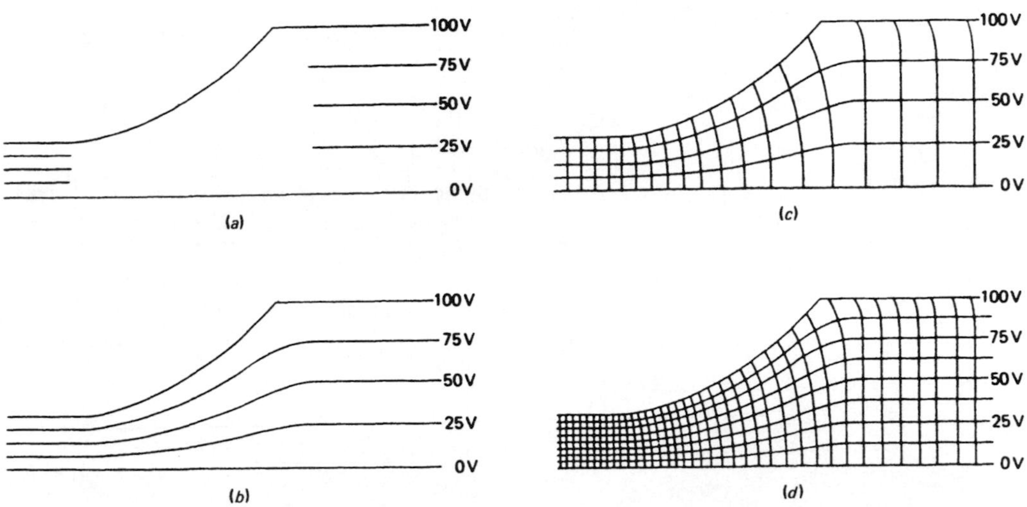

Fig. 12.6

The electric field distribution is then given by the streamlines drawn in Fig. 12.6(*d*).

Calculations

Once the mapping of a field has been completed it can be used to determine some of the parameters of the system. The field really exists in three dimensions so that each curvilinear square in the map is actually a cube, the third dimension being the depth d metres. Hence, if m is the number of squares measured along each equi-potential, n is the number of squares measured along each streamline and d is the depth of field in metres, then

- Conduction field:

  Conductance $G = \sigma dm/n$ S (12.11)

  where σ is the conductivity of the material.
- Electric field:

  Capacitance $C = \varepsilon dm/n$ F (12.12)

  where ε is the permittivity of the material.
- Magnetic field:

  $1/\text{reluctance} = 1/S = \mu dm/n$ Wb/A (12.13)

  where μ is the permeability of the material.
- Thermal field:

  Thermal conductance $= C_{th}\ dm/n$ W/°C (12.14)

  where C_{th} is the thermal conductivity of the material.

In all four cases the value for a single square is found by putting $m = n = d = 1$.

Example 12.3

The mapping of the electric field between two concentric cylinders contains 4 rows of squares with 26 squares in each row. (*a*) Calculate the capacitance of the cylinders if the relative permittivity of the dielectric is 4. Take $\varepsilon_0 = 8.854 \times 10^{-12}$ F/m. (*b*) Calculate the inductance if the permeability of the material between the cylinders is 600. (*c*) Calculate the resistance if the conductivity of the material is 1×10^{-6} S/m.

Solution

(*a*) $C = (4 \times 8.854 \times 10^{-12} \times 26)/4 = 0.23 \times 10^{-9}$ F/m (*Ans.*)

(*b*) $L = (600 \times 4\pi \times 10^{-7} \times 26)/4 = 4.9$ mH/m (*Ans.*)

(*c*) $G = (1 \times 10^{-6} \times 26)/4 = 6.5 \times 10^{-6}$ S
$R = 1/G = 153.8$ kΩ (*Ans.*)

Example 12.4

Determine the capacitance of a parallel-plate capacitor whose 8 cm × 5 cm plates are 1 mm apart if $\varepsilon_r = 3$.

Solution
Divide the 1 mm distance into equal parts, giving $n = 2$, and 0.5 mm squares. Then the number of squares M in an 8 cm distance is $80/0.5 = 160$. Therefore, since $d = 5 \times 10^{-2}$

$$C = [(160 \times 8.854 \times 10^{-12} \times 3)/2] \times 5 \times 10^{-2} = 106 \text{ pF} \quad (\textit{Ans.})$$

Example 12.5

The two concentric cylinder system shown in Fig. 12.1(*b*) is mapped with curvilinear squares. If there are 6 equi-potentials between the inner and outer cylinders, and 16 streamlines determine the capacitance per metre. $\varepsilon_r = 3$.

Solution
From equation (12.12),

$$C = [(3 \times 16)/6] \times 8.854 \times 10^{-12} = 70.83 \text{ pF/m} \quad (\textit{Ans.})$$

Capacitance of systems

Capacitance of a two-wire line

Figure 12.7 shows two long parallel conductors each of radius r metres which are spaced d metres apart. Suppose the charge per unit length on each conductor is Q coulombs and the permittivity of the dielectric is $\varepsilon = \varepsilon_r\varepsilon_0$. Then, from equation

(12.7) and since $D = \varepsilon E$, the electric field strength at distance x from a conductor is given by equation (12.15)

$$E = Q/2\pi\varepsilon x \text{ V/m} \tag{12.15}$$

Therefore at the point distance x from the left-hand conductor,

$$E = Q/(2\pi\varepsilon x) - (-Q)/[2\pi\varepsilon(d-x)]$$
$$= (Q/2\pi\varepsilon)[1/x + 1/(d-x)]$$

The capacitance C of the line is $C = Q/V = Q/\int E\,\mathrm{d}x$.

$$\int_r^{d-r} E\,\mathrm{d}x = \int_r^{d-r} [Q/(2\pi\varepsilon)][1/x + 1/(d-x)]\,\mathrm{d}x$$
$$= [Q/(2\pi\varepsilon)][\log_e x - \log_e(d-x)]_r^{d-r}]$$
$$= [Q/(\pi\varepsilon)]\{\log_e[(d-r)/r] - \log_e[r/(d-r)]\}$$
$$= [Q/(2\pi\varepsilon)]\log_e[(d-r)/r]$$

Therefore,

$$C = Q/V = \pi\varepsilon/\{\log_e[(d-r)/r]\} \text{ F/m} \tag{12.16}$$

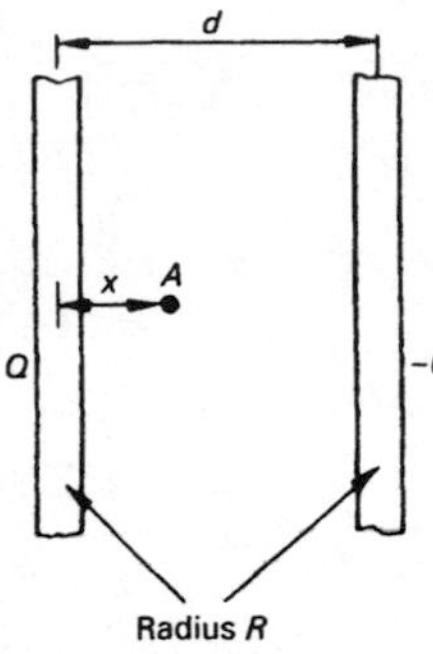

Fig. 12.7 Two parallel conductors

Example 12.6

A two-wire line uses conductors of 4.8 mm diameter which are held 20 mm apart by insulating material whose relative permittivity is 2.1. Calculate the capacitance of a 100 m length of the line.

Solution

$$C = [\pi \times 8.854 \times 10^{-12} \times 2.1]/\{\log_e[(20 - 4.8)/4.8]\}$$
$$= 50.68 \text{ pF/m}$$

Therefore,

Capacitance of 100 m length = 5.07 nF (*Ans.*)

Electric field between two conductors

The electric field between the two conductors is given by equation (12.15), i.e.

$$E = Q/(2\pi\varepsilon x) = (V\pi\varepsilon)[1/(2\pi\varepsilon x)]/\{\log_e[(d-r)/r]\}$$
$$E = V/\{2x \log_e[(d-r)/r]\} \tag{12.17}$$

Capacitance of a coaxial line

Figure 12.8 shows two concentric cylinders or a coaxial cable. The inner conductor has a radius of r while the inner radius of

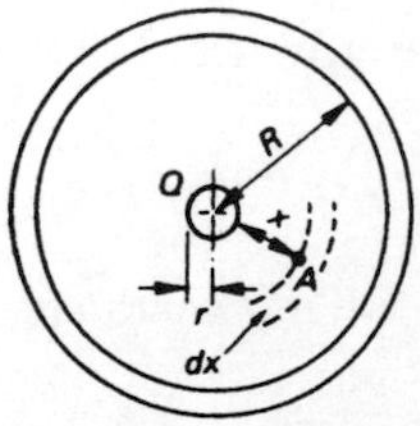

Fig. 12.8 Two concentric cylindrical conductors

the outer conductor is R. The electric field strength at a point distance x from the centre of the inner conductor is

$$E = Q/(2\pi\varepsilon x)\ \text{V/m} \qquad (12.18)$$

The voltage between the conductors is

$$V = \int_r^R E\,dx = [Q/(2\pi\varepsilon)]\int_r^R (1/x)\,dx = [Q/(2\pi\varepsilon)][\log_e x]_r^R$$

$$V = [Q/(2\pi\varepsilon)]\log_e(R/r)\ \text{V}$$

$$\text{Capacitance } C = Q/V = (2\pi\varepsilon)/[\log_e(R/r)]\ \text{F/m} \qquad (12.19)$$

Electric field between a pair of coaxial conductors

$$E = Q/(2\pi\varepsilon x) = \{2\pi\varepsilon V/[\log_e(R/r)]\}[1/(2\pi\varepsilon x)]$$

$$E = V/[x\log_e(R/r)]\ \text{V/m} \qquad (12.20)$$

The electric field strength varies inversely with the distance x from the inner conductor and it has its maximum value at the surface of the inner conductor, i.e. when $x = r$. Thus,

$$E_{max} = V/[r\ \log_e(R/r)] \qquad (12.21)$$

Transposing equation (12.21) gives

$$R = r\,e^{V/rE_{max}} \qquad (12.22)$$

The best cable has the minimum outer diameter for given values of applied voltage and permissible maximum electric stress. To determine the minimum radius for the outer conductor differentiate equation (12.22) with respect to r and equate the result to zero. Then,

$$dR/dr = 0 = e^{V/rE_{max}} - (r\,e^{V/rE_{max}} \cdot V/r^2E_{max})$$

$$r = V/E_{max}$$

Substituting this result into equation (12.22), the minimum value for the diameter of the outer conductor is

$$R_{min} = r\,e^{VE_{max}/VE_{max}} = r\,e^1 = 2.7182r \qquad (12.23)$$

Example 12.7

A coaxial cable has an inner diameter of 0.26 cm and an outer diameter of 0.95 cm. Calculate its capacitance and the maximum potential gradient when the applied voltage is 250 V, given that $\varepsilon_r = 1$.

Solution

From equation (12.19)

$$C = (2\pi \times 8.854 \times 10^{-12})/[\log_e(0.95/0.26)] = 43\ \text{pF/m} \quad (\textit{Ans.})$$

From equation (12.21),

$$E_{max} = 250/[2.6 \times 10^{-3} \log_e(0.95/0.26)] = 74.21 \text{ kV/m} \quad (Ans.)$$

Example 12.8

Re-calculate the capacitance of the cable in example 12.7 by mapping the electric field.

Solution

Consider a segment θ of the inter-conductor space, see Fig. 12.9. Its capacitance is $C = C\theta/2\pi$, where C is the total capacitance. The capacitance of a square is ε and hence

$$\varepsilon = [(2\pi\varepsilon)/\log_e(R/r)] \times (\theta/2\pi) = \varepsilon\theta/[\log_e(R/r)].$$

$$\theta = \log_e(R/r) \quad \text{or} \quad e^{\theta} = R/r$$

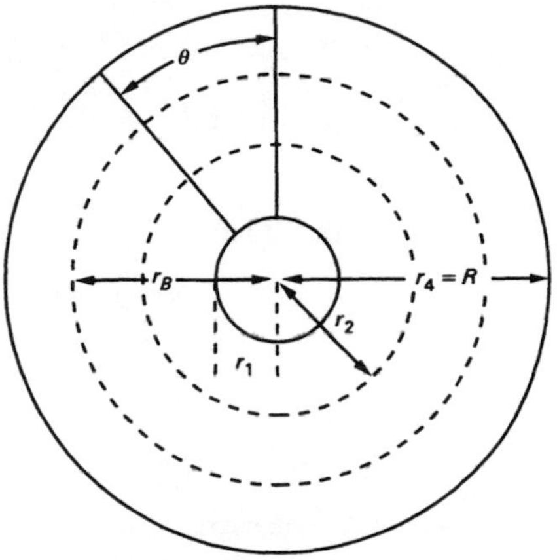

Fig. 12.9

If the segment is divided into, say, three parts as in the figure, then for each part $e^{\theta} = r_n/r_{n-1}$, so that $e^{3\theta} = R/r = 0.95/0.26$ and $\theta = 0.432$ radians. Therefore, there are $2\pi/14.55$ segments and

$$C = (8.854 \times 10^{-12} \times 14.55)/3 \approx 43 \text{ pF/m} \quad (Ans.)$$

Example 12.9

A concentric cylindrical capacitor has an outer conductor whose inner diameter is 60 mm. Calculate (*a*) the diameter of the inner conductor to give minimum electric stress at the surface of the inner conductor and (*b*) the maximum voltage that can be applied to the capacitor if the breakdown voltage of the dielectric is 6 kV/mm.

Solution

(*a*) From equation (12.23),

$$r = R_{min}/e = 60/2.718 = 22.08 \text{ mm} \quad (Ans.)$$

(*b*) From equation (12.21)

$$6000 = V_{max}/(22.08 \ \log_e e)$$
$$V_{max} = 6000 \times 22.08 \times 1 = 132.48 \text{ kV} \quad (\textit{Ans.})$$

Concentric sphere capacitors

A capacitor may be formed by two concentric metal spheres of radii R and r, where $R > r$. The flux density at a distance x from a sphere is $Q/4\pi x^2$ and $E = Q/4\pi\varepsilon x^2$ V/m, where Q is the charge on the inner sphere. The voltage between the two spheres is

$$V = \int_r^R E \, dx = (Q/4\pi\varepsilon) \int_r^R dx/x^2$$
$$= (Q/4\pi\varepsilon)(-1/x)_r^R$$
$$= (Q/4\pi\varepsilon)(1/r - 1/R) \text{ V} \quad \text{and}$$

$$C = Q/V = 4\pi\varepsilon/(1/r - 1/R) \tag{12.24}$$

Electric field

$$E = Q/4\pi\varepsilon x^2 = CV/4\pi\varepsilon x^2 = [4\pi\varepsilon/(1/r - 1/R)][V/4\pi\varepsilon x^2]$$
$$= V/[x^2(1/r - 1/R)] \tag{12.25}$$

Example 12.10

A concentric sphere capacitor has an inner radius of 80 mm and an outer radius of 400 mm. It has a voltage of 80 kV applied across its terminals. Calculate the maximum value of the electric field strength in the capacitor.

Solution

The electric field strength is at its maximum value when $x = r$. Then,

$$E_{max} = 80 \text{ kV}/(0.08 - 0.016) = 80 \text{ kV}/0.064$$
$$= 1.25 \text{ MV/m} \quad (\textit{Ans.})$$

Resistance of systems

The analogy between electric and conduction fields provides a means of determining the resistance of a system. The capacitance C between two conductors separated by a dielectric of permittivity $\varepsilon = \varepsilon_0\varepsilon_r$, and the resistance R between the same two conductors separated by a medium of resistivity ρ are related by:

$$C/\varepsilon = \rho/R = 1/\sigma R \tag{12.26}$$

where σ is the conductivity of the medium.

This relationship may be employed to determine the resistance of a system using one of the expressions derived for capacitance suitably modified using equation (12.26). Hence the equation (12.16) for the capacitance between two conductors can be modified to give the insulation resistance between the conductors.

$$R = \{\rho \log_e[(d - r)/r]\}/\pi \ \Omega/\text{m} \tag{12.27}$$

and equation (12.19) becomes

$$R = \rho[\log_e(R/r)]/2\pi \tag{12.28}$$

Example 12.11

Calculate the insulation resistance between two conductors of radius 1.5 cm which are spaced 15 cm apart. The resistivity of the insulating medium is 10^{13} Ωm.

Solution

$$R = 10^{13} \log_e[(15 - 1.5)/1.5]/\pi = 7 \times 10^{12} \ \Omega/\text{m} \quad (\textit{Ans.})$$

Inductance of systems

Inductance of a two-wire line

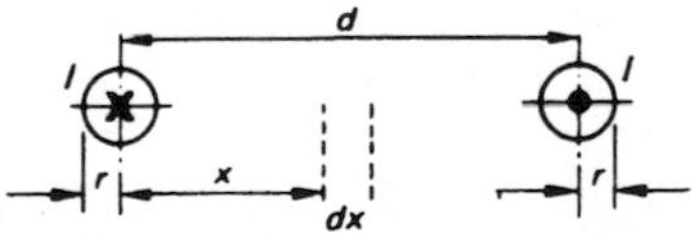

Fig. 12.10 Two parallel conductors

Figure 12.10 shows a twin line whose conductors are of radius r separated by a spacing of d metres. Suppose the left-hand conductor carries a current of I amperes in the direction into the paper, while the right-hand conductor carries the same current out of the paper.

The magnetic field strength H at distance x from the left-hand conductor is $H = I/2\pi x$ A/m. The flux passing through the small distance dx is

$$\Phi = [(\mu_0 I/2\pi x) + (\mu_0 I/2\pi(d - x))] \ dx \text{ per metre length}$$

The total flux linking the two conductors is

$$\int_r^{d-r} \Phi \ \mathrm{d}x = (\mu_0 I/2\pi) \int_r^{d-r} [1/x + 1/(\mathrm{d} - x)] \ \mathrm{d}x$$
$$= (\mu_0 I/\pi) \log_e[(d - r)/r]$$

Now $L = N\Phi/I = \Phi/I$ so that

$$L = (\mu_0/\pi) \log_e[(d - r)/r] \ \text{H/m} \tag{12.29}$$

There is a further component to the inductance because of the flux linkages inside the two conductors themselves. This extra component is equal to $\mu_0/8\pi$ per conductor. For a two-wire line there are two conductors, so the increase is $\mu_0/4\pi$ H/m.

Example 12.12

Calculate the inductance of the two-wire line of example 12.6.

Solution

$$L = [(4\pi \times 10^{-7})/\pi] \log_e[(20 - 4.8)/4.8] + (4\pi \times 10^{-7})/4\pi$$
$$= 4.61 \times 10^{-7} + 1 \times 10^{-7}$$

For a 100 m length of line

$$L = 5.61 \times 10^{-7} \times 100 = 56.1\ \mu\text{H} \quad (\textit{Ans.})$$

Inductance of a coaxial pair

Suppose the outer conductor shown in Fig. 12.8 carries a current I amperes in one direction while the inner conductor carries the same current in the opposite direction. The magnetic field strength at the point marked as A due to the current flowing in the inner conductor is $H = I/2\pi x$ AT/m. The flux passing through the small element dx is (since $BA = \mu HA = \mu I\, dx$ since $N = 1$).

$$d\Phi = [(\mu_0 I/2\pi x) + (\mu_0 I)/(2\pi(R - x))]\ dx \text{ per metre length}$$

The total flux linking the two conductors is

$$\Phi = (\mu_0 I/2\pi) \int_r^R [(1/x) + 1/(R - x)]\ dx$$
$$= (\mu_0 I/2\pi)\ \log_e(R/r)$$

and so,

$$L = \Phi/I = [\mu_0\ \log_e(R/r)]/2\pi \qquad (12.30)$$

As for the two-wire line there is an extra component to the inductance equal to $\mu_0/4\pi$ because of flux linkages inside the conductors.

Example 12.13

An air-spaced coaxial cable has an inner conductor of radius 0.16 cm and an outer conductor of inner radius 0.55 cm. Calculate its inductance.

Solution

$$L = [4\pi \times 10^{-7}\ \log_e(0.55/0.16)]/2\pi + (4\pi \times 10^{-7})/4\pi$$
$$= (247 + 100) \times 10^{-9} = 347\ \text{nH} \quad (\textit{Ans.})$$

Exercises 12

12.1 A capacitor consists of two 400 mm long concentric cylinders. The capacitance of the component is 60 pF and the maximum electric field strength is 3.5 MV/m when the applied voltage is 235 kV r.m.s. Calculate the inner and outer radii of the two conductors.

12.2 Show that the voltage at any point distance x from the centre of two coaxial cylinders is given by

$$V = [V_{rR} \log_e(R/x)]/\log_e(R/r)$$

where R and r are the radii of the outer and the inner cylinders respectively.

12.3 A dielectric of relative permittivity 5 is 5 cm thick and has an area of 20 cm^2. The applied voltage is 1000 V. Calculate the energy stored in the dielectric. Calculate also the capacitance of a parallel-plate capacitor made using this dielectric.

12.4 Calculate (a) the capacitance and (b) the inductance per metre of a two-wire line that consists of two 1.5 cm diameter conductors which are spaced 15 cm apart. Assume the dielectric to be air.

12.5 A coaxial cable pair has an inner conductor of radius 0.15 cm and an outer conductor of inner radius 0.5 cm. Calculate the capacitance of the pair if the relative permittivity of the dielectric is 2.5.

12.6 Write down the expression for the capacitance per metre of two concentric cylinders. Use the analogy between electric and conduction fields to calculate the insulation resistance between the two conductors shown in Fig. 12.11 if the radius of the inner is 6 mm, the inner radius of the outer is 22 mm, and the resistivity of the insulating material is 9×10^{12} Ωm.

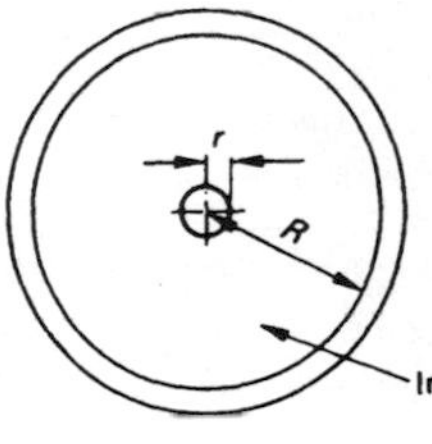

Fig. 12.11

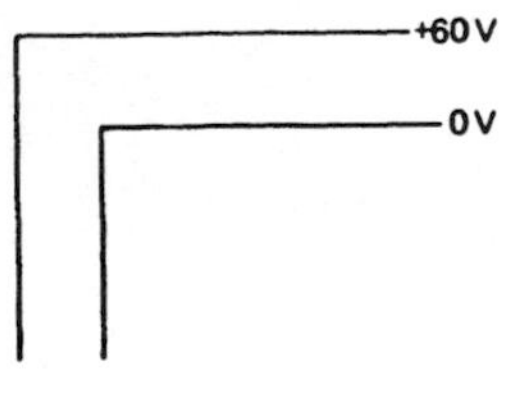

Fig. 12.12

12.7 Map the electric field in the system shown in Fig. 12.12

12.8 The voltage at any point at radius x from the centre of the inner of two concentric cylinders is given by

$$V_x = V \log_e[(R/x)/(R/r)]$$

Calculate the voltage midway between the cylinders if the applied voltage is 100 kV and $R/r = 3$.

12.9 The mapping of the current conduction field of a particular system was achieved using 5 rows of curvilinear squares each of which contained 12 squares. If the conductivity of the material is 60×10^6 siemen/metre calculate the current that would flow when 120 mV are applied to the system.

12.10 For the system shown in Fig. 12.13 map the electric field and estimate the potentials at D, E and F.

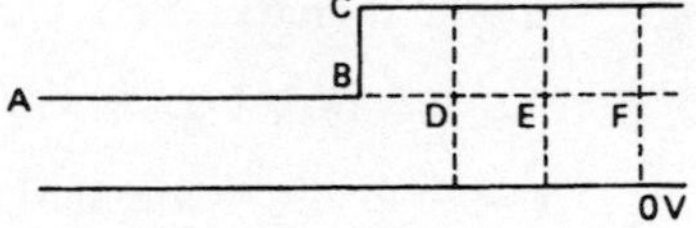

Fig. 12.13

12.11 Show that the extra component of inductance due to flux linkages inside the conductor, equations (12.29) and (12.30), is $\mu_0/4\pi$.

12.12 A coaxial cable uses a dielectric with $\varepsilon_r = 4$ and a maximum electric field strength of 14.14 μV/m. The applied voltage is to be 100 kV. Determine the dimensions of the best cable.

12.13 Two air-spaced conductors of radius 1 cm have a capacitance of 12 pF/m. Determine their distance apart.

12.14 Two conductors are spaced 25 cm apart. Each conductor has a radius of 1.2 cm and the insulation between the conductors has $\varepsilon_r = 2.3$. Calculate the voltage applied across the conductors that will produce an electric field of 22 kV/m between them.

12.15 Two conductors have a diameter of 1 cm and are 5 cm apart. Calculate their inductance.

12.16 A conductor carries a current of 22 A. Determine the field strength and the flux density produced at a distance of (a) 1 m and (b) 2 m.

12.17 A pair of concentric cylinders has an outer radius of 24 mm and an inner radius of 8 mm. A voltage of 50 V is applied across the cylinders. Determine the voltage at a radius of 20 mm.

12.18 The region between the inner and outer conductors of a coaxial cable is mapped into curvilinear squares. If the squares are drawn so that there are 7 equi-potentials and 28 streamlines, calculate the capacitance of the cable. $\varepsilon_r = 3.8$.

13 Energy transfer

Energy is the capability to do work. Any machine that possesses some energy is able to perform work, converting energy from one form to another. Conversely, work must have been done in order to provide the machine with energy in the first place. A machine will accept input energy in one form and convert some of it into output energy in some other form. Some of the output energy will not be useful since it will be provided in the form of heat. The part of the input energy that does not appear at the output is stored within the machine. This means that work done and energy are two aspects of the same thing and they are both measured in terms of the same unit; namely the *Joule*.

> 1 joule is the work done when a force of 1 newton acts over a distance of 1 metre in the direction of the force.

There are several different forms of energy; those considered in this chapter are electrical, mechanical, chemical, heat, light and sound.

Principle of energy conversion

The principle of energy conversion states that energy can be neither created nor destroyed, but only changed from one form to another. This means that all the input energy to a machine either must appear at the output of the machine or must be stored within it.

Several examples of energy conversion are well known:

- In both primary and secondary cell batteries, chemical energy is changed into electrical energy.
- In power stations, chemical energy is obtained from the burning of coal and oil and is changed first into heat, then into mechanical energy, and, finally, by means of a generator into electrical energy.

- In a telephone transmitter sound energy is changed into electrical energy, while in the receiver the opposite process takes place; that is, electrical energy is converted into sound energy.
- In an electric light bulb, electrical energy is changed into both light and heat energy.
- In an electric fire, electrical energy is changed into heat energy with some light energy also.
- Electric motors and generators of many kinds convert electrical energy into mechanical energy or vice versa.

It is well known, in some of these cases at least, that not all of the input energy is converted into the required form of output energy. Take, for example, the electric bulb and fire. The electric light bulb produces quite a lot of unwanted heat energy as well as the wanted light energy, while an electric fire generally produces some light energy as well as heat.

This is always true; some of the input energy is always used for the production of energy in one, or more, unwanted forms. As a result the efficiency of any energy conversion device or transducer always falls short of 100 per cent.

An electric field is a region of space within which any electric charges are subject to a force. Work is done in an electrical field when a number of electrons move from one point to another. The work done is equal to QV where Q is the total charge in coulombs that is transferred and V is the potential difference between the two points. Since $Q = It$, where I is the current in amperes and t is the time in seconds, the energy is given by

$$W = QV = IVt \text{ joules} \tag{13.1}$$

$$= I^2Rt = V^2t/R \text{ joules} \tag{13.2}$$

Power is the rate of doing work and its unit is the watt. 1 watt is the power dissipated when energy is used at the rate of 1 J/s, i.e. energy = power × time.

Energy in an electric field

Capacitance

If a capacitance of C μF is charged to have a voltage of V volts across its terminals, the charge Q stored is

$$Q = CV \text{ coulombs} \tag{13.3}$$

If the voltage across the capacitor is then increased to $V + \mathrm{d}V$ volts in a time of $\mathrm{d}t$ seconds the current taken must be equal to $I = C\,\mathrm{d}V/\mathrm{d}t$. The energy supplied to the capacitor is $VI\,\mathrm{d}t = VC(\mathrm{d}V/\mathrm{d}t) \times \mathrm{d}t = VC\,\mathrm{d}V$ joules.

The total energy W supplied as the capacitance is charged from zero volts to V volts terminal voltage is

$$W = C\int_0^V V\,\mathrm{d}V = CV^2/2 \text{ joules} \tag{13.4}$$

or $$W = QV/2 \text{ joules} \tag{13.5}$$

Example 13.1

(*a*) A 1 μF capacitor is charged from a 24 V d.c. supply and is then disconnected from the supply. Calculate the energy stored in the capacitor.

(*b*) Two uncharged capacitors, one of 2 μF and one of 5 μF, are now connected in series with one another and then the series combination is connected in parallel with the 1 μF capacitor. Calculate the energy stored in each capacitor and the total energy stored. Assume no loss of charge occurs at any time.

Solution

(*a*) $W = 0.5 \times 1 \times 10^{-6} \times 24^2 = 288 \times 10^{-6}$ J (*Ans.*)

(*b*) When the 2 μF and 5 μF capacitors are connected in series the total capacitance is

$1 + [5 \times 2/(5+2)] = 2.43$ μF

The voltage V across the combination (Fig 13.1) is

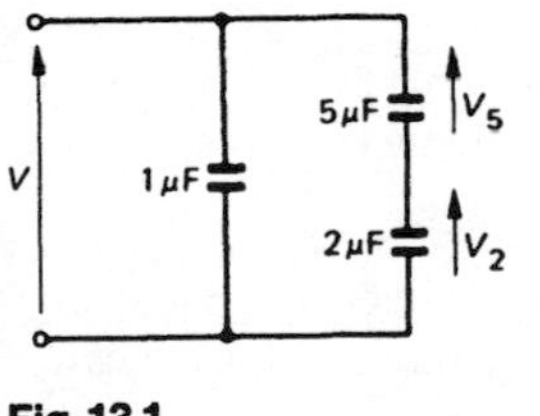

Fig.13.1

$V = Q/C = (24 \times 1 \times 10^{-6})/(2.43 \times 10^{-6}) = 9.88$ V

The energy stored in the 1 μF capacitor is

$0.5 \times 1 \times 10^{-6} \times 9.88^2 = 48.8 \times 10^{-6}$ J (*Ans.*)

The voltage V_5 across the 5 μF capacitor $= 9.88 \times 2/7 = 2.82$ V. Hence, the energy stored is

$0.5 \times 5 \times 10^{-6} \times 2.82^2 = 19.9 \times 10^{-6}$ J (*Ans.*)

The voltage V_2 across the 2 μF capacitor $= 9.88 \times 5/7 = 7.06$ V. So, the energy stored in this capacitor is

$0.5 \times 2 \times 10^{-6} \times 7.06^2 = 49.8 \times 10^{-6}$ J (*Ans.*)

The total energy stored is

$(48.8 + 19.9 + 49.8) \times 10^{-6} = 118.5 \times 10^{-6}$ J (*Ans.*)

The total energy stored is less than that which was originally stored in the 1 μF capacitor. This is because some energy has been dissipated in the form of heat (I^2R) when current flowed to re-distribute the charge when the new capacitors were connected into the circuit.

Example 13.2

A 2 μF capacitor is connected in series with a 100 kΩ resistor and a 120 V d.c. supply. Calculate the rate at which energy is being stored in the capacitor when its terminal voltage is 40 V.

Solution

When $V_C = 40$ volts the voltage across the resistor must be 80 volts and so the current flowing is $80/100 = 0.8$ mA. At this instant therefore, the power supplied to the circuit is $120 \times 0.8 \times 10^{-3} = 96$ mW, and the power dissipated in the resistor is $80 \times 0.8 \times 10^{-3} = 64$ mW. Therefore, the rate at which energy is stored in the capacitor is

$$96 - 64 = 32 \text{ mW} \quad (\textit{Ans.})$$

Dielectrics

If a very small portion, dx deep and dy wide of length 1 metre, of dielectric is considered, then for this small volume, regardless of the shape of the dielectric, both the lines of force and the equipotential lines are parallel. The minute portion of the dielectric therefore acts just like a parallel plate capacitor.

The capacitance of a parallel plate capacitor is $C = \varepsilon A/d$ F and hence for the very small piece of dielectric, $C = \varepsilon\, dx/dy$ and the energy stored is $W = [(\varepsilon\, dx/dy)/2](E\, dy)^2 = (\varepsilon\, dx\, dy\, E^2)/2$ J.

The energy stored in the dielectric per cubic metre is

$$W = \varepsilon E^2/2 \text{ J/m}^3 \qquad (13.6)$$

or $$W = \varepsilon D^2/2\varepsilon^2 = D^2/2\varepsilon \text{ J/m}^3 \qquad (13.7)$$

$$= DE/2 \text{ J/m}^3 \qquad (13.8)$$

Example 13.3

A rectangular slab of a dielectric material is 2 cm thick and has a voltage of 600 V applied across it. The energy stored is then 2 μJ. Calculate the flux density if the area of the slab is 24 cm^2.

Solution

The volume of the dielectric is

$$2 \times 10^{-2} \times 24 \times 10^{-4} = 48 \times 10^{-6} \text{ m}^3$$

and so the energy stored per cubic metre is

$$(2 \times 10^{-6})/(48 \times 10^{-6}) = 41.67 \text{ mJ/m}^3$$

Therefore,

$$D = (0.0417 \times 4 \times 10^{-2})/600 = 2.78 \times 10^{-6} \text{ C/m}^2 \quad (\textit{Ans.})$$

Force between two charged plates

Suppose the plates are disconnected from both the supply voltage and the external circuit so that charge can neither be supplied nor lost. If, then, one of the plates is fixed in position and the other plate is moved away through a distance of dx metres, the stored charge will not change. The capacitance C of the plates will be reduced and so the voltage V between the plates must increase. Since C is inversely proportional to d, and V is inversely proportional to C, the ratio V/d will remain constant. Consequently the energy stored per cubic metre is unchanged. This means that the energy stored in the extra volume of dielectric must have been supplied by the work done in moving the plate. Therefore, if the plate area is A

$$F\,\mathrm{d}x = (\varepsilon V^2/2d^2)A\,\mathrm{d}x$$

Hence, the force F between two charged plates is

$$F = (\varepsilon A V^2)/(2d^2) \text{ joules} \tag{13.9}$$

or $$F = QV/2d = W/d \text{ joules} \tag{13.10}$$

Example 13.4

Calculate the force of attraction between two air-spaced metal plates; each of area 100 cm^2 and 2 mm apart, which have a voltage of 3 kV applied across them.

Solution
From equation (13.9),

$$F = W = [(8.854 \times 10^{-12}) \times 100 \times 10^{-4} \times (3000)^2]/[2 \times (2 \times 10^{-3})^2] = 99.6 \text{ mN} \quad (Ans.)$$

Energy in a Magnetic Field

When the current flowing in an inductance of L henrys increases in value, the e.m.f. induced into the inductance is equal to $-L\,\mathrm{d}i/\mathrm{d}t$ volts. The energy stored in the magnetic field of the inductance in a time of dt seconds will be

$$iv\,\mathrm{d}t = i\,\mathrm{d}tL\,\mathrm{d}i/\mathrm{d}t = Li\,\mathrm{d}i \text{ joules}$$

The total energy absorbed when the current increases from zero to I amperes is

$$W = L\int_0^I i\,\mathrm{d}i = \tfrac{1}{2}LI^2 \text{ joules} \tag{13.11}$$

provided the inductance is constant.

Mutual Inductance

Consider two coils of self-inductances L_1 and L_2 and suppose that L_2 is initially open circuited. If a current I_1 flows in L_1 the energy stored is $W_1 = L_1 I_1^2/2$ joules. If, now, the current I_1 is kept constant while another current I_2 is passed through L_2 the extra energy which is stored is $W_2 = L_2 I_2^2/2 + W_M$, where W_M is the energy that is required to keep current I_1 constant as the current I_2 tends to reduce I_1.

$$W_M = \int_0^t (M\,dI_2/dt)I_1\,dt = \int_0^{I_2} MI_1\,dI_2 = MI_1I_2$$

Therefore,

$$W = L_1 I_1^2/2 + L_2 I_2^2/2 \pm MI_1I_2 \qquad (13.12)$$

Example 13.5

Two inductors of 4 H and 2 H have a mutual inductance of 1 H. The inductors are connected in series aiding and have a current 0.5 A passing through them. Calculate the stored energy.

Solution

$$W_T = 4 \times 0.25/2 + 2 \times 0.25/2 + 1 \times 0.25 = 1\ \text{J} \quad (\textit{Ans.})$$

Example 13.6

A coil has an inductance of 2 H and an effective resistance of 10 Ω and it is connected across a 12 V d.c. supply. Calculate the rate at which energy is being stored in the magnetic field when the current is 0.4 A. Also find the maximum energy stored.

Solution

When the current is 0.4 A the voltage dropped across the series resistor is $0.4 \times 10 = 4$ V. The input power is $12 \times 0.4 = 4.8$ W and the power dissipated is $4 \times 0.4 = 1.6$ W. Therefore, the rate at which energy is being stored is

$$4.8 - 1.6 = 3.2\ \text{W} \quad (\textit{Ans.})$$

Maximum current = final current = $12/10 = 1.2$ A. Therefore, maximum energy stored is

$$0.5 \times 1.2^2 \times 2 = 1.44\ \text{joules} \quad (\textit{Ans.})$$

Energy stored in a magnetic material

The maximum energy stored in an inductance L was shown on page 227 to be equal to $LI^2/2$ joules. But $L = N\Phi/I$ and $H = NI/l$ and therefore

$$W = \Phi I/2 = (BAHl)/2 \quad (\text{since } N = 1)$$

$$W = BH/2 = \mu H^2/2\ \text{J/m}^3 \qquad (13.13)$$

Example 13.7

Calculate the energy stored in a magnetic material of relative permeability 1000 when the magnetizing force is 2000 A/m.

Solution

From equation (13.13),

$$W = (10^3 \times 4\pi \times 10^{-7} \times 4 \times 10^6) = 2513.3 \text{ J/m}^3 \quad (Ans.)$$

Magnetic circuit equations

When energy calculations are to be performed upon a magnetic circuit, a knowledge of the various magnetic circuit equations is required. These should have been studied at an earlier stage and so this chapter will only quote the relevant formulae. (l is the length of the magnetic path in metres.)

- Magnetomotive force (m.m.f.) = NI ampere (A) (13.14)
- Magnetizing force $H = NI/l$ ampere/metre (A/m) (13.15)
- Magnetic flux Φ = m.m.f./reluctance webers (Wb) (13.16)
- Flux density B = flux/area = Φ/A tesla (T) (13.17)
- Absolute permeability ($\mu = \mu_r\mu_0$)

$$= \frac{\text{Flux density}}{\text{magnetizing force}}$$

$$= B/H \text{ henry/metre (H/m)} \quad (13.18)$$

- Relative permeability μ_r

$$= \frac{\text{Flux density in material}}{\text{Flux density in air}} \quad (13.19)$$

(The flux density must be caused by the same magnetizing force.)

- μ_0 = permeability of air (strictly a vacuum)

$$= 4\pi \times 10^{-7} \text{ H/m} \quad (13.20)$$

- Reluctance S = m.m.f./flux = $NI/\Phi = Hl/BA$

$$= l/\mu A \text{ A/Wb} \quad (13.21)$$

- Inductance = flux linkages per ampere = $N\Phi/I$ (13.22)

$$= (N/I)(NI/S) = N^2/S \quad (13.23)$$

The energy stored in an inductance is given by equation (13.11). Hence,

$$W = 0.5I^2N^2/S = (0.5I^2N^2\mu A)/l = [(NI)^2/2l^2]\mu Al$$
$$= \mu H^2Al/2 \text{ joules}$$

Therefore, the energy stored per cubic metre is

$$W = \mu H^2/2 = BH/2 = B^2/2\mu \text{ J/m}^3 \quad (13.24)$$

When this expression is used to determine the force of attraction between two magnetic surfaces it is the energy stored in the field within the air gap that is important and hence equation (13.24) becomes

$$W = B^2/2\mu_0 \text{ J/m}^3 \qquad (13.25)$$

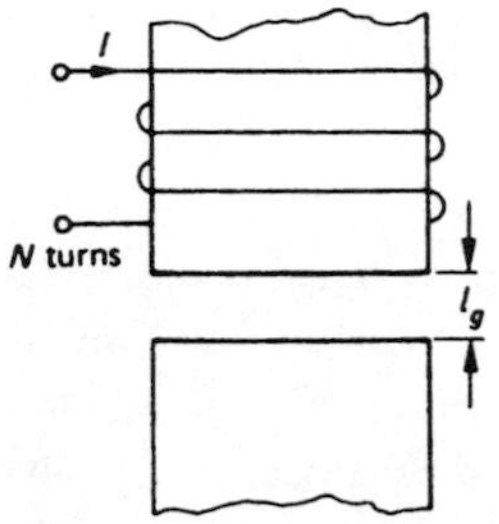

Fig. 13.2

Consider Fig. 13.2 which shows two parallel flat surfaces, made from a magnetic material, and separated from one another by an air gap of length l_g metres. The upper surface has a coil of N turns wound around it that carries a current of I amperes. Suppose a force of F newtons is exerted upon the lower surface to move it away from the upper surface through a distance of dl_g. The reluctance of the air gap will then increase and, if the flux density in the air gap is to be maintained constant at its original value, the current I must also increase by the appropriate amount. The flux linkages will then be unchanged and so there will be zero induced e.m.f. in the winding. This means that zero electrical energy will have been supplied to the circuit. Therefore the extra energy stored in the air gap must have been derived from the work done by the applied force. Hence,

$$F\, dl_g = B^2/2\mu_0 \cdot A\, dl_g, \quad \text{and} \quad F = B^2A/2\mu_0 \text{ N} \qquad (13.26)$$

The various energy changes that take place during the operation of the relay are as follows:

- When a current first flows in the coil winding, the input electrical energy is converted mainly into magnetic energy but also into heat energy (I^2R dissipation in the winding resistance). During this time an attractive force is exerted upon the armature but its magnitude is not large enough for it to be able to overcome the mechanical forces (mainly exerted by the contact springs) opposing any movement of the armature.
- When the magnetic flux density in the air gap has built up sufficiently, the attractive force exerted upon the armature becomes large enough to balance the opposing mechanical forces and the armature starts to move. As the armature moves, mechanical energy is needed to lift the contact springs and to supply kinetic energy to the moving parts. This energy is supplied by the electrical circuit. As the armature moves the inductance of the relay falls.
- When the armature is brought to a halt at the end of its travel, all of the kinetic energy it possesses is dissipated.

- Once the armature has fully operated, electrical energy must still be supplied to the relay to keep the armature in its operated position. This energy must provide a sufficiently large attractive force to balance the restoring force exerted by the contact springs.
- When the current in the coil is reduced, the point will be reached at which the mechanical force exerted by the springs will exceed the magnetic force trying to keep the armature operated. As soon as this point is reached the armature will release and return to its non-operated position.

The energy and voltage equations for the relay are:

$$dW/dt = LI\,dI/dt + (I^2 dL/dt)/2$$

$$= LI\,dI/dt + I^2(dL/dx)(dx/dt)/2 \qquad (13.27)$$

and $V = L\,dI/dt + I\,dL/dt$

$$= L\,dI/dt + (I\,dL/dx)(dx/dt) \qquad (13.28)$$

Example 13.8

A relay has an inductance of 2 H when its armature is in the non-operated position. At the moment the relay operates the current in the coil is 50 mA and is increasing at 2 A/s, the armature position is changing at 0.2 m/s, and the inductance is increasing at 500 H/m. (*a*) Calculate the rate at which the energy stored in the air gap changes. (*b*) Calculate the voltage induced into the coil. (*c*) Determine the force exerted upon the armature.

Solution

(*a*) From equation (13.27),

$$dW/dt = 2 \times 50 \times 10^{-3} \times 2 + 0.5 \times (50 \times 10^{-3})^2 \times 500 \times 0.2$$

$$= 0.325\ \text{W} \quad (\textit{Ans.})$$

(*b*) From equation (13.28),

$$v = (2 \times 2) + (50 \times 10^{-3} \times 500 \times 0.2) = 9\ \text{V} \quad (\textit{Ans.})$$

(*c*) Force on armature $= 0.325/0.2 = 1.625$ N (*Ans.*)

Example 13.9

Data for the relay shown in Fig. 13.3 is

Mean length of magnetic circuit = 32 cm
Total length of air gaps = 3.5 mm
Mean cross-sectional area of magnetic circuit = 0.6 cm^2
Number of turns = 8000 coil current = 45 mA $\mu_r = 700$

Calculate the force exerted on the armature.

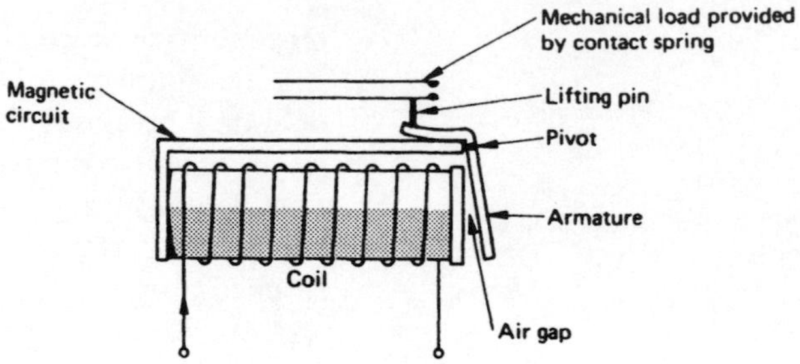

Fig.13.3 Basic relay

Solution
Total reluctance is

$$S = 0.32/(700 \times 4\pi \times 10^{-7} \times 60 \times 10^{-6})$$
$$+ (3.5 \times 10^{-3})/(4\pi \times 106^{-7} \times 60 \times 10^{-6})$$
$$= 52.48 \times 10^{6} \text{ A/Wb}$$

Flux $\Phi = (8000 \times 45 \times 10^{-3})/(52.48 \times 10^{6}) = 6.86\ \mu\text{Wb}$

Flux density $B = (6.86 \times 10^{-6})/(60 \times 10^{-6}) = 0.114$ T

Therefore, from equation (13.26) the force F exerted on the armature is

$$F = (0.114^2 \times 60 \times 10^{-6})/(2 \times 4\pi \times 10^{-7}) = 0.31 \text{ N} \quad (\textit{Ans.})$$

Example 13.10

Figure 13.4 shows a section of a brake on a train. The effective length of the magnet is 35 cm and its cross-sectional area is 70 cm^2. Irregularities at the points where the magnet meets the track cause air gaps of 0.5 mm to exist. The track has a cross-sectional area of 80 cm^2 and the mean length between the poles of the magnet is 5 cm. Calculate (a) the current that must flow in the winding of 150 turns to produce a magnetic flux in the air gaps of 5 mWb, (b) the attractive force between the track and the magnet. Data for the magnetic material employed is given in Table 13.1.

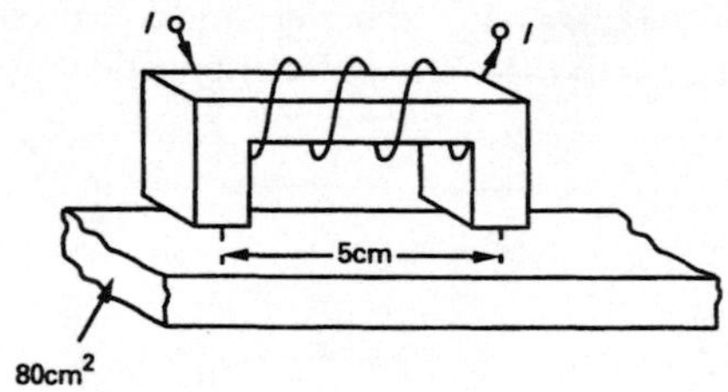

Fig.13.4

Table 13.1

H(AT)	200	300	400	500	600	700	800
B(T)	0.25	0.55	0.7	0.8	0.9	0.97	1.04

Solution

(a) Flux density in air gaps $= \varphi/A = (5 \times 10^{-3})/(70 \times 10^{-4}) = 0.714$ T

$H_g = B_g/\mu_0 = (5 \times 10^{-3})/[(4\pi \times 10^{-7})(70 \times 10^{-4})] = 568\,411$ A

$B_b \approx 0.714$ T and $H_b = 420$ A from the tabled data. Also, $B_t = (5 \times 10^{-3})/(80 \times 10^{-4}) = 0.625$ T, and $H_t = 350$ A from the tabled data. Hence the required m.m.f. is

$$(568\,411 \times 2 \times 0.5 \times 10^{-3}) + (420 \times 35 \times 10^{-2})$$
$$+ (350 \times 5 \times 10^{-2}) = 732.9 \text{ A.}$$

Therefore, the required current is

$I = 732.9/150 = 4.89$ A (*Ans.*)

(*b*) $F = (2 \times 0.714^2 \times 70 \times 10^{-4})/(2 \times 4\pi \times 10^{-7}) = 2840$ N (*Ans.*)

Example 13.11

A horseshoe magnet holds a rectangular block of iron. The force which must be applied to free the iron block is 10 N. The effective length of the horseshoe is 500 mm and its cross-sectional area is 200 mm². The iron block is 150 mm in length and it has a cross-sectional area of 200 mm². If the relative permeability of the magnetic material is 800, and there are 120 turns in the winding, calculate the required current.

Solution

Since there are two air gaps, $F = B^2A/\mu_0 = 10$ N.

$B = \sqrt{[(10 \times 4\pi \times 10^{-7})/(200 \times 10^{-6})]} = 0.25$ T

$H = B/(\mu_0\mu_r) = 0.25/(4\pi \times 10^{-7} \times 800) = 248.7$ A

m.m.f. $= 248.7 \times 650 \times 10^{-3} = 161.7$ A (*Ans.*)

Forces in inductive circuits

When a current flows in a conductor, or a system of conductors, it sets up a magnetic field around the conductor. This field will exert a force upon the conductor(s) that will tend to move the conductor, or to deform the conductor system. The deformation is always such that the number of flux linkages is increased. The increase in flux linkages produces an increase in the stored energy; the increase being equal to the work done in changing the dimensions of the circuit. The change in the number of flux linkages will induce an e.m.f. $N\,d\Phi/dt$ into the circuit. If the current flowing in the circuit is maintained at its original value, electrical energy must be supplied either *to* or *by* the circuit to satisfy equation (13.29).

$$\text{Energy supplied} = \text{Work done} + \text{Change in stored energy} \qquad (13.29)$$

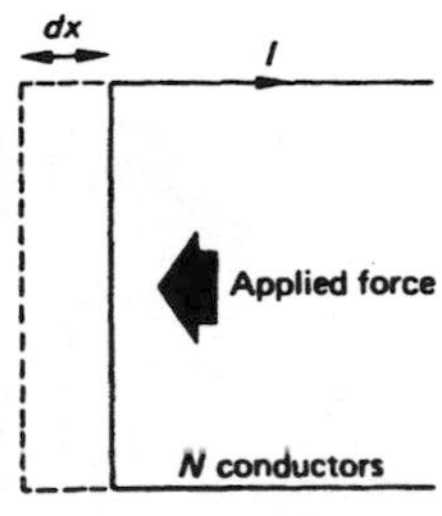

Fig.13.5

Suppose that a force of F newtons is exerted upon the N current-carrying conductors shown in Fig. 13.5 with the result that the conductors are moved through a distance of dx metres. Then the work done is $F\,dx$ joules. If the change in the position of the conductors reduces the number of flux linkages, the e.m.f. V which is induced will act in such a direction that the current is kept constant at its original value. This means that energy is supplied *by* the electrical source *to* the circuit and is equal to $IV\,dt$ joules. Now

$$IV\,dt = IN(d\Phi/dt)dt = I\,d(N\Phi) = I\,d(LI) = I^2\,dL \text{ joules}$$

Since the current is maintained constant at I amperes, the change in the stored magnetic energy will be $(I^2\,dL)/2$, where dL is the change in the inductance of the circuit. The increase in the electrical input energy is equal to the sum of the work done and the change in the stored magnetic energy. Therefore,

$$I^2\,dL = F\,dx + (I^2\,dL)/2$$

$$F = (I^2\,dL/dx)/2 \text{ newtons} \qquad (13.30)$$

If the change in the dimensions of the current-carrying conductors is such that the number of flux linkages is increased, energy will be delivered *by* the circuit *to* the electrical source. In this case $IV\,dt - LI^2/2 = F\,dx$, which leads to equation (13.30) again,

If the system of conductors is rotational and is subject to an applied torque T, the corresponding equation is

$$T = (I^2\,dL/d\theta)/2 \qquad (13.31)$$

where $d\theta$ is the angular distance through which the conductors move.

Example 13.12

A wire of 1 cm radius is bent into the shape of a circle of radius 100 cm and a current of 14.14 kA peak value flows in it. The inductance of the circle of wire is $L = \mu_0 R[\log_e(8R/r) - 1.75]$, where R is the radius of the circle and r is the radius of the wire. Calculate (a) the peak and (b) the average tensile force in the wire.

Solution

(a) Figure 13.6(a) shows the wire circle. The forces exerted upon the wire when a current flows in it are exerted outwards and are equal to $(I^2\,dL/dR)/2$ N. The forces will alter the dimensions of the wire and hence it is necessary to find dL/dR.

$$dL/dR = \mu_0[\log_e(8R/r) - 1.75] + \mu_0 R[(r/8R)(8/r)]$$

$$= \mu_0[\log_e 800 - 1.75 + 1] = 5.94\mu_0$$

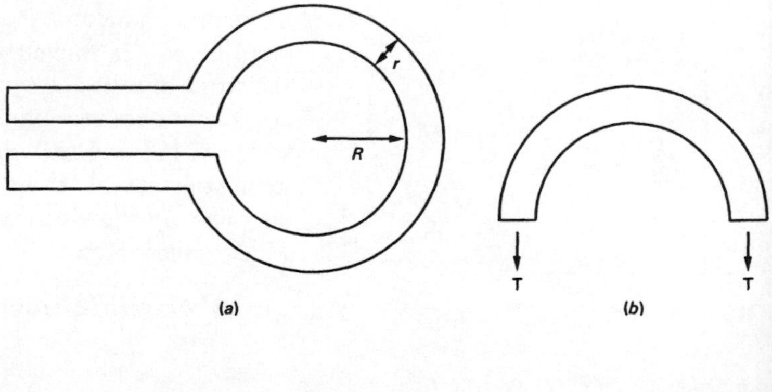

Fig. 13.6

To determine the tension in the wire assume it to be cut in half as shown by Fig 13.6(*b*). The total force exerted is F and the force exerted per metre is $F/2\pi R$. Hence, $2T = 2RF/2\pi R = F/\pi$, and

$$T = F/2\pi = (I^2 \, \mathrm{d}L/\mathrm{d}R)/2\pi = (I^2 \times 5.94\mu_0)/2\pi$$
$$= (14.14 \times 10^3)^2 \times (5.94 \times 4\pi \times 10^{-7})/2\pi = 237.5 \text{ T} \quad (\textit{Ans.})$$

(*b*) Average tension $= T/2 = 118.8\text{T}$ (*Ans.*)

Non-electrical energy

Mechanical energy

Energy in a mechanical system exists by virtue of either the *position* or the *velocity* of a body. If a body is positioned above the Earth it will possess *potential energy*. A body of mass m travelling with a velocity v m/s possesses a kinetic energy of $mv^2/2$ joules.

If a body is moved through a distance of x metres by the application of a force of F newtons, the work done is Fx joules. In the case of a body suspended above the Earth the applied force is the force of gravity. For a rotating body with an angular velocity of ω radians per second the kinetic energy is given by $J\omega^2/2$ joules, where J is the moment of inertia.

A spring possesses potential energy whenever it is deformed. For a linear spring the potential energy is $W = kf^2/2$ where k is the compliance and f is the restoring force. For a rotary spring $W = kM^2/2$ where M is the torque.

Heat, sound and light energy

Three other forms of energy, which are commonly met in everyday life, are heat, sound and light. Whenever a current flows in a conductor, or a component, electrical energy is converted into heat energy. In some cases, such as an electric fire, the heat produced is the wanted output but in many other cases it is an embarrassment. The increase in the temperature of a conductor, or a component, is not proportional to the heat dissipated since some of the generated heat is lost by convection, conduction or radiation, or a combination of them. When an increase in the temperature is undesirable, as in a transistor for example, the component is often mounted on a 'heat sink' to assist in the removal of the unwanted heat as rapidly as possible.

For electrical energy to be converted into sound or light energy, a *transducer* is required. Some possibilities have been mentioned earlier in this chapter. Others are to be found in the fields of radio, television and electronics: for example, photodiodes and light-emitting diodes.

Electro-mechanical energy transfer

Electric machines, of many kinds, are widely used for the conversion of either electrical energy into mechanical energy (e.g. a motor or a relay) or mechanical energy into electrical energy (a generator). In either case the machine must be supplied with energy from some external source. Some of this input energy is lost, mainly in the form of heat energy, but most of it is converted into another desired form.

Thus, in the case of an electric motor, the electrical input energy is used to drive a rotating shaft which, when coupled to an external mechanical load, will produce output mechanical energy. Conversely, a generator is provided with input mechanical energy, in the form of a rotating shaft, and this is converted into electrical energy. In both cases the available output energy is always smaller than the input energy so that the conversion efficiency is always less than 100%.

Energy balance equation

The *energy balance equation* for an electrical machine is

$$\begin{bmatrix}\text{Electrical input energy}\\ +\\ \text{Mechanical input energy}\end{bmatrix} = \begin{bmatrix}\text{Stored magnetic}\\ \text{field energy}\end{bmatrix}$$

$$+\begin{bmatrix}\text{Stored}\\ \text{mechanical}\\ \text{energy}\end{bmatrix} + \begin{bmatrix}\text{Energy}\\ \text{dissipation}\end{bmatrix} \qquad (13.32)$$

The energy dissipation is nearly all heat energy.

If the machine is operated as a motor its output will consist of mechanical energy and hence this will be *negative* in equation (13.32). Conversely, the output energy of a generator is electrical and then the electrical *input* energy in equation (13.32) is *negative*.

The energy balance equation can be expressed more succinctly using symbols, as shown by equation (13.33).

$$W_E + W_M = W_F + W_S + W_C \qquad (13.33)$$

where W_E is the electrical input energy
W_M is the input mechanical energy
W_F is the stored magnetic energy
W_S is the stored mechanical energy
W_C is the total energy dissipation.

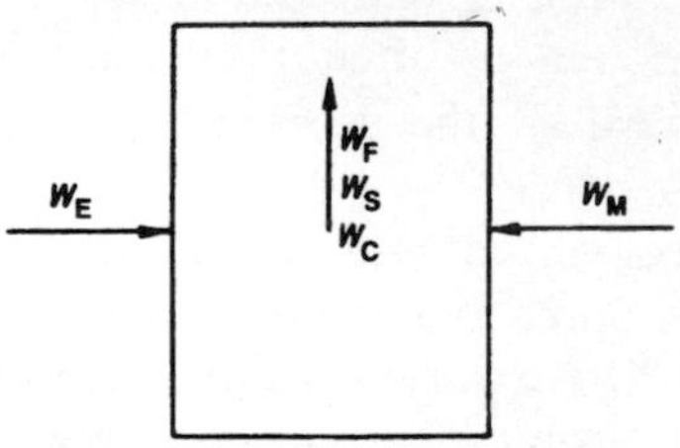

Fig. 13.7 Energy balance diagram

The energy balance diagram corresponding to equation (13.33) is shown in Fig. 13.7.

Power balance equation

The energy balance concept is still applicable if either *changes* of energy or *rates of change* of energy are considered. Equations (13.34) and (13.35) respectively apply.

$$\delta W_E + \delta W_M = \delta W_F + \delta W_S + \delta W_C \quad (13.34)$$

$$\delta W_E/\delta t + \delta W_M/\delta t = \delta W_F/\delta t + \delta W_S/\delta t + \delta W_C/\delta t \quad (13.35)$$

$$P_E + P_M = \delta W_F/\delta t + \delta W_S/\delta t + \delta W_C/\delta t \quad (13.36)$$

Equation (13.36) is generally known as the *power balance equation*. It states that the rate of change of stored energy in a machine is equal to the total power input.

When the machine first starts to run, its electrical and mechanical losses are unequal and the machine is able to increase its speed. At some particular speed the electrical and mechanical losses become equal to one another and then the machine operates under steady-state conditions. Before steady state conditions prevail the question as to which of the two types of loss is the greater depends upon the type of machine. For a generator, the mechanical losses are greater than the electrical losses until the steady running speed is reached. For a motor, on the other hand, the electrical losses are the greater until loss balance is achieved. Once steady-state conditions exist, the currents and velocities at various points in the machine are constant and there are then *no* changes in either electrical or mechanical energy storage to consider.

The energy balance and power balance equations are difficult to solve since they include transient terms, i.e. terms that exist only while the machine is either speeding-up or is slowing-down. When a machine is in one of its steady-state conditions (either stationary or running at a constant speed) there are no changes in the stored energy and so

$$\delta W_F/\delta t = \delta W_S/\delta t = 0$$

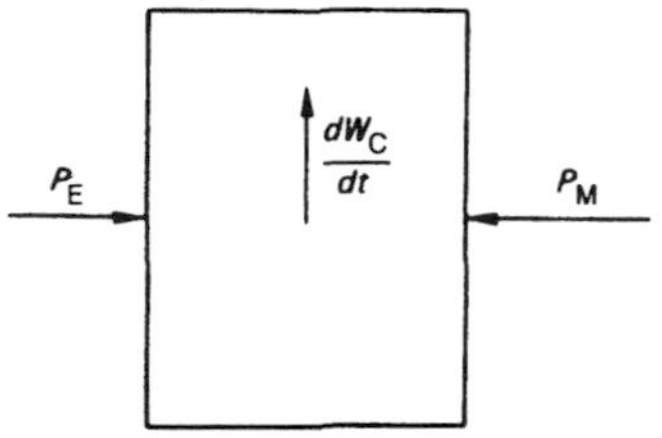

Fig. 13.8 Simplified energy balance diagram

Then equation (13.36) can be written in the form

$$P_E + P_M = \delta W_C/\delta t = P_C \quad (13.37)$$

The energy balance diagram is shown by Fig. 13.8.

[*Note:* the field input is a loss and appears on both sides of the balance equation.]

Example 13.13

A shunt-wound machine is rated at 240 V, 3 kW and 2500 r.p.m. An input power of 200 W is needed to provide the field current. When the machine is operated as a motor with no load, a further input power of 150 W is needed to supply the armature losses. Assuming the power losses are constant calculate the efficiency of the machine when it is used as a generator producing the rated output power. The armature has a resistance of 1.2 Ω.

Solution
At full load the armature current is 3000/240 = 12.5 A and so the armature loss is $12.5^2 \times 1.2 = 187.5$ W. The energy balance equation becomes

$$-3000 + 200 + W_M = 150 + 187.5 + 200$$
$$W_M = 3000 + 150 + 187.5 = 3337.5 \text{ W}$$

The efficiency is $3000/(3337.5 + 200) \times 100 = 84.8\%$ (*Ans.*)

Power losses in electrical motors

In the calculation of the efficiency of an electric motor the various sources of loss, both electrical and mechanical, must be identified. The losses can be illustrated by a *power flow diagram.*

D.C Motor

The power flow diagram of a d.c. motor is shown in Fig. 13.9. Some of the losses shown vary with the load on the motor and hence with the input current I, and some of them are relatively constant. In a shunt-wound motor the field current, and hence the field loss, is constant. The armature loss varies with the load. The friction loss varies with the speed of the machine but not with the current taken from the supply. The iron losses are approximately constant with load.

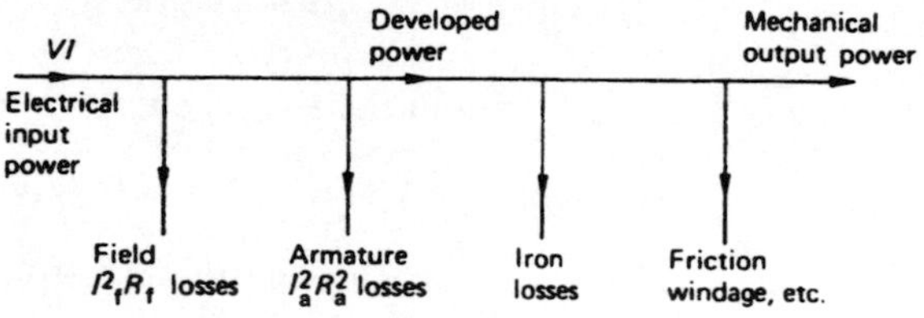

Fig.13.9 Power flow diagram of a d.c. shunt motor

The efficiency η of a d.c. machine is 100 times the ratio of the output power to the input power. Thus,

$$\eta = (VI - I_a^2R_a - I_f^2R_f - F)/VI \times 100\% \qquad (13.38)$$

where F represents the sum of the iron, friction and windage, etc. losses.

If it assumed that the field current is much smaller than the armature current so that $I_a \approx I$, then η can be written as

$$\eta = [V - IR_a - (I_f^2R_f + F)/I]/V \times 100\% \qquad (13.39)$$

The maximum value of the motor's efficiency can be found by differentiating equation (13.39) with respect to the input current I and then equating the result to zero. Thus:

$$d[V - IR_a - (I_f^2 R_f + F)/I]/dI = 0$$

$$-R_a + (I_f^2 R_f + F)/I^2 = 0 \quad \text{and} \quad I^2 R_a = I_f^2 R_f + F$$

Thus, the efficiency of the motor is at its maximum possible value when the load (i.e. the input current) is such that the variable losses are equal to the fixed losses.

In the case of a series-wound motor the field losses will also vary with the load.

Cage rotor induction motor

The power flow diagram for a cage rotor induction motor is shown in Fig. 13.10. Some of the electrical input power is lost in the form of heat energy dissipated in the effective resistances of the rotor and stator windings and some (small) iron losses. The remainder of the input energy is converted into mechanical form and then some of this is, in turn, lost because of friction, etc. The fixed losses are the stator core loss and friction/windage losses. The variable losses consist of the I^2R losses in both the rotor and the stator and the core loss in the rotor. However, the stator core loss is usually very small.

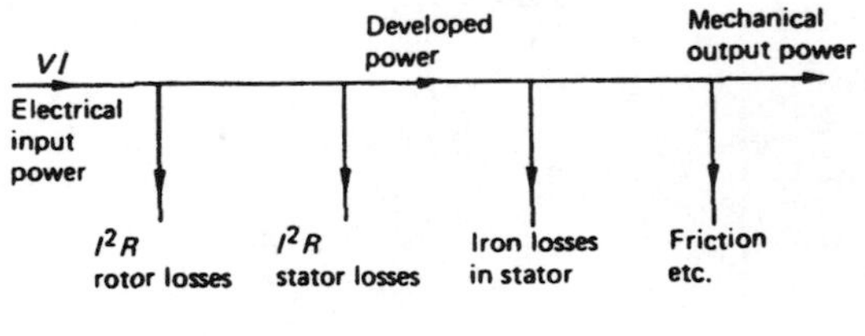

Fig. 13.10 Power flow diagram of an induction motor

A.C. motor

Figure 13.11 shows the power flow diagram for an a.c. motor. For a motor of this type the fixed losses are (a) the core loss and (b) friction and windage losses. All the other losses vary with the load on the armature.

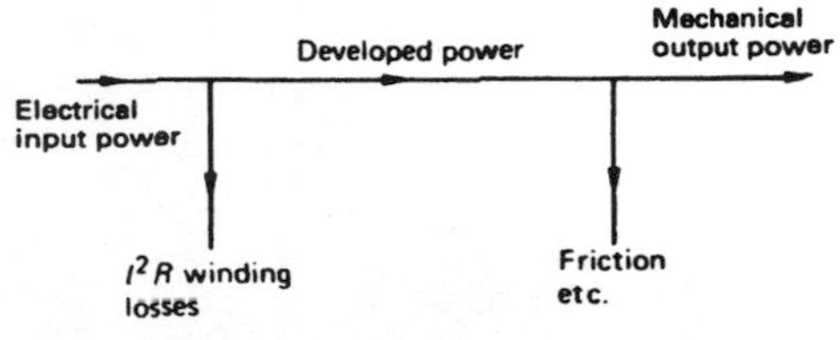

Fig. 13.11 Power flow diagram of an a.c. motor

Exercises 13

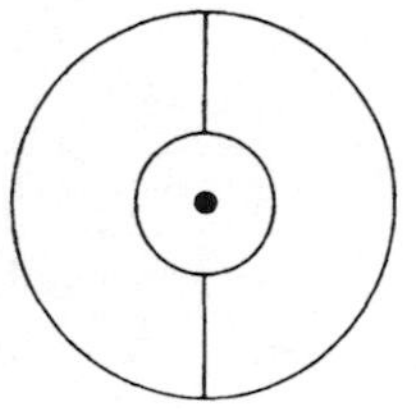

Fig. 13.12

13.1 Figure 13.12 shows a conductor carrying a current I amps that passes axially through the centre of an iron ring. The ring consists of two halves held together by a magnetic force of attraction. The ring has a mean diameter of 45 cm and a cross-sectional area of 50 cm^2. Calculate the force that must be exerted to pull the two halves of the ring apart when the current is 160 A. (Take $\mu_r = 400$.)

13.2 A conductor has a diameter ($2R$) of 1.6 cm and it is bent into a circular shape. The inductance of the wire is then

$$L = \mu_0 R[\log_{10}(R/0.015)]$$

Calculate the average force exerted on the wire when the current flowing in it is 125 A peak.

13.3 The basic construction of a relay is shown in Fig. 13.3. The coil has 1000 turns and the air gap has a length of 0.6 cm and a cross-sectional area of 1.4 cm^2. Calculate the current needed to operate the relay if the force exerted on the armature by the contact springs is 4.8 N.

13.4 A d.c. machine is rated at 240 V, 2000 W at 2000 r.p.m. The power dissipated in the field circuit is 100 W. When the machine is operated as a motor the armature losses are 200 W. Calculate the efficiency of the machine when it is used as a generator. State any assumptions made but take armature resistance as 1 Ω.

13.5 A capacitor is charged with 20 mC and it is then disconnected from the voltage supply. The energy stored in the capacitor is then 2.9 J. Calculate (a) the capacitance and (b) the voltage across the capacitor. This capacitor is then connected across a 10 μF capacitor. Calculate the voltage across each capacitor.

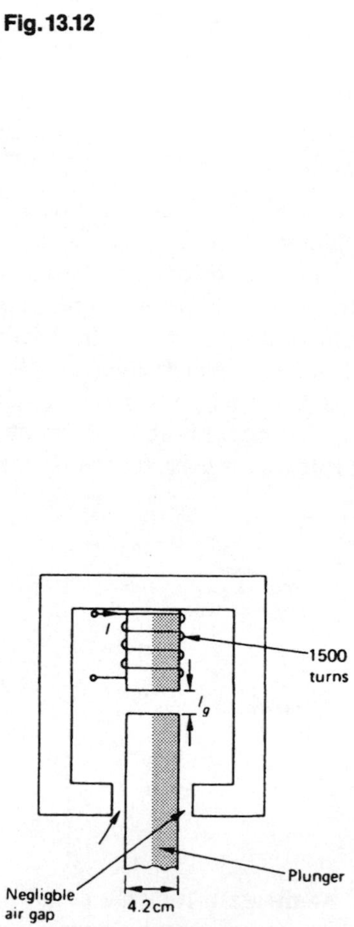

Fig. 13.13

13.6 Figure 13.13 shows an electromagnetic device. When current flows in the coil the magnetic force developed attracts the armature upwards. The length of the air gap varies from 2 cm when the device is non-operated to 0.76 cm when it is operated. There are 1500 turns and the current passed through the coil is 4 A. Calculate (a) the flux density in the air gap, (b) the inductance of the coil when the plunger is (i) un-operated, (ii) operated and (c) the work done as the armature moves from its non-operated to its operated positions.

13.7 Figure 13.14 represents an electromagnet holding up a block of iron. The mean length of the magnetic circuit is 0.76 m and its cross-sectional area is 4 cm^2. Calculate the force exerted on the iron block when a current of 536 mA flows through the coil. (Take $\mu_r = 700$.)

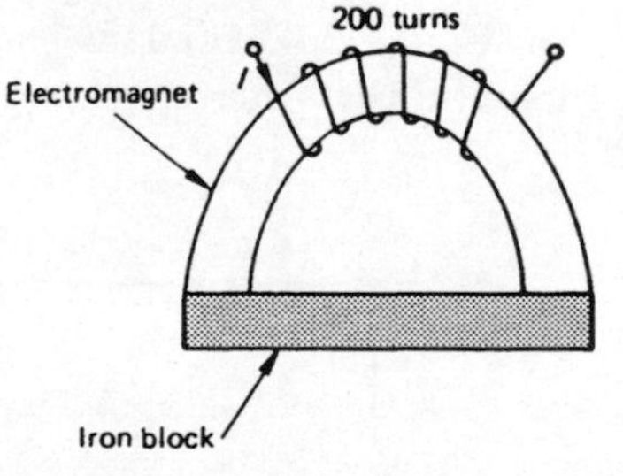

Fig. 13.14

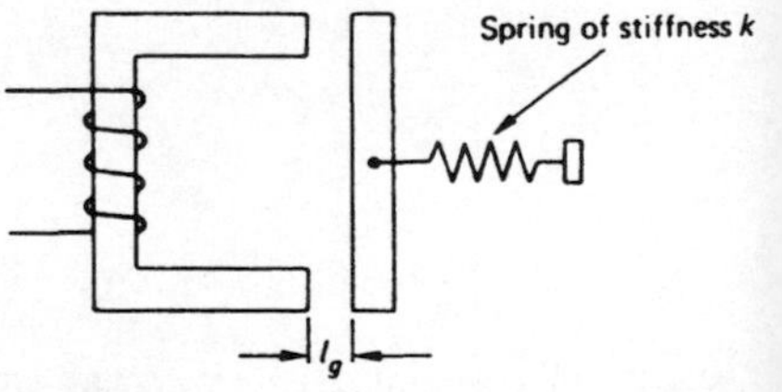

Fig. 13.15

13.8 Figure 13.15 shows an actuator with a spring-loaded armature. When the device is non-activated the length of the air gap is l_0. Show that when a current I flows in the coil to operate the device and reduce the air gap to l_1, the electrical input energy needed is given by $W_E = k(l_0^2 - l_1^2)/2$.

13.9 A capacitor is connected to a d.c. supply when it receives a charge of 1.8 mC. The energy stored is then 1.56 J. The voltage source is then disconnected. Calculate the capacitance of the capacitor and the voltage across its terminals. Two 2 μF capacitors are now connected in series with each other and in series with the first capacitor. The end capacitors are linked to form a complete circuit. Calculate the charge in each capacitor and the voltage across its terminals.

13.10 Calculate the pull on the armature of a relay having the following data: Mean length of magnetic circuit = 30 cm, total length of air gaps = 3 mm, mean cross-sectional area of both the magnetic circuit and the air gaps = 0.6 cm^2, number of turns in coil = 10 000, current flowing in coil = 50 mA, and $\mu_r = 600$.

13.11 A horseshoe-shaped magnet has a cross-sectional area of 1000 mm^2 and a length of 0.6 m. The magnetic material has $\varepsilon_r = 1000$ and there are 400 turns in the winding. Each air gap is 1 mm wide and the leakage coefficient is 1.2. Calculate the load that the magnet is able to lift.

13.12 A m.m.f. of 1600 A is applied to a magnetic circuit. Calculate the energy stored if data for the circuit is: length = 0.75 m, cross-sectional area = 1000 mm^2, $\mu_r = 1800$.

13.13 Calculate the energy stored in an air gap 1.5 mm wide and 8 cm^2 area when the magnetic flux density in the gap is 0.85 T.

13.14 A magnetic system has a m.m.f. of 200 A and a total reluctance of $S = 200 \times 10^3$ A/Wb. Calculate the energy stored in the air gap.

13.15 A magnetic circuit has 0.35 joules of energy stored in its air gap when the m.m.f. is 300 A. If there are 800 turns calculate the inductance of the circuit.

13.16 The energy stored in an air gap is 4 J. What will be the energy stored if the length of the air gap is doubled? Assume that the flux density remains constant.

13.17 A relay has an inductance of 3 H. Calculate the rate of change of the energy stored in the air gap as the current increases from 20 mA at the rate of 15 A/s. Assume the inductance to remain constant.

13.18 A 10 μF capacitor is charged to 50 V and it is then disconnected from the supply. A 2 μF, initially discharged, capacitor is then connected across the 10 μF capacitor. Calculate (a) the voltage across the combination and (b) the energy stored.

13.19 An inductance of $L = 5.5$ H and $r = 75\ \Omega$ is connected across a d.c. voltage of 150 V. Calculate the energy stored in the magnetic field of the inductor when the current is (a) at its initial value, (b) at 50 per cent of its final value and (c) at its final value.

13.20 Two capacitors are connected across a 24 V supply and then have stored charges of 0.22 mC and 0.24 mC respectively. Calculate the energy stored in each capacitor.

13.21 Repeat exercise 13.1 if the diameter of the ring is 250 mm, the cross-sectional area = 500 mm^2 and $\mu_r = 500$. Now calculate the current required to produce a force of 25 N between the two halves of the ring.

13.22 The magnetic device shown in Fig. 13.13 has its air gap l_g varied from 0.5 cm to 1.5 cm. The coil has 2000 turns and a current of 2.5 A flows in it. Calculate (*a*) the inductance of the coil for the two extreme positions of the armature and (*b*) the work done by the armature as it moves from one extreme position to the other.

13.23 Figure 13.16 shows a relay whose armature travel is restricted to give an air gap of 5 mm to 15 mm. The winding has 240 turns and carries a current of 3 A. Calculate (*a*) the inductance of the winding for the extreme armature positions, (*b*) the work done as the armature moves between these positions and (*c*) the average force exerted.

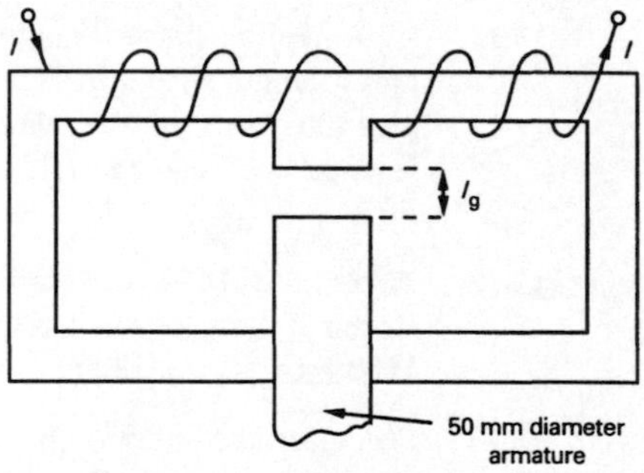

Fig. 13.16

13.24 For a device similar to that shown in Fig. 13.14, the mean length of the magnetic circuit is 500 mm and its cross-sectional area is 200 mm^2. The iron block is 150 mm in length and its cross-sectional area is also 200 mm^2. If $\mu_r = 800$ and there are 120 turns in the coil, calculate the current needed for a force of 10 N to be exerted on the iron block.

14 Network parameters

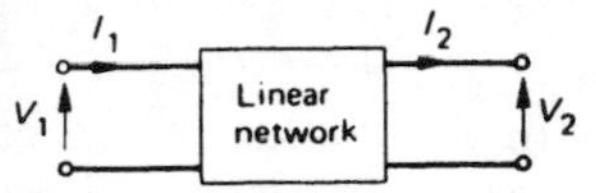

Fig.14.1 Four-terminal network

When a voltage is applied to the input terminals of a four-terminal (or two-port) network, input and output currents will flow and an output voltage will appear across the output terminals. Figure 14.1 shows a four-terminal network with input and output currents I_1 and I_2 respectively and input and output voltages V_1 and V_2 respectively. It is assumed that the impedances within the network are *linear*, i.e. that they obey Ohm' law. The directions of current and voltage shown in the figure are conventional and any one, or more, of them may be the other way around in any particular case.

Any two of the four variables can be taken as the *independent variables*, so that the other two automatically become the *dependent variables*. This leads to *six* different ways in which the terminal performance of the network can be described. However, not all of these sets of variables are used in practice.

Parameters

h Parameters

Suppose that the input current I_1 and the output voltage V_2 are taken to be the independent variables. Then the input voltage V_1 and the output current I_2 can be written down in terms of the output voltage, the input current and four circuit parameters. Thus

$$V_1 = h_{11}I_1 + h_{12}V_2 \tag{14.1}$$

$$I_2 = h_{21}I_1 + h_{22}V_2 \tag{14.2}$$

$$\begin{bmatrix} V_1 \\ I_2 \end{bmatrix} = \begin{bmatrix} h_{11} & h_{12} \\ h_{21} & h_{22} \end{bmatrix} \begin{bmatrix} I_1 \\ V_2 \end{bmatrix} \tag{14.3}$$

Equation (14.1) states that the input voltage V_1 is equal to a parameter h_{11} times the input current I_1 *plus* another parameter h_{12} times the output voltage V_2. The right-hand side of the equation must have the dimensions of a voltage and hence h_{11}

must be an impedance while h_{12} must be a dimensionless quantity. Similarly the right-hand side of equation (14.2) must have the dimensions of a current; therefore h_{21} is dimensionless and h_{22} is an admittance.

If the output terminals of the network are short circuited, $V_2 = 0$ and then

$$V_1 = h_{11}I_1 \quad \text{or} \quad h_{11} = V_1/I_1\ \Omega \quad (V_2 = 0)$$

$$\text{and} \quad I_2 = h_{21}I_2 \quad \text{or} \quad h_{21} = I_2/I_1 \quad (V_2 = 0)$$

Similarly, if $I_1 = 0$,

$$V_1 = h_{12}V_2 \quad \text{or} \quad h_{12} = V_1/V_2 \quad (I_1 = 0)$$

$$\text{and} \quad I_2 = h_{22}V_2 \quad \text{or} \quad h_{22} = I_2/V_2\ \text{S} \quad (I_1 = 0)$$

Because one of the parameters has the dimensions of impedance, another parameter has the dimensions of admittance, and the other two parameters are dimensionless, this set of parameters is known as the *hybrid parameters*. The *h* parameters are particularly appropriate for the representation of the a.c. performance of the bipolar transistor and they are widely employed for this purpose. When the *h* parameters are used in transistor work, the four *h* parameters are generally re-labelled to indicate both the transistor configuration and the nature of the parameter. Thus, h_{11} becomes h_{ie}, where the suffix *i* indicates 'input' and the suffix *e* indicates the common-emitter configuration. The other parameters are: the forward current gain h_{fe}, the output admittance h_{oe}, the reverse voltage ratio h_{re}, and h_{ie} the input impedance of the transistor.

y Parameters

With *y* parameters the input and output currents of a network are expressed in terms of the input and output voltages. Thus:

$$I_1 = y_{11}V_1 + y_{12}V_2 \tag{14.4}$$

$$I_2 = y_{21}V_1 + y_{22}V_2 \tag{14.5}$$

$$\begin{bmatrix} I_1 \\ I_2 \end{bmatrix} = \begin{bmatrix} y_{11} & y_{12} \\ y_{21} & y_{22} \end{bmatrix} \begin{bmatrix} V_1 \\ V_2 \end{bmatrix} \tag{14.6}$$

In this case, all four of the parameters have the dimensions of *admittance*. y_{11} and y_{21} can be determined by short circuiting the output terminals to make $V_2 = 0$, while y_{12} and y_{22} can be found by setting V_1 to zero by short circuiting the input terminals. Then,

$$y_{11} = I_1/V_1 \text{ with } V_2 = 0 \qquad y_{12} = I_1/V_2 \text{ with } V_1 = 0$$

$$y_{21} = I_2/V_1 \text{ with } V_2 = 0 \qquad y_{22} = I_2/V_2 \text{ with } V_1 = 0$$

z Parameters

With the z parameters of a network the input and output voltages are expressed as functions of the input and output currents.

$$V_1 = z_{11}I_1 + z_{12}I_2 \tag{14.7}$$

$$V_2 = z_{21}I_1 + z_{22}I_2 \tag{14.8}$$

$$\begin{bmatrix} V_1 \\ V_2 \end{bmatrix} = \begin{bmatrix} z_{11} & z_{12} \\ z_{21} & z_{22} \end{bmatrix} \begin{bmatrix} I_1 \\ I_2 \end{bmatrix} \tag{14.9}$$

All four parameters have the dimensions of impedance and their values can be determined with first I_1 and then I_2 set to zero by open circuiting the appropriate terminals. z Parameters are used to obtain the equivalent circuit of a coupled circuit (p. 179) or of a transformer (p. 195).

With $I_1 = 0$, $V_1 = Z_{12}I_2$ or $Z_{12} = V_1/I_2$, and $V_2 = Z_{22}I_2$ or $Z_{22} = V_2/I_2$. With $I_2 = 0$, $V_1 = Z_{11}I_1$ or $Z_{11} = V_1/I_1$, and $V_2 = Z_{21}I_i$ or $Z_{21} = V_2/I_1$.

General circuit or transmission parameters

The general circuit, or transmission, parameters are often employed to describe the operation of four-terminal networks of all kinds, including transformers and transmission lines.

$$V_1 = AV_2 + BI_2 \tag{14.10}$$

$$I_1 = CV_2 + DI_2 \tag{14.11}$$

$$\begin{bmatrix} V_1 \\ I_1 \end{bmatrix} = \begin{bmatrix} A & B \\ C & D \end{bmatrix} \begin{bmatrix} V_2 \\ I_2 \end{bmatrix} \tag{14.12}$$

In this case the parameters A and D are dimensionless, B has the dimensions of an impedance and C is an admittance.

g Parameters

With this type of parameter the input current and the output voltage are expressed in terms of the input voltage and the output current.

$$I_1 = g_{11}V_1 + g_{12}I_2 \tag{14.13}$$

$$V_2 = g_{21}V_1 + g_{22}I_2 \tag{14.14}$$

$$\begin{bmatrix} I_1 \\ V_2 \end{bmatrix} = \begin{bmatrix} g_{11} & g_{12} \\ g_{21} & g_{22} \end{bmatrix} \begin{bmatrix} V_1 \\ I_2 \end{bmatrix} \tag{14.15}$$

The choice of parameters

The choice of which set of parameters to use in a particular case is determined by the type of network and the way in which it is interconnected with other networks. It has already been mentioned that the h parameters are widely employed in the analysis of bipolar transistor circuits at audio frequencies. Some difficulty is experienced however in obtaining the values of the h parameters at high frequencies and for this reason (and others) high-frequency transistor analysis is generally carried out using y parameters.

The transmission or general circuit parameters are commonly used for the analysis of heavy-current circuits and for power frequency transmission lines. These parameters are also employed when two or more networks are connected in cascade, as in Fig. 14.2, since the overall performance of the cascade can be described by the product of their parameter matrices. This is possible because the output current and voltage of the first network are the input current and voltage of the second network.

$$\begin{bmatrix} V_1 \\ I_1 \end{bmatrix} = \begin{bmatrix} A & B \\ C & D \end{bmatrix} \begin{bmatrix} A' & B' \\ C' & D' \end{bmatrix} \begin{bmatrix} V_3 \\ I_3 \end{bmatrix}$$

$$= \begin{bmatrix} AA'+BC' & AB'+BD' \\ CA'+DC' & CB'+DD' \end{bmatrix} \begin{bmatrix} V_3 \\ I_3 \end{bmatrix} \qquad (14.16)$$

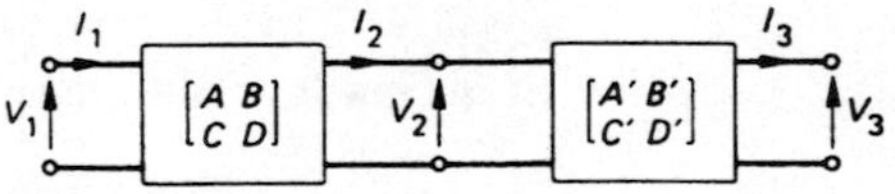

Fig.14.2 Networks connected in cascade

Similarly, a particular set of parameters is best suited to the overall behaviour of two networks connected in one of the other possible ways. Figure 14.3(a) shows two networks connected in parallel; for this connection the y parameters are the most convenient because the overall matrix is equal to the sum of the individual y matrices of the two networks. Thus,

$$\begin{bmatrix} I_1 \\ I_2 \end{bmatrix} = \begin{bmatrix} [y_a] + [y_b] \end{bmatrix} \begin{bmatrix} V_1 \\ V_2 \end{bmatrix} \qquad (14.17)$$

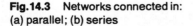

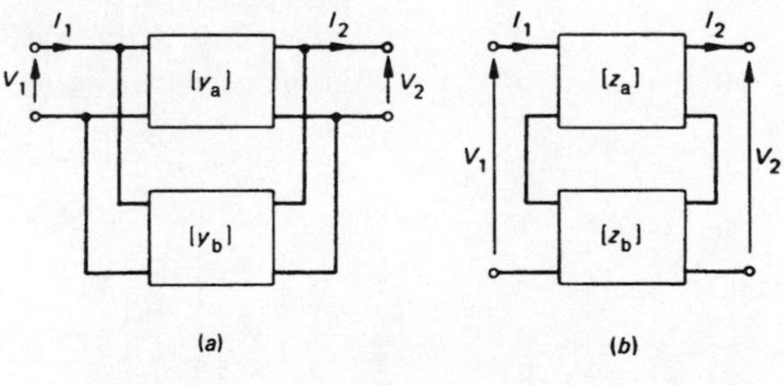

Fig.14.3 Networks connected in: (a) parallel; (b) series

When two networks are connected in series (Fig. 14.3(*b*)), the *z* parameters become the most convenient set and the overall matrix is

$$\begin{bmatrix} V_1 \\ V_2 \end{bmatrix} = \begin{bmatrix} [z_a] + [z_b] \end{bmatrix} \begin{bmatrix} I_1 \\ I_2 \end{bmatrix} \tag{14.18}$$

Conversion from one set of parameters to another

It is always possible to convert from one set of parameters to another set and the necessary relationships are given in Table 14.1.

Table 14.1 Relationships between parameters

To \ From	$[z]$	$[y]$	$[h]$	$[g]$	$\begin{bmatrix} A & B \\ C & D \end{bmatrix}$
Note: $\lvert y\rvert = y_{11}y_{22} - y_{12}y_{21}$, $\lvert a\rvert = AD - BC$, $\lvert z\rvert = z_{11}z_{22} - z_{12}z_{21}$, $\lvert h\rvert = h_{11}h_{22} - h_{12}h_{21}$, $\lvert g\rvert = g_{11}g_{22} - g_{12}g_{21}$					
$[z]$	$\begin{bmatrix} z_{11} & z_{12} \\ z_{21} & z_{22} \end{bmatrix}$	$\frac{1}{\lvert y\rvert}\begin{bmatrix} y_{22} & -y_{12} \\ -y_{21} & y_{11} \end{bmatrix}$	$\frac{1}{h_{22}}\begin{bmatrix} \lvert h\rvert & h_{12} \\ -h_{21} & 1 \end{bmatrix}$	$\frac{1}{g_{11}}\begin{bmatrix} 1 & -g_{12} \\ g_{21} & \lvert g\rvert \end{bmatrix}$	$\frac{1}{C}\begin{bmatrix} A & \lvert a\rvert \\ 1 & D \end{bmatrix}$
$[y]$	$\frac{1}{\lvert z\rvert}\begin{bmatrix} z_{22} & -z_{12} \\ -z_{21} & z_{11} \end{bmatrix}$	$\begin{matrix} y_{11} & y_{12} \\ y_{21} & y_{22} \end{matrix}$	$\frac{1}{h_{11}}\begin{bmatrix} 1 & -h_{12} \\ h_{21} & \lvert h\rvert \end{bmatrix}$	$\frac{1}{g_{22}}\begin{bmatrix} \lvert g\rvert & g_{12} \\ -g_{21} & 1 \end{bmatrix}$	$\frac{1}{B}\begin{bmatrix} D & -\lvert a\rvert \\ -1 & A \end{bmatrix}$
$[h]$	$\frac{1}{z_{22}}\begin{bmatrix} \lvert z\rvert & z_{12} \\ -z_{21} & 1 \end{bmatrix}$	$\frac{1}{y_{11}}\begin{bmatrix} 1 & -y_{12} \\ y_{21} & \lvert y\rvert \end{bmatrix}$	$\begin{matrix} h_{11} & h_{12} \\ h_{21} & h_{22} \end{matrix}$	$\frac{1}{\lvert g\rvert}\begin{bmatrix} g_{22} & -g_{12} \\ -g_{21} & g_{11} \end{bmatrix}$	$\frac{1}{\lvert D\rvert}\begin{bmatrix} B & \lvert a\rvert \\ -1 & C \end{bmatrix}$
$[g]$	$\frac{1}{z_{11}}\begin{bmatrix} 1 & -z_{12} \\ z_{21} & \lvert z\rvert \end{bmatrix}$	$\frac{1}{y_{22}}\begin{bmatrix} \lvert y\rvert & y_{12} \\ -y_{21} & 1 \end{bmatrix}$	$\frac{1}{\lvert h\rvert}\begin{bmatrix} h_{22} & -h_{12} \\ -h_{21} & h_{11} \end{bmatrix}$	$\begin{matrix} g_{11} & g_{12} \\ g_{21} & g_{22} \end{matrix}$	$\frac{1}{A}\begin{bmatrix} C & -\lvert a\rvert \\ 1 & B \end{bmatrix}$
$\begin{bmatrix} A & B \\ C & D \end{bmatrix}$	$\frac{1}{z_{21}}\begin{bmatrix} z_{11} & \lvert z\rvert \\ 1 & z_{22} \end{bmatrix}$	$\frac{1}{y_{21}}\begin{bmatrix} -y_{22} & -1 \\ -\lvert y\rvert & -y_{11} \end{bmatrix}$	$\frac{1}{h_{21}}\begin{bmatrix} -\lvert h\rvert & -h_{11} \\ -h_{22} & -1 \end{bmatrix}$	$\frac{1}{g_{21}}\begin{bmatrix} 1 & g_{22} \\ g_{11} & \lvert g\rvert \end{bmatrix}$	$\begin{matrix} A & B \\ C & D \end{matrix}$

Example 14.1

A four-terminal network has the transmission matrix $\begin{bmatrix} 150 & 60 \\ 8 & 150 \end{bmatrix}$. Determine the *y* matrix for the network. What is the overall *y* matrix for two such networks connected in parallel?

Solution

From Table 14.1

$$[y] = \frac{1}{60}\begin{bmatrix} 150 & -(150^2 - 480) \\ -1 & 150 \end{bmatrix}$$

or $$[y] = \begin{bmatrix} 2.5 & 367 \\ 0.0167 & 2.5 \end{bmatrix} \quad (\textit{Ans.})$$

Also $$[y_{ov}] = \begin{bmatrix} 5 & 734 \\ 0.0334 & 5 \end{bmatrix} \quad (\textit{Ans.})$$

The main application for the h and the y parameters is the calculation of the input and output impedances and the current and voltage gains of a bipolar transistor amplifier. Both sets of parameters *can* be applied to the solution of network problems but such usage is not common.

Transmission parameters

The transmission, or general circuit, parameters are

$$V_1 = AV_2 + BI_2 \tag{14.19}$$

$$I_1 = CV_2 + DI_2 \tag{14.20}$$

or, in matrix form

$$\begin{bmatrix} V_1 \\ I_1 \end{bmatrix} = \begin{bmatrix} A & B \\ C & D \end{bmatrix} \begin{bmatrix} V_2 \\ I_2 \end{bmatrix} \tag{14.21}$$

Measurement of parameters

The values of the transmission parameters of a network may be determined by first short circuiting, and then open circuiting, the output terminals of the network.

If V_2 is made equal to zero,

$$V_1 = BI_2 \quad \text{or} \quad B = V_1/I_2\ \Omega, \quad \text{and}$$

$$I_1 = DI_2 \quad \text{or} \quad D = I_1/I_2 \quad \text{(dimensionless)}$$

If I_2 is made equal to zero,

$$V_1 = AV_2 \quad \text{or} \quad A = V_1/V_2 \quad \text{(dimensionless), and}$$

$$I_1 = CV_2 \quad \text{or} \quad C = I_1/V_2\ \text{S}$$

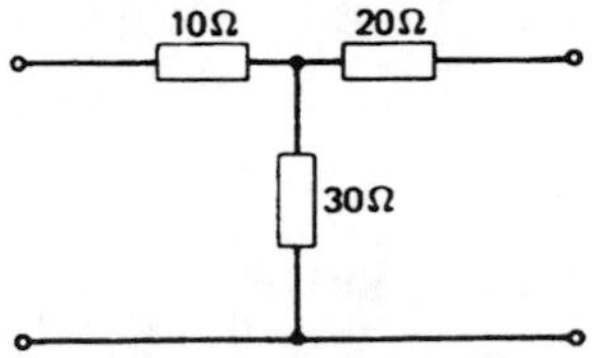

Fig. 14.4

Example 14.2

Determine the transmission parameters of the network shown in Fig. 14.4.

Solution

$$A = V_1/V_2 \quad (I_2 = 0). \quad \text{Since} \quad V_2 = 30V_1/(10+30) = 3V_1/4$$

$$A = 4/3 = 1.33$$

$$C = I_1/V_2 \quad (I_2 = 0)$$

Since $I_1 = V_1/(10+30) = V_1/40$ and $V_2 = 3V_1/4$

$$C = (V_1/40)(4/(3V_1)) = 1/30\ \text{S} = 33.33 \times 10^{-3}\ \text{S}$$

$$B = V_1/I_2 \quad (V_2 = 0) \quad \text{Since}$$

$$I_1 = V_1/\{10 + [(20 \times 30)/(20+30)]\} = V_1/22 \quad \text{and}$$

$$I_2 = (V_1/22)(30/50)$$

$$B = (22 \times 50)/30 = 36.7\ \Omega$$

$$D = I_1/I_2 \quad (V_2 = 0)$$

Since $I_2 = I_1 \times (30/50)$ then $D = 5/3 = 1.67$

Thus the transmission matrix is

$$\begin{bmatrix} 1.33 & 36.7 \\ 33.33 \times 10^{-3} & 1.67 \end{bmatrix} \quad (Ans.)$$

$AD - BC = 1$

If the output terminals of a passive four-terminal network are short circuited and a voltage V is applied to the input terminals then,

$$V = V_1 = BI_{2sc}, \quad \text{and} \quad I_{1sc} = DI_{2sc}.$$

If, now, the input terminals are short circuited and the same voltage V is applied to the output terminals, then

$$V_1 = V_1' = 0 \quad \text{and} \quad 0 = AV_2 + BI_2 = AV_1 - DI_2' \quad \text{or}$$
$$I_2' = AV_1/B.$$

Hence $-I_1' = CV_2 + DI_2 = CV_1 - DI_2'$ since the current is now flowing in the opposite direction. The network contains no current or voltage sources, hence

$$I_1' = I_{2sc} \quad \text{and} \quad -I_{2sc} = CV_1 - DAV_1/B$$

Now,

$$I_{2sc} = V_1/B \quad \text{and so} \quad -V_1/B = CV_1 - DAV_1/B$$

Finally,

$$AD - BC = 1 \tag{14.22}$$

This relationship is true for all linear passive networks. For a symmetrical network $A = D$ and the relationship becomes $A^2 - BC = 1$.

Note that in Example 14.2,

$$AD - BC = (1.33 \times 1.67) - (36.7 \times 33 \times 10^{-3}) = 1$$

Transmission parameters of networks

Series impedance

From Fig. 14.5 suppose that a current $I_1 = I_2$ flows into and out of the circuit. Then

$$V_1 = V_2 + I_2 Z \quad \text{and} \quad I_1 = I_2$$

Therefore,

$$\begin{bmatrix} A & B \\ C & D \end{bmatrix} = \begin{bmatrix} 1 & Z \\ 0 & 1 \end{bmatrix}$$

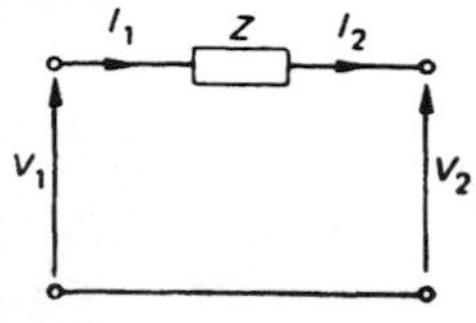

Fig. 14.5 Series impedance

Alternatively, if the output terminals of the network are first open circuited and then short circuited, $I_2 = I_1 = 0$ and $V_1 = V_2$. Hence

$$V_1/V_2 = A = 1 \quad \text{and} \quad I_1/V_2 = C = 0$$

Also, if $V_2 = 0$, $V_1 = I_2 Z$ and $I_1 = I_2$. Hence

$$V_1/I_2 = B = Z \quad \text{and} \quad I_1/I_2 = D = 1$$

Shunt admittance

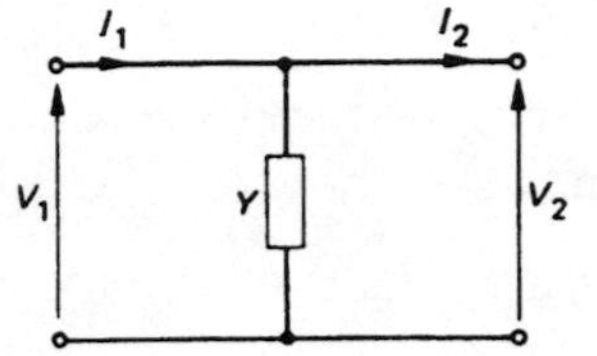

Fig.14.6 Shunt admittance

From Fig. 14.6, assuming I_2 exists,

$$V_1 = V_2 \quad \text{and} \quad I_1 = V_2 Y + I_2$$

Therefore,

$$\begin{bmatrix} A & B \\ C & D \end{bmatrix} = \begin{bmatrix} 1 & 0 \\ Y & 1 \end{bmatrix}$$

Alternatively, when $V_2 = 0$, $V_1 = 0$ and $I_1 = I_2$. Hence,

$$V_1/I_2 = B = 0 \quad \text{and} \quad I_1/I_2 = D = 1$$

When $I_2 = 0$, $V_1 = V_2$ and $I_1 = V_2 Y$. Hence,

$$V_1/V_2 = A = 1 \quad \text{and} \quad I_1/V_2 = C = Y$$

L Network

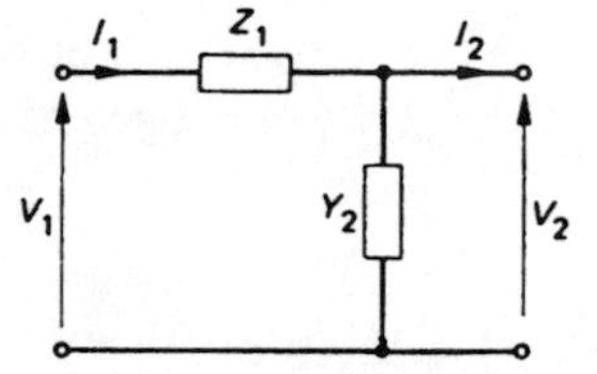

Fig.14.7 L network

From Fig. 14.7

$$V_1 = V_2 + I_1 Z_1 \quad \text{and} \quad I_1 = V_2 Y_2 + I_2$$

Hence

$$V_1 = V_2 + (V_2 Y_2 + I_2) Z_1 = V_2(1 + Y_2 Z_1) + I_2 Z_1$$

Therefore,

$$\begin{bmatrix} A & B \\ C & D \end{bmatrix} = \begin{bmatrix} 1 + Y_2 Z_1 & Z_1 \\ Y_2 & 1 \end{bmatrix}$$

The L network can, alternatively, be regarded as the cascade connection of a series impedance Z_1 and a shunt admittance Y_2. In this case,

$$\begin{bmatrix} A & B \\ C & D \end{bmatrix} = \begin{bmatrix} 1 & Z_1 \\ 0 & 1 \end{bmatrix} \begin{bmatrix} 1 & 0 \\ Y_2 & 1 \end{bmatrix} = \begin{bmatrix} 1 + Z_1 Y_2 & Z_1 \\ Y_2 & 1 \end{bmatrix}$$

(as before)

If the L network is reversed so that the shunt admittance is across the input terminals of the network the transmission matrix becomes

$$\begin{bmatrix} A & B \\ C & D \end{bmatrix} = \begin{bmatrix} 1 & Z_1 \\ Y_2 & 1 + Z_1 Y_2 \end{bmatrix}$$

T Network

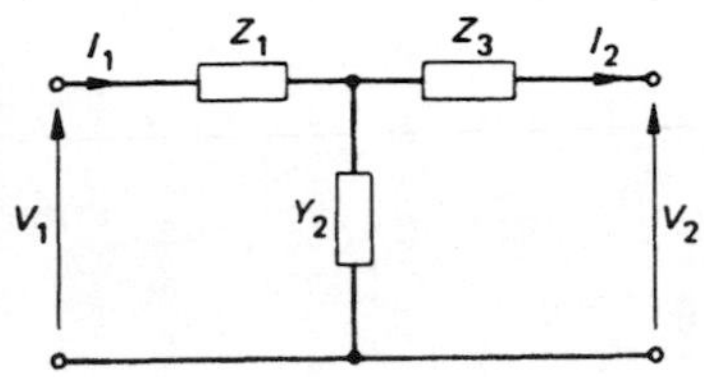

Fig. 14.8 T network

Treating the T network (Fig. 14.8) as the cascade connection of a series impedance Z_1, a shunt admittance Y_2 and a series impedance Z_3 gives

$$\begin{bmatrix} A & B \\ C & D \end{bmatrix} = \begin{bmatrix} 1 & Z_1 \\ 0 & 1 \end{bmatrix} \begin{bmatrix} 1 & 0 \\ Y_2 & 1 \end{bmatrix} \begin{bmatrix} 1 & Z_3 \\ 0 & 1 \end{bmatrix}$$

$$= \begin{bmatrix} 1 + Z_1 Y_2 & Z_1 \\ Y_2 & 1 \end{bmatrix} \begin{bmatrix} 1 & Z_3 \\ 0 & 1 \end{bmatrix}$$

$$= \begin{bmatrix} 1 + Z_1 Y_2 & Z_3(1 + Z_1 Y_2) + Z_1 \\ Y_2 & 1 + Y_2 Z_3 \end{bmatrix}$$

π Network

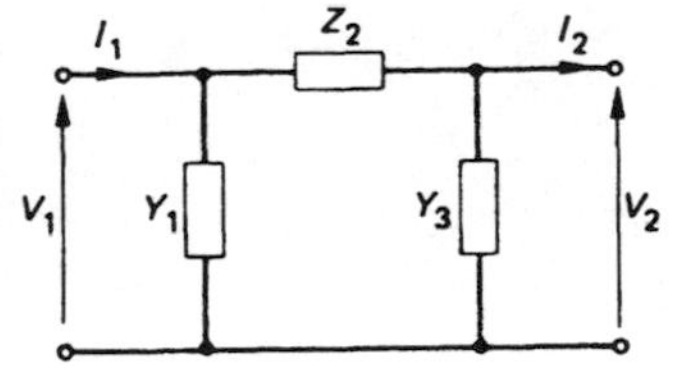

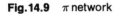

Fig. 14.9 π network

The π network of Fig. 14.9 can be regarded as the cascade connection of a shunt admittance Y_1, a series impedance Z_2 and a shunt admittance Y_3, and hence its transmission matrix is

$$\begin{bmatrix} A & B \\ C & D \end{bmatrix} = \begin{bmatrix} 1 & 0 \\ Y_1 & 1 \end{bmatrix} \begin{bmatrix} 1 & Z_2 \\ 0 & 1 \end{bmatrix} \begin{bmatrix} 1 & 0 \\ Y_3 & 1 \end{bmatrix}$$

$$= \begin{bmatrix} 1 & Z_2 \\ Y_1 & 1 + Y_1 Z_2 \end{bmatrix} \begin{bmatrix} 1 & 0 \\ Y_3 & 1 \end{bmatrix}$$

$$= \begin{bmatrix} 1 + Z_2 Y_3 & Z_2 \\ Y_1 + Y_3(1 + Y_1 Z_2) & 1 + Y_1 Z_2 \end{bmatrix}$$

Example 14.3

A symmetrical T network has series impedances of $500\angle 50°\ \Omega$ and a shunt admittance of $10\angle 30°$ mS. (*a*) Obtain the transmission matrix. (*b*) Calculate the input voltage and current needed to produce 1 V across a 100 Ω load resistor.

Solution

(*a*) The transmission matrix is

$$\begin{bmatrix} 1 + 500\angle 50^\circ \times 0.01\angle 30^\circ & 500\angle 50^\circ(1 + 500\angle 50^\circ \times 0.01\angle 30^\circ) + 500\angle 50^\circ \\ 0.01\angle 30^\circ & 1 + 0.01\angle 30^\circ \times 500\angle 50^\circ \end{bmatrix}$$

$$= \begin{bmatrix} 1 + 5\angle 80^\circ & 500\angle 50^\circ(1.87 + j4.92) + 321.4 + j383 \\ 0.01\angle 30^\circ & 1.87 + j4.92 \end{bmatrix}$$

$$= \begin{bmatrix} 5.27\angle 69.2^\circ & 2849\angle 109.8^\circ \\ 0.01\angle 30^\circ & 5.27\angle 69.2^\circ \end{bmatrix} \quad (Ans.)$$

(*b*) Load current $I_2 = 1/100 = 0.01$ A.

$$V_1 = AV_2 + BI_2 = 5.27\angle 69.2^\circ \times 1 + 2849\angle 109.8^\circ \times 0.01$$
$$= -7.76 + j31.72 = 32.66\angle 103.8^\circ \text{ V} \quad (Ans.)$$
$$I_1 = 0.01\angle 30^\circ \times 1 + 5.27\angle 69.2^\circ \times 0.01 \text{ A}$$
$$= 27.48 + j44.5 \text{ mA} = 52.3\angle 58.3^\circ \text{ mA} \quad (Ans.)$$

Lattice Network

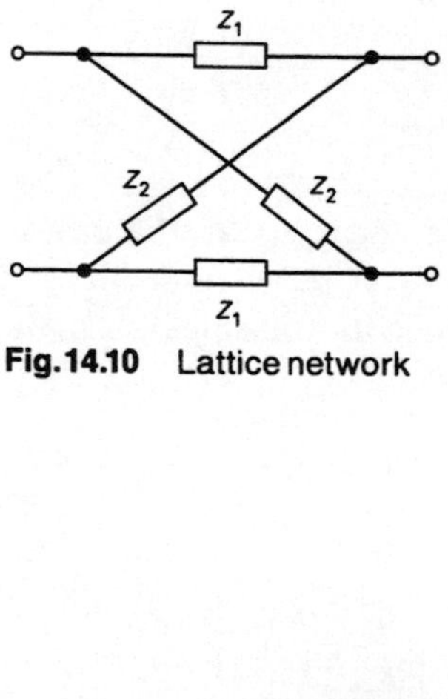

Fig.14.10 Lattice network

Figure 14.10 shows a lattice network with series impedances Z_1 and shunt impedances Z_2. When the output terminals are open circuited the current $I_2 = 0$ and then

$$V_1 = AV_2 = DV_2$$
$$A = D = V_1/[Z_2/(Z_1 + Z_2) - Z_1/(Z_1 + Z_2)]$$
$$= (Z_1 + Z_2)/(Z_1 - Z_2)$$

Also,

$$I_1 = CV_2,\ C = I_1/V_2 = AI_1/V_1$$
$$= [2/(Z_1 + Z_2)][(Z_1 + Z_2)/(Z_2 - Z_1)] = 2/(Z_2 - Z_1)$$
$$B = (A^2 - 1)/C = (2Z_1Z_2)/(Z_2 - Z_1)$$

Transformer

For an ideal transformer of turns ratio $n :: 1$, $V_1 = nV_2$ and $I_1 = I_2/n$. Hence,

$$\text{Transmission matrix} = \begin{bmatrix} n & 0 \\ 0 & 1/n \end{bmatrix}$$

The losses in a transformer are represented by a series impedance Z_1 and a shunt admittance Y_2 in the primary circuit, and by a series impedance Z_3 in the secondary circuit. The transmission matrix of a practical transformer is therefore:

$$\begin{bmatrix} 1 & Z_1 \\ 0 & 1 \end{bmatrix} \begin{bmatrix} 1 & 0 \\ Y_2 & 1 \end{bmatrix} \begin{bmatrix} n & 0 \\ 0 & 1/n \end{bmatrix} \begin{bmatrix} 1 & Z_3 \\ 0 & 1 \end{bmatrix}$$

$$= \begin{bmatrix} n(1+Z_1Y_2) & nZ_3(1+Z_1Y_2)+Z_1/n \\ nY_2 & nZ_3Y_2+1/n \end{bmatrix}$$

Example 14.4

A 20 :: 1 turns ratio transformer has a primary resistance of 4 Ω, a secondary resistance of 0.05 Ω and primary shunt admittance of $1\angle 90°$ mS. Determine its transmission matrix.

Solution

$$\begin{bmatrix} 20(1+\mathrm{j}4\times 10^{-3}) & 20\times 0.05(1+\mathrm{j}4\times 10^{-3})+4/20 \\ \mathrm{j}20\times 10^{-3} & \mathrm{j}20\times 0.05\times 10^{-3}+1/20 \end{bmatrix}$$

$$= \begin{bmatrix} 20+\mathrm{j}80\times 10^{-3} & 1.2+\mathrm{j}4\times 10^{-3} \\ \mathrm{j}20\times 10^{-3} & 0.05+\mathrm{j}1\times 10^{-3} \end{bmatrix} \quad (\textit{Ans.})$$

Output voltage and current

The basic equations are repeated here:

$$V_1 = AV_2 + BI_2 \qquad I_1 = CV_2 + DI_2$$

From these equations, $I_2 = (I_1 - CV_2)/D$ and therefore

$$V_1 = AV_2 + B(I_1 - CV_2)/D$$

or $DV_1 - BI_1 = V_2(AD - BC) = V_2$

Hence

$$V_2 = DV_1 - BI_1 \tag{14.23}$$

Similarly

$$I_2 = AI_1 - CV_1 \tag{14.24}$$

To obtain V_2 purely in terms of V_1 so that the voltage ratio of the network can be obtained, recourse must be made to the equation for V_1. This can be written as $V_1 = AV_2 + (BV_2/Z_L)$, and therefore the voltage ratio is

$$V_2/V_1 = V_2/[AV_2 + (BV_2/Z_L)]$$
$$= Z_L/(AZ_L + B) \tag{14.25}$$

The input impedance of the network is the ratio $Z_{in} = V_1/I_1$. Therefore,

$$\begin{aligned} Z_{in} &= (AV_2 + BI_2)/(CV_2 + DI_2) \\ &= [AV_2 + (BV_2/Z_L)]/[CV_2 + (DV_2/Z_L)] \\ &= (AZ_L + B)/(CZ_L + D) \qquad (14.26) \end{aligned}$$

Example 14.4

A network has the following transmission parameters:

$$A = D = 1\angle 60^\circ \qquad B = 200\angle 70^\circ\ \Omega \qquad C = 1.5\times 10^{-3}\angle -90^\circ\ \text{S}$$

Calculate its input impedance when its output terminals are (*a*) open circuited and (*b*) short-circuited.

Solution

(*a*) When $Z_L = \infty$, the input impedance Z_{in} is given by $A/C\ \Omega$. Therefore

$$Z_{in} = (1\angle 60^\circ)/[(1.5\times 10^{-3})\angle -90^\circ] = 666.7\angle 150^\circ\ \Omega \quad (\textit{Ans.})$$

(*b*) When $Z_L = 0$, $Z_{in} = B/D\ \Omega$. Therefore

$$Z_{in} = (200\angle 70^\circ)/(1\angle 60^\circ) = 200\angle 10^\circ\ \Omega \quad (\textit{Ans.})$$

Example 14.5

Fig. 14.11

Figure 14.11 shows an L network whose transmission parameters are

$$\begin{bmatrix} 1 + (r + j\omega L)/R & r + j\omega L \\ 1/R & 1 \end{bmatrix}$$

If $r = 20\ \Omega$ and $\omega L = 30\ \Omega$ calculate the value of R for two such networks connected in cascade to introduce an overall phase shift of 90°.

Solution

When the output of the circuit is open circuit, equation (14.19) becomes $V_1 = AV_2$. This means that only the A parameter of the cascade is required. This is

$$\begin{aligned} &[1 + (r + j\omega L)/R][1 + (r + j\omega L)/R] + (r + j\omega L)/R \\ &= [R^2 + r^2 - \omega^2 L^2 + j2r\omega L + 3rR]/R^2 + j3\omega L/R \end{aligned}$$

For the network to introduce a 90° phase shift the real part of this equation must be equal to zero. Therefore, $R^2 + 3rR + r^2 - \omega^2 L^2 = 0$, $R^2 + 60R + 400 - 900 = 0$

$$R = [-60 \pm \sqrt{(3600 + 2000)}]/2 = 7.42\ \Omega \quad (\textit{Ans.})$$

(ignoring the negative result).

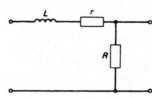

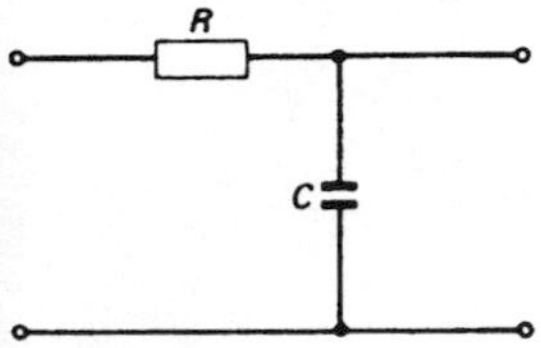

Fig. 14.12

Example 14.6

Three of the circuits shown in Fig. 14.12 are connected in cascade. Derive an expression for the frequency at which the combination will introduce a phase shift of 180° when terminated in an open circuit.

Solution

For a single network $Z_1 = R$ and $Y_2 = \mathrm{j}\omega C$. Hence the transmission matrix is

$$\begin{bmatrix} 1+\mathrm{j}\omega CR & R \\ \mathrm{j}\omega C & 1 \end{bmatrix}$$

For three such networks in cascade the overall matrix is

$$\begin{bmatrix} 1+\mathrm{j}\omega CR & R \\ \mathrm{j}\omega C & 1 \end{bmatrix}^3$$

but as in the previous example only the A term is needed. This is

$$(1+\mathrm{j}\omega CR)^3 + \mathrm{j}\omega CR(1+\mathrm{j}\omega CR) + \mathrm{j}\omega CR(1+\mathrm{j}\omega CR) + \mathrm{j}\omega CR$$
$$= 1 - 5\omega^2C^2R^2 + \mathrm{j}6\omega CR - \mathrm{j}\omega^3C^3R^3$$

For V_2 to be 180° out of phase with V_1, the j terms must sum to zero. Therefore

$$6\omega_0 CR = \omega_0^3 C^3 R^3 \quad \text{or} \quad f_0 = \sqrt{6}/2\pi CR \quad (Ans.)$$

Transmission parameters may be employed to solve a circuit whose analysis is normally carried out using a network theorem or Kirchhoff's laws. This is demonstrated by the following example (Example 14.1 repeated).

Example 14.7

For the circuit given in Fig 14.2 use transmission parameters to find the current that flows in the 10 Ω resistor.

Solution

$$Z_{\text{in}} = 10 + (20 \times 40)/60 = 23.33\ \Omega \quad \text{and}$$
$$I_{\text{in}} = 6/23.33 = 0.257\ \text{A}$$

With the output terminals short circuited

$$I_1 = 6/[10 + (30 \times 20)/50] = 6/22\ \text{A}$$

Hence

$$I_2 = (6/22 \times 20/50) = 109.1\ \text{mA}$$
$$B = V_1/I_2 = 6/0.1091 = 55\ \Omega$$
$$D = I_1/I_2 = (6/22)/0.1091 = 2.5$$

With the output terminals open circuited

$$V_2 = (6 \times 20)/30 = 4\ \text{V}$$

Therefore,

$A = V_1/V_2 = 6/4 = 1.5 \qquad I_1 = 6/30 = 0.2\ \text{A}$ and so

$C = I_1/V_2 = 0.2/4 = 0.05\ \text{S}$

$$\begin{bmatrix} V_1 \\ I_1 \end{bmatrix} = \begin{bmatrix} 1.5 & 55 \\ 0.05 & 2.5 \end{bmatrix} \begin{bmatrix} V_2 \\ I_2 \end{bmatrix}.$$

The inverse matrix is $\begin{bmatrix} 2.5 & -0.05 \\ -55 & 1.5 \end{bmatrix} \Big/ (3.75 - 2.75)$

Therefore,

$$\begin{bmatrix} V_2 \\ I_2 \end{bmatrix} = \begin{bmatrix} 6 \\ 0.257 \end{bmatrix} \begin{bmatrix} 2.5 & -0.05 \\ -55 & 1.5 \end{bmatrix} = \begin{bmatrix} 0.87 \\ 0.086 \end{bmatrix}$$

Hence,

$I_2 = 0.086\ \text{A}$ (*Ans.*)

Before, $I_{in} = I_1 - I_2 = 0.257 - 0.171 = 0.086\ \text{A}$

Star–delta transform

The *star–delta transform* has already been discussed in Chapter 9. The transmission parameters provide an alternative method of deriving the necessary equations.

Consider the T and π networks shown in Fig. 14.13. The transmission matrices are

$$\begin{bmatrix} 1 + (Z_2/Z_1) & Z_3[1 + (Z_2/Z_1)] + Z_2 \\ 1/Z_1 & 1 + (Z_3/Z_1) \end{bmatrix}$$

and

$$\begin{bmatrix} 1 + (Z_C/Z_A) & Z_C \\ (1/Z_B) + (1/Z_A)[1 + (Z_C/Z_B)] & 1 + (Z_C/Z_B) \end{bmatrix}$$

Hence

$$Z_C = (Z_1 Z_3 + Z_2 Z_3 + Z_1 Z_2)/Z_1 \tag{14.27}$$

$Z_2/Z_1 = Z_C/Z_A$ or

$$Z_A = Z_C Z_1/Z_2 = (Z_1 Z_3 + Z_2 Z_3 + Z_1 Z_2)/Z_2 \tag{14.28}$$

and $Z_3/Z_1 = Z_C/Z_B$ or

$$Z_B = Z_C Z_1/Z_3 = (Z_1 Z_3 + Z_2 Z_3 + Z_1 Z_2)/Z_3 \tag{14.29}$$

Similarly,

$$\begin{aligned} 1/Z_1 &= (1/Z_B) + (1/Z_A)[1 + (Z_C/Z_B)] \\ &= (1/Z_B) + Z_C/(Z_A Z_B) + (1/Z_A) \\ &= (Z_A + Z_B + Z_C)/Z_A Z_B \end{aligned}$$

or $$Z_1 = Z_A Z_B/(Z_A + Z_B + Z_C) \tag{14.30}$$

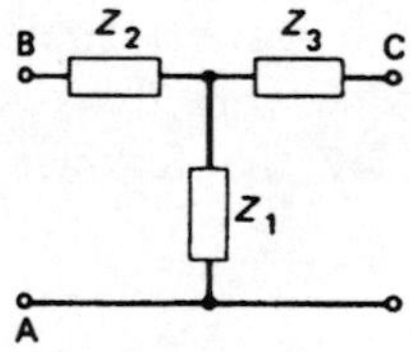

(a)

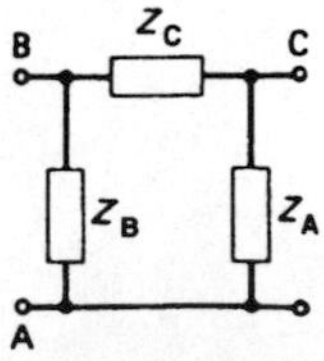

(b)

Fig. 14.13

$$Z_3/Z_1 = Z_C/Z_B \quad \text{or}$$

$$Z_3 = Z_1 Z_C/Z_B = Z_A Z_C/(Z_A + Z_B + Z_C) \qquad (14.31)$$

Lastly, $Z_2/Z_1 = Z_C/Z_A$ or

$$Z_2 = Z_1 Z_C/Z_A = Z_B Z_C/(Z_A + Z_B + Z_C) \qquad (14.32)$$

y Parameters

When networks which are connected in parallel are to be analysed it is more convenient to use y parameters. The methods to be used to determine the values of the y parameters of a particular network are very similar to those already described. For this reason, only one example is given and then the results for the more commonly used networks are quoted in Table 14.2.

Table 14.2

Figure no.	*y matrix*
14.6	$\begin{bmatrix} \infty & \infty \\ \infty & \infty \end{bmatrix}$
14.7	$\begin{bmatrix} 1/Z_1 & -1/Z_1 \\ -1/Z_1 & -(1+Z_1Y_2)/Z_1) \end{bmatrix}$
14.7 (reversed)	$\begin{bmatrix} (1+Z_1Y_2)/Z_1 & -1/Z_1 \\ 1/Z_1 & -1/Z_1 \end{bmatrix}$
14.8	$\begin{bmatrix} (1+Z_3Y_2)/(Z_1+Z_3+Z_1Y_2Z_3) & -1/(Z_1+Z_3+Z_1Y_2Z_3) \\ 1/(Z_1+Z_3+Z_1Y_2Z_3) & -(1+Z_3Y_2)/(Z_1+Z_3+Z_1Y_2Z_3) \end{bmatrix}$
14.9	$\begin{bmatrix} (1+Z_1Y_2)/Z_1 & -Y_2 \\ -Y_2 & (1+Y_2Z_3)/Z_3 \end{bmatrix}$

Consider the series impedance of Fig. 14.5 again. Since $I_2 = I_1$ and $V_1 = V_2 + I_2 Z$ then $I_1 = (V_1/Z) - (V_2/Z)$. Hence, $y_{11} = 1/Z$ and $y_{12} = -1/Z$. Also $I_2 = (V_1/Z) - (V_2/Z)$ so that $y_{21} = 1/Z$ and $y_{22} = -1/Z$. Alternatively, with the output terminals short circuited, $V_2 = 0$ and then $V_1 = I_1 Z_1$ so that $I_1/V_1 = y_{11} = 1/Z$. Also, $V_1 = I_2 Z$ (since $I_2 = I_1$) and so $I_2/V_1 = y_{21} = 1/Z$. With the input terminals short circuited, $V_1 = 0$ and $I_1 = -y_{12} V_2 = -V_2/Z$ and $y_{12} = 1/Z$. Finally, $I_2 = -V_2/Z$ or $I_2/V_2 = y_{22} = -1/Z$

Exercises 14

14.1 Determine the transmission parameters of the network shown in Fig. 14.14

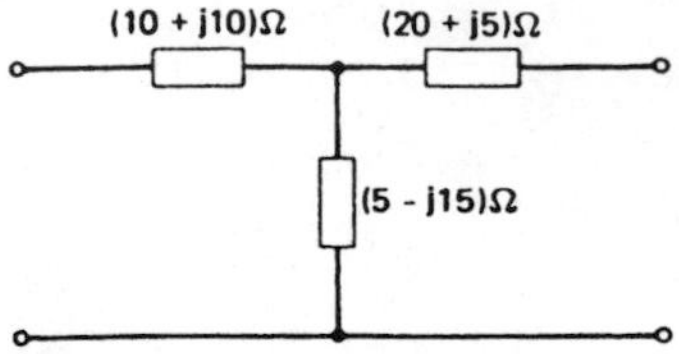

Fig. 14.14

14.2 A network has the following transmission parameters:

$$A = 1\angle 30^\circ \qquad B = 150\angle 50^\circ\ \Omega \qquad C = 0.02\angle -6.8^\circ\ \text{S}$$
$$D = 3\angle 0^\circ$$

Calculate the input impedance of the network when its output terminals are (*a*) open circuited and (*b*) short circuited.

14.3 A network has the transmission parameters $A = D = 14.14\angle 0^\circ$, $B = 50\angle 60^\circ\ \Omega$ and $C = 4\angle -60^\circ$ S. Calculate (*a*) the ratio V_{out}/V_{in} and (*b*) the input impedance of the network when a 600 Ω resistor is connected across the output terminals.

14.4 Determine (*a*) the *y* parameters and (*b*) the *h* parameters of the network shown in Fig. 14.15.

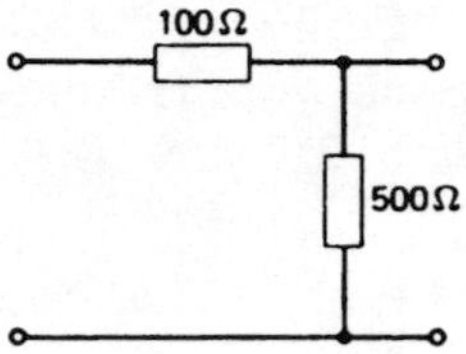

Fig. 14.15

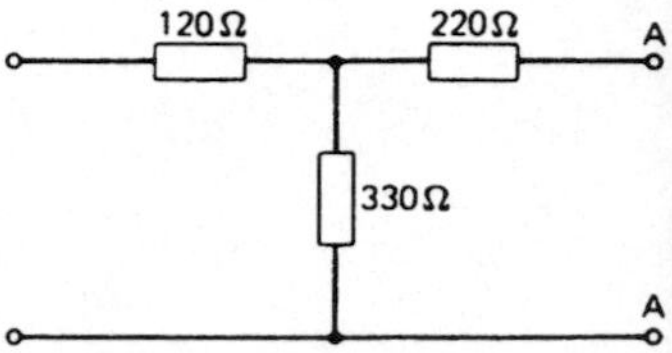

Fig. 14.16

14.5 A network has the following *h* parameters: $h_{11} = 1000\ \Omega$, $h_{12} = h_{21} = 0.5$ and $h_{22} = 0.015$. Determine its transmission parameters.

14.6 For the network in Fig. 14.16 determine (*a*) the transmission parameters and (*b*) the voltage ratio V_2/V_1 when a 1200 Ω resistor is connected across the terminals AA.

14.7 Determine the overall transmission matrix of the two cascaded networks shown in Fig. 14.17.

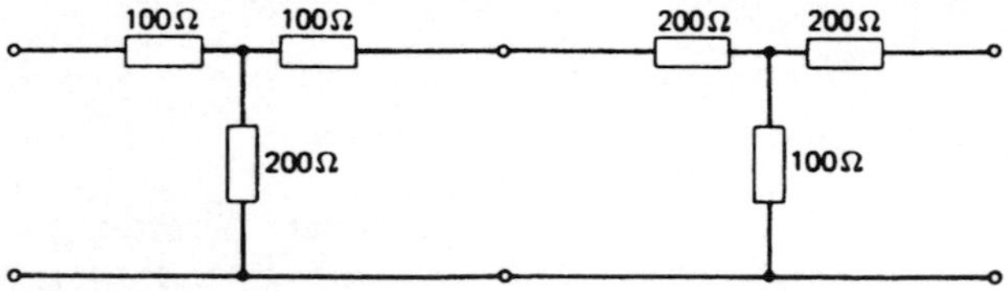

Fig. 14.17

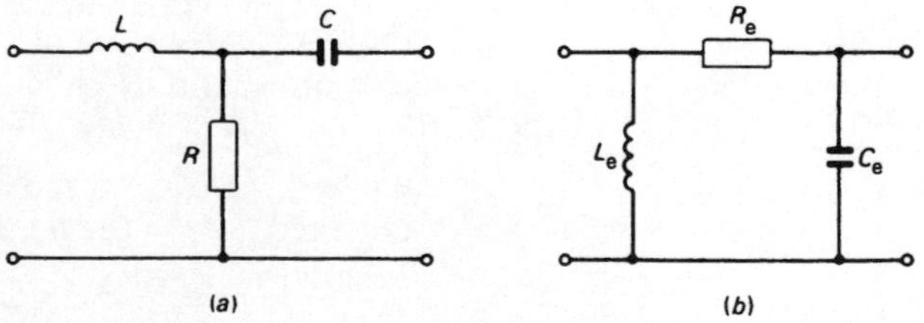

Fig. 14.18

14.8 Calculate the transmission parameters of the circuit given in Fig. 14.18. If, when the circuit is terminated in a 500 Ω resistor, the output voltage is 100 V calculate the input voltage.

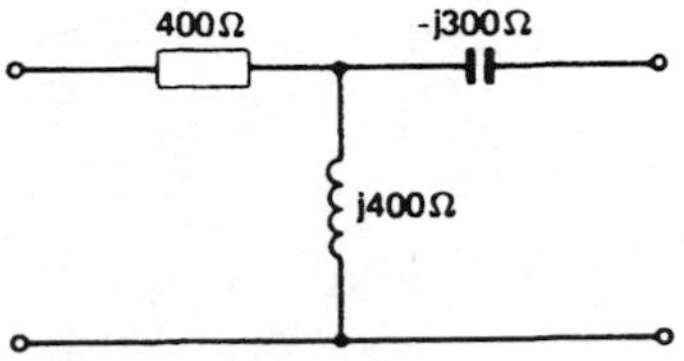

Fig. 14.19

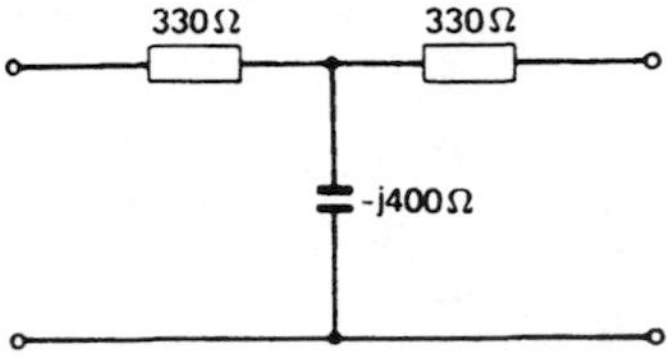

Fig. 14.20

14.9 Find the transmission parameters of a symmetrical T network which has series arms of (250 + j350) Ω and a shunt arm of (−j180) Ω.

14.10 Calculate the transmission parameters of the network of Fig. 14.20. Calculate the supply voltage and current required to give an output of 100 V and 50 mA at unity power factor to a load of R ohms connected across the output terminals.

14.11 Calculate the transmission parameters of the circuit shown in Fig. 14.21. If the output voltage is $25\angle 0°$ volts and the output current is $16\angle 0°$ mA calculate the input voltage.

Fig. 14.21

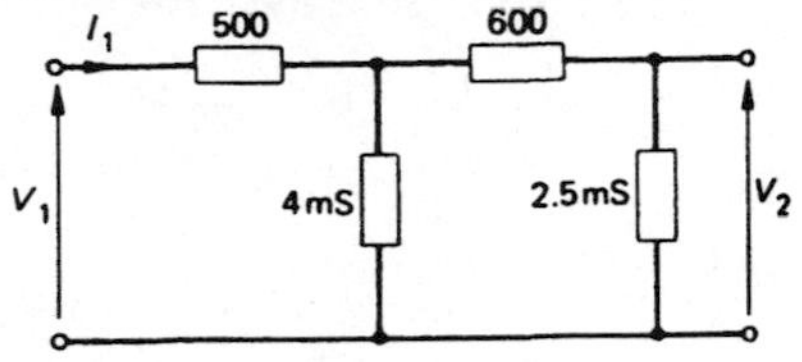

14.12 A 6 V d.c. source is connected to a circuit that consists of two 12 Ω resistors connected in series. Use transmission parameters to calculate the voltage across either of the resistors.

14.13 The expressions representing the behaviour of a 4-terminal network are

$$10 = 500I_1 + 0.02V_2 \qquad 5 = 50I_1 + 0.04V_2$$

What kind of parameters have been used and for what purpose are they best suited?

14.14 A network has the impedance matrix

$$\begin{bmatrix} 100 & 10 \\ 20 & 40 \end{bmatrix}$$

and is connected in series with another network. The overall impedance matrix is then

$$\begin{bmatrix} 150 & 50 \\ 30 & 80 \end{bmatrix}.$$

Obtain the impedance matrix of the second network.

14.15 One set of parameters has not been given in this chapter, i.e. those for which I_2 and V_2 are the independent variables. Write down equations for this set of parameters, using coefficients E, F, G and H and deduce the dimensions of each parameter.

14.16 A passive network has the following transmission parameters: $A = D = 100$, $B = 500\ \Omega$. Calculate the value of C.

14.17 A 4-terminal network has the h parameter matrix

$$\begin{bmatrix} 1000 & 1 \times 10^{-6} \\ 120 & 4 \times 10^{-4} \end{bmatrix}$$

Determine the y matrix for the network.

14.18 A network has the transmission matrix

$$\begin{bmatrix} 1.5 & 40 \\ 0.0313 & 1.5 \end{bmatrix}.$$

Two such networks are connected in cascade. Determine the transmission matrix of the combination.

14.19 A symmetrical T network has two series impedances of $50\angle 60°\ \Omega$ and a shunt admittance of $1.2 \times 10^{-3}\angle 90°$ S. Calculate the transmission parameters of the network.

14.20 Show that the transmission matrix of a pure mutual inductance is

$$\begin{bmatrix} 0 & j\omega M \\ 1/j\omega M & 0 \end{bmatrix}$$

14.21 A symmetrical π network has series impedance of j100 Ω and two shunt impedances of −j80 Ω each. Calculate its ABCD parameters.

14.22 A two-port network has its output terminals first open circuited and then short circuited. The input impedances are then $Z_{oc} = 800\ \Omega$ and $Z_{sc} = 550 - j250\ \Omega$. When the input terminals are open-circuited the output impedance of the network is $Z_0 = 500 - j500\ \Omega$. Calculate the transmission parameters of the network.

14.23 A π network has a series impedance of $50 + j200\ \Omega$ and shunt admittances $Y_1 = 5 \times 10^{-4}$ S and $Y_3 = 1 \times 10^{-3}$ S. Determine the transmission parameters of the network.

14.24 A T network has $Z_1 = -j30\ \Omega$, $Z_3 = j60\ \Omega$ and $Y = j0.015$ S. Determine its ABCD parameters.

15 Network theory

A 4-terminal, or two-port, network is one that has two input and two output terminals. Very often one input and one output terminal are common. The four basic versions of networks are known as the T, the π, the L and the lattice networks and they are shown, respectively, in Figs 15.1(a), (b), (c) and (d). When a T or a π network is *symmetrical* it is customary to label the components to make the total series impedance equal to Z_1 and the total shunt impedance equal to Z_2. This practice is followed in this chapter (see Fig. 15.7).

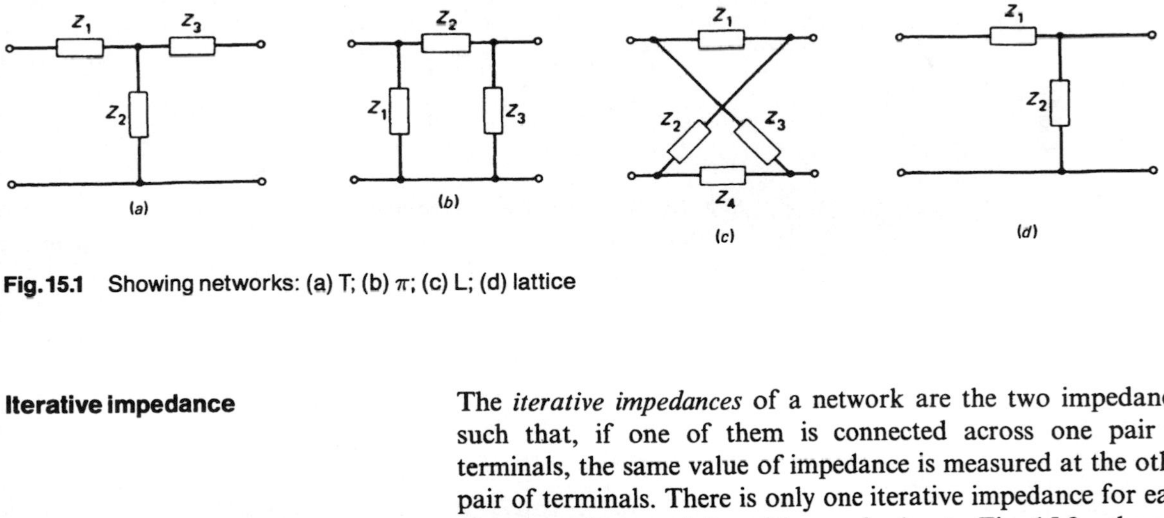

Fig. 15.1 Showing networks: (a) T; (b) π; (c) L; (d) lattice

Iterative impedance

The *iterative impedances* of a network are the two impedances such that, if one of them is connected across one pair of terminals, the same value of impedance is measured at the other pair of terminals. There is only one iterative impedance for each direction of transmission. Thus, referring to Fig. 15.2, when an impedance Z_A is connected across the terminals BB, the input

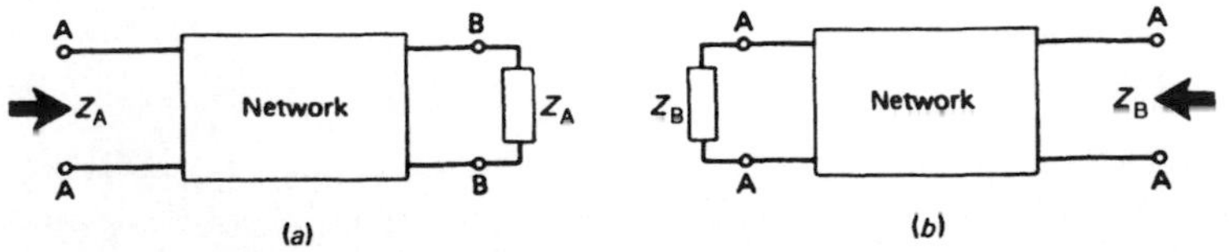

Fig. 15.2 Iterative impedances

impedance at the terminals AA is also Z_A. Conversely, when an impedance Z_B is connected across the terminals AA the impedance measured at terminals BB is also Z_B.

Example 15.1

Calculate the iterative impedances of the network shown in Fig. 15.3(*a*).

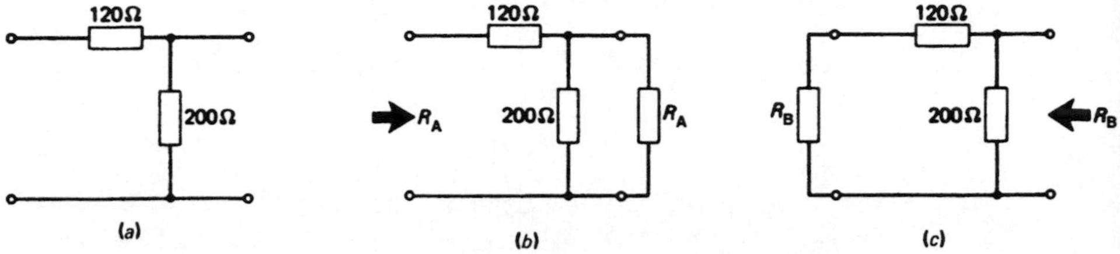

Fig.15.3

Solution
From the definition of an iterative impedance, when a resistor R_A is connected across the output terminals of the network the input impedance will also be R_A. Therefore, from Fig. 15.3(*b*),

$$R_A = 120 + 200R_A/(200 + R_A)$$
$$= (24\,000 + 120R_A + 200R_A)/(200 + R_A)$$

Hence $R_A^2 - 120R_A - 24\,000 = 0$ or

$$R_A = [120 \pm \sqrt{(120^2 + 4 \times 24\,000)}]/2 = 226\ \Omega \quad \text{or} \quad -106\ \Omega \quad (Ans.)$$

Clearly, the negative answer is inadmissible.
Similarly, from Fig. 15.3(*c*),

$$R_B = [200(120 + R_B)]/(200 + 120 + R_B)$$
$$320R_B + R_B^2 = 24\,000 + 200R_B$$

or $R_B^2 + 120R_B - 24\,000 = 0$

Solving

$$R_B = 106\ \Omega \quad \text{or} \quad -226\ \Omega \quad (Ans.)$$

Again, the negative answer is inadmissible. Note that the negative result obtained for R_A is actually the value of R_B and vice versa. This means, of course, that for a resistive network only one iterative impedance calculation need be carried out.

Example 15.2

A T network has series resistors of 300 Ω and 500 Ω and a shunt resistor of 100 Ω. Calculate its iterative impedances.

Solution

$$R_A = 300 + [(500 + R_A)100]/(600 + R_A),$$
$$600R_A + R_A^2 = 18 \times 10^4 + 300R_A + 5 \times 10^4 + 100R_A$$
$$R_A^2 + 200R_A - 23 \times 10^4 = 0$$

and $R_A = [-200 \pm \sqrt{(4 \times 10^4 + 9.2 \times 10^5)}]/2 \approx 390\ \Omega$ (*Ans.*)

Taking the negative value

$$R_B \approx 590\ \Omega \quad (Ans.)$$

Image impedance

The two *image impedances* of a network are such that, when one of them is connected across one pair of terminals, the other is measured at the other pair of terminals and vice versa. The concept of image impedance is illustrated by Fig. 15.4.

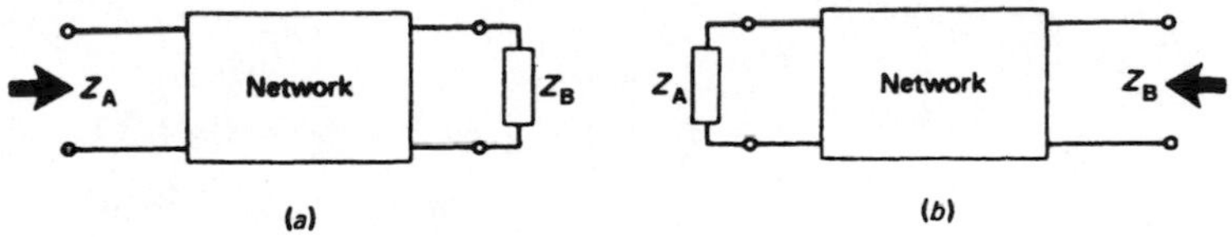

Fig. 15.4 Image impedances

The two image impedances of a network can be calculated using the method employed for the determination of the iterative impedances.

Example 15.3

An L network has a series resistor of 2 Ω and a shunt resistor of 6 Ω. Calculate its image resistances.

Solution

$$R_A = 2 + 6R_B/(6 + R_B),\ 6R_A + R_AR_B = 12 + 8R_B \quad (15.1)$$
$$R_B = 6(2 + R_A)/(6 + 2 + R_A), -6R_A + R_AR_B = 12 - 8R_B \quad (15.2)$$

Adding equations (15.1) and (15.2) gives,

$$2R_AR_B = 24, \quad R_AR_B = 12 \quad \text{and} \quad R_A = 12/R_B$$

Substitute into equation (15.1)

$$72/R_B + 12 = 12 + 8R_B, \quad \text{or} \quad R_B = 3\ \Omega \quad (Ans.)$$
$$R_A = 12/3 = 4\ \Omega \quad (Ans.)$$

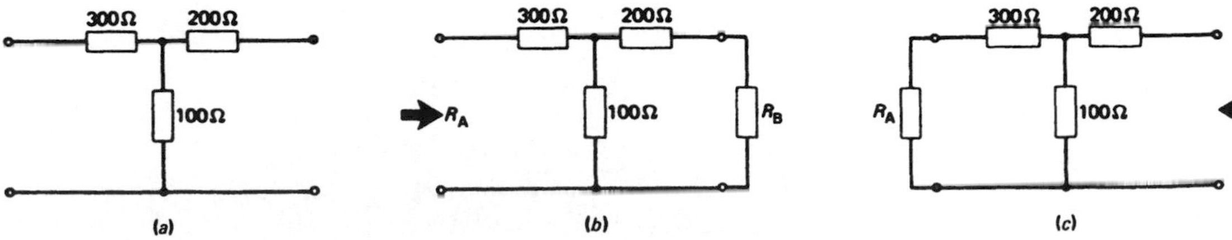

Fig. 15.5

Example 15.4

Calculate the image impedances of the network shown in Fig. 15.5(a).

Solution
From Fig. 15.5(b)

$$R_A = 300 + [100(200 + R_B)]/(100 + 200 + R_B)$$

$$300R_A + R_AR_B = (9 \times 10^4) + 300R_B + (2 \times 10^4) + 100R_B$$

$$300R_A + R_AR_B = (11 \times 10^4) + 400R_B \quad (15.3)$$

From Fig. 15.5(c)

$$R_B = 200 + [100(300 + R_A)]/(100 + 300 + R_A)$$

$$400R_B + R_AR_B = (8 \times 10^4) + 200R_A + (3 \times 10^4) + 100R_A$$

$$-300R_A + R_AR_B = (11 \times 10^4) - 400R_B \quad (15.4)$$

Adding equations (15.3) and (15.4),

$$2R_AR_B = 22 \times 10^4 \qquad R_A = (11 \times 10^4)/R_B$$

Substituting into equation (15.3)

$$(300 \times 11 \times 10^4)/R_B + (11 \times 10^4) = (11 \times 10^4) + 400R_B$$

$$R_B^2 = (3 \times 11 \times 10^4)/4 \quad \text{and} \quad R_B = 287.2\ \Omega \quad (\textit{Ans.})$$

Therefore

$$R_A = (11 \times 10^4)/287.2 = 383\ \Omega \quad (\textit{Ans.})$$

$Z_{im} = \sqrt{(Z_{oc}Z_{sc})}$

Alternatively, and somewhat more simply, the expression

$$Z_{im} = \sqrt{(Z_{oc}Z_{sc})} \quad (15.5)$$

can be used to calculate the image impedances of a network, where Z_{oc} and Z_{sc} are the impedances measured at a pair of terminals when the other pair of terminals is first open circuited and then short circuited.

For the network in example 15.3,

$$Z_{A(oc)} = 2 + 6 = 8\ \Omega, \quad Z_{A(sc)} = 2\ \Omega \quad \text{and}$$

$$Z_A = \sqrt{(2 \times 8)} = 4\ \Omega$$

as before. Also,

$$Z_{B(oc)} = 6\ \Omega, \quad Z_{B(sc)} = (6 \times 2)/(6 + 2) = 1.5\ \Omega \quad \text{and}$$

$$Z_B = \sqrt{(6 \times 1.5)} = 3\ \Omega$$

as before. For the circuit of example 15.4, looking into the input terminals of the network,

$$R_{oc} = 300 + 100 = 400\ \Omega$$

Also

$$R_{sc} = 300 + (200 \times 100)/(200 + 100) = 366.7\ \Omega$$

Hence

$$R_A = \sqrt{(400 \times 366.7)} = 383\ \Omega \quad \text{(as before)}$$

Looking into the output terminals, $R_{oc} = 300\ \Omega$, $R_{sc} = [200 + (300 \times 100)]/400 = 275\ \Omega$. Hence $R_B = \sqrt{(300 \times 275)} = 287.5\ \Omega$.

Example 15.5

For the circuit shown in Fig. 15.6 calculate (*a*) the image impedances, (*b*) the input and output currents and (*c*) the input power to the network and the power dissipated in the load.

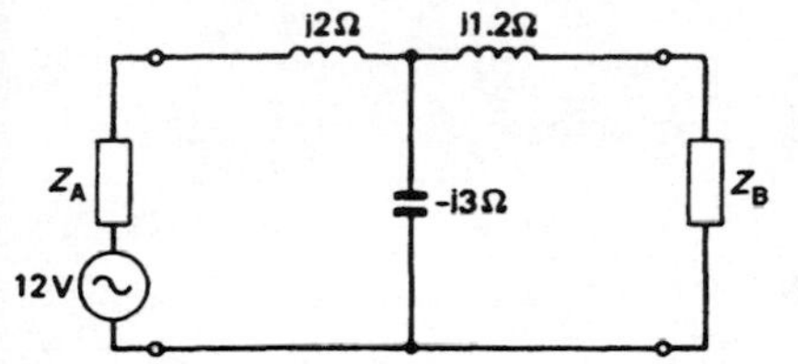

Fig. 15.6

Solution

(*a*) $Z_A = \sqrt{(Z_{oc}Z_{sc})}$
$= \sqrt{\{(j2 - j3)[j2 + (-j3 \times j1.2)]/(j1.2 - j3)\}}$
$= 2\ \Omega$ (*Ans.*)
$Z_B = \sqrt{\{(j1.2 - j3)[j1.2 + (j3 \times j3)]/(j2 - j3)\}} = 3.6\ \Omega$ (*Ans.*)

(*b*) $V_{in} = 12/2 = 6$ V. Therefore $I_{in} = 6/2 = 3$ A (*Ans.*)
$I_2 = (3 \times -j3)/(3.6 + j1.2 - j3) = 1 - j2$ A (*Ans.*)

(*c*) $P_{in} = 6^2/2 = 18$ W (*Ans.*)
$P_{out} = |1 - j2|^2 \times 3.6 = 2.2362^2 \times 3.6 = 18$ W (*Ans.*)

There is no loss in the network because it contains no resistances.

Characteristic impedance

For a symmetrical network the iterative impedances and the image impedances are equal to one another and have a common value for both directions of transmission. The common value of impedance is known as the *characteristic impedance* of the network.

Figures 15.17(*a*) and (*b*) show, respectively, a symmetrical T and a symmetrical π network.

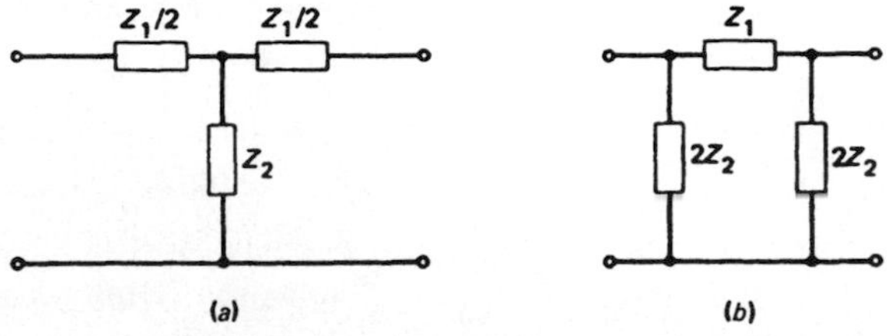

Fig. 15.7 Symmetrical T and π networks

T network

From Fig. 15.7(a)

$$Z_{0T} = Z_1/2 + [Z_2(Z_1/2 + Z_{0T})]/(Z_1/2 + Z_2 + Z_{0T})$$

$$Z_{0T}Z_1/2 + Z_{0T}Z_2 + Z_{0T}^2$$

$$= Z_1^2/4 + Z_1Z_2/2 + Z_1Z_{0T}/2 + Z_1Z/2 + Z_{0T}Z_2$$

$$Z_{0T}^2 = Z_1^2/4 + Z_1Z_2$$

$$Z_{0T} = \sqrt{(Z_1^2/4 + Z_1Z_2)} \qquad (15.6)$$

Alternatively, using $Z_0 = \sqrt{(Z_{oc}Z_{sc})}$

$$Z_{oc} = Z_1/2 + Z_2 \quad \text{and}$$

$$Z_{sc} = Z_1/2 + (Z_1Z_2/2)/(Z_1/2 + Z_2)$$

$$= (Z_1^2/4 + Z_1Z_2)/(Z_1/2 + Z_2)$$

Therefore

$$Z_{0T} = \sqrt{(Z_1^2/4 + Z_1Z_2)} \quad \text{as before}$$

Bartlett's Bisection theorem

When a network is symmetrical the work involved in its analysis can be greatly reduced by the use of *Bartlett's Bisection theorem.* The theorem states that a symmetrical network can be cut in half and the open-circuit and short-circuit impedances of either half obtained. Then the image/characteristic impedance is given by $Z_0 = \sqrt{(Z_{oc} \times Z_{sc})}$ applied to either half of the network. Consider the symmetrical T network shown in Fig. 15.7(a). Bisecting the network gives two L networks each with a series impedance of $Z_1/2$ and a shunt impedance of $2Z_2$. Then, $Z_{oc} = Z_1/2 + 2Z_2$, and $Z_{sc} = Z_1/2$.

Hence

$$Z_0 = \sqrt{[(Z_1/2 + 2Z_2)(Z_1/2)]}$$

$$= \sqrt{(Z_1^2/4 + Z_1Z_2)} \quad \text{(which is equation (15.6))}$$

Example 15.6

A T network has series impedances of $100 + j100\ \Omega$ and a shunt impedance of $-j100\ \Omega$. Calculate its characteristic impedance, (a) using equation (15.6), (b) using $\sqrt{(Z_{oc}Z_{sc})}$ and (c) using Bartlett's Bisection theorem.

Solution

(a) $Z_0 = \sqrt{[(200 + j200)^2/4 + (200 + j200)(-j100)]} = 141\ \Omega$ (*Ans.*)

(b) $Z_{oc} = (100 + j100) + (-j100) = 100\ \Omega$

$$Z_{sc} = (100 + j100) + (100 + j100)(-j100)/100$$

$$= 100 + j100 + 100 - j100 = 200\ \Omega$$

Therefore,

$$Z_0 = \sqrt{(100 \times 200)} = 141\ \Omega \quad (\textit{Ans.})$$

(*c*) $Z_{oc} = 100 - j100$ and $Z_{sc} = 100 + j100$. Hence,

$$Z_0 = \sqrt{[(141\angle -45°) \times (141\angle 45°)]} = 141\ \Omega \quad (\textit{Ans.})$$

π Network

For the symmetrical π network of Fig. 15.7(*b*)

$$Z_{oc} = [2Z_2(Z_1 + 2Z_2)]/(4Z_2 + Z_1) \quad \text{and}$$
$$Z_{sc} = (2Z_2Z_1)/(2Z_2 + Z_1)$$
$$Z_{oc} \times Z_{sc} = (4Z_2^2Z_1)/(4Z_2 + Z_1)$$

Multiplying both the numerator and the denominator by $Z_1/4$ gives

$$Z_{oc}Z_{sc} = Z_1^2Z_2^2/(Z_1^2/4 + Z_1Z_2)$$

and so

$$Z_{0\pi} = Z_1Z_2/\sqrt{(Z_1^2/4 + Z_1Z_2)}$$

or $$Z_{0\pi} = Z_1Z_2/Z_{0T} \qquad (15.7)$$

Applying Bartlett's Bisection theorem to Fig. 15.7(*b*) gives a series impedance of $Z_1/2$ and a shunt impedance of $2Z_2$. Then

$$Z_{oc} = 2Z_2 \quad \text{and} \quad Z_{sc} = (2Z_2 \times Z_1/2)/(Z_1/2 + 2Z_2) = Z_1Z_2/(Z_1/2 + 2Z_2)$$

Hence

$$Z_{0\pi} = \sqrt{[(2Z_1Z_2^2)/(Z_1/2 + 2Z_2)]}$$

Multiply numerator and denominator by $Z_1/2$ to get

$$Z_{0\pi} = \sqrt{[(Z_1^2Z_2^2)/(Z_1^2/4 + Z_1Z_2)]} = Z_1Z_2/Z_{0T} \qquad (15.8)$$

Image transfer coefficient

The *image transfer coefficient* γ of a network is defined by

$$\gamma = 0.5\ \log_e[(I_SV_S)/(I_RV_R)] \qquad (15.9)$$
$$= \log_e[(V_S/V_R)\sqrt{(Z_B/Z_A)}]$$
$$= \log_e[(I_S/I_R)\sqrt{(Z_A/Z_B)}] \qquad (15.10)$$

where I_S and V_S are the current and voltage at the input terminals and I_R and V_R are the current and voltage at the output terminals. The term 'image' implies that the network is connected between its image impedances.

The real part of γ is known as the *image attenuation coefficient* α, expressed in nepers. The imaginary part of γ is known as the image phase-change coefficient β, expressed in radians. Thus,

$$\gamma = \alpha + j\beta \tag{15.11}$$

In the case of a symmetrical network, the ratio of the input and output currents is equal to the ratio of the input and output voltages and then

$$\gamma = 0.5 \log_e(I_S^2/I_R^2) = \log_e(I_S/I_R) \tag{15.12}$$

or $$\gamma = \log_e(V_S/V_R) \tag{15.13}$$

γ is then known as the *propagation coefficient* of the network.

Example 15.7

Calculate the image attenuation coefficient of the network shown in Fig. 15.8(*a*).

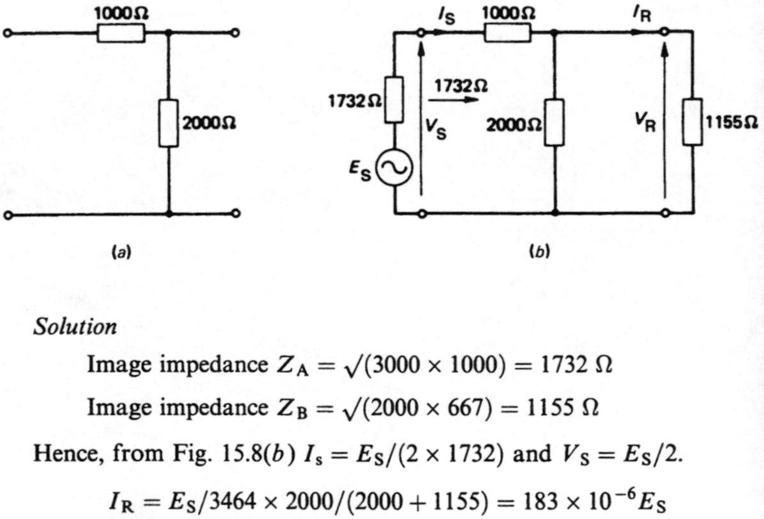

Fig. 15.8

Solution

$$\text{Image impedance } Z_A = \sqrt{(3000 \times 1000)} = 1732\ \Omega$$

$$\text{Image impedance } Z_B = \sqrt{(2000 \times 667)} = 1155\ \Omega$$

Hence, from Fig. 15.8(*b*) $I_s = E_S/(2 \times 1732)$ and $V_S = E_S/2$.

$$I_R = E_S/3464 \times 2000/(2000 + 1155) = 183 \times 10^{-6} E_S$$

$$V_R = I_R Z_B = 183 \times 10^{-6} E_S \times 1155 = 0.211 E_S$$

Therefore, from equation (15.9)

$$\gamma = 0.5 \log_e\{[(E_S/3464)(E_S/2)]/(183 \times 10^{-6} E_S \times 0.211 E_S)\}\ \text{N}$$

$$= 0.5 \log_e 3.738 = 0.66\ \text{N} \quad (\textit{Ans.})$$

$$= 0.66 \times 8.686 = 5.73\ \text{dB} \quad (\textit{Ans.})$$

Example 15.8

A T network has series resistors $R_1 = 300\ \Omega$ and $R_3 = 200\ \Omega$ and a shunt resistor $R_2 = 100\ \Omega$. Calculate its image attenuation coefficient.

Solution

This is the network of example 15.4, from this the image impedances are $R_A = 383\ \Omega$ and $R_B = 287.2\ \Omega$. Hence,

$$I_S = E_S/(2 \times 383) = E_S/766$$
$$I_R = (E_S/766) \times (100/587.2) = 22.223 \times 10^{-4} E_S$$
$$V_S = E_S/2$$

and

$$V_R = 287.2 \times 2.223 \times 10^{-4} E_S$$

Therefore,

$$\gamma = 0.5\ \log_e [E_S{}^2/(766 \times 2)]/[(2.223 \times 10^{-4})^2 E_S^2 \times 287.2]$$
$$= 0.5\ \log_e (1/0.022) = 1.908$$

Image attenuation coefficient

$$\alpha = 1.908\ \text{N} = 16.57\ \text{dB} \quad (\textit{Ans.})$$

Note that the image phase-change coefficient is zero.

Example 15.9

A T network has series resistors of 1000 Ω and a shunt capacitor of 0.1 μF. for a frequency of 5000/2π Hz calculate (*a*) the characteristic impedance, (*b*) the image transfer coefficient and (*c*) the output voltage when the input voltage is 1 V and two such networks are cascaded.

Solution

$$Z_1 = 2000\ \Omega \quad \text{and} \quad Z_2 = -\text{j}2000\ \Omega$$

(*a*) $Z_0 = \sqrt{[(4 \times 10^6)/4 - \text{j}4 \times 10^6]} = 10^3\sqrt{(1 - \text{j}4)}$
$= 2031\ \angle{-38^\circ}\ \Omega \quad (\textit{Ans.})$

Or, using Bartlett's Bisection theorem,

$Z_{oc} = 1000 - \text{j}4000\ \Omega,\ Z_{sc} = 1000\ \Omega$ and
$Z_0 = 10^3\sqrt{(1 - \text{j}4)} = 2031\ \angle{-38^\circ}\ \Omega$ as before.

(*b*) $\gamma = \log_e[1 + 500/{-\text{j}2000} + (2031\ \angle{-38^\circ})/{-\text{j}2000}]$
$= \log_e[1 + \text{j}0.25 + 1.016\ \angle 52^\circ] = \log_e[1.626 + \text{j}1.051]$
$= \log_e[1.936\ \angle 32.9^\circ] = \log_e[\text{e}^{(0.66 + \text{j}0.57)}]$
$= 0.66 + \text{j}0.57$

Therefore,

$\alpha = 0.66$ N and $\beta = 0.57$ radians (*Ans.*)

(*c*) For two sections, $\alpha = 1.32$ N and $\beta = 1.14$ rad. Hence,

$V_R/V_S = \text{e}^{-(1.32 + \text{j}1.14)} = 0.267\ \angle{-65.3^\circ}$ (*Ans.*)

Example 15.10

A T network has series impedances $Z_1 = 5 + j20\ \Omega$ and $Z_3 = 4 + j16\ \Omega$, and a shunt impedance Z_2 of $2 - j15\ \Omega$. Calculate (*a*) its image transfer coefficient, (*b*) its image attenuation coefficient and (*c*) its phase change coefficient.

Solution

$$Z_{A(oc)} = (5 + j20) + (2 - j15) = 7 + j5\ \Omega = 8.6\angle 35.5°\ \Omega$$

$$Z_{A(sc)} = 5 + j20 + (2 - j15)(4 + j16)/(6 + j1)$$

$$= 44.46 + j8.76 = 45.32\angle 11.1°\ \Omega$$

Hence

$$Z_A = \sqrt{(8.6\angle 35.5° \times 45.32\angle 11.1°)} = 19.74\angle 23.3°\ \Omega$$

$$Z_{B(oc)} = 4 + j16 + (2 - j15) = 6 + j1 = 6.08\angle 9.5°\ \Omega$$

$$Z_{B(sc)} = 4 + j16 + (5 + j20)(2 - j15)/(7 + j5)$$

$$= 31 - j8.23 = 32.1\angle -14.9°\ \Omega$$

Hence

$$Z_B = \sqrt{(6.08\angle 9.5° \times 32.1\angle -14.9°)} = 13.97\angle -2.7°\ \Omega$$

(*a*) $I_S = E_S/2Z_A = E_S/39.48\angle 23.3°$

$$I_R = I_S(2 - j15)/(6 + j1 + 13.96\angle -2.7°) - 0.759\angle -81.4° I_S$$

$$\gamma = \log_e[1/(0.759\angle -81.4°)] \times \sqrt{[(19.74\angle 23.3°)/(13.96\angle -2.7°)]}$$

$$= \log_e[1.318\angle 81.4° \times 1.189\angle 13°] = \log_e[1.567\angle 94.4°]$$

$$= \log_e[e^{(0.449 + j1.648)}] = 0.45 + j1.65 \quad (\textit{Ans.})$$

(*b*) $\alpha = 0.45$ N

(*c*) $\beta = 1.65$ rad (*Ans.*)

Propagation coefficient of a symmetrical T network

Consider Fig. 15.9 which shows a symmetrical T network fed by a source of e.m.f. E_S and internal impedance Z_{0T} and terminated by a load of impedance Z_{0T}.

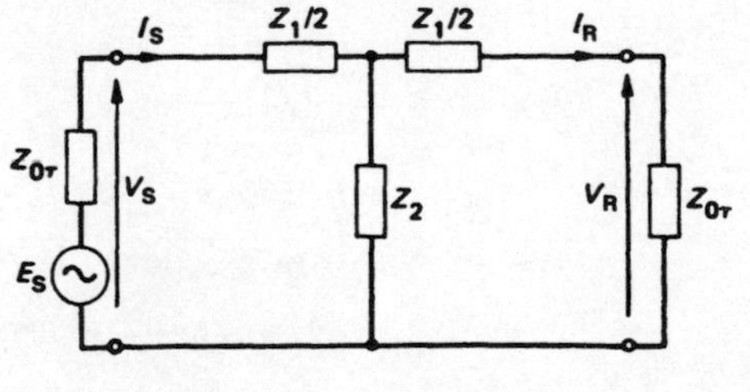

Fig. 15.9 Calculation of the propagation coefficient of a T network

Since $V_S = I_S Z_{0T}$ and $V_R = I_R Z_{0T}$

$$V_S I_S / V_R I_R = I_S^2 Z_{0T} / I_R^2 Z_{0T} = I_S^2 / I_R^2$$

and $\gamma = \log_e I_S / I_R$.

From the figure

$$I_R = I_S Z_2 / (Z_2 + Z_1/2 + Z_{0T}) \qquad (15.14)$$

$$e^{\gamma} = I_S / I_R = [Z_2 + (Z_1/2) + Z_{0T}]/Z_2$$

$$= 1 + Z_1/2Z_2 + Z_{0T}/Z_2 \qquad (15.15)$$

and $\gamma = \log_e[1 + Z_1/2Z_2 + Z_{0T}/Z_2]$ (15.16)

Example 15.11

The network shown in Fig. 15.9 has $Z_1 = 100 + j50\ \Omega$ and $Z_2 = 10 - j20\ \Omega$. Calculate (*a*) its characteristic impedance, (*b*) its propagation coefficient, (*c*) its attenuation coefficient and (*d*) its phase change coefficient.

Solution

(*a*) $Z_{0T} = \sqrt{[(100 + j50)^2/4 + (100 + j50)(10 - j20)]}$

$= \sqrt{(3124.8\angle 53.2° + 2500\angle -36.8°)}$

$= \sqrt{(1871.8 + j2502.1 + 2001.8 - j1497.5)}$

$= \sqrt{(3873.7 + j1004.6)} = \sqrt{(4001.8\angle 14.5°)}$

$= 63.3\angle 7.3°\ \Omega$ (*Ans.*)

(*b*) $\gamma = \log_e[1 + (100 + j50)/(20 - j40) + (63.26\angle 7.3°)/(10 - j20)]$

$= \log_e[1.94 + j5.17] = \log_e[5.52\angle 69.4°] = \log_e e^{(1.708 + j1.211)}$

$= 1.71 + j1.21$ (*Ans.*)

(*c*) $\alpha = 1.71$ N (*Ans.*)

(*d*) $\beta = 1.21$ rad (*Ans.*)

Propagation coefficient of a symmetrical π network

Figure 15.10 shows a symmetrical π network connected between source and load impedances which are both equal to the characteristic impedance of the network. From the figure,

$$V_R = [V_S 2Z_2 Z_{0\pi}/(2Z_2 + Z_{0\pi})]/\{Z_1 + [(2Z_2 Z_{0\pi}/(2Z_2 + Z_{0\pi})]\}$$

$$= (V_S 2Z_2 Z_{0\pi})/(2Z_1 Z_2 + Z_1 Z_{0\pi} + 2Z_2 Z_{0\pi})$$

$$e^{\gamma} = V_S/V_R = 1 + (Z_1/2Z_2) + (Z_1/Z_{0\pi}) \qquad (15.17)$$

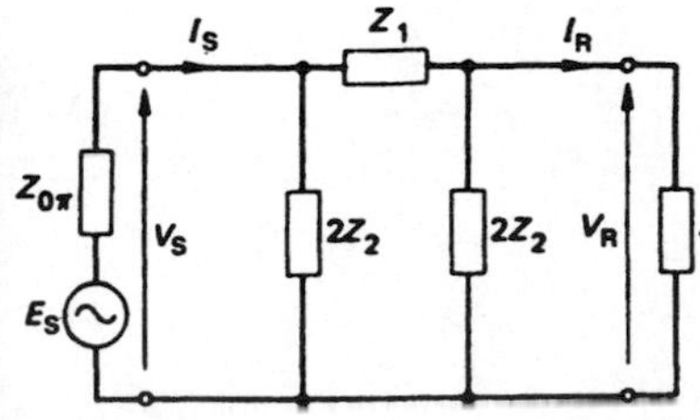

Fig. 15.10 Calculation of the propagation coefficient of a π network

Since $Z_{0\pi} = Z_1Z_2/Z_{0T}$ the equation for γ can be written as

$$\gamma = \log_e[1 + Z_1/2Z_2 + Z_{0T}/Z_2]$$

which is the same as equation (15.16). This means that symmetrical T and π networks having the same values of total series and shunt impedances have the same propagation coefficient.

Cosh$\gamma = 1 + Z_1/2Z_2$

Since $e^{\gamma} = 1 + (Z_1/2Z_2) + (Z_{0T}/Z_2)$

$$e^{-\gamma} = 1/[1 + (Z_1/2Z_2) + (Z_{0T}/Z_2)]$$

Now $\cosh\gamma = (e^{\gamma} + e^{-\gamma})/2$ and therefore,

$$\cosh\gamma = \left[\frac{Z_2 + Z_1/2 + Z_{0T}}{Z_2} + \frac{Z_2}{Z_2 + Z_1/2 + Z_{0T}}\right]/2$$

or $\cosh\gamma = 1 + (Z_1/2Z_2)$ (15.18)

Alternatively, consider Fig. 15.11 which shows two symmetrical T networks connected in cascade. Applying Kirchhoff's law to the middle loop,

$$0 = I_S\,e^{-\gamma}Z_1 + (I_S\,e^{-\gamma} - I_S\,e^{-2\gamma})Z_2 - (I_S - I_S\,e^{-\gamma})Z_2$$
$$I_SZ_2 = I_S\,e^{-\gamma}(Z_1 + 2Z_2) - e^{-2\gamma}Z_2I_S$$
$$Z_2\,e^{\gamma} + Z_2\,e^{-\gamma} = Z_1 + 2Z_2$$

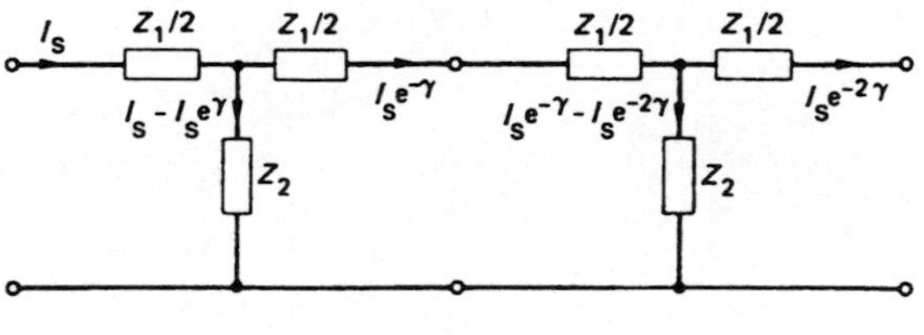

Fig. 15.11 Determination of $\cosh\gamma = 1 + (Z_1/Z_2)$

and so

$$(e^{\gamma} + e^{-\gamma})/2 = \cosh\gamma = 1 + (Z_1/2Z_2) \quad \text{as before.}$$

Similarly, for a symmetrical π network

$\cosh\ \gamma = 1 + (2Z_2/Z_1)$ (15.19)

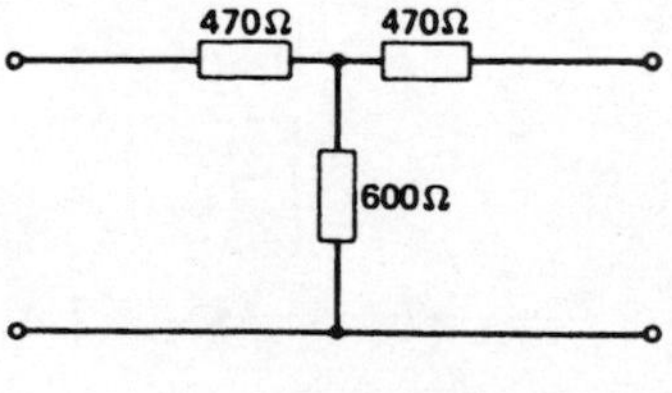

Fig. 15.12

Example 15.12

Calculate the attenuation of the circuit shown in Fig. 15.12.

Solution

From equation (15.18)

$$\cosh\gamma = 1 + (470/600) = 1.783$$

$$\cosh\alpha\cos\beta + j\sinh\alpha\sin\beta = 1.783$$

There is no imaginary part and hence $\sin\beta = 0$ and $\cos\beta = 1$. Therefore $\cosh\alpha = (e^{\alpha} + e^{-\alpha})/2 = 1.783$. Solving for e^{α} gives $e^{\alpha} = 3.26$ or 0.31. Therefore

$$\alpha = 1.182 \text{ N} = 10.26 \text{ dB} \quad (\textit{Ans.})$$

Alternatively, the characteristic impedance of the network is

$$Z_{0T} = \sqrt{[940^2/4 + (940 \times 600)]} = 886\ \Omega$$

Hence, from Fig. 15.13,

$$I_R = I_S 600/1956 \quad \text{and} \quad I_S/I_R = e^{\alpha} = 3.26 \quad \text{as before}$$

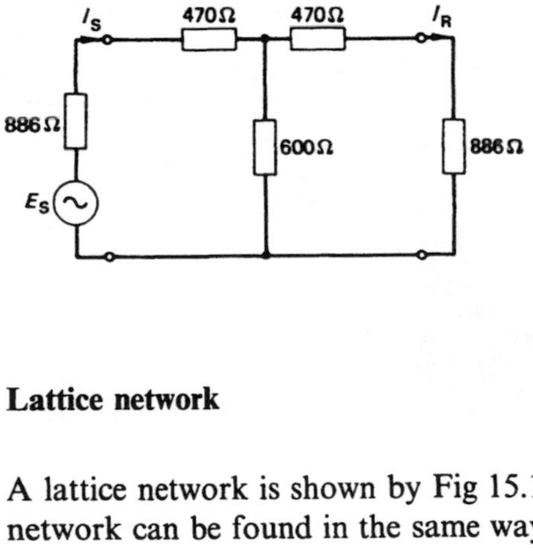

Fig. 15.13

Lattice network

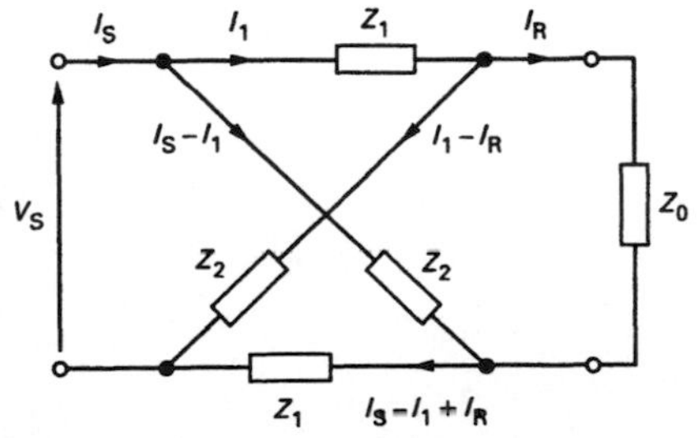

Fig. 15.14 Lattice network

A lattice network is shown by Fig 15.14. The parameters of the network can be found in the same way as for a T network.

Characteristic impedance

The characteristic impedance can best be found using $Z_0 = \sqrt{Z_{oc}Z_{sc}}$. From the figure,

$$Z_{oc} = (Z_1 + Z_2)/2 \quad \text{and} \quad Z_{sc} = (2Z_1Z_2)/(Z_1 + Z_2)$$

Therefore,

$$Z_0 = \sqrt{\{[Z_1 + Z_2)/2][2Z_1Z_2/(Z_1 + Z_2)]\}}$$
$$= \sqrt{(Z_1Z_2)} \qquad (15.20)$$

Propagation coefficient

Applying Kirchhoff's voltage law to the network,

$$V_S = I_1Z_1 + I_RZ_0 + (I_S - I_1 + I_R)Z_1$$
$$= I_SZ_1 + I_R(Z_0 + Z_1)$$

Now, $V_S = I_SZ_0$ and hence

$$I_S(Z_0 - Z_1) = I_R(Z_0 + Z_1) \quad \text{and}$$
$$I_S/I_R = (Z_0 + Z_1)(Z_0 - Z_1)$$

Therefore

$$\gamma = \log_e[(Z_0 + Z_1)/(Z_0 - Z_1)]$$
$$= \log_e[(Z_2 + Z_0)/(Z_2 - Z_0)] \qquad (15.21)$$

Example 15.13

The lattice network shown in Fig. 15.14 has $Z_1 = 1000\ \Omega$ and $Z_2 = 360\ \Omega$. Calculate (*a*) the characteristic impedance and (*b*) the attenuation coefficient of the network.

Solution

(*a*) $R_{oc} = 1360/2 = 680\ \Omega$

$R_{sc} = (2 \times 1000 \times 360)/(1000 + 360) = 529\ \Omega$ (*Ans.*)

(*b*) $\alpha = \log_e[(529 + 360)/(529 - 360)] = \log_e[5.25] = 1.66$ N

$= 14.1$ dB (*Ans.*)

Example 15.14

(*a*) If, for the lattice network shown in Fig. 15.14, $Z_1 = X_1$ and $Z_2 = X_2$ show that $\tanh(\gamma/2) = \sqrt{(X_1/X_2)}$. (*b*) Confirm that if X_1 and X_2 are of opposite sign the network will have zero loss at all frequencies.

Solution

(*a*) $Z_0 = \sqrt{(X_1X_2)}$ and

$\gamma = \log_e[\sqrt{[(X_1X_2) + X_1]}/(\sqrt{(X_1X_2)} - X_1)]$ and

$e^\gamma = [\sqrt{(X_1X_2)} + X_1]/[\sqrt{(X_1X_2)} - X_1]$

Hence, $e^\gamma\sqrt{(X_1X_2)} - e^\gamma X_1 = \sqrt{(X_1X_2)} + X_1$

$\sqrt{X_1X_2}(e^\gamma - 1) = X_1(e^\gamma + 1)$

$X_1/(\sqrt{(X_1X_2)} = \sqrt{(X_1/X_2)} = (e^\gamma - 1)/(e^\gamma + 1) = \tanh(\gamma/2)$

(*b*) $\tanh(\gamma/2) = \sqrt{(jX_1/-jX_2)} = j\sqrt{(X_1/X_2)}$

This means that $\tanh(\gamma/2)$ is purely imaginary and hence the network is lossless.

Insertion loss

The *insertion loss* of a network is the ratio, expressed in dB, of the powers dissipated in a load before and after the insertion of the network in between the source and the load.

When (Fig. 15.15(*a*)) a source of e.m.f. E_S and internal impedance Z_S is directly connected to a load Z_L, the current that flows in the load is I_1. When (Fig. 15.15(*b*)) the network is inserted in between the source and the load, the load current is reduced to a new value I_2. The insertion loss of the network is then equal to

$$10\ \log_{10}[(|I_1|^2 R_L)/(|I_2|^2 R_L)]\ \text{d}B$$

or $\quad$ Insertion loss $= 20\ \log_{10}[|I_1|/|I_2|]$ dB $\qquad$ (15.22)

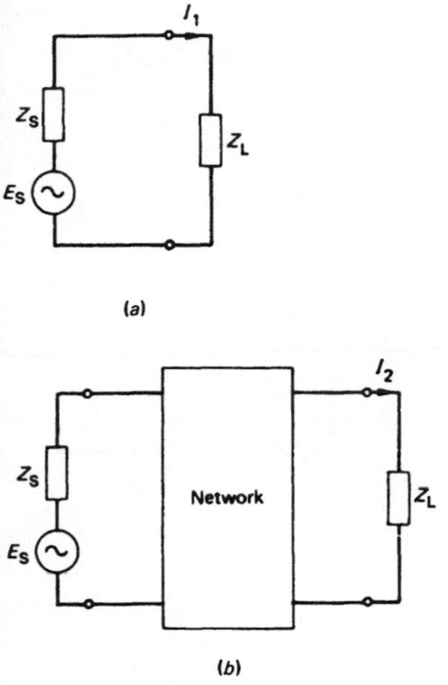

Fig. 15.15 Insertion loss

Example 15.15

When a voltage source is connected to a load the current in the load is 5 mA. When a network is connected in between the source and the load current falls to 2 mA. Calculate the insertion loss of the network.

Solution

Insertion loss $= 20\ \log_{10}(5/2) = 7.96$ dB $\quad$ (*Ans.*)

Exercise 15.16

Calculate the insertion loss of the network shown in Fig. 15.12 when it is connected (*a*) between 600 Ω impedances and (*b*) between its image impedances.

Solution

(*a*) From Fig. 15.16(*a*), the current supplied by the source is

$$E_S/\{600 + 470 + [600(470 + 600)]/(600 + 470 + 600)\}$$
$$= E_S/1454$$

and hence

$$I_2 = (E_S/1454)(600/1670)$$

From Fig. 15.16(*b*) the current supplied to the directly connected load is $I_1 = E_S/1200$. Therefore, the insertion loss is

$$20\ \log_{10}[(E_S \times 1454 \times 1670)/(1200 \times E_S \times 600)]$$
$$= 10.56 \text{ dB} \quad (\textit{Ans.})$$

(*b*) From Fig. 15.16(*c*) the current supplied by the source to the network is $E_S/1772$. Hence

$$I_4 = (E_S/1772) \times (600/1956)$$

From Fig. 15.16(*d*) $I_3 = E_S/1772$. Therefore, the insertion loss is

$$20\ \log_{10}[(E_S \times 1772 \times 1956)/(1772 \times E_S \times 600)]$$
$$= 10.26 \text{ dB} \quad (\textit{Ans.})$$

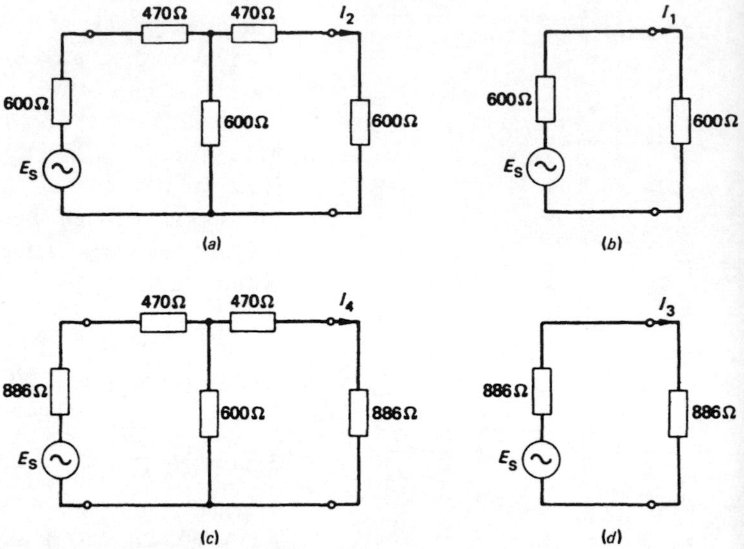

Fig. 15.16

When a network is image-matched at both input and output terminals its insertion loss is at its minimum possible value and is equal to its image attenuation.

Example 15.17

Calculate the insertion loss of a T network with series resistors of 60 Ω and a shunt resistor of 100 Ω when it is connected between a 40 Ω voltage source and a 110 Ω load.

Solution

When directly connected $I_1 = V/150$. With the network in circuit its open-circuit voltage is $V/2$ and the output resistance is 110 Ω. Hence

$$I_2 = V/(2 \times 220)$$

$$\text{Insertion loss} = 20 \log_{10}(440/150) = 9.35 \text{ dB} \quad (Ans.)$$

Example 15.18

Calculate the insertion loss of the 0.1 μF capacitor shown in Fig. 15.17.

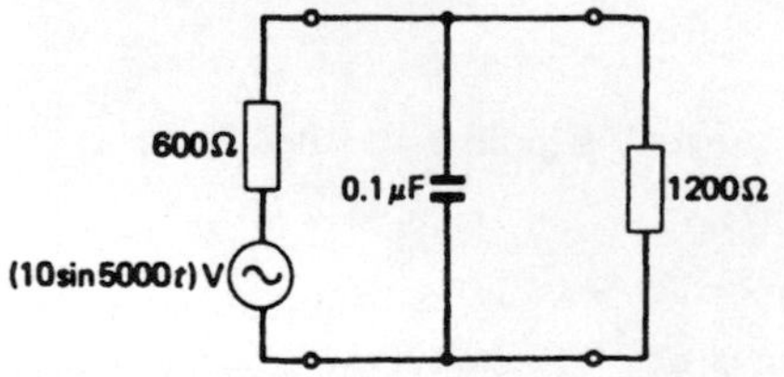

Fig. 15.17

Solution

Without the capacitor in circuit the current flowing in the 1200 Ω resistor is $I_1 = 10/1800$ A. The reactance of the capacitor is $-j2000$ Ω. Thus, when the capacitor is connected in circuit, the current taken from the source is

$$I_S = 10/[600 + (1200 \times -j2000)/(1200 - j2000)]$$
$$= 10/(1482 - j529)$$

The load current is

$$I_2 = [10/(1482 - j529)][-j2000/(1200 - j2000)]$$
$$|I_2| = 5.44 \times 10^{-3}\ \text{A}$$

Therefore, the insertion loss of the capacitor is

$$20\ \log_{10}[10/(1800 \times 5.45 \times 10^{-3})] = 0.18\ \text{dB} \quad (\textit{Ans.})$$

Insertion gain of a transformer

If a transformer is used to match a load R_L to a source of e.m.f. E_S and resistance R_S its turns ratio is $n = \sqrt{(R_L/R_S)}$. Then the reflected resistance into the primary winding is R_L/n^2 and the voltage across this reflected resistance is $E_S/2$ volts. The voltage across the load $= (E_S/2)\sqrt{(R_L/R_S)}$ and the current flowing in the load is

$$I_{L1} = (E_S/2R_L)\sqrt{(R_L/R_S)} = E_S/[2\sqrt{(R_L R_S)}]$$

When the source is directly connected to the load the current in the load is $I_{L2} = E_S/(R_S + R_L)$. Therefore

$$\text{Insertion gain} = 20\ \log_{10}(I_{L1}/I_{L2})$$
$$= 20\ \log_{10}\left\{\left[\frac{E_S}{2\sqrt{(R_L R_S)}}\right]\left[\frac{R_S + R_L}{E_S}\right]\right\}$$
$$= 20\ \log_{10}\left[0.5\left(\sqrt{\frac{R_S}{R_L}} + \sqrt{\frac{R_L}{R_S}}\right)\right]\ \text{dB} \quad (15.23)$$

Example 15.19

Calculate the insertion gain of a transformer that is used to match a 150 Ω voltage source to a 600 Ω load.

Solution
Insertion gain $= 20\ \log_{10}\{0.5[\sqrt{(150/600)} + \sqrt{(600/150)}]\}$
$= 20\ \log_{10}\ 1.25 = 1.94\ \text{dB}$ (*Ans.*)

Complete insertion loss

Sometimes it is necessary to determine the insertion loss of a network whose internal configuration is not known. If the input and output image impedances are matched to the source and to the load respectively, the insertion loss will be equal to the image attenuation α. If, however, the input and output terminals are not matched to the source and the load respectively, then the total insertion loss is the sum of the following terms:

- Reflection loss at the input terminals due to mismatch between the source impedance Z_S and the input image impedance Z_A. The input current is then $I_1 = E_S/(Z_S + Z_A)$ and the input power is

  $$P_{in} = |I_1|^2 |Z_A| \cos\varphi$$
  $$= [|E_S|^2/(|Z_S + Z_A|)^2] \times [|Z_A| \cos\varphi]$$

  where φ is the phase angle of Z_A. For resistive impedances $P_{in} = E_S^2 Z_A/(Z_S + Z_A)^2$. If the input were matched, i.e. $Z_A = Z_S$, the input power would be $P'_{in} = E_S^2/4Z_S$. Therefore the reflection loss is

  $$10 \log_{10}(P'_{in}/P_{in})$$
  $$= 10 \log_{10}\{[E_S^2/4Z_S][(Z_S + Z_A)^2/(E_S^2 Z_A)]\}$$
  $$= 10 \log_{10}[(Z_S + Z_A)^2/(4Z_S Z_A)]$$
  $$= 20 \log_{10}[(Z_S + Z_A)/(2\surd(Z_S Z_A))] \text{ dB} \qquad (15.24)$$

- Reflection loss at the output terminals because of mismatch between the load impedance Z_L and the output image impedance Z_B. This is equal to

  $$20 \log_{10}[(Z_L + Z_B)/(2\surd(Z_L Z_B))] \text{ dB} \qquad (15.25)$$

- The attenuation α of the network. In dB this is equal to $20 \times 0.4343\alpha$.
- Losses due to multiple reflections within the network because of the input and output mismatches. This contribution to the total loss is usually fairly small. The *interaction loss* is given by

  Interaction loss
  $$= 20 \log_{10}\{1 - [(Z_A - Z_S)/(Z_A + Z_S)]$$
  $$[(Z_B - Z_L)/(Z_B + Z_L)]e^{-2\alpha}\} \text{ dB} \qquad (15.26)$$

  Usually the interaction loss is small and it is often neglected.
- From the listed above terms must be subtracted the reflection loss between the source and the load when they are directly connected together. This is equal to

  $$20 \log_{10}[(Z_S + Z_L)/(2\surd(Z_L Z_S))] \text{ dB} \qquad (15.27)$$

Example 15.20

A two-pole network has image impedances $Z_A = 600\ \Omega$, $Z_B = 750\ \Omega$ and an image attenuation coefficient of 1.5 N. It is connected between a source of 500 Ω resistance and a 500 Ω load. Calculate its insertion loss.

Solution

$$\begin{aligned}\text{Insertion loss} &= 20\{\log_{10}[100/(2\sqrt{(30\times10^4)})]\\ &\quad+\log_{10}[1250/(2\sqrt{(375\times10^3)})]\\ &\quad+\log_{10}[1-(100/1100)(-250/1250)\times e^{-3}]\\ &\quad+\log_{10}[2\sqrt{(25\times10^4)}/1000]+0.4343\times1.5\}\\ &=20\{0.0018+0.0088-0.0004+0.6514\}\\ &=13.23\text{ dB}\quad(\textit{Ans.})\end{aligned}$$

Attenuators

An *attenuator* is a purely resistive network designed to introduce a specified loss into a circuit usually without a change in impedance. The majority of attenuators employ either the T or the π configuration although other arrangements may also be met.

Symmetrical attenuators

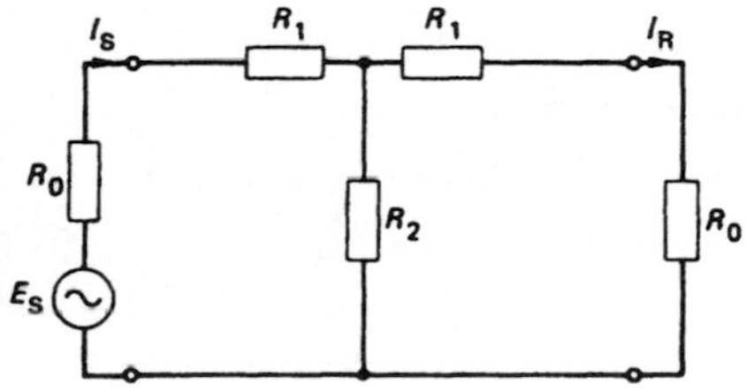

Fig. 15.18 T attenuator

Figure 15.18 shows a T attenuator connected between a source and a load both of which have an impedance equal to the characteristic impedance of the network. From the figure,

$$I_R = I_S R_2/(R_1+R_2+R_0)$$

Therefore,

$$I_S/I_R = M = (R_1+R_2+R_0)/R_2 \tag{15.28}$$

Also,

$$\begin{aligned}R_0 &= R_1+[R_2(R_1+R_0)]/(R_1+R_2+R_0)\\ &= R_1+(R_1+R_0)/M\\ R_0M &= R_1M+R_1+R_0 \qquad R_0(M-1)=R_1(M+1)\\ R_1 &= [R_0(M-1)]/(M+1)\end{aligned} \tag{15.29}$$

From equation (15.28)

$$\begin{aligned}R_2M &= R_1+R_2+R_0\\ R_2(M-1) &= R_1+R_0=[R_0(M-1)]/(M+1)+R_0\\ R_2(M-1)(M+1) &= R_0(M-1)+R_0(M+1)=2R_0M\end{aligned}$$

and so

$$R_2=(2R_0M)/(M^2-1) \tag{15.30}$$

Example 15.21

Design a T attenuator to have a characteristic impedance of 600 Ω and an attenuation of 20 dB.

Solution
20 dB is a voltage or current ratio of 10 : 1, and therefore,

$$R_1 = [600(10-1)]/(10+1) = 491\ \Omega \quad (Ans.)$$

and $R_2 = (2 \times 600 \times 10)/(100-1) = 121\ \Omega \quad (Ans.)$

π Attenuator

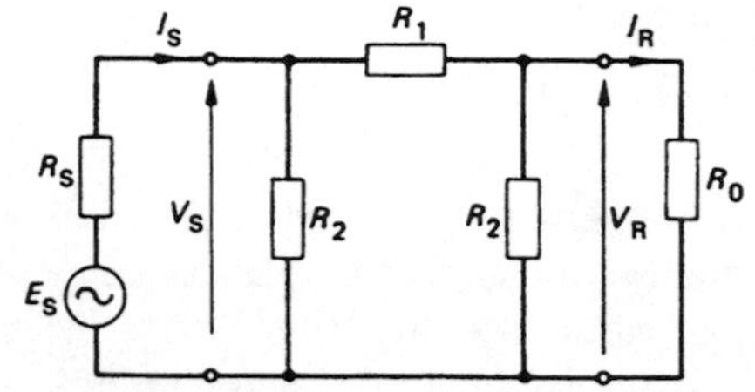

Fig. 15.19 π attenuator

Figure 15.19 shows a π attenuator. For this circuit

$$I_S = V_S/R_0 = (V_S/R_2) + (V_R/R_2) + (V_R/R_0)$$

Dividing throughout by V_R gives

$$M/R_0 = (M/R_2) + (1/R_2) + (1/R_0)$$
$$(1/R_0)(M-1) = (1/R_2)(M+1)$$

or $R_2 = [R_0(M+1)]/(M-1)$ (15.31)

Also

$$I_S = V_S/R_0 = (V_S/R_2) + [(V_S - V_R)/R_1]$$
$$M/R_0 = (M/R_2) + (M/R_1) - (1/R_1)$$
$$\begin{aligned}(1/R_1)(M-1) &= (M/R_0) - (M/R_2)\\ &= M/R_0 - M/[R_0(M+1)/(M-1)]\\ &= [M - M(M-1)/(M+1)]/R_0\\ &= (1/R_0)[2M/(M+1)]\end{aligned}$$
$$1/R_1 = [2M/(M+1)(M-1)]/R_0$$
$$R_1 = [R_0(M^2-1)/2M] \qquad (15.32)$$

Example 15.22

Design a π attenuator to have a characteristic impedance of 600 Ω and an attenuation of 20 dB.

Solution
From equation (15.31),

$$R_2 = [600(10+1)]/(10-1) = 733\ \Omega \quad (Ans.)$$

From equation (15.32)

$$R_1 = [600(100-1)]/(2 \times 10) = 2970\ \Omega \quad (Ans.)$$

If a number of attenuators of identical characteristic impedances are connected in cascade, the overall attenuation is the *sum* of the attenuations of the individual attenuators.

Matching networks

A network is often employed to match a source to a load. If the minimum loss is required and the match to occur only at a

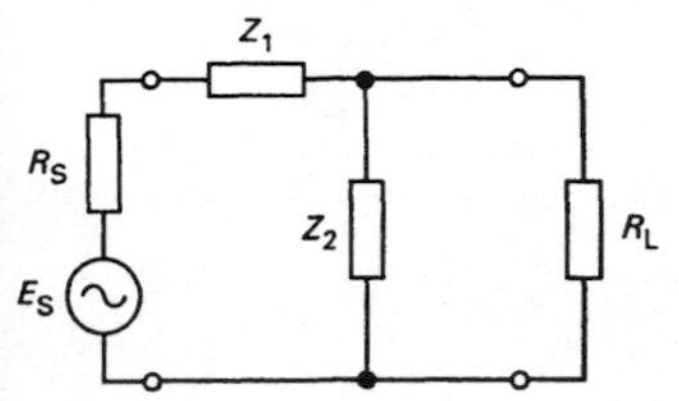

Fig.15.20 L matching network

certain frequency the components of the network will be lossless inductors and capacitors. Figure 15.20 shows an L network connected between a voltage source of e.m.f. E_S and resistance R_S and a load of resistance R_L. The shunt impedance is always connected in parallel with the higher value of the source and load impedances. For the source to be matched to the load,

$$R_s = Z_1 + R_L Z_2/(Z_2 + R_L),$$

$$R_S R_L + R_S Z_2 = Z_1 Z_2 + R_L Z_1 + R_L Z_2 \qquad (15.33)$$

$$R_L = [Z_2(R_S + Z_1)]/(Z_1 + Z_2 + R_S),$$

$$R_L Z_1 + R_L Z_2 + R_L R_S = Z_2 R_S + Z_1 Z_2 \qquad (15.34)$$

Add equations (15.33) and (15.34) to get,

$$2R_S R_L + R_S Z_2 + R_L Z_1 + R_L Z_2$$
$$= 2Z_1 Z_2 + R_S Z_2 + R_L Z_1 + R_L Z_2$$

Therefore,

$$2R_S R_L = 2Z_1 Z_2 \quad \text{or} \quad Z_1 = R_S R_L / Z_2 \qquad (15.35)$$

Subtract equation (15.34) from (15.33) to obtain,

$$R_S Z_2 - R_L Z_1 - R_L Z_2 = R_L Z_1 + R_L Z_2 - R_S Z_2,$$

$$Z_2(R_S - R_L) = R_L Z_1$$

Therefore,

$$Z_1 = Z_2(R_S - R_L)/R_L \qquad (15.36)$$

Equate (15.35) and (15.36),

$$R_S R_L / Z_2 = [Z_2(R_S - R_L)]/R_L \quad \text{or}$$

$$Z_2 = \surd(R_L^2 R_S)/(R_S - R_L) \qquad (15.37)$$

From equation (15.35),

$$Z_1^2 = R_S^2 R_L^2 / Z_2^2 = (R_S - R_L) R_S \quad \text{and}$$

$$Z_1 = \surd[R_S(R_S - R_L)] \qquad (15.38)$$

Lastly,

$$Z_2 = (R_S R_L)/Z_1 \qquad (15.39)$$

Example 15.23

An L network is used to match a 650 Ω source to a 150 Ω load. Determine the necessary values of Z_1 and Z_2 if the match is to be (a) at all frequencies and (b) at 100 kHz only.

Solution

(*a*) $R_1 = \sqrt{(650 \times 500)} = 570\ \Omega$,

$R_2 = \sqrt{[(150^2 \times 650)/(650 - 150)]} = 171\ \Omega$ (*Ans.*)

(*b*) $Z_1 = \sqrt{[(150(150 - 650)]} = j274 = j2\pi \times 10^5 L$ or

$L = 437\ \mu H$ (*Ans.*)

$Z_2 = (650 \times 150)/j274 = -j356 = 1/\omega C$

Hence $C = 4.47$ nF (*Ans.*)

Voltage and current ratios

Consider the L network calculated in Example 15.23 and suppose that a voltage source of 12 V e.m.f. and 650 Ω resistance is applied to it. Then

$$I_S = 12/(650 + 650) = 9.23 \text{ mA} \quad \text{and} \quad V_S = 6 \text{ V}$$

$$I_R = 9.23 \times 171/(171 + 570 + 150) = 1.77 \text{ mA}$$

$$V_R = 1.77 \times 10^{-3} \times 150 = 0.27 \text{ V}$$

The current ratio is $9.23/1.77 = 5.215$ and the voltage ratio is $6/0.27 = 22.2$. Clearly the two ratios are different. This means that the loss of a non-symmetrical network must be expressed in terms of its image attenuation coefficient (equation 15.9). Thus, in this case,

$$\alpha = 0.5 \log_e[(6 \times 9.23)/(0.27 \times 1.77)$$

$$= 2.376 \text{ N} \quad \text{or} \quad 20.64 \text{ dB}$$

Equivalent T networks

Any network, of whatever configuration, can always be represented by an *equivalent T network*. The component values of the equivalent network can be determined by means of three measurements followed by some calculations.

The necessary measurements are of

- The input impedance with the output terminals open circuited, Z_{oc1}
- The input impedance with the output terminals short circuited, Z_{sc1}
- The output impedance with the input terminals open circuited, Z_{oc2}

Referring to Fig. 15.1(*a*)

$$Z_{oc1} = Z_1 + Z_2 \qquad (15.40)$$

$$Z_{sc1} = Z_1 + Z_2 Z_3/(Z_2 + Z_3) \qquad (15.41)$$

$$Z_{oc2} = Z_2 + Z_3 \qquad (15.42)$$

From equations (15.40) and (15.42)

$$Z_1 = Z_{oc1} - Z_2 \quad \text{and} \quad Z_3 = Z_{oc2} - Z_2$$

Substituting into equation (15.41) gives

$$Z_{sc1} = Z_{oc1} - Z_2 + [(Z_{oc2} - Z_2)Z_2]/(Z_{oc2} - Z_2 + Z_2)$$
$$Z_{sc1}Z_{oc2} = Z_{oc1}Z_{oc2} - Z_2Z_{oc2} + Z_{oc2}Z_2 - Z_2^2$$
$$Z_2 = \pm\sqrt{[Z_{oc2}(Z_{oc1} - Z_{sc1})]} \qquad (15.43)$$

Z_1 and Z_3 can then be obtained using equations (15.40) and (15.42).

The components of the equivalent T network can also be expressed in terms of the image impedances and the image transfer coefficient of the network. The derivation of the necessary equations is complex in the case of a non-symmetrical network and it will not be attempted here. The results are

$$Z_1 = Z_A \coth\gamma - Z_3 \qquad Z_2 = Z_B \coth\gamma - Z^3$$
$$Z_3 = \sqrt{(Z_A Z_B)}/\sinh\gamma \qquad (15.44)$$

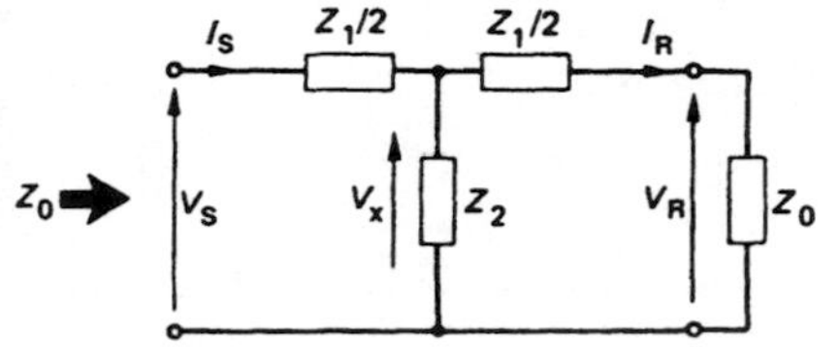

Fig. 15.21

For a symmetrical network the equivalent T network is shown in Fig. 15.21. From the figure,

$$I_R = V_x/(Z_1/2 + Z_0) = [V_S - (I_S Z_1/2)]/(Z_1/2 + Z_0)$$
$$= [I_S Z_0 - (I_S Z_1/2)]/(Z_1/2 + Z_0)$$
$$I_R/I_S = e^{-\gamma} = (Z_0 - Z_1/2)/(Z_0 + Z_1/2)$$
$$(Z_1/2)(1 + e^{-\gamma}) = Z_0(1 - e^{-\gamma})$$
$$Z_1/2 = [Z_0(1 - e^{-\gamma})]/(1 + e^{-\gamma})$$
$$= [Z_0(e^{\gamma/2} - e^{-\gamma/2})]/(e^{\gamma/2} + e^{-\gamma/2})$$
$$Z_1/2 = Z_0 \tanh\gamma/2 \qquad (15.45)$$

Also

$$I_R = I_S Z_2/(Z_1/2 + Z_2 + Z_0)$$
$$I_R/I_S = e^{-\gamma} = Z_2/(Z_1/2 + Z_2 + Z_0)$$
$$e^{-\gamma}[Z_1/2 + Z_0] = Z_2(1 - e^{-\gamma})$$
$$e^{-\gamma}Z_0[2/(1 + e^{-\gamma})] = Z_2(1 - e^{-\gamma})$$
$$Z_2 = Z_0 2e^{-\gamma}/[(1 + e^{-\gamma})(1 - e^{-\gamma})] = 2Z_0 e^{-\gamma}/(1 - e^{-2\gamma})$$
$$= Z_0/[(e^{\gamma} - e^{-\gamma})/2]$$

or $$Z_2 = Z_0/\sinh\gamma \qquad (15.46)$$

Example 15.24

A symmetrical network has a characteristic impedance of 600 Ω and an image transfer coefficient of $2 + j0$. Calculate the component values of the equivalent T network.

Solution
From equation (15.45) $Z_1/2 = Z_0 \tanh \gamma/2$. Here $\gamma = \alpha + j\beta = 2 + j0$. Hence

$$Z_1/2 = 600 \tanh 1$$

$$\tanh 1 = (e^1 - e^{-1})/(e^1 + e^{-1})$$

$$= (2.7183 - 0.3679)/(2.7183 + 0.3679) = 0.76$$

Therefore

$$Z_1/2 = 600 \times 0.76 = 456\ \Omega \quad (\textit{Ans.})$$

From equation (15.46), $Z_2 = 600/\sinh 2$

$$\sinh 2 = (e^2 - e^{-2})/2 = (7.389 - 0.1353)/2 = 3.63$$

Therefore

$$Z_2 = 600/3.63 = 165\ \Omega \quad (\textit{Ans.})$$

Use of transmission parameters

Equations can be derived for the iterative and image impedances, and the image transfer coefficient, of a network using the transmission parameters of that network.

Image impedance

Referring to equation (14.26) and letting $Z_{in} = Z_A$ and $Z_L = Z_B$, gives

$$Z_A = (AZ_B + B)/(CZ_B + D) \tag{15.47}$$

If a voltage V_2 is applied to the output terminals with such a polarity that current flows in the same direction as before, then $V_1 = AV_2 + BI_2$ becomes

$$-I_1 Z_A = -AI_2 Z_B + BI_2 = I_2(-AZ_B + B) \tag{15.48}$$

Also, $I_1 = CV_2 + DI_2$ becomes

$$I_1 = -CI_2 Z_B + DI_2 = I_2(-CZ_B + D) \tag{15.49}$$

Dividing equation (15.48) by (15.49) gives,

$$(-I_1 Z_A)/I_1 = [I_2(-AZ_B + B)]/[I_2(-CZ_B + D)]$$

or $$-Z_A = (-AZ_B + B)/(-CZ_B + D) \tag{15.50}$$

Adding equation (15.47) to (15.50) gives

$$\frac{AZ_B + B}{CZ_B + D} + \frac{-AZ_B + B}{-CZ_B + D} = 0$$

Hence,

$$-2ACZ_B^2 = -2BD \quad \text{or} \quad Z_B = \sqrt{(BD/AC)} \qquad (15.51)$$

Substitute equation (15.51) into (15.47) to get

$$Z_A = [A\sqrt{(BD/AC)} + B]/[C\sqrt{(BD/AC)} + D]$$

$$= [A\sqrt{(BD)} + B\sqrt{(AC)}]/[C\sqrt{(BD)} + D\sqrt{(AC)}] \quad \text{or}$$

$$Z_A = \sqrt{(AB/CD)} \qquad (15.52)$$

Example 15.25

(*a*) Use ABCD parameters to determine the image impedances of the T network shown in Fig. 14.4.
(*b*) Confirm the answers using $\sqrt{(Z_{oc}Z_{sc})}$.

Solution

(*a*) Using the values on page 248,

$$Z_A = \sqrt{[(1.33 \times 36.7)/(33.33 \times 10^{-3} \times 1.67)]} = 29.6\ \Omega \quad (Ans.)$$

$$Z_B = [(36.7 \times 1.67)/(1.33 \times 33.33 \times 10^{-3})] = 37\ \Omega \quad (Ans.)$$

(*b*) $Z_{A(oc)} = 40\ \Omega$, $Z_{A(sc)} = 10 + (20 \times 30)/50 = 22\ \Omega$,

$$Z_A = \sqrt{(40 \times 22)} = 29.6\ \Omega \quad (Ans.)$$

$$Z_{B(oc)} = 50\ \Omega, \quad Z_{B(sc)} = 20 + (30 \times 10)/40 = 27.5\ \Omega.$$

$$Z_B = \sqrt{(50 \times 27.5)} = 37\ \Omega \quad (Ans.)$$

Characteristic impedance

For a symmetrical network $A = D$ and then

$$Z_A = Z_B = Z_0 = \sqrt{(B/C)} \qquad (15.53)$$

Example 15.26

A symmetrical 4-terminal network has the following transmission parameters: $A = D = 1\angle 45°$, $B = 80\angle 60°\ \Omega$, and $C = 10\angle 90°$ mS. The network is connected to a resistive load of 90 Ω. Determine (*a*) its characteristic impedance and (*b*) its output voltage and current when a voltage of $22\angle 0°$ V is applied to the input terminals.

Solution

(*a*) $Z_0 = \sqrt{(80\angle 60°)/(10 \times 10^{-3}\angle 90°)} = 89.4\angle -15°\ \Omega$ (*Ans.*)

(*b*) $V_2 = 90I_2$. Hence, $22 = 90AI_2 + BI_2$, or

$$I_2 = 22/(90 \times 1\angle 45° + 80\angle 60°)$$

$$= 0.08 - j0.1 = 0.128\angle -51.3°\ \text{A} \quad (Ans.)$$

$$V_2 = 0.128\angle -51.3° \times 90 = 11.53\angle -51.3°\ \text{V} \quad (Ans.)$$

Image transfer coefficient

The expression for the image transfer coefficient of a network is given by equation (15.9), i.e. $\gamma = 0.5 \log_e[I_S V_S / I_R V_R]$. Using transmission parameters,

$$\begin{aligned} V_S I_S / V_R I_R &= [(AV_R + BI_R)(CV_R + DI_R)]/V_R I_R \\ &= ACV_R/I_R + AD + BC + BDI_R/V_R \\ &= ACZ_B + AD + BC + BD/Z_B \\ &= AC\sqrt{\frac{BD}{AC}} + AD + BC + BD\sqrt{\frac{AC}{BD}} \\ &= \surd(AC)\surd(BD) + AD + BC + \surd(BD)\surd(AC) \\ &= AD + 2\surd(AC)\surd(BD) + BC \\ &= AD + 2\surd(AD)\surd(BC) + BC \\ &= [\surd(AD) + \surd(BC)]^2 \end{aligned}$$

Therefore

$$\gamma = \log_e[\surd(AD) + \surd(BC)) \tag{15.54}$$

Example 15.27

A network has the following transmission parameters. Calculate (*a*) its image impedances and (*b*) its image transfer coefficient.

$$A = 1.6\angle{-150°}, \quad B = 200\angle{-143°}\ \Omega,$$
$$C = 6.32 \times 10^{-3}\angle{-71.6°}\ \text{S}, \quad D = 0.45\angle{-117°}$$

Solution

(*a*) $$Z_A = \sqrt{\frac{1.6\angle{-150°} \times 200\angle{-143°}}{6.32 \times 10^{-3}\angle{-71.6°} \times 0.45\angle{-117°}}}$$
$$= 335.4\angle{-52.2°}\ \Omega \quad (\textit{Ans.})$$

$$Z_B = \sqrt{\frac{200\angle{-143°} \times 0.45\angle{-117°}}{1.6\angle{-150°} \times 6.32 \times 10^{-3}\angle{-71.6°}}}$$
$$= 94.3\angle{-19.2°}\ \Omega \quad (\textit{Ans.})$$

(*b*) $$\begin{aligned} \gamma &= \log_e[\surd(1.6\angle{-150°} \times 0.45\angle{-117°}) \\ &\qquad + \surd(200\angle{-143°} \times 6.32 \times 10^{-3}\angle{-71.6°})] \\ &= \log_e[\surd(0.72\angle{-267°}) + \surd(1.264\angle{-214.6°})] \\ &= \log_e[0.849\angle{-133.5°} + 1.124\angle{-107.3°}] \\ &= \log_e[1.922\angle{-118.5°}] = \log_e[e^{(0.653 - j2.068)}]. \end{aligned}$$

Therefore

$$\gamma = 0.653 - j2.068 \quad (\textit{Ans.})$$

Constant-*k* filters

First-order *RC* filters provide adequate selectivity for many applications. Greater selectivity can be obtained if an *LC* filter is employed. The basic *LC* filter is known as a constant-*k* filter. Modern circuitry generally employs *active filters* which employ an *RC* network in conjunction with one, or more, op-amps.

A *prototype* or *constant*-k *filter* consists of a number of inductors and capacitors connected as either a T or a π network. The term 'constant-*k*' indicates that the product of the series impedance and the shunt impedance is a constant at all frequencies. Assuming that the components employed in the filter have zero resistance, the characteristic impedance of the network must always be either wholly real or wholly imaginary. When the characteristic impedance is real the filter will be able to accept power from a source and, since it is supposed to contain zero resistance, *none* of this power will be dissipated within the filter. This means that all the input power is transmitted through the filter and delivered to the load. The filter therefore has zero attenuation in its pass-band. On the other hand, when the characteristic impedance of the filter is imaginary, the filter will not accept power from the source and so there can be no output power. Thus, the pass- and stop-bands of a filter are determined by whether or not its characteristic impedance is real or imaginary.

The T low-pass filter

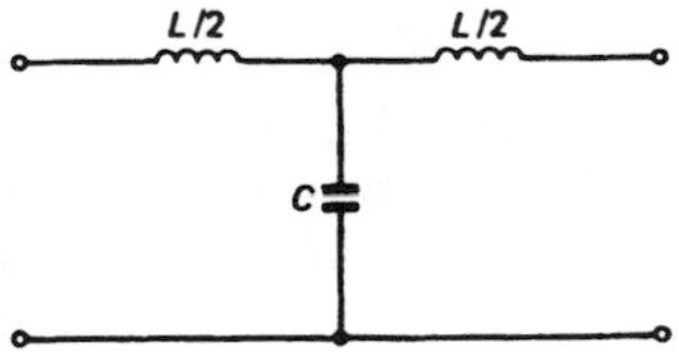

Fig.15.22 T constant-k low-pass filter

A low-pass filter should be able to transmit, with the minimum attenuation, all frequencies from 0 Hz up to the *cut-off frequency* f_{co}. At frequencies greater than f_{co} the attenuation of the filter will increase with increase in frequency up to a high value. The circuit of a T constant-*k* low-pass filter is given in Fig. 15.22. The total series impedance is $Z_1 = j\omega L$ and the total shunt impedance is $Z_2 = 1/j\omega C$. The product of the series and shunt impedances is to be constant at all frequencies, i.e. $Z_1 Z_2 = R_0^2$ where R_0 is the design impedance. Therefore,

$$R_0^2 = j\omega L \times 1/j\omega C = L/C \quad \text{and} \quad R_0 = \sqrt{(L/C)}\ \Omega$$

The characteristic impedance Z_0 of the filter is, from equation (15.6),

$$Z_{0T} = \sqrt{(Z_1^2/4 + Z_1 Z_2)} = \sqrt{[Z_1 Z_2(1 + Z_1/4Z_2)]}$$
$$= \sqrt{[(L/C)(1 + j\omega L/(4/j\omega C))]}$$

$$\text{or} \quad Z_{0T} = R_0\sqrt{(1 - \omega^2 LC/4)} \qquad (15.55)$$

At all frequencies where $1 > \omega^2 LC/4$, Z_{0T} will be a real quantity and the frequencies lie in the pass band of the filter. Conversely, if $\omega^2 LC/4 > 1$, then Z_{0T} will be an imaginary quantity and this specifies the stop-band of the filter.

The cut-off point of the filter occurs at the frequency at which the changeover from the pass-band to the stop-band takes place. This is the frequency f_{co} at which $1 = \omega_{co}^2 LC/4$ or

$$f_{co} = 1/[\pi\sqrt{(LC)}] \text{ Hz} \tag{15.56}$$

When the frequency is 0 Hz the input impedance of the filter is equal to the design impedance R_0, but at the cut-off frequency Z_{0T} is zero. In between these two frequencies Z_{0T} varies in the manner shown by Fig. 15.23.

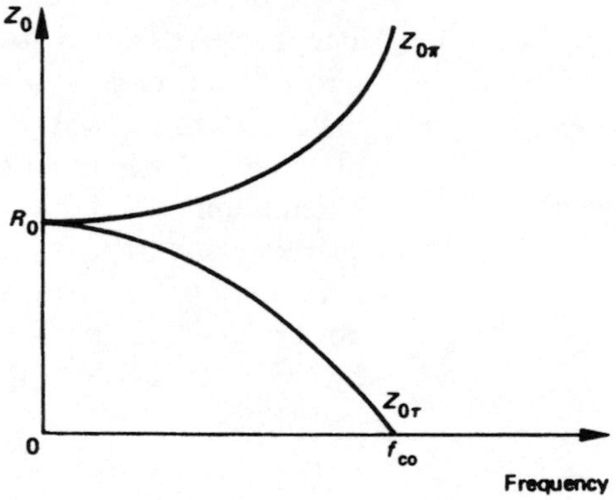

Fig. 15.23 Characteristic impedance-frequency characteristic of a constant-k low-pass filter

The π low-pass filter

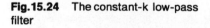

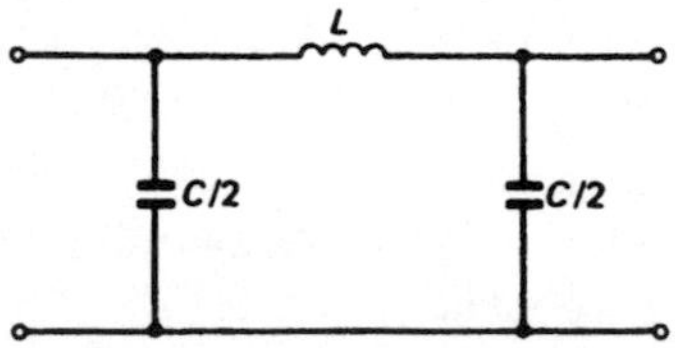

Fig. 15.24 The constant-k low-pass filter

The circuit of a constant-*k* π low-pass filter is shown by Fig. 15.24. Once again the total series impedance is $Z_1 = j\omega L$ and the total shunt impedance is $Z_2 = 1/j\omega C$ so that the design impedance is unchanged at $R_0 = \sqrt{(L/C)}\ \Omega$. From equation 15.8

$$Z_{0\pi} = Z_1 Z_2 / Z_{0T} = R_0^2 / Z_{0T}$$
$$= R_0/\sqrt{(1 - \omega^2 LC/4)} \tag{15.57}$$

At zero frequency $Z_{0\pi} = R_0$, as for the T filter, but as the frequency is increased $Z_{0\pi}$ rises and becomes equal to $R_0/0$ or ∞ at the cut-off frequency (see Fig. 15.23).

Example 15.28

Calculate the component values for a constant-*k* low-pass filter if the design impedance is to be 600 Ω and the cut-off frequency is 3000 Hz.

Solution

$$600 = R_0 = \surd(L/C) \quad \text{and} \quad 3000 = 1/\pi\surd(LC)$$

Therefore,

$$600 \times 3000 = 1/\pi C \quad \text{or}$$
$$C = 1/(600 \times 3000 \times \pi) = 0.177\ \mu\text{F} \quad (\textit{Ans.})$$

Also

$$600/3000 = \pi L \quad \text{or}$$
$$L = 600/3000\pi = 63.7\ \text{mH} \quad (\textit{Ans.})$$

Attenuation and phase shift

For a low-pass filter the expression $\cosh\gamma = 1 + (Z_1/2Z_2)$

$$\cosh\gamma = 1 + [\text{j}\omega L/(2/\text{j}\omega C)] = 1 - (\omega^2 LC/2)$$

Therefore

$$\cosh(\alpha + \text{j}\beta) = 1 - (\omega^2 LC/2)$$
$$\cos, \alpha\cos\beta + \text{j}\sinh\alpha\sin\beta = 1 - (\omega^2 LC/2)$$

In the pass-band, $\alpha = 0$. Hence $\cosh\alpha = 1$ and $\sinh\alpha = 0$, Hence,

$\cos\beta = 1 - (\omega^2 LC/2)$ or the phase shift β in the pass-band is $\beta = \cos^{-1}[1 - 2f^2/f_{\text{co}}^2]$ (15.58)

In the stop-band, $\beta = 180°$, $\sin\beta = 0$ and $\cos\beta = -1$. Hence

$-\cosh\alpha = 1 - (\omega_2 LC/2)$, $\cosh\alpha = 2\omega^2/\omega_{\text{co}}^2 - 1$, and $\cosh^2\alpha/2 = \omega^2/\omega_{\text{co}}^2$

Therefore

$$\alpha = 2\cosh^{-1}(f/f_{\text{co}})\ \text{N} \qquad (15.59)$$

The high-pass filter

Figure 15.25 shows the arrangements of a T and a π high-pass filter. In each case the total series impedance is $Z_1 = 1/\text{j}\omega C$ and the total shunt impedance is $Z_2 = \text{j}\omega L$. The design impedance is $R_0 = \surd(Z_1 Z_2) = \surd(L/C)$, as for the low-pass filter. The characteristic impedance $Z_{0\text{T}}$ of the T filter is

$$Z_{0\text{T}} = \surd(Z_1^2/4 + Z_1 Z_2) = \surd[Z_1 Z_2(1 + Z_1/4Z_2)]$$
$$= \surd\{(L/C)[1 - 1/(4\omega^2 LC)]\}$$

or $\quad Z_{0\text{T}} = R_0\surd[1 - 1/(4\omega^2 LC)] \qquad (15.60)$

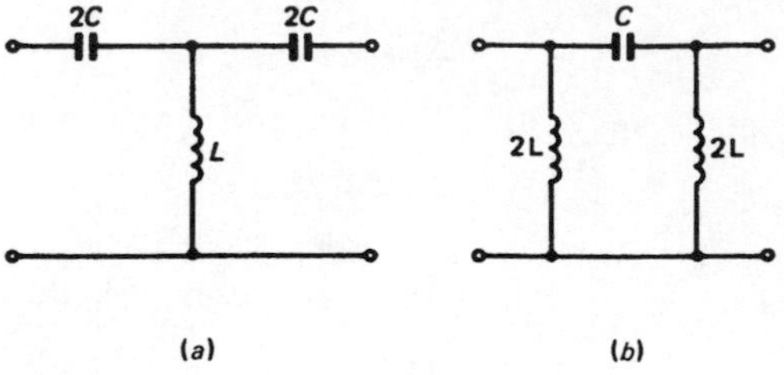

Fig. 15.25 The constant-k high-pass filter

In this case Z_{0T} will be imaginary for all frequencies low enough to ensure that $1/4\omega^2 LC > 1$ and real for all high frequencies at which $1 > 1/4\omega^2 LC$. The cut-off frequency f_{co} is the frequency at which Z_{0T} changes over from being imaginary to being real. At this point $1 = 1/4\omega_{co}^2 LC$ or

$$f_{co} = 1/[4\pi\sqrt{(LC)}] \text{ Hz} \tag{15.61}$$

At the cut-off frequency, $Z_{0T} = 0\ \Omega$ and then increases with increase in frequency, becoming equal to the design impedance R_0 when the frequency is very high. Since $Z_{0T} = Z_1 Z_2 / Z_{0T}$ the impedance $Z_{0\pi}$ of a constant-k π high-pass filter varies from infinity $(Z_1 Z_2/0)$ at the cut-off frequency to R_0 at very high frequencies.

The phase-shift in the pass-band and the attenuation in the stop-band can be determined from the equation $\cosh\gamma = 1 + (Z_1/2Z_2)$. Following the same steps as before gives

$$\beta = \cos^{-1}[1 - 2f_{co}^2/f^2] \tag{15.62}$$

$$\alpha = 2\cosh^{-1}(f_{co}/f) \tag{15.63}$$

Active filters

Active filters are considered in Chapter 20.

Exercises 15

15.1 For the network shown in Fig. 15.26 calculate the image impedances and the image transfer coefficient. Calculate the insertion loss of the network when it is connected in between its image impedances.

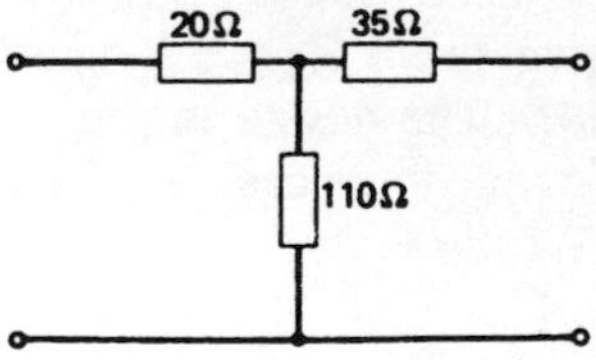

Fig. 15.26

15.2 Measurements on a network produced the following data:
input resistance with output open circuited = 1100 Ω
input resistance with output short circuited 550 Ω
output resistance with input open circuited = 1650 Ω
Calculate the component values of the equivalent T network.

15.3 For the network shown in Fig. 15.27 calculate (a) its image impedances, (b) its iterative impedances and (c) its insertion loss when connected between its image impedances.

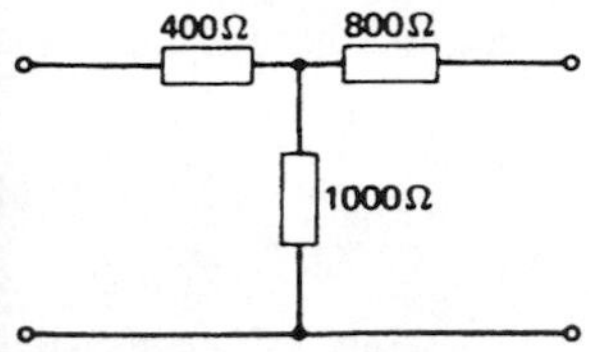

Fig. 15.27

15.4 A T network has a total series impedance Z_1 of 2000 Ω and a shunt impedance Z_2 of 5600 Ω. Calculate (*a*) the characteristic impedance and (*b*) the propagation coefficient of the network. (*c*) Calculate the current in a matched load when five such networks are connected in cascade if the input current is 10 mA.

15.5 Explain with the aid of a sketch the difference between an image impedance and an iterative impedance. The T network shown in Fig. 15.28 is connected between its image impedances. Calculate its insertion loss.

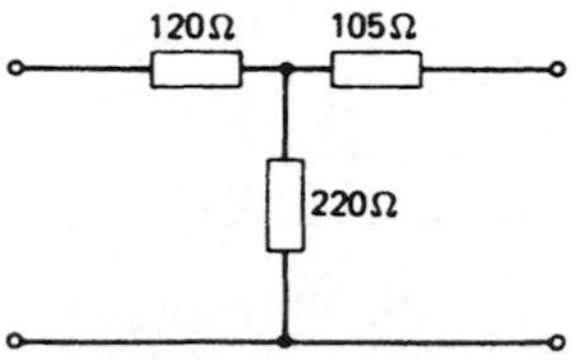

Fig. 15.28

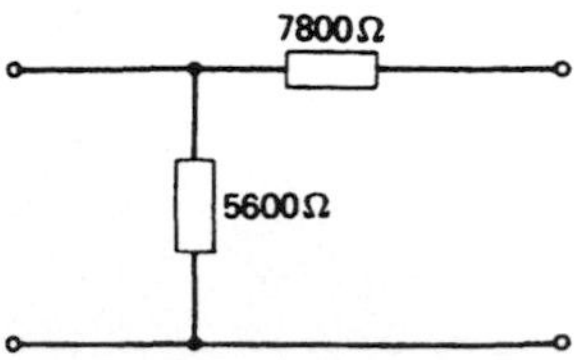

Fig. 15.29

15.6 Calculate the insertion loss of the network given in Fig. 15.29 when it is connected between (*a*) its image impedances and (*b*) its iterative impedances.

15.7 Derive an expression for the characteristic impedance of a symmetrical T network. A symmetrical T attenuator has a total series impedance of 400 Ω and a shunt impedance of 380 Ω. Calculate the characteristic impedance of the network and the overall loss of four such sections connected in tandem.

15.8 Prove that the two image impedances of an L network may both be calculated using the expression $Z_{im} = \sqrt{(Z_{oc}Z_{sc})}$. An L network has a series resistance of 1200 Ω and a shunt impedance of 3300 Ω. Calculate its insertion loss when it is connected between its image impedances.

15.9 A 4-terminal network has an input resistance of 1200 Ω when the output terminals are open circuited. When the output terminals are short circuited the input impedance is 480 Ω. Lastly, when the input terminals are short-circuited the output impedance is 560 Ω. Calculate the component values of the equivalent T network.

15.10 Design a T attenuator to have a characteristic impedance of 140 Ω and an attenuation of 15 dB. Derive any formulae used.

15.11 A non-symmetrical passive network has input and output image impedances of 1000 Ω and 750 Ω respectively and an image attenuation coefficient of 1.5 nepers. The network is connected between a source of impedance 600 Ω and a load of 500 Ω. Determine the insertion loss of the network.

15.12 A T network has $Z_0 = 600$ Ω and series resistors of 480 Ω. Calculate (*a*) the shunt resistor, (*b*) the insertion loss when connected between 600 Ω resistances and (*c*) the insertion loss if the load is reduced to 300 Ω.

15.13 A low-pass filter section has a total series inductance of 2 H and a total shunt capacitance of 2 μF. Calculate the loss of the filter at $2500/2\pi$ Hz when it is connected between a source and load impedance equal to the design impedance.

15.14 Design a low-pass filter to operate between a source and a load impedance of 600 Ω. The cut-off frequency is to be $25\,000/2\pi$ Hz.

15.15 Calculate the iterative impedances of an L network having a series resistor of 20 Ω and a shunt resistor of 120 Ω.

15.16 A network is connected between its image impedances. The source has an e.m.f. of 6 V and an internal impedance of 350 Ω and the load is 400 Ω. (*a*) What is the input impedance of the network? (*b*) Is the network symmetrical? (*c*) What is the input current to the network? (*d*) If the load voltage is 0.5 V what is the image attenuation of the network? (*e*) What is the insertion loss of the network?

15.17 The input and output currents of a non-symmetrical network are 80 mA and 10 mA respectively. The input and output voltages are 100 V and 25 V respectively Calculate (*a*) the input and load impedances and (*b*) the image transfer coefficient of the network. State whether the input and load impedances are equal to the image impedances of the network.

15.18 A load $R_L = 70$ Ω is matched to a source $R_S = 600$ Ω at 200 kHz. Find the values of the reactances X_1 and X_2 in the matching L network.

15.19 A T network has series resistors $R_1 = 20$ Ω and $R_3 = 35$ Ω and a shunt resistor $R_2 = 110$ Ω. Calculate its insertion loss when it is connected between its iterative impedances.

15.20 Use ABCD parameters to show that for a symmetrical T network $\cosh \gamma = A$, where γ is the propagation coefficient.

15.21 The symmetrical T network given in Fig. 15.7(*a*) has an impedance Z_A connected between the input and output terminals to form a bridged-T network. Show that the characteristic impedance is given by

$$Z_0 = \sqrt{[Z_A Z_1(Z_1 + 4Z_2)]/[4(Z_1 + Z_A)]},$$

and the propagation coefficient is

$$\gamma = \log_e\{[Z_0(Z_1 + Z_A) + Z_1 Z_A/2]/[Z_0(Z_1 + Z_A) - Z_1 Z_A/2]\}$$

15.22 A symmetrical network has $Z_1/2 = 175$ Ω and $Z_2 = 350$ Ω. Calculate (*a*) its characteristic impedance and (*b*) its loss.

15.23 A T network has series impedances $Z_1 = 40$ Ω, $Z_3 = 72$ Ω, and a shunt impedance $Z_2 = 360$ Ω. Calculate its image transfer coefficient.

15.24 A resistive T network has series resistances of 400 Ω and 800 Ω and a shunt resistance of 1000 Ω. Calculate its image impedances and its insertion loss when connected between these impedances.

15.25 An L network has a series impedance of j900 Ω and a shunt impedance of −j1300 Ω. It is connected between a voltage source of e.m.f. 156 V and impedance Z_S and a load of impedance Z_L. Z_S and Z_L are equal to the image impedances of the network. Calculate (*a*) Z_S and Z_L, (*b*) the phase shift produced by the network, and (*c*) the ratio (power supplied to networks)/(power in load).

15.26 A T network has equal series resistors of 200 Ω and a shunt resistor of 400 Ω. (*a*) Determine its ABCD parameters. (*b*) Calculate the image impedances. (*c*) If two such networks are connected in cascade find the transfer matrix of the combination.

15.27 A voltage source of resistance 140 Ω is to be matched by an L network to a load of resistance 75 Ω. Calculate the values of Z_1 and Z_2 if the match is to be (*a*) at all frequencies and (*b*) at 48 kHz only.

15.28 Derive an expression for the characteristic impedance of a T network that has series arms L_1 in series with C_1 and a shunt arm of $2C_2$.

15.29 Show that if a transformer that is used to match a source resistance R_S to a load R_L is connected the wrong way around the insertion loss is

$$20\ \log_{10}[\sqrt{(R_S/R_L)} + \sqrt{(R_L/R_S)} - 2(\sqrt{(R_S R_L)}/(R_S + R_L)]\ \text{dB}$$

15.30 An L network has 10 dB attenuation and an input iterative resistance of 500 Ω. (*a*) Calculate its resistors R_1 and R_2. (*b*) Calculate its insertion loss when connected between a source of resistance 100 Ω and a load of 100 Ω.

15.31 A network has image impedances of $60\sqrt{30}\angle 0°$ Ω and $40\sqrt{30}\angle 0°$ Ω and an image transfer coefficient of 1.5433. Calculate its insertion loss when it is connected between resistances of 60 Ω.

15.32 Calculate the insertion loss of a network with a characteristic impedance of 600 Ω and a loss of 5 dB, when it is connected between a source of 150 Ω and a load of 200 Ω.

15.33 A T network has $Z_1 = -j30$ Ω, $Z_3 = j60$ Ω and $Y_2 = j0.15$ S. Use ABCD parameters to determine (*a*) its image impedances and (*b*) the phase shift it introduces to a voltage when connected between its image impedances.

16 Matched transmission lines

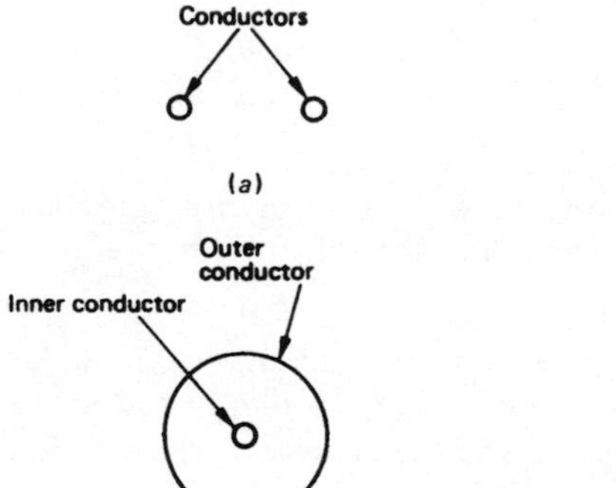

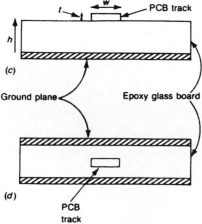

Fig. 16.1 Types of line: (a) twin; (b) coaxial; (c) microstrip; (d) stripline

Transmission lines are widely used in both telecommunication engineering and electric power engineering for the transmission of electrical energy from one point to another. A transmission line may operate in any frequency band from zero hertz to a few gegahertz. Lines employed in power transmission systems generally operate at 50 Hz and lines used to transmit telecommunication signals may need to work from zero hertz to several megahertz. In radio engineering, at VHF and higher frequency bands, transmission lines are also used to simulate the action of a component, such as an inductor or a capacitor, or a tuned circuit. In such cases the frequencies involved may be as high as a few GHz.

In electronic circuits the wiring and/or printed circuit track used to interconnect components may have to be treated as a transmission line. A conductor linking two components will have to be considered as a transmission line and not as a short circuit: (*a*) for an analogue system when the electrical length (in wavelengths) is of the same order, or exceeds, the signal wavelength; (*b*) in a digital system, the propagation delay along the conductor determines whether or not transmission line effects need to be taken into account. If the risetime of the shortest pulse is five times smaller than the round-trip propagation delay, then the conductor must be regarded as a transmission line.

Four main types of transmission line are in common usage: the two-wire or twin line, the coaxial line, the strip line and the microstrip line. The four types of line are shown in Fig. 16.1. The coaxial line is only used at frequencies higher than about 60 kHz. Strip line and microstrip are formed by a conductor mounted on a printed circuit board (PCB) and are the kind of line employed with high-speed digital circuits and at microwave frequencies.

The physical length of a line may be kilometres in the case of power lines and audio-frequency telecommunication lines; and only a few tens of metres for some radio applications of lines such as a feeder to connect an aerial to a receiver; or even just a fraction of a metre when a line is employed to simulate a component, or to inter-connect components in an electronic circuit.

Example 16.1

A length of conductor has a phase velocity of 200×10^6 m/s. It connects an electronic device with a risetime of 14 ns to another device. Calculate the maximum length of the conductor that may be considered as a short circuit and not as a transmission line.

Solution

Round-trip delay $= (14 \times 10^{-9})/10 = 1.4 \times 10^{-9}$ s. Therefore the maximum length of conductor that acts like a short circuit is

$$1.4 \times 10^{-9} \times 200 \times 10^6 = 0.28 \text{ m} \quad (\textit{Ans.})$$

Primary coefficients of a line

Resistance

The resistance R of a line is the sum of the resistances of the two conductors comprising a pair. The unit of resistance is ohms per kilometre loop.

At zero frequency the resistance of a pair is merely the d.c. resistance R_{dc} but at frequencies greater than a few kilohertz a phenomenon known as *skin effect* comes into play. This effect ensures that current flows only in a thin layer or 'skin' at the surface of the conductors. The thickness of this layer reduces as the frequency is increased and this means that the effective cross-sectional area of the conductor is reduced. Since resistance is equal to $\rho l/a$ the a.c. resistance of a conductor will increase with increase in frequency. While the skin effect is developing, the relationship between the a.c. resistance and the frequency is rather complicated, but once skin effect is fully developed (usually at about 12 kHz) the a.c. resistance becomes directly proportional to the square root of the frequency, i.e.

$$R_{ac} = k_1\sqrt{f} \qquad (16.1)$$

where k_1 is a constant.

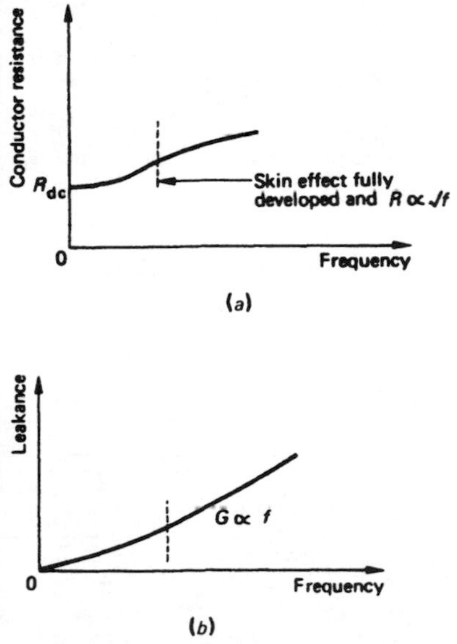

Fig. 16.2 Variation with frequency of: (a) line resistance; (b) line leakance

Figure 16.2(a) shows how the resistance of a line varies with frequency. Initially, little variation from the d.c. value is observed but at higher frequencies the shape of the graph is determined by equation (16.1).

Inductance

The inductance of a line, in henrys/kilometre, depends upon the dimensions and the spacing of its two conductors and the permeability of the material between them, see equations (12.29) and (12.30).

The permeability is not a function of frequency and consequently the inductance of a line is not a frequency-dependent parameter.

Capacitance

The capacitance of a line, in μF/km, is determined by the dimensions and spacing of the conductors and the permittivity of the insulating material. Expressions for the capacitance are given by equations (12.16) and (12.19). The permittivity of the dielectric is not frequency dependent and so the capacitance of a line can also be considered to be reasonably constant with change in frequency.

Leakance

The leakance G of a line, in siemen per kilometre, represents the leakage of current between the conductors via the dielectric separating them. The leakage current has two components: one of these passes through the insulation resistance between the conductors, while the other supplies the energy losses in the dielectric as the line capacitance repeatedly charges and discharges.

Leakance increases with increase in frequency and at the higher frequencies it becomes directly proportional to frequency:

$$G = k_2 f \tag{16.2}$$

where k_2 is another constant.

Figure 16.2(*b*) shows how the leakance of a line varies with frequency. Some typical figures for the primary coefficients of a line are given in Table 16.1. Clearly, the values of the four primary coefficients may differ considerably between pairs in different types of cable.

Table 16.1 Line primary coefficients

Line type	*R* (Ω/km)	*L* (mH/km)	*C* (μF/km)	*G* (μS/km)
Twin (800 Hz)	55	0.6	0.033	0.6
Coaxial (1 MHz)	34	0.28	0.05	1.4

Secondary coefficients of a line

The secondary coefficients of a transmission line are its *characteristic impedance* Z_0, *propagation coefficient* γ and *velocity of propagation* v_p.

The propagation coefficient has both a real part and an imaginary part; the former is known as the *attenuation coefficient* α and the latter is known as the *phase-change coefficient* β.

Characteristic impedance

When a symmetrical T network is terminated in its characteristic impedance Z_0, the input impedance of the network is also equal to Z_0 (p. 265). Similarly, if five or more identical T networks are connected in cascade, the input impedance of the combination will also be equal to Z_0. Since a transmission line can be considered to consist of the tandem connection of a very large number of very short lengths δl of line, as shown by Fig. 16.3, the concept of characteristic impedance is also applicable to a line. Each elemental length δl of line has a total series impedance of $(R+\mathrm{j}\omega L)\delta l$ and a total shunt admittance of $(G+\mathrm{j}\omega C)\delta l$. The *characteristic impedance* Z_0 of a line can therefore be defined as being the input impedance of a long length of that line. Figure 16.4 shows a very long length of line; its input impedance is the ratio of the voltage V_S impressed across the sending-end terminals of the line to the current I_S that flows into those terminals, i.e.

$$Z_0 = V_S/I_S \ \Omega \tag{16.3}$$

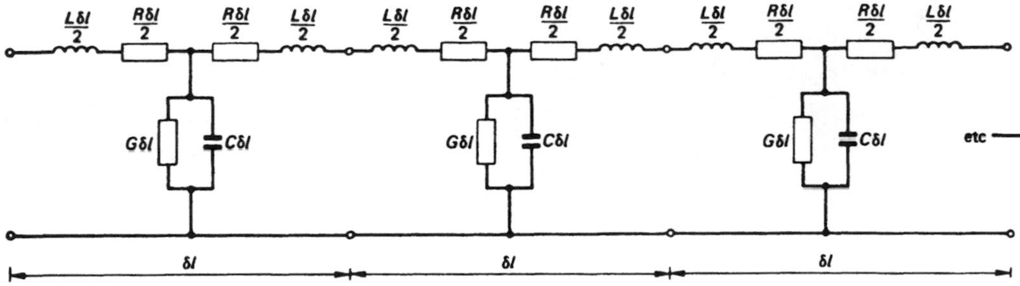

Fig.16.3 Representation of a line by the cascade connection of δl sections

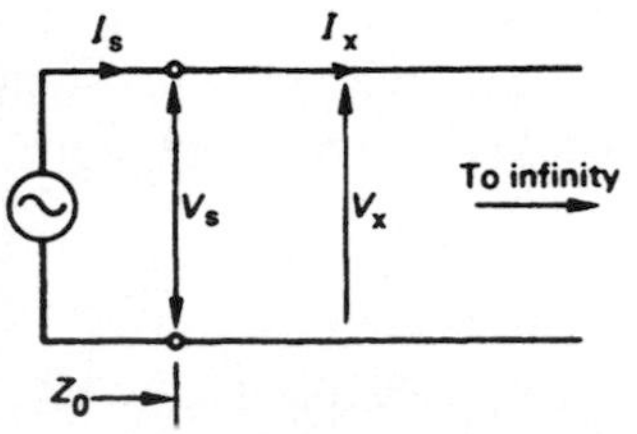

Fig.16.4 Definition of the characteristic impedance of a line

Similarly, at any point x along the line the ratio V_x/I_x is always equal to Z_0. Suppose that the line is cut a finite distance from its sending-end terminals as shown by Fig. 16.5(a). The remainder of the line is still very long and so the impedance measured at terminals 2-2 is equal to the characteristic impedance. Thus, before the line was cut, terminals 1-1 were effectively terminated in impedance Z_0. The conditions at the input terminals will not be changed if the terminals 1-1 are closed in a physical impedance equal to Z_0, as in Fig. 16.5(b). This leads to a more practical definition: the characteristic impedance of a transmission line is

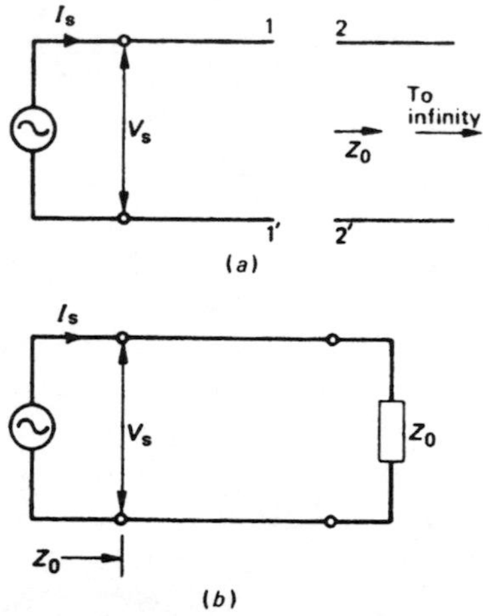

Fig.16.5 Alternative definition of the characteristic impedance of a line

the input impedance of a line that is itself terminated in the characteristic impedance. This result can be used to derive an expression for the characteristic impedance. A line that is terminated in its charac-teristic impedance is generally said to be *correctly terminated*, and all the incident energy is absorbed by the load. If a line is in-correctly terminated, i.e. $Z_L \neq Z_0$, some of the incident energy will be *reflected* from the load. Reflection is considered in Chapter 17.

The characteristic impedance of a single T network is given by equation (15.6).

For the elemental length of transmission line shown in Fig. 16.6, $Z_1 = (R + j\omega L)\delta l$ and $Z_2 = 1/(G + j\omega C)\delta l$. Therefore

$$Z_0 = \sqrt{\left[\frac{(R + j\omega L)^2 \delta l^2}{4} + \frac{R + j\omega L}{G + j\omega C}\right]}$$

$$\approx \sqrt{[(R + j\omega L)/(G + j\omega C)]} \qquad (16.4)$$

since δl^2 is *very* small.

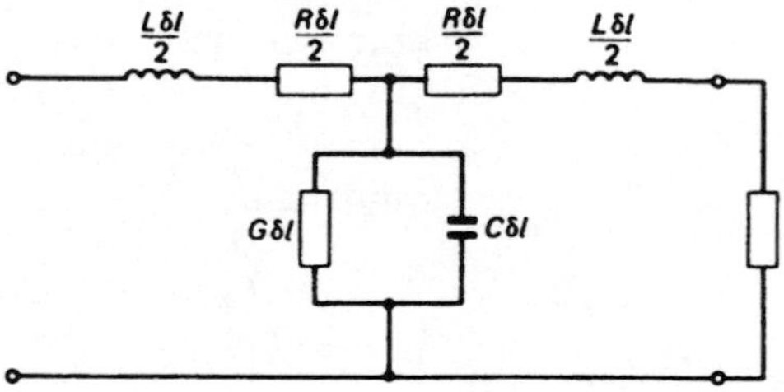

Fig.16.6 Calculation of Z_0

At zero, and very low, frequencies, $R \gg \omega L$ and $G \gg \omega C$ and then $Z_0 \approx \sqrt{(R/G)}\ \Omega$. With increase in frequency both the magnitude and the phase angle of Z_0 decrease in value and, when the frequency is reached at which the statements $\omega L \gg R$ and $\omega C \gg G$ are valid, then

$$Z_0 = \sqrt{(L/C)}\ \Omega \qquad (16.5)$$

Since both L and C are both very nearly constant with frequency, Z_0 remains at this value with any further increase in frequency. Furthermore, Z_0 is now a purely resistive quantity. The approximate expression (16.5) for Z_0 is always used for radio-frequency lines since these are only used at fairly high frequencies. Fig. 16.7 shows how the magnitude of the characteristic impedance of a line varies with frequency.

Example 16.2

A transmission line has the following primary coefficients: $R = 40\ \Omega/\text{km}$, $L = 48\ \text{mH/km}$, $C = 0.06\ \mu\text{F/km}$ and $G = 4 \times 10^{-6}\ \text{S/km}$. Calculate its characteristic impedance at a frequency of $5000/2\pi$ Hz.

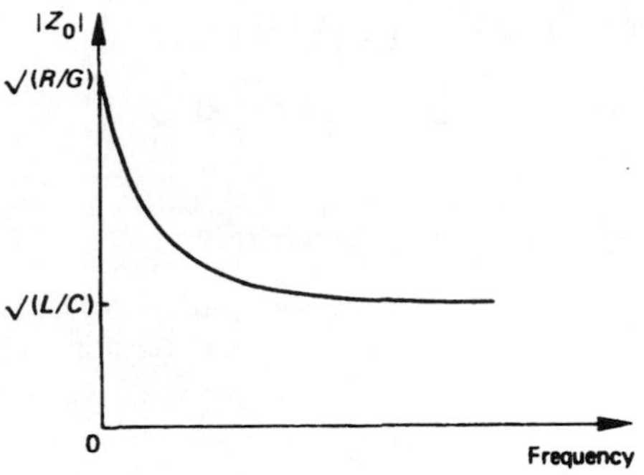

Fig.16.7 Variation of Z_0 with frequency

Solution
From equation (16.4)

$$Z_0 = \sqrt{\{(40 + j5000 \times 48 \times 10^{-3})/[(4 + j5000 \times 0.06] \times 10^{-6})\}}$$
$$= \sqrt{\{(40 + j240)/[(4 + j300) \times 10^{-6}]\}}$$

Clearly, $G \ll \omega C$ and so it may be neglected with very little error. Therefore,

$$Z_0 = \sqrt{[(243.3 \angle 80.5°)/(300 \angle 90° \times 10^{-6})]}$$
$$\approx 900 \angle -4.8° \ \Omega \quad (Ans).$$

The characteristic impedance of a transmission line is a function of the physical dimensions of the line and the permittivity of the dielectric. At higher frequencies when the approximate expression (16.5) for Z_0 is valid, the characteristic impedance can be written as

- Twin line

$$Z_0 = \sqrt{(L/C)} = \sqrt{\left[\frac{\mu_0 \log_e d/r}{\pi} \times \frac{\log_e d/r}{\pi\varepsilon}\right]}$$
$$= \frac{\log_e d/r}{\pi}\sqrt{\frac{\mu_0}{\varepsilon_0\varepsilon_r}} = 120 \log_e \frac{d}{r}\sqrt{\frac{1}{\varepsilon_r}} \qquad (16.6)$$
$$= [276 \log_{10}(d/r)]/\sqrt{\varepsilon_r} \qquad (16.7)$$

- Coaxial line

$$Z_0 = \sqrt{(L/C)} = \sqrt{\left[\frac{\mu_0 \log_e R/r}{2\pi} \times \frac{\log_e R/r}{2\pi\varepsilon}\right]}$$
$$= \frac{\log_e R/r}{2\pi}\sqrt{\frac{\mu_0}{\varepsilon_0\varepsilon}} = [60 \log_e(R/r)]/\sqrt{\varepsilon_r} \qquad (16.8)$$
$$Z_0 = [138 \log_{10}(R/r)]/\sqrt{\varepsilon_r} \qquad (16.9)$$

- Microstrip:

$$Z_0 = [87/\sqrt{(\varepsilon_r + 1.41)}]\{\log_e[(5.98\ \mathrm{H})/(0.8\ \mathrm{W} + \mathrm{T})]\} \qquad (16.10)$$

- Stripline:

$$Z_0 = (60/\sqrt{\varepsilon_r})\{\log_e[4\ B/(0.7\pi\ W(0.8 + T/W))]\} \qquad (16.11)$$

Example 16.3

A coaxial pair is terminated by an annular shaped resistor of inner radius r and outer radius R. The resistor is made from a material whose resistivity is $\rho\,\Omega/\square$. If $\varepsilon_r = 2.8$ calculate the required value for ρ for a matched termination.

Solution
For a matched termination

$$R_L = Z_0, \text{ or } (\rho/2\pi)\log_e(R/r) = [60\ \log_e(R/r)]/\sqrt{\varepsilon_r}).$$

Hence, $\rho = 120\pi/\sqrt{2.8} = 225.3\ \Omega/\square$ (*Ans.*)

Choice of Z_0 *for a coaxial line*

The characteristic impedance of a coaxial cable depends upon its dimensions and this means that its value is determined by its main requirements.

- Minimum attenuation. For the minimum attenuation which is required for a long-distance telephone cable, the ratio (outer radius R)/(inner radius r) should be equal to 3.59 and this gives a characteristic impedance of 76 Ω.
- Maximum power handling capability. The power handling capability of a coaxial cable is limited by the breakdown voltage of its dielectric. The ratio R/r giving the maximum breakdown voltage is 2.72 and this results in a characteristic impedance of 60 Ω.

These figures are reduced if a dielectric other than air is employed, the reduction being $1/\sqrt{\varepsilon_r}$. Most coaxial cables are manufactured with a characteristic impedance of either 50 Ω or 75 Ω as a reasonable compromise.

Propagation coefficient

The *propagation coefficient* γ of a 1 km length of correctly terminated transmission line is defined as

$$\gamma = \log_e V_S/V_R = \log_e I_S/I_R \qquad (16.12)$$

or $$V_R = V_S e^{-\gamma} \qquad (16.13)$$

For a line which is 2 km in length the received voltage will be

$$V_R = V_S e^{-\gamma} e^{-\gamma} = V_S e^{-2\gamma}$$

Similarly, for a line of length l kilometres

$$V_R = V_S e^{-\gamma l} \tag{16.14}$$

In similar fashion the current received at the end of a line of length l can be written as

$$I_R = I_S e^{-\gamma l} \tag{16.15}$$

For an elemental length of line

$$Z_1 = (R + j\omega L)\delta l \qquad Z_2 = 1/(G + j\omega C)\delta l$$

and, of course,

$$Z_0 = \sqrt{(R + j\omega L)/(G + j\omega C)}$$

Hence, substituting into equation (15.15),

$$\begin{aligned} e^{\gamma} &= 1 + \frac{(R + j\omega L)\delta l(G + j\omega C)\delta l}{2} + \sqrt{\left(\frac{R + j\omega L}{G + j\omega C}\right)}\,(G + j\omega C) \\ &= 1 + \sqrt{[(R + j\omega L)(G + j\omega C)]}\delta l \\ &\quad + [(R + j\omega L)(G + j\omega C)\delta l^2]/2 \end{aligned} \tag{16.16}$$

The series form of e^x is

$$e^x = 1 + x + x^2/2! + x^3/3! + \cdots \tag{16.17}$$

Comparing equations (16.16) and (16.17) term by term it is evident that for an elemental length of line

$$\gamma = \sqrt{[(R + j\omega L)(G + j\omega C)]}\delta l$$

In a 1 kilometre length of line there are $1/\delta l$ such elemental sections and so

$$\gamma = \sqrt{[(R + j\omega L)(G + j\omega C)]} \text{ per km} \tag{16.18}$$

Note that the propagation coefficient is dimensionless.

The propagation coefficient γ is a complex quantity and it therefore has both a real part and an imaginary part. The real part of γ is known as the *attenuation coefficient* α, measured in units of nepers/kilometre. The imaginary part of γ is called the *phase-change coefficient* β, measured in terms of units of radians/kilometre.

Thus $\gamma = \alpha + j\beta$. (16.19)

The voltage V_x at any point along a line, distance x from the sending end, can be written as

$$\begin{aligned} V_x &= V_S e^{-\gamma x} = V_S e^{-(\alpha + j\beta)x} \\ &= V_S e^{-\alpha x} \cdot e^{-j\beta x} = V_S e^{-\alpha x} \angle -\beta x \end{aligned} \tag{16.20}$$

The voltage V_R received at the far end of the line can now be written as

$$V_R = V_S e^{-\alpha l} \angle -\beta l \qquad (16.21)$$

Equation (16.21) demonstrates that the magnitude of the line voltage (or current) decreases exponentially with the distance x from the sending end of the line. The phase angle of the line voltage is always a lagging angle and this angle increases in direct proportion to the length of the line.

Expressions for the attenuation coefficient and for the phase-change coefficient can be derived from equation (16.18) but the results are unwieldy and not worthwhile. When the values of α and β are required it is better to use equation (16.18) and take the real and imaginary parts of the result.

Example 16.4

Calculate, for a frequency of $5000/2\pi$ Hz, the propagation coefficient of a line whose primary coefficients are: $R = 55\ \Omega$/km, $L = 28$ mH/km, $C = 0.07\ \mu$F/km and $G = 2.0\ \mu$S/km. Determine also the values of α in dB and β in degrees.

Solution

$$\gamma = \sqrt{[(55 + j5000 \times 28 \times 10^{-3})(2 \times 10^{-6} + j5000 \times 0.07 \times 10^{-6})]}$$

$$= \sqrt{[(55 + j140)(2 \times 10^{-6} + j3.5 \times 10^{-4})]}$$

$$= \sqrt{[150.4 \angle 68.6° \times 3.5 \times 10^{-4} \angle 89.7°]}$$

$$= 0.229 \angle 79.2°/\text{km} \quad (\textit{Ans.})$$

$$\alpha = 0.229 \cos 79.2° = 0.043 \text{ N/km}$$

$$= 0.043 \times 8.686 = 0.37 \text{ dB/km} \quad (\textit{Ans.})$$

$$\beta = 0.229 \sin 79.2° = 0.225 \text{ R/km} = 13°/\text{km} \quad (\textit{Ans.})$$

Example 16.5

A 16 km length of transmission line has primary coefficients $R =$ 22.5 Ω/loop km, $L = 175\ \mu$H/loop km, $C = 65$ nF/km, and $G \approx 0$. Calculate (*a*) its characteristic impedance, (*b*) its propagation coefficient. (*c*) If the line is terminated in its characteristic impedance and supplied by a 400 V matched voltage source at a frequency of 5 kHz, calculate the current flowing in the load.

Solution

$$(a)\quad Z_0 = \sqrt{\frac{22.5 + j2\pi \times 5000 \times 175 \times 10^{-6}}{j2\pi \times 5000 \times 65 \times 10^{-9}}}$$

$$= 100\sqrt{(22.5 + j5.5)/(j20.4)} = 106.6 \angle -38°\ \Omega \quad (\textit{Ans.})$$

$$(b)\quad \gamma = \sqrt{(22.5 + j5.5)(j20.4 \times 10^{-4})} = 0.22 \angle 51.9° \quad (\textit{Ans.})$$

(c) $I_S = 200/(106.6\angle -38°) = 1.88\angle 38°$ A.
$I_R = 1.88\angle 38° \times e^{-(0.14+j0.17)16} = 0.22\angle -118.8°$ A (*Ans.*)

Approximate expressions for γ and β

Approximate expressions for the attenuation coefficient and the phase-change coefficient can be obtained at (*a*) very low frequencies where $R \gg \omega L$ and $G \gg \omega C$, (*b*) low frequencies where $R \gg \omega L$ and $\omega C \gg G$, and (*c*) high frequencies where $\omega L \gg R$ and $\omega C \gg G$.

- Very low frequencies. If $R \gg \omega L$ and $G \gg \omega C$ then

  $\gamma = \sqrt{(RG)}$ and is wholly real (16.22)

- Low frequencies. If $R \gg \omega L$ and $\omega C \gg G$ then

  $$\gamma \approx \sqrt{(j\omega CR)} = \sqrt{(\omega CR\angle 90°)} = \sqrt{(\omega CR)}\angle 45°$$

  Taking the real and the imaginary parts,

  $$\gamma = \sqrt{(\omega CR/2)} + j\sqrt{(\omega CR/2)} \qquad (16.23)$$

- High frequencies. If $\omega L \gg R$ and $\omega C \gg G$ then

  $$\begin{aligned}\gamma &= \sqrt{[(1 + R/j\omega L)j\omega L(1 + G/j\omega C)j\omega C]}\\ &= j\omega\sqrt{(LC)}[1 + R/j\omega L]^{1/2}[1 + G/j\omega C]^{1/2}\\ &= j\omega\sqrt{(LC)}[1 + R/2j\omega L][1 + G/2j\omega C]\\ &\approx j\omega\sqrt{(LC)}[1 + R/2j\omega L + G/2j\omega C]\\ &= (R/2)\sqrt{(C/L)} + (G/2)\sqrt{(L/C)} + j\omega\sqrt{(LC)}\end{aligned}$$

  Hence

  $$\alpha = R/2Z_0 + GZ_0/2 \text{ N/km} \qquad (16.24)$$

  and

  $$\beta = \omega\sqrt{(LC)} \text{ R/km} \qquad (16.25)$$

Example 16.6

A radio-frequency line has an attenuation of 2.5 dB/km at 2 MHz, 10 per cent of which is dielectric loss. Determine the attenuation of the line at 4 MHz.

Solution
At 2 MHz the conductor loss is 2.25 dB/km and the dielectric loss is 0.25 dB/km. From equation (16.24) the conductor loss is proportional to the resistance of the line and so it is proportional to the square root of the frequency. Therefore, the conductor loss at 4 MHz is $2.25\sqrt{(4/2)} = 3.18$ dB/km.

The dielectric loss is proportional to the leakance of the line and hence to the frequency. Therefore, the dielectric loss at 4 MHz is $0.25 \times 4/2 = 0.5$ dB/km.

Therefore, the attenuation of the line at 4 MHz is

$$3.18 + 0.5 = 3.68 \text{ dB/km} \quad (\textit{Ans.})$$

Example 16.7

An air-spaced line has a characteristic impedance of 70 Ω and an attenuation coefficient of 0.2 dB/m at a frequency of 1.1 GHz. The dielectric is changed from air to a material with $\varepsilon_r = 3$ and a power factor of 5×10^{-4}. Calculate the new value of the attenuation coefficient.

Solution

In nepers $\alpha = 0.2/8.686 = 0.023$ N. For the air-spaced line $G \approx 0$ and $\alpha = 0.023 = R/(2 \times 70)$, or $R = 3.22\ \Omega$. With the dielectric material inserted between the conductors $Z_0 = 70/\sqrt{3} = 40.41\ \Omega$ and now $G = \omega C \times$ power factor $= 2\pi \times 1.1 \times 10^9 C \times 5 \times 10^{-4} = 3.456 \times 10^6 C$. Hence,

$$\begin{aligned}\alpha &= 3.22/(2 \times 40.41) + 3.456 \times 10^6 C Z_0/2 \\ &= 0.04 + 1.728 \times 10^6 C\sqrt{(L/C)} \\ &= 0.04 + 1.728 \times 10^6 \times \sqrt{(LC)} \\ &= 0.04 + 1.728 \times 10^6 \times \sqrt{3}/(3 \times 10^8) = 0.05 \text{ N/m} \\ &= 0.43 \text{ dB/km} \quad (\textit{Ans.})\end{aligned}$$

When a digital signal is transmitted over a line the transitions at the leading and trailing edges of the pulses are attenuated more than the relatively flat tops and bottoms of the pulses because they contain higher-frequency components.

Velocity of Propagation

The *phase velocity* v_p of a line is the velocity with which a sinusoidal wave travels along that line. Any sinusoidal wave travels with a velocity of one wavelength per cycle. There are f cycles per second and so a wave travels with a velocity of λf kilometres per second, i.e.

$$v_p = \lambda f \text{ kilometres/second} \tag{16.26}$$

In one wavelength, a phase change of 2π radians occurs. Hence the phase change per kilometre is $2\pi/\lambda$ radians and this is also equal to the phase-change coefficient β of the line. Thus $\beta = 2\pi/\lambda$ or $\lambda = 2\pi/\beta$ and

$$v_p = 2\pi f/\beta = \omega/\beta \text{ km/sec} \tag{16.27}$$

The *phase delay* of a line is the product of the line length and the reciprocal of its phase velocity.

For most cable insulating materials the relative permittivity ε_r is approximately 3 to 4 and this value gives a phase velocity

$1/\sqrt{(\varepsilon_r \mu)}$ of about 50–60 per cent of the free-space velocity $c = 3 \times 10^8$ m/s.

Any repetitive non-sinusoidal waveform contains components at a number of different frequencies. Each of these components will be propagated along a line with a phase velocity given by equation (18.27). For all of these components to travel with the same velocity and arrive at the far end of the line together it is necessary for β to be a *linear* function of frequency. Unfortunately, it is only at the higher frequencies where $\omega L \gg R$ and $\omega C \gg G$ that this requirement is satisfied. At these frequencies $\beta = \omega\sqrt{(LC)}$ and hence

$$v_p = \omega/\beta = 1/\sqrt{(LC)} \text{ km/sec} \qquad (16.28)$$

When the component frequencies of a complex wave travel with different velocities, their relative phase relationships will be altered and the resultant waveform will be *distorted.*

The change in phase velocity from ω/β at low frequencies to $1/\sqrt{(LC)}$ at high frequencies is known as *frequency dispersion.* Frequency dispersion causes the transitions in a digital signal to have a smaller delay than the flatter tops and bottoms of the pulses. The combined effect of attenuation and frequency dispersion on a digital waveform is to round off the pulse waveshape.

Example 16.8

A transmission line is 10 km in length. When a voltage of 6 V at 2000 Hz is applied to the sending-end terminals of the line a voltage of 1 V lagging by 200° appears across the matched load. Calculate (a) the attenuation and phase-change coefficients of the line, (b) the wavelength of the signal and (c) the phase velocity of propagation.

Solution

(a) $1 = 6e^{-10\alpha}$, so $6 = e^{10\alpha}$ and

$$\alpha(= \log_e 6)/10 = 0.179 \text{ N/km} = 1.56 \text{ dB/km} \quad (\textit{Ans.})$$

and $10\beta = 200°$ so that $\beta = 20°/\text{km} = 0.349$ R/km (*Ans.*)

(b) $\lambda = 2\pi/\beta = 360°/20° = 18$ km (*Ans.*)

(c) $v_p = \omega/\beta = (2\pi \times 2000)/0.349 = 36 \times 10^3$ km/sec (*Ans.*)

Example 16.9

A line has an inductance of 200 nH/m and a capacitance of 80 pF/m. Calculate (a) its characteristic impedance, (b) its phase change coefficient, (c) its phase velocity, if the frequency is 50 MHz.

Solution

(a) $Z_0 = \sqrt{(200 \times 10^{-9})/(80 \times 10^{-12})} = 50\ \Omega$ (*Ans.*)

(b) $\beta = \omega\sqrt{(LC)} = 2\pi \times 50 \times 10^6\sqrt{(200 \times 10^{-9} \times 80 \times 10^{-12})}$

$= 1.257$ rad/m $= 72°$/m (*Ans.*)

(*c*) $v_p = 1/\sqrt{(LC)} = 1/\sqrt{(200 \times 10^{-9} \times 80 \times 10^{-12})}$
$= 2.5 \times 10^8$ m/s (*Ans.*)

Group velocity

It is customary to consider the *group velocity* of a complex wave rather than the phase velocities of its individual components. Group velocity is the velocity with which the envelope of a complex wave is propagated along a line. The *group delay* of a line is the product of the length of the line and the reciprocal of its group velocity. Group delay measures the time taken for the envelope of a complex wave to propagate over a transmission line.

The *group velocity* of a complex wave is the velocity with which the envelope of the wave is transmitted. Consider a wave that contains components at two frequencies, $\omega/2\pi$ and $(\omega + \delta\omega)/2\pi$. The phase velocity of the lower frequency component is $v_{p1} = \omega/\beta_1$ and the phase velocity of the other component is $v_{p2} = (\omega + \delta\omega)/(\beta + \delta\beta)$. At any point along the line distance x from the sending end the instantaneous voltages of the components are

$$v_1 = V_1 \cos(\omega t - \beta x) \quad \text{and}$$
$$v_2 = V_2 \cos[(\omega + \delta\omega)t - (\beta + \delta\beta)x].$$

The two components move into and out of phase with one another as they travel along the line and they are in phase after a time t_1 when the peak value of the envelope occurs. At time t_1,

$$\omega t_1 - \beta x = (\omega + \delta\omega)t_1 - (\beta + \delta\beta)x, \quad \text{or}$$
$$\delta\omega t_1 = \delta\beta x \tag{16.29}$$

At some later time t_2 and distance y from the sending end the two components are again in phase with one another. Now

$$\omega t_2 - \beta y = (\omega + \delta\omega)t_2 - (\beta + \delta\beta)y, \quad \text{or}$$
$$\delta\omega t_2 = \delta\beta y \tag{16.30}$$

In a time $t_2 - t_1$ the peak value of the complex waveform has travelled a distance $y - x$, which is one wavelength. Therefore, the group velocity is

$$v_g = (y - x)/(t_2 - t_1) = \delta\omega t_2/\delta\beta - \delta\omega t_1/\delta\beta$$
$$= \delta\omega/\delta\beta \tag{16.31}$$

If the phase change coefficient β of the line is directly proportional to frequency then $\omega/\beta = (\omega + \delta\omega)/(\beta + \delta\beta)$, or $\omega/\beta = \delta\omega/\delta\beta$. This means that the phase velocity is equal to the group velocity.

Relationship between v_g, v_p, and λ

The phase velocity is $v_p = \omega/\beta$, or $\omega = \beta v_p$. Then $\delta\omega = \delta(\beta v_p) = v_p\delta\beta + \beta\delta v_p$,

$$\delta\omega/\delta\beta = v_g = v_p + \beta\delta v_p/\delta\beta = v_p + (\beta\delta\lambda/\delta\beta)(\delta v_p/\delta\lambda).$$

But $\beta = 2\pi/\lambda$. Hence, $\delta\beta/\delta\lambda = -2\pi/\lambda^2$ and $v_g = v_p + (2\pi/\lambda)(-\lambda^2/2\pi)(\delta v_p/\delta\lambda)$.

$$\text{Therefore, } v_g = v_p - \lambda\delta v_p/\delta\lambda \tag{16.32}$$

Received current and voltage

If a transmission line is correctly terminated, i.e. the load impedance is equal to the characteristic impedance Z_0, then the input impedance of the line will also be equal to Z_0. This point is emphasized by Fig. 16.8 and it is true regardless of the length of the line.

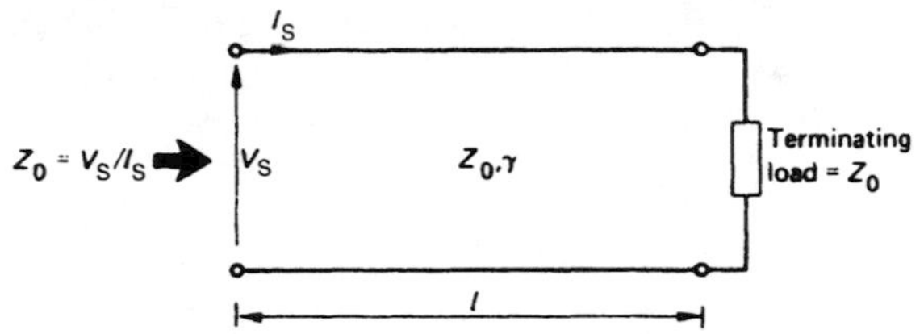

Fig.16.8 Matched line

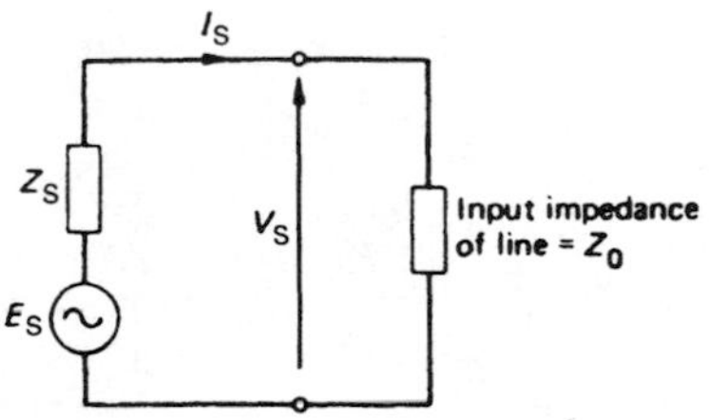

Fig.16.9 Input circuit of a matched line

Figure 16.9 shows a source of e.m.f E_S and internal impedance Z_S connected across the input terminals of a line. The current I_S flowing into the line is

$$I_S = E_S/(Z_S + Z_0)$$

and the voltage appearing at the input terminals is

$$V_S = E_S Z_0/(Z_S + Z_0)$$

Note that $V_S/I_S = Z_0$.

The input current and voltage propagate along the line to its far end with a phase velocity v_p. The waves are subject to both attenuation and phase shift (lagging) as they travel. At any point along the line, distance x from the sending end, the current and voltage are equal to

$$I_x = I_S e^{-\gamma x} \quad \text{and} \quad V_x = V_S e^{-\gamma x}$$

respectively. Note that once again the ratio V_x/I_x is equal to Z_0.

At the far end of the line the received current I_R is equal to $I_S e^{-\gamma l}$ and the received voltage is $V_R = V_S e^{-\gamma l}$. Once again the ratio V_R/I_R is equal to Z_0. The power dissipated in the load can be worked out in two different ways, e.g.

$$P_R = |I_R|^2 \times (\text{real part of } Z_0) \quad \text{or}$$
$$P_R = |V_R||I_R| \cos\theta,$$

where θ is the phase angle between the received current and voltage.

A *polar diagram* can be drawn to show the magnitude and the phase, relative to the sending end, of the line voltage (or current) at any point along a line. Such a diagram consists of a series of phasors drawn to represent the line voltage at various points along the line. Each phasor is drawn with an appropriate length and angle to indicate the magnitude and the phase of the voltage at that point.

Consider an 8 km length of line that has an attenuation of 2 dB/km and a phase-change coefficient of 45°/km at a particular frequency. If the sending-end voltage is 1 volt then the magnitude and phase of the line voltage at 1 km steps from the sending end of the line are given in Table 16.1.

The polar diagram is shown plotted in Fig. 16.10.

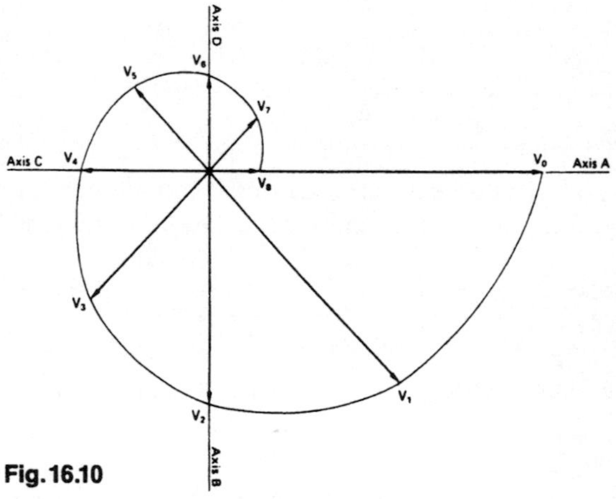

Fig. 16.10

Table 16.1

Distance from sending end	*Attenuation (dB)*	*Voltage (V)*	*Phase lag (deg)*
0	0	1	0
1	2	0.79	45
2	4	0.63	90
3	6	0.5	135
4	8	0.4	180
5	10	0.32	215
6	12	0.25	270
7	14	0.2	315
8	16	0.16	360

At the instant when the sending-end voltage is zero and just about to go positive, the line voltage at various points along the line can be plotted by projecting from the tips of the phasors to the corresponding points on axis A.

A quarter of the period time later, the sending-end voltage is instantaneously at its peak positive value of 1 V. The line voltages at this instant in time can be plotted by projecting from the tips of the phasors onto axis B.

The waveforms of the line voltages at the instants in time when the input voltage is, first, zero and about to go negative and, secondly, at its peak negative value are plotted on axes C and D respectively.

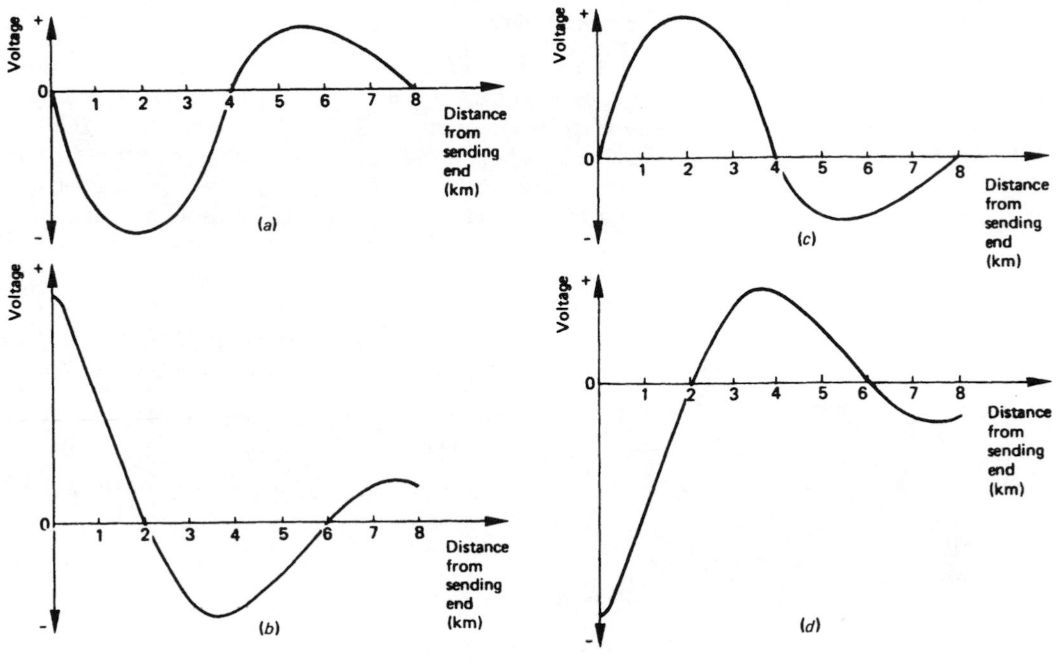

Fig. 16.11

Figure 16.11 shows the four line voltage waveforms and it illustrates clearly that a particular part of the applied voltage waveform moves along the line. For example, note that the negative peak voltage moves progressively through distances of approximately 1.5 km, 3.5 km, 5.5 km and 7.5 km from the sending-end of the line.

Example 16.10

A transmission line has a characteristic impedance of $600\angle-20°\ \Omega$ and a propagation coefficient of $0.04 + j0.35$ per kilometre. A voltage source of e.m.f. 12 V r.m.s. and impedance $600\angle-20°\ \Omega$ is connected across the sending end terminals of the line. If the line is 5 km long and correctly terminated calculate (*a*) the current in the load, (*b*) the power dissipated in the load, and (*c*) the distance from the sending end of the line at which the voltage is 5 V.

Solution

(*a*) Sending-end current $I_S = 12/(1200\angle-20°) = 10\angle 20°$ mA
Received current $I_R = I_S e^{-\gamma l} = 10e^{-0.2}\angle(20° - 100°)$

$I_R = 8.19\angle-80°$ mA (*Ans.*)

(*b*) Power in load $= (8.19 \times 10^{-3})^2 600\ \cos(-20°) = 37.82$ mW (*Ans.*)

(*c*) Sending-end voltage $E_S/2 = 6$ V. Hence,

$5 = 6e^{-0.04x}$ or $x = 4.56$ km (*Ans.*)

Example 16.11

A 3 km length of transmission line is terminated in its characteristic impedance of $Z_0 = 1200\angle -30°\ \Omega$. At a certain frequency the attenuation coefficient of the line is 3 dB/km and its phase change coefficient is 0.2 radians/km. If the sending-end voltage is 1 V calculate (*a*) the amplitude of the received current, (*b*) its phase relative to the sending end voltage, and (*c*) the power dissipated in the load.

Solution

(*a*) Input current $= I_S = 1/(1200\angle -30°) = 833\angle 30°\ \mu A$. Line loss $= 3 \times 3 = 9$dB which is a current ratio of 2.818. Therefore, the amplitude of the received current $= 833/2.818 \approx 296\ \mu A$ (*Ans.*)

(*b*) Phase shift $= 0.2 \times 3 = 0.6$ radians $= 34.4°$. Therefore I_R lags V_S by $(+30° - 34.4°) = 4.4°$ (*Ans.*)

(*c*) Load power $= I_R{}^2|Z_L|\cos\theta = (296 \times 10^{-6})^2 \times 1200\cos(-30°) = 91\ \mu W$ (*Ans*).

Example 16.12

A 3 km length of transmission line has a characteristic impedance of $600\angle -15°\ \Omega$ and a propagation coefficient of $0.25 + j0.4$ per km. The power developed in the matched load is 15.2 mW. Determine (*a*) the sending-end current, and (*b*) the e.m.f. of the applied voltage source.

Solution

(*a*) Power in load $= 15.2 \times 10^{-3} = I_R{}^2|Z_L|\cos\theta$,

$I_R{}^2 = (15.2 \times 10^{-3})/[600\cos(-15°)] = 26.227 \times 10^{-6}$,

and $I_R = 5.12$ mA. Hence, $5.12 \times 10^{-3} = I_S e^{-3(0.25+j0.4)}$,

$I_S = e^{0.75} \times (5.12 \times 10^{-3}\angle 3 \times 0.4) = 10.84\angle 68.8°$ mA (*Ans.*)

(*b*) $V_S = I_S Z_0 = 10.84 \times 10^{-3}\angle 68.8° \times 600\angle -15° = 6.5\angle 53.8$ V.

Therefore $E_S = 2V_S = 13\angle 53.8$ V (*Ans.*)

[Angles quoted with respect to the received current.]

Nominal T and π circuits

Series impedance circuit

If a transmission line is of very short electrical length, its capacitance and conductance may be neglected since only the series resistance and inductance will introduce noticeable loss. The input current I_S then is equal to the received current I_R. The received voltage V_R is equal to the input current I_S minus the voltage drop $I_R(R + jX_L)$ across the series impedance of the line. 50 Hz power lines are regarded as being very short whenever their length is less than about 80 km. The representation of an electrically short line is shown by Fig. 16.12(*a*) and its phasor diagram is given in Fig. 16.12(*b*).

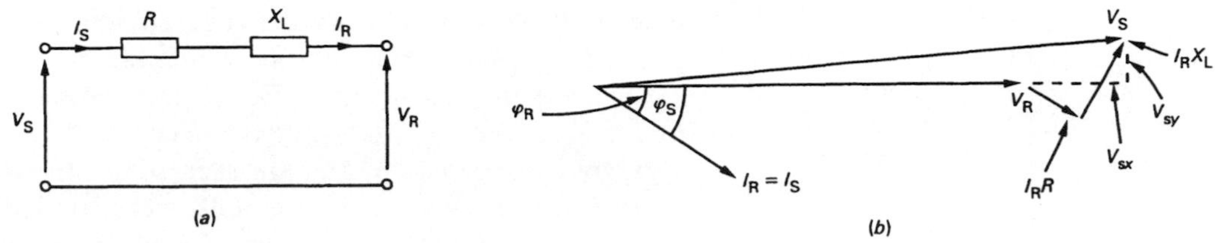

Fig. 16.12 Series impedance circuit (a) of a line; (b) phasor diagram

From Fig. 16.12(*a*), $I_S = I_R$, and

$$V_S = V_R + I_S(R + jX_L) \qquad (16.33)$$

From the phasor diagram,

$$V_{sx} = V_R + I_S R\cos\varphi_R + I_R X_L \sin\varphi_R, \text{ and}$$
$$V_{sy} = I_R X_L \cos\varphi_R - I_R R \sin\varphi_R.$$
$$V_S = \sqrt{(V_{sx}^2 + V_{sy}^2)}$$
$$\approx V_R + I_R R\cos\varphi_R + I_R X_L \sin\varphi_R \qquad (16.34)$$

The voltage regulation of the line is given by

$$(V_S - V_R)/V_R$$
$$\approx I_R(R\cos\varphi_R + X_L\sin\varphi_R)/V_R \qquad (16.35)$$

Example 16.13

A 50 Hz 3-phase line delivers 4 MW at 12 kV to a star-connected load over a distance of 15 km. The power factor of the load is 0.8 lagging, and the primary coefficients are $R = 2.6\ \Omega$ and $X_L = 5.8\ \Omega$. Calculate (*a*) the sending-end voltage, and (*b*) the voltage regulation.

Solution

(*a*) $V_R = 12\,000/\sqrt{3} = 6928$ V.

$I_R = (4 \times 10^6)/(\sqrt{3} \times 12 \times 10^3 \times 0.8) = 240.6$ A.

$V_S \approx 6928 + (240.6 \times 2.6 \times 0.8) + (240.6 \times 5.8 \times 0.6)$

$= 8265.8$ V per phase $= 14\,317$ V line (*Ans.*)

(*b*) Voltage regulation

$= [(8265.8 - 6928)/8265.8] \times 100 = 16.19\%$ (*Ans.*)

Nominal T And π Circuits

For lines of somewhat greater electrical length the shunt capacitance and conduotance can no longer be neglected. The four primary coefficients may be considered to exist at a few points along the line allowing the line to be represented by either a *nominal T* or a *nominal π* circuit. These circuits are shown by Figs. 16.13(*a*) and (*b*) respectively. In the nominal T circuit all of the capacitance and conductance is assumed to be concentrated at the mid-point of the line. In the nominal π circuit the total capacitance and conductance of the line are each split into two equal parts located at each end of the line. Often the conductance is small enough to be neglected. Problems involving 50 Hz power lines can be solved using either the nominal T or the nominal π circuit whenever the line length is somewhere in the range 75 to 250 km.

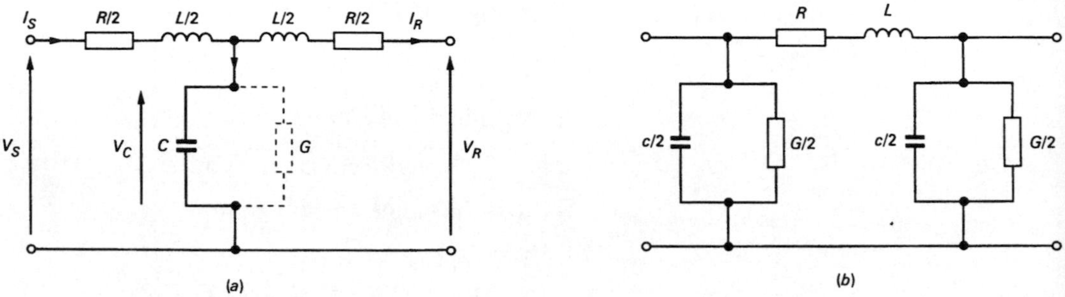

Fig. 16.13 Nominal circuits: (a) T; (b) π

The phasor diagram of the nominal T circuit is shown in Fig. 16.14 (neglecting the conductance to simplify the drawing), where V_C is the voltage across the capacitance C. From the diagram,

$$V_C = V_R + I_R Z_1/2, \quad \text{(where } Z_1 = R + j\omega L\text{)},$$
$$I_C = V_C/Z_2 = V_C Y, \quad \text{(where } Z_2 = 1/j\omega C\text{)}$$
$$I_S = I_R + I_C = I_R + V_C Y = I_R + (V_R + I_R Z_1/2)Y$$
$$= V_R Y + I_R(1 + YZ_1/2) \tag{16.36}$$

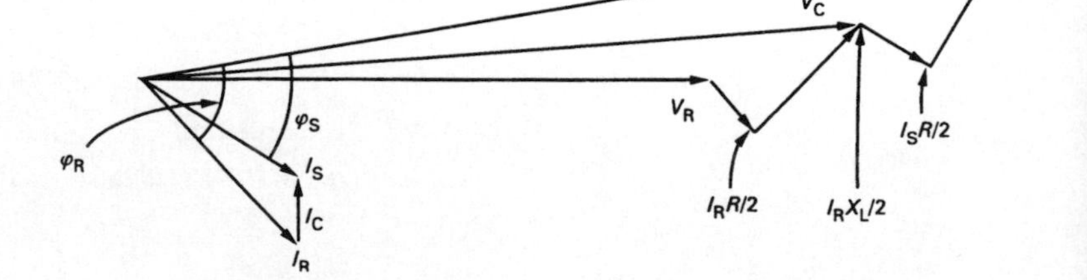

Fig. 16.14 Phasor diagram of nominal T circuit

Also,

$$V_S = V_C + I_S Z_1/2$$
$$= V_R + I_R Z_1/2 + [V_R Y + I_R(1 + YZ_1/2)]Z_1/2$$

or $$V_S = V_R(1 + YZ_1/2) + I_R Z_1(1 + YZ_1/4) \tag{16.37}$$

Example 16.14

A 75 km long three-phase line delivers 6 MW power at 50 kV to a load with a power factor of 0.8 lagging. The components of the nominal T circuit of the line are: $X_L = 30\ \Omega$, $Y_C = 2 \times 10^{-4}$ S, and $R = 75\ \Omega$. Calculate (*a*) the sending-end current, (*b*) the sending-end voltage and (*c*) the power factor at the sending end of the line.

Solution

(*a*) $V_R = (50 \times 10^3)/\sqrt{3} = 28.868$ kV.

$I_R = [(6 \times 10^6)/(\sqrt{3} \times 50 \times 10^3 \times 0.8)][0.8 - j0.6]$

$= 86.603(0.8 - j0.6) = 69.28 - j51.96$ A

$Z_1 = 75 + j30\ \Omega$, so $Z_1/2 = 37.5 + j15\ \Omega$,

$V_C = 28868 + (37.5 + j15)(69.28 - j51.96)$

$= 32245.4 - j909.3\ VI_C = V_C Y$

$= (j2 \times 10^{-4})(32245.4 - j909.3) = 0.182 + j6.45$ A.

$I_S = I_C + I_R = (0.182 + j6.45) + (69.28 - j51.96)$

$= 69.462 - j45.51$ A $= 83\sqrt{-33.2°}$ A (*Ans.*)

(*b*) $V_S = V_C + I_S Z_1/2$

$= 32245.4 - j909.3 + (69.462 - j45.51)(37.5 + j15)$

$= 35533 - j1574 = 35568\angle{-2.5°}$ V (*Ans.*)

(*c*) Sending-end power factor $= \cos[(-33.2 - 2.5)°] = 0.81$ (*Ans.*)

Example 16.15

A three-phase line supplies a balanced load of 15 MVA at 88 kV, 50 Hz with a power factor of 0.9 lagging. The primary coefficients of the line per phase are: $R = 50\ \Omega$, $X_L = 80\ \Omega$ and $X_C = j5 \times 10^{-4}$ S. Use a nominal T circuit to calculate (*a*) the sending-end voltage, and (*b*) the efficiency of the line.

Solution

(*a*) kVA per phase $= (15 \times 10^3)/3 = 5000$.

Phase voltage $= 88/\sqrt{3} = 50.8$ kV.

Load current

$I_L = (5 \times 10^6)/(50.8 \times 10^3) = 98.5\angle{-25.9°}$

$= 88.6 - j42.8$ A.

$V_C = (88.6 - j42.8)(25 + j40) + 50800$

$= 54790 + j2470 = 55\angle{2.5°}$ kV.

$I_C = (55 \times 10^3\angle{2.5°})(5 \times 10^{-4}\angle{90°}) = -1.2 + j27.5$ A.

$I_S = I_C + I_R = -1.2 + j27.5 + 88.6 - j42.8 = 87.4 - j15.3$

$= 88.7\angle{-9.9°}$ A.

$V_S = V_C + I_S Z_1/2 = 54790 + j2470 + 88.7\angle{-9.9°} \times (25 + j40)$

$= 57584 + j5584$

$= 57.854\angle{5.4°}$ kV per phase $= 100.21$ kV line (*Ans.*)

(*b*) Input power $= 57.854 \times 10^3 \times 88.7 \cos [5.4 - (-9.9)]°$

$= 4.9497$ MW per phase $= 14.849$ MW total.

Line efficiency $= [(15 \times 0.9)/14.849] \times 100 = 90.92\%$ (*Ans.*)

Example 16.16

Use the nominal π circuit to calculate (*a*) the sending-end voltage, (*b*) the sending-end current and (*c*) the sending-end power factor, of a single phase line that supplies a load of 15 MVA, 0.8 lagging power factor at 76.21 kV. The components of the π circuit are $Z = 25 + j62.8\ \Omega$ and $Y_1 = Y_2 = j392\ \mu S$.

Solution

(*a*) Load current

$I_R = (15 \times 10^6)/(76.21 \times 10^3) = 197\angle -36.9° = 158 - j118$ A

$I_{C2} = 76.21 \times 10^3 \times j392 \times 10^{-6} = j29.87$ A

$I_x = I_{C2} + I_R = j29.87 + 158 - j118 = 158 - j88.2 = 179\angle -29°$ A

$V_x = (25 + j62.8)(158 - j88) = 942 + j775 = 1220\angle 39.5°$ V

$V_S = V_R + V_x = 76.21 \times 10^3 + 942 + j775 = 77142 + j775$

$= 77.8\angle 6°$ kV (*Ans.*)

(*b*) $I_{C2} = 77.8 \times 10^3 \angle 6° \times 392 \times 10^{-6} \angle 90°$

$= 30.5\angle 96° = -3 + j30.2$ A

$I_S = I_{C1} + I_x = -3 + j30.2 + 158 - j88.2 = 155 - j58$

$= 164\angle -20.5°$ A (*Ans.*)

(*c*) Regulation $= (77.8 - 76.2)/77.8 = 0.21$ (*Ans.*)

Equivalent circuit of a line

When an electrically long transmission line is considered it is necessary to take into account the distributed nature of the primary coefficients. This requirement applies to power lines longer than about 250 km and lines that are required to transmit signals at audio frequencies and higher. Figure 16.15(*a*) shows a length l of transmission line and Fig. 16.15(*b*) shows its equivalent circuit.

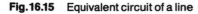

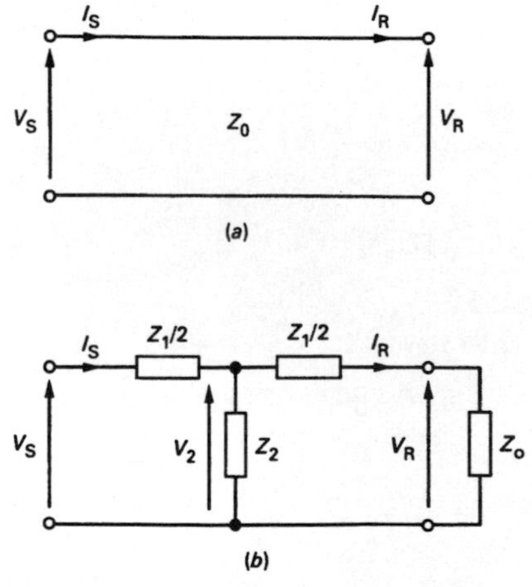

Fig. 16.15 Equivalent circuit of a line

$$V_S = I_S Z_0. \quad V_2 = I_S Z_0 - I_S Z_1/2,$$

$$I_R = I_S(Z_0 - Z_1/2)/(Z_0 + Z_1/2),$$

$$I_R/I_S = e^{-\gamma l} = (Z_0 - Z_1/2)/(Z_0 + Z_1/2),$$

$$e^{-\gamma l}(Z_0 + Z_1/2) = Z_0 - Z_1/2,$$

$$(Z_1/2)(1 + e^{\gamma l}) = Z_0(1 - e^{-\gamma l})$$

and

$$Z_1/2 = Z_0(1 - e^{-\gamma l})/(1 + e^{-\gamma l}) = Z_0 \tanh(\gamma l/2) \quad (16.38)$$

Also,

$$I_R = I_S Z_2/(Z_1/2 + Z_2 + Z_0),$$
$$I_R/I_S = e^{-\gamma l} = Z_2/(Z_1/2 + Z_2 + Z_0)$$
$$Z_2 e^{\gamma l} = Z_1/2 + Z_2 + Z_0, \quad Z_1/2 + Z_0 = Z_2(e^{\gamma l} - 1),$$
$$Z_0[1 + (1 - e^{-\gamma l})/(1 + e^{-\gamma l})] = 2Z_0/(1 + e^{-\gamma l}) = Z_2(e^{\gamma l} - 1)$$

or $$Z_2 = 2Z_0/(1 + e^{-\gamma l})(e^{\gamma l} - 1)$$
$$= 2Z_0/(e^{\gamma l} - e^{\gamma l}) = Z_0/\sinh \gamma l \qquad (16.39)$$

Similarly, for the equivalent π circuit,

$$Z_1 = Z_0 \sinh \gamma l \qquad (16.40)$$

and $$2Z_2 = Z_0 \coth \gamma l \qquad (16.41)$$

Transmission parameters

Nominal T and π circuits

To obtain the ABCD parameters of the nominal T (or π) circuit of a transmission line (*a*) use the parameters for a T or a π circuit given on page 251, and (*b*) substitute $Z = R + j\omega L$ and $Y = G + j\omega C$.

Equivalent T Circuit

For the equivalent T circuit the transmission parameters are: $A = D = \cosh \gamma$, $B = Z_0 \sinh \gamma$, and $C = \sinh \gamma / Z_0$.

Example 16.17

A transmission line has a characteristic impedance of $600\angle{-25^\circ}$ Ω and a propagation coefficient of $0.4 + j0.3$. Calculate its ABCD parameters.

Solution

$$A = D = \cosh \gamma = \cosh(0.4 + j0.3)$$
$$= \cosh 0.4 \cos 0.3 + j \sinh 0.4 \sin 0.3$$
$$= [(e^{0.4} + e^{-0.4})/2] \cos\ 17.2^\circ + j[(e^{0.4} - e^{-0.4})/2] \sin 17.2^\circ$$
$$= 1.031 + j0.121 = 1.038\angle 6.7^\circ \quad (\textit{Ans.})$$
$$B = Z_0 \sinh \gamma = 600\angle{-25^\circ}[\sinh 0.4 \cos 0.3 + j \cosh 0.4 \sin 0.3]$$
$$= 600\angle{-25^\circ}[0.41 \times 0.955 + j1.08 \times 0.296]$$
$$= 600\angle{-25^\circ} \times 0.506\angle 39.2^\circ = 303.6\angle 14.2^\circ \quad (\textit{Ans.})$$
$$C = \sinh \gamma / Z_0 = (0.506\angle 39.3^\circ)/(600\angle{-25^\circ})$$
$$= 843 \times 10^{-6}\angle 64.2^\circ\ \mathrm{S} \quad (\textit{Ans.})$$

Example 16.18

A 320 km long three-phase line delivers 10 MW power at a lagging power factor of 0.8 and 120 kV to a load. The ABCD parameters of the line are: $A = D = 0.94 + j0.039$, $B = 86.3 + j135.1\ \Omega$, and $C = (-1.48 + j8.62) \times 10^{-5}$ S. Calculate (*a*) the sending-end voltage, (*b*) the sending-end current, and (*c*) the sending-end power factor.

Solution

(*a*) Power factor $= 0.8 = \cos\theta$, $\theta = 36.9°$.

$$10 \times 10^6 = \sqrt{3} \times 120 \times 10^3 \times 0.8 I_R, \quad I_R = 60.14 \text{ A}$$

$$V_S = AV_R + BI_R$$

$$= (0.94 + j0.039)(69.28 \times 10^3) + (86.3 + j135.1)(60.14\angle -36.9°)$$

$$= 74.4\angle 4.7° \text{ kV} \quad (Ans.)$$

(*b*) $I_S = CV_R + DI_R = [(-1.48 + j8.62) \times 10^{-5} \times 69.28 \times 10^3]$

$$+ (0.94 + j0.039)(60.14\angle -36.9°)$$

$$= 52.53\angle -29.9° \text{ A} \quad (Ans.)$$

(*c*) Power factor $= \cos[4.7 - (-29.9)]° = 0.823$ (*Ans.*)

Exercises 16

16.1 A transmission line 22 km long has a propagation coefficient of $(0.05 + j\pi/10)$ per kilometre and a characteristic impedance of $600\angle -30°$ ohms. The line is connected to a generator of e.m.f. 10 V and internal impedance $600\angle -30°$ ohms and is correctly terminated at the far end. Calculate (*a*) the distance from the sending end at which the line voltage is 3.5 V, (*b*) the distance from the sending end at which the line current is in phase with the sending-end current, and (*c*) the power dissipated in the load.

16.2 Explain what is meant by the propagation coefficient of a line. A transmission line has a propagation coefficient of $(0.015 + j\pi/60)$ per kilometre and a characteristic impedance of $550\angle -25°$ ohms. The line is terminated in its characteristic impedance 20 km from the sending end. The voltage across the sending end terminals is 12 V. Calculate (*a*) the power dissipated in the load and (*b*) the phase difference between the sent and received voltages.

16.3 A voltage source of e.m.f. 20 V and internal impedance $500\angle 0°$ ohms is connected across the input terminals of a loss-free line of $Z_0 = 500\angle 0°$ ohms. (*a*) Calculate the power dissipated in the correctly terminated load. (*b*) If the phase-change coefficient of the line is $\pi/3$ radians/m calculate the phase velocity on the line when the frequency of the signal is 30 MHz.

16.4 A transmission line has a characteristic impedance of $600\angle -30°$ ohms and a propagation coefficient of $(0.1 + j\pi/2)$ per km. If the line is terminated in its characteristic impedance calculate the current in the load when the sending-end voltage is 10 V and the line is 12 km long.

16.5 A correctly terminated transmission line is 10 km long and has the following primary coefficients: $R = 20\ \Omega/\text{km}$, $L = 16\ \text{mH/km}$, $C = 0.045\ \mu\text{F/km}$ and $G \approx 0$ at $\omega = 5000$ R/s. A voltage generator with an e.m.f. of 12 V and an internal impedance of 500 Ω is connected across the sending-end terminals of the line. Calculate the magnitude and phase, relative to the sending-end voltage, of the load current.

16.6 Explain what is meant by the term 'characteristic impedance' when applied to a transmission line. A 2 m length of line is terminated in its characteristic impedance of 75 ohms and has a voltage of 10 V at 100 MHz applied to the sending-end terminals. If the line has negligible loss calculate (*a*) the wavelength of the signal on the line, (*b*) the power dissipated in the load, and (*c*) the current 1 m from the sending-end terminals.

16.7 A 3 km loss-free line has a characteristic impedance of 50 Ω. It is fed by a source of e.m.f. 6 V and impedance 50 Ω and is terminated by a 50 Ω resistor. If the wavelength of the signal on the line is 200 m calculate (*a*) the sent voltage, (*b*) the received voltage, (*c*) the power in the 50 Ω resistor, and (*d*) the phase-change coefficient of the line.

16.8 A transmission line has the following primary coefficients: $R = 25\ \Omega/\text{km}$, $L = 1\ \text{mH/km}$, $C = 0.05\ \mu\text{F/km}$ and $G \approx 0$. If the frequency is $10\,000/2\pi$ Hz calculate (*a*) the characteristic impedance, (*b*) the attenuation coefficient, and (*c*) the phase-change coefficient of the line.

16.9 A line has an attenuation coefficient of 6 dB/km, a phase-change coefficient of 0.62 R/km and a characteristic impedance of $1800\angle{-28^\circ}\ \Omega$ at a particular frequency. A voltage of 3 V is maintained across the input terminals of a 2 km length of this line. Calculate (*a*) the current in the load and (*b*) the length of the line in wavelengths.

16.10 A transmission line has the following primary coefficients: $R = 28\ \Omega/\text{km}$, $L = 0.6\ \text{mH/km}$, $C = 0.055\ \mu\text{F/km}$ and $G \approx 0$ at $\omega = 18\,000$ R/s. Calculate its velocity of propagation.

16.11 For any line $Z_0 = \sqrt{Z_{oc}Z_{sc}}$ and $\tanh \gamma l = \sqrt{(Z_{sc}/Z_{oc})}$. (*a*) Show that $\gamma l = 0.5 \log_e[(\sqrt{Z_{oc}} + \sqrt{Z_{sc}})/(\sqrt{Z_{oc}} - \sqrt{Z_{sc}})]$. (*b*) Show that $\gamma Z_0 = R + j\omega L$ and $\gamma/Z_0 = G + j\omega C$.

16.12 Calculate the components of the equivalent T network of a 10 km length of transmission line having a characteristic impedance of $280\angle{-30^\circ}\ \Omega$ and a propagation coefficient of $0.08\angle 40^\circ$ per km at a frequency of $5000/2\pi$ Hz.

16.13 A 50 Hz line can be represented by a nominal T circuit with $Z_1/2 = 35\angle 60^\circ\ \Omega$ and $Z_2 = 2000\angle 90^\circ\ \Omega$. Calculate (*a*) the sending-end current, (*b*) the sending-end voltage and (*c*) the sending-end power factor when the load is 270 A, 80 kV, with a lagging power factor of 0.8.

16.14 A three-phase 50 Hz line is 95 km long and has the following primary coefficients per phase: $R = 18\ \Omega$, $X_L = 38\ \Omega$, $X_C = 2400\ \Omega$, and $G \approx 0$. Calculate its transmission parameters.

16.15 A line has a characteristic impedance of $650\angle{-20^\circ}$ ohms and a propagation coefficient of $(0.025 + j\pi/3)$/km. If the line is 10 km long calculate the power dissipated in its correctly terminated load when the sending-end voltage is 60 V. State the phase difference between (*a*) the voltages at each end of the line and (*b*) the current at the load and the sending-end voltage.

16.16 A line has the following primary coefficients: $R = 40\ \Omega/\text{km}$, $L = 1.2\ \text{mH/km}$, $C = 0.1\ \mu\text{F/km}$, and $G = 1.3 \times 10^{-3}\ \text{S/km}$ at $\omega = 10\ 000\ \text{R/s}$. Calculate its characteristic impedance and its propagation coefficient.

16.17 A line has negligible losses and inductance and capacitance values of 1.5 μH and 12.2 pF per metre respectively. If this line is terminated by its characteristic impedance calculate (*a*) the power in the load when the sending-end voltage is 30 V and (*b*) the velocity of propagation.

16.18 A line has $Z_0 = 500\angle -15°$ ohms at 1500 Hz and is correctly terminated. If a voltage of 10 V is maintained across the sending-end terminals calculate the input power to the line. If the load power is 10 mW calculate the attenuation of the line.

16.19 A line has a propagation coefficient of $0.5 + j0.26$ and a characteristic impedance of $150\angle -30°\ \Omega$. Calculate its transmission parameters.

16.20 The primary coefficients of a transmission line are $R = 44\ \Omega/\text{km}$, $C = 0.07\ \mu\text{F/km}$, $L = 1\ \text{mH/km}$, and $G \approx 0$. Calculate (*a*) the characteristic impedance, (*b*) the propagation coefficient, (*c*) the attenuation coefficient and (*d*) the phase change coefficient at a frequency of $10\ 000/2\pi$ Hz.

16.21 A transmission line has a characteristic impedance of $2000\angle -30°\ \Omega$, an attenuation coefficient of 6 dB/km, and a phase change coefficient of 0.6 rad/km. Draw polar diagrams to show how the current and voltage vary along 3 km of this line when it is correctly terminated and 2 V is applied to its sending-end terminals.

16.22 A line has transmission parameters $A = D = 1\angle 1°$, $B = 40\angle 58°$ and $C = 0.4\angle 90°$ mS. Calculate (*a*) its sending-end voltage, (*b*) its sending-end current and (*c*) its efficiency, when the load is supplied with 100 MW power at 0.82 lagging power factor and a line voltage of 220 kV.

16.23 A line has $\gamma = 0.04 + j0.5$ and $Z_0 = 500\angle -12°\ \Omega$. Determine its ABCD parameters.

16.24 The nominal π circuit of a line has $Z = 120\angle 60°\ \Omega$ and $Y/2 = 2.5 \times 110^{-3}\angle 90°$ S. Calculate the characteristic impedance of the line.

16.25 A balanced star load of $300 + j100\ \Omega$ is connected to a three-phase line which has a total impedance per phase of $24 + j28\ \Omega$. Calculate (*a*) the total impedance per phase of the system, (*b*) the line current, (*c*) the load line voltage and (*d*) the phase angle between the load voltage and the applied voltage of 33 kV.

16.26 A transmission line has an input impedance of $348\angle 32°\ \Omega$ when the far-end terminals are open circuited and an input impedance of $1775\angle -45°\ \Omega$ when the far terminals are short circuited. Calculate (*a*) the characteristic impedance and (*b*) the propagation coefficient of the line.

17 Mismatched transmission lines

Very often a transmission line is operated with a load impedance that is not equal to its characteristic impedance. In some instances this state of affairs is intentional but in many others it is because the correct load impedance is not available for one reason or another. When a line is *mismatched*, the load impedance is unable to absorb all of the power incident upon it and so some of this power is reflected back towards the sending end of the line. If the sending-end terminals are matched to the source impedance, all of the reflected energy will be absorbed by the source impedance and there will be no further reflections. On the other hand, if the sending end of the line is also mismatched, some of the energy reflected by the load will be further reflected at the sending-end terminals and will be returned towards the load again. In such situations multiple reflections will take place.

Whether reflections on a line are desirable or not depends upon the intended use of the line. If the line is to be employed for the transmission of energy from one point to another, reflections are undesirable for a number of reasons that will be discussed later in this chapter. If, however, the intended application of the line is to simulate a component of some kind, reflections will be essential.

Reflected waves

Open-circuited loss-free line

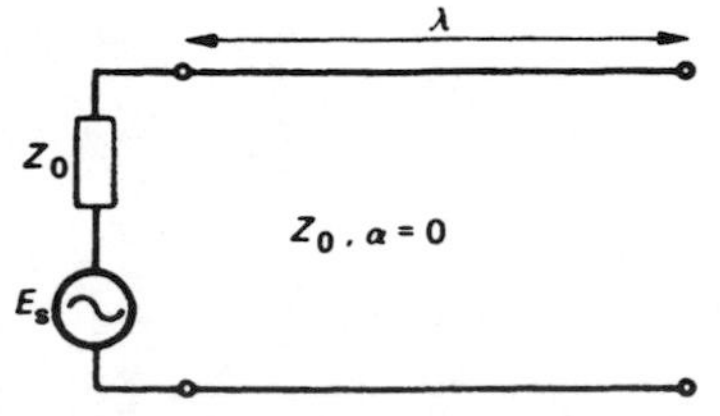

Fig. 17.1 Open-circuited loss-free line

Consider Fig. 17.1 which shows a loss-free line whose output terminals are open circuited. The line has an electrical length of one wavelength and its input terminals are connected to a source of e.m.f. E_S volts and impedance Z_0 ohms.

When the voltage source is first connected to the line, the input impedance of the line is equal to its characteristic impedance Z_0. An *incident* current of $E_S/2Z_0$ then flows into the line and an *incident* voltage $E_S/2$ appears across the input terminals. These are, of course, the same values of sending-end current and voltage that flow into a correctly terminated line. The incident current and voltage waves propagate along the line, being phase-shifted as they travel. Since the electrical length of the line is one wavelength, the overall phase shift experienced is 360°.

Since the output terminals of the line are open circuited, no current can flow between them. This means that all of the incident current must be reflected at the open circuit. The total current at the open circuit is the phasor sum of the incident and reflected currents, and since this must be zero the current must be reflected with 180° phase shift. The incident voltage is also totally reflected at the open circuit but with the zero phase shift. The total voltage across the open-circuited terminals is twice the voltage that would exist if the line were correctly terminated, and this means that the incident voltage is totally reflected with zero degrees phase shift. The reflected current and voltage waves propagate along the line towards its sending end, being phase shifted as they go. When the reflected waves reach the sending end, they are completely absorbed by the impedance of the matched source.

At any point along the line, the total current and voltage is the phasor sum of the incident and reflected currents and voltages. Consider Fig. 17.2(*a*). At the open circuit the phasors representing the incident and reflected currents are of equal length (since all the incident current is reflected) and point in opposite directions. The current flowing in the open circuit is the sum of these two phasors and is zero (as expected). At a distance of $\lambda/8$ from the open circuit, the incident current phasor is 45° leading, and the reflected current phasor is 45° lagging, the

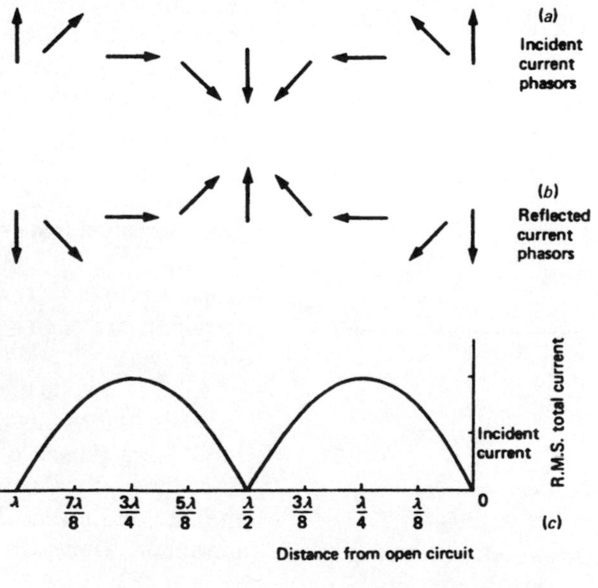

Fig. 17.2 R.M.S. current on an open-circuited loss-free line

open-circuit phasors. The lengths of the two phasors are equal since the line loss is zero but they are now 90° out of phase with one another. The total current at this point is $\sqrt{2}$ times the incident current. Moving a further $\lambda/8$ along the line, the incident and reflected current phasors have rotated, in opposite directions, through another 45° and are now again in phase with one another. The total current $\lambda/4$ from the open circuit is equal to twice the incident current. A further $\lambda/8$ along the line finds the two phasors once again at right angles to one another so that the total line current is again $\sqrt{2}$ incident current. At a point $\lambda/2$ from the end of the line, the incident and reflected current phasors have rotated to become in antiphase with one another and so the total line current is zero. Over the next half-wavelength of line the phasors continue to rotate in opposite directions, by 45° in each $\lambda/8$ distance, and the total line current is again determined by their phasor sum.

It is usual to consider the r.m.s. values of the total line current and then its phase need not be considered. The way in which the r.m.s. line current varies with distance from the open circuit is shown by Fig. 17.2(*c*) The points at which maxima (*antinodes*) and minima (*nodes*) of current occur are always the same and do not vary with time. Because of this the waveform shown in Fig. 17.2(*c*) is said to be a *standing wave*.

If, now, the voltages existing on the line of Fig. 17.1 are considered, the phasors shown in Fig. 17.3(*a*) are obtained. At the open-circuited output terminals, the incident and reflected voltage phasors are in phase with one another and the total voltage is twice the incident voltage. Moving from the open circuit towards the sending end of the line, the phasors rotate through an angle of 45° in each $\lambda/8$ length of line; the incident

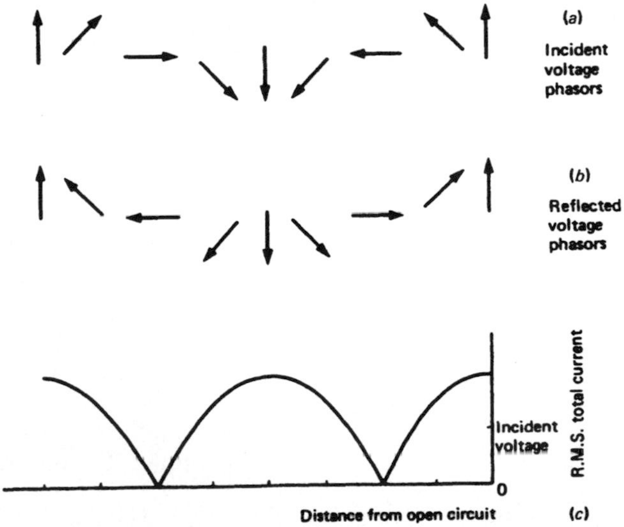

Fig. 17.3 R.M.S. voltage on an open-circuited loss-free line

voltage phasors rotate in the anticlockwise direction and the reflected voltage phasors rotate clockwise. The total voltage at any point along the line is the phasor sum of the incident and reflected voltages and its r.m.s. value varies in the manner shown in Fig. 17.3(*c*).

Two things should be noted from Figs 17.2 and 17.3. Firstly, the *voltage standing-wave* pattern is displaced by $\lambda/4$ from the current standing wave pattern, i.e. a current antinode occurs at the same place as a voltage node and vice versa. Secondly, the current and voltage values at the open circuit are repeated at $\lambda/2$ intervals along the length of the line; this remains true for any longer length of loss-free line.

Short-circuited loss-free line

When the output terminals of a loss-free line are short circuited, the conditions at the load are reversed. There can be no voltage across the short-circuited output terminals but the current that flows is twice as large as the current that would flow in a matched load. This means that at the short circuit the incident current is totally reflected with zero phase shift, and the incident voltage is totally reflected with 180° phase shift. Thus, Fig. 17.2(*c*) shows how the r.m.s. voltage on a short-circuited line varies with distance from the load, and Fig. 17.3(*c*) shows how the r.m.s. current varies along the line.

Line terminated in $Z_L \neq Z_0$

If the loss of the line is not negligibly small, the incident and reflected current and voltage waves will be attenuated as they propagate along the line. At any point distant x from the load the incident current and voltage can be written as $I_i e^{\gamma x}$ and $V_i e^{\gamma x}$ respectively, where I_i and V_i are their values at the receive terminals of the line. The magnitudes of the two waves increase exponentially with distance because the distance is measured from the load and, of course, the waves actually travel in the opposite direction.

The reflected current and voltage at any point distance x from the load can be written as $I_r e^{-\gamma x}$ and $V_r e^{-\gamma x}$ respectively, where I_r and V_r are their values at the load.

The total voltage at any point on the line is the sum of the incident and the reflected waves. Hence,

$$V_x = V_i e^{\gamma x} + V_r e^{-\gamma x} \tag{17.1}$$

Similarly, the total current at any point on the line is given by

$$I_x = I_i e^{\gamma x} + I_r e^{-\gamma x} \qquad (17.2)$$

The incident current is equal to V_i/Z_0 and the reflected current is equal to $-V_r/Z_0$. The minus sign is necessary because the reflected current is always in antiphase with the reflected voltage. This statement can be confirmed by looking at the phasors drawn in Figs 17.2(*b*) and 17.3(*b*). Therefore, equation (17.2) can be written as

$$I_x = V_i e^{\gamma x}/Z_0 - (V_r e^{-\gamma x})/Z_0 \qquad (17.3)$$

Open- and short-circuit terminations are the two extreme cases of a mismatched load and in most instances the load will have an impedance that is somewhere in between these values. Then, not all of the incident current and voltage will be reflected but only some fractional value. The fraction of the incident current and voltage reflected by the mismatched load is determined by the *reflection coefficient* of the load.

Reflection coefficients

Voltage reflection coefficient

The *voltage reflection coefficient* ρ_v of a mismatched load is the ratio

$$\rho_v = \text{(Reflected voltage)/(Incident voltage)} \qquad (17.4)$$

At the load $x = 0$ so that $V_R = V_i + V_r$ and $I_R = V_i/Z_0 - V_r/Z_0$. The load impedance Z_L is equal to the ratio (Load voltage)/(Load current), i.e.

$$Z_L = V_R/I_R = (V_i + V_r)/(V_i/Z_0 - V_r/Z_0)$$
$$= [(1 + V_+/V_i)Z_0](1 - V_r/V_i)$$

i.e. $Z_L/Z_0 = (1 + \rho_v)/(1 - \rho_v)$

Re-arranging $\rho_v = (Z_L - Z_0)/(Z_L + Z_0)$ (17.5)

Current reflection coefficient

The *current reflection coefficient* ρ_i is the ratio

$$\rho_i = \text{(Reflected current)/(Incident current)} \qquad (17.6)$$

Always $\rho_i = -\rho_v$ Hence

$$\rho_i = (Z_0 - Z_L)/(Z_0 + Z_L) \qquad (17.7)$$

Equations (17.5) and (17.7) give $\rho_v = +1$ and $\rho_i = -1$ for an open-circuited line and $\rho_v = -1, \rho_i = +1$ for a short-circuited line.

Thevenin's theorem

The expression for the current reflection coefficient can also be derived by applying Thevenin's theorem to the end of the line.

Assume that the line is matched at its sending end terminals so that the impedance measured at the open-circuited output terminals is the characteristic impedance of the line. Then, from Fig. 17.4(a) the current I_L flowing in the mismatched load Z_L is $V_{oc}/(Z_0 + Z_L)$. If the load impedance were equal to the characteristic impedance of the line, the incident current would flow in it. Hence, from Fig. 17.4(b),

$$I_i = V_{oc}/2Z_0$$

The load current is the phasor sum of the incident and the reflected currents at the load, i.e. $I_L = I_i + I_r$. Hence,

$$I_r = I_L - I_i = V_{oc}/(Z_0 + Z_L) - V_{oc}/2Z_0$$
$$= [V_{oc}(Z_0 - Z_L)]/[2Z_0(Z_0 + Z_L)]$$

Therefore,

$$\rho_i = I_r/I_i = (Z_0 - Z_L)/(Z_0 + Z_L) \qquad (17.7) \text{ (again)}$$

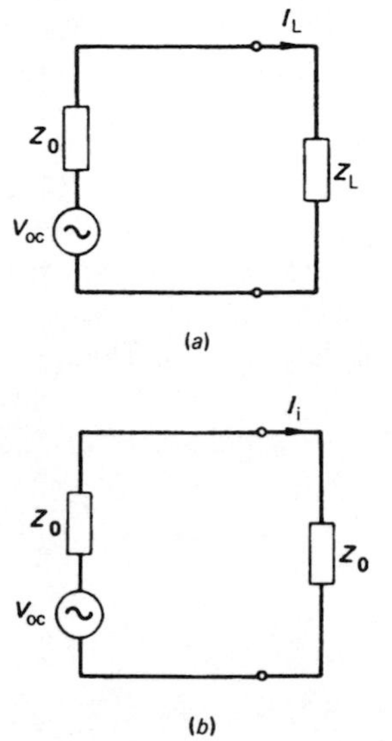

Fig. 17.4 Calculation of current reflection coefficient

Example 17.1

A mismatched line has a load voltage reflection coefficient of $0.5\angle -120°$ and an attenuation of 1.2 dB per wavelength at a certain frequency. Calculate the positions, measured from the load, and relative amplitudes of (a) the first two voltage minimums (nodes), and (b) the first two voltage maximums (antinodes).

Solution

(a) The first voltage minimum occurs when the incident and reflected voltage phasors are in antiphase with one another, i.e. when $180° = \beta x + 120° + \beta x$, or $\beta x = 30°$ or $\lambda/12$. The next node occurs 180° or $\lambda/2$ further down the line, i.e. when $\beta x = 210°$ or $7\lambda/12$. (*Ans.*)

Voltage maximums occur 90° from a voltage minimum, i.e. when $\beta x = 120°$ or $\lambda/3$ and 300° or $5\lambda/6$. (*Ans.*)

(b) Let the incident voltage at the load be 1 V. $30° = \lambda/12$, hence loss $= 1.2/12 = 0.1$ dB $= 1.012$ voltage ratio. $V_i = 1.012$ and $V_r = 0.989 \times 0.5 = 0.495$.

Amplitude $= 1.012 - 0.495 = 0.517$

$120° = \lambda/3$, loss $= 1.2/3 = 0.4$ dB $= 1.047$ voltage ratio. $V_i = 1.047$ and $V_r = 0.955 \times 0.5 = 0.478$

Amplitude $= 1.047 + 0.478 = 1.525$

$210° = 7\lambda/12$, loss $= (7 \times 1.2)/12 = 0.7$ db $= 1.084$ voltage ratio. $V_i = 1.084$ and $V_r = 0.923 \times 0.5 = 0.462$.

Amplitude $= 1.084 - 0.462 = 0.622$

$300° = 10\lambda/12$, loss $= (10 \times 1.2)/12 = 1$ dB $= 1.122$ voltage ratio. $V_i = 1.122$ and $V_r = 0.891 \times 0.5 = 0.446$.

Amplitude $= 1.122 + 0.446 = 1.568$ (*Ans.*)

Example 17.2

A line has a characteristic impedance of $600\angle{-20°}$ Ω and is terminated in a load of $300\angle 0°$ Ω. Calculate its current and voltage reflection coefficients.

Solution

$$600\angle-20° = 563.8 - j205.2$$

Therefore,

$$\rho_v = (300 - 563.8 + j205.2)/(300 + 563.8 - j205.2)$$
$$= (-263.8 + j205.2)/(863.8 - j205.2) = 0.38\angle 155.5° \quad \text{(Ans.)}$$
$$\rho_i = 0.38\angle(155.5 + 180)° = 0.38\angle{-24.5°} \quad (\textit{Ans.})$$

Voltage transmission coefficient

Reflections do not occur only at the mismatched load of a line but also at any discontinuity along that line. A discontinuity may arise at the junction of two lines of different characteristic impedances, or at a point at which an impedance has been connected across the line. Figure 17.5(*a*) and (*b*) shows examples of each form of discontinuity. Consider Fig. 17.5(*a*). The voltage at the discontinuity is equal to the sum of the incident and reflected voltages,

$$V_d = V_{i1} + V_{r1} = V_{i1} + \rho_v V_{i1}$$
$$= V_{i1} + [(Z_{02} - Z_{01})V_{i1}]/(Z_{01} + Z_{02})$$
$$= V_{i1}[2Z_{02}/(Z_{01} + Z_{02})]$$

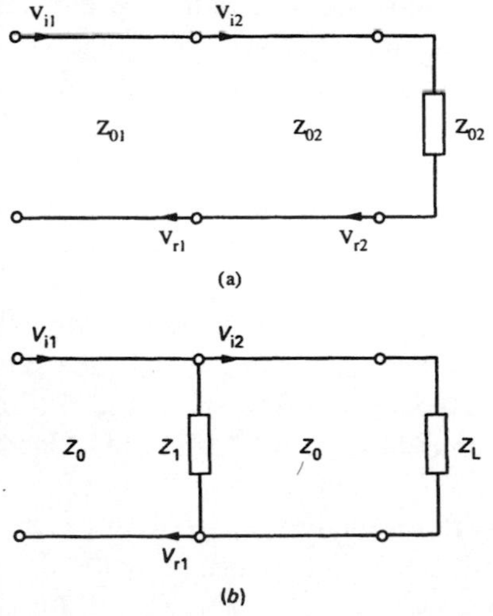

Fig. 17.5 Voltage transmission coefficient

This voltage is transmitted over the second line to the distant load.

The *voltage transmission coefficient* V_d/V_{i1} is

$$T_v = 2Z_{02}/(Z_{01} + Z_{02}) \tag{17.8}$$

The current transmitted past the junction of the two cables is

$$I_t = V_t/Z_{02} = 2V_{i1}/(Z_{01} + Z_{02}) = (2Z_{01}I_{i1})/(Z_{01} + Z_{02})$$

The *current transmission coefficient* I_t/I_{i1} is

$$T_i = (2Z_{01})/(Z_{01} + Z_{02}) \tag{17.9}$$

Example 17.3

A 3 km length of a line having a characteristic impedance of 1000 Ω and 2 dB/km attenuation is connected to another line, 1 km in length, that has a characteristic impedance of 600 Ω and a loss of 3 dB/km. 6 V is applied to the input terminals of the first line and the second line is correctly terminated in 600 Ω. Determine the voltage across the 600 Ω load.

Solution

At the junction of the two lines, the voltage transmission coefficient T_v is $(2 \times 600)/1600 = 0.75$. The loss of the first line is $2 \times 3 = 6$ dB and hence the incident voltage at the junction is $6/2 = 3$ V. Therefore

$$\text{Transmitted voltage} = 3 \times 0.75 = 2.25 \text{ V}$$

and Load voltage $= 2.25/\sqrt{2} = 1.59$ V (*Ans.*)

Voltage and current on a mismatched line

Figure 17.6 shows a line of length l with secondary coefficients Z_0 and γ that is terminated in a load impedance Z_L that is not equal to Z_0.

The line is matched to its source so that reflections will not occur at the sending end of the line. The incident voltage V_i at the sending-end terminals is the voltage that would appear if the line were correctly matched at the far end. (Since there are initially no reflections present at the sending end the input impedance is initially determined solely by the physical dimensions of the line, i.e. it is Z_0.) Therefore,

$$V_i = E_S Z_0 / 2Z_0 = E_S/2 \quad \text{and} \quad I_i = E_S/2Z_0$$

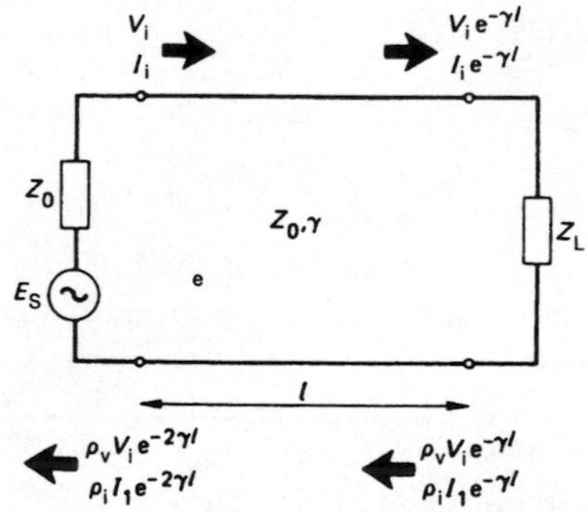

Fig. 17.6 Currents and voltages on a mismatched line

As the incident current and voltage waves travel along the line they are each subjected to both attenuation and phase lag. At the far end of the line the incident current and voltage waves are given by $I_i\, e^{-\gamma l}$ and $V_i\, e^{-\gamma l}$ respectively.

At the load $Z_L \neq Z_0$ and so both waves are reflected. The reflected voltage wave is found by merely multiplying the received incident voltage by the voltage reflection coefficient (see equation (17.4)). Hence the reflected voltage at the load is equal to $\rho_v V_i\, e^{-\gamma l}$. Similarly, the reflected current at the load is given by $\rho_i I_i\, e^{-\gamma l}$.

The reflected current and voltage waves now travel along the line towards the sending end and are subject to exactly the same attenuation and phase shift as the incident waves. At the sending end of the line the reflected current is $\rho_i I_i\, e^{-2\gamma l}$ and the reflected voltage is $\rho_v V_i\, e^{-2\gamma l}$.

The total voltage at any point on the line is the phasor sum of the incident and reflected voltages at that point. Hence

Sending-end voltage

$$V_S = V_i + \rho_v V_i\, e^{-2\gamma l} = V_i(1 + \rho_v\, e^{-2\gamma l}) \qquad (17.10)$$

Load voltage

$$V_R = V_i\,e^{-\gamma l} + \rho_v V_i\,e^{-\gamma l} = V_i\,e^{-\gamma l}(1+\rho_v) \qquad (17.11)$$

Voltage at distance x from the sending end:

$$V_x = V_i\,e^{-\gamma x} + \rho_v V_i\,e^{-\gamma(2l-x)} \qquad (17.12)$$

This can be written as

$$V_x = V_i\,e^{-\gamma x} + V_r\,e^{\gamma x} \qquad (17.13)$$

where V_r is the voltage reflected by the load.

Similarly, the current at any point along the line distance x from the receiving end is

$$I_x = I_i\,e^{-\gamma x} + I_r\,e^{\gamma x} = V_i\,e^{-\gamma x}/Z_0 - V_r\,e^{\gamma x}/Z_0 \qquad (17.14)$$

Example 17.4

At a particular frequency a transmission line is $\lambda/2$ long and has an attenuation of 3 dB. The characteristic impedance of the line is $600\angle 0^\circ$ ohms. A voltage source of e.m.f. 60 V and $600\angle 0^\circ$ ohms impedance is connected across the input terminals of the line. Calculate (*a*) the voltage across the 300 ohm load, (*b*) the voltage across the sending-end terminals.

Solution

(a) *Method A:* 3 dB = 0.345 nepers = αl

$\rho_v = (300-600)/(300+600) = -1/3$

$V_i = 60/2 = 30$ V. Substituting into equation (17.11),

$V_R = 30\,e^{-0.345}(1-1/3) = 14.16$ V (*Ans.*)

Method B: 3 dB is a voltage ratio of $1/\sqrt{2}$ and hence, referring to Fig. 17.7

$V_R = 30/\sqrt{2} - 30/(3\sqrt{2}) = 14.14$ V(*Ans.*)

The slight difference in the results arises because 3 dB is not exactly equal to 0.345 N and neither is it exactly equal to $1/\sqrt{2}$.

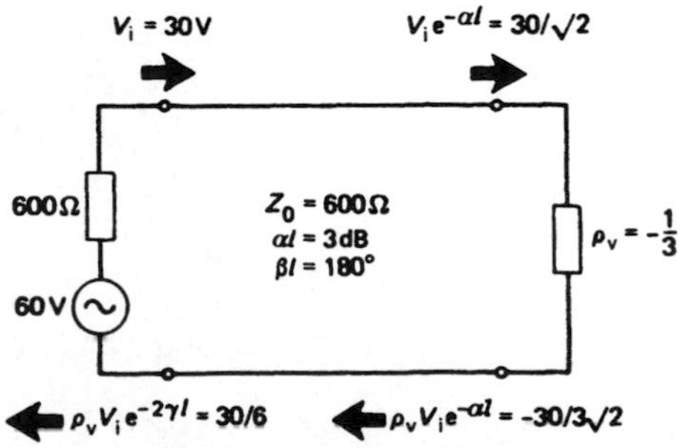

Fig. 17.7

(*b*) *Method A:* Substituting into equation (17.10),

$V_S = 30[1 - (1/3)e^{-0.69}] = 25$ V (*Ans.*)

Method B: From Fig. 17.7,

$V_S = 30 - 30/6 = 25$ V (*Ans.*)

General line equations

At the sending end of a line $x = 0$ in equations (17.13) and (17.14) and hence

$$V_S = V_i + V_r \quad \text{and} \quad I_S = V_i/Z_0 - V_r/Z_0$$
$$\text{or} \quad I_s Z_0 = V_i - V_r$$

Adding these two equations together gives $V_i = (V_S - I_S Z_0)/2$ and subtracting gives $V_r = (V_S - I_S Z_0)/2$. Equation (17.13) can now be written as

$$V_x = [(V_S + I_S Z_0)/2]\,e^{-\gamma x} + [(V_S - I_S Z_0)/2]\,e^{\gamma x}$$
$$= V_S[(e^{\gamma x} + e^{-\gamma x})/2] - I_S Z_0[(e^{\gamma x} - e^{-\gamma x})/2], \text{ or}$$
$$V_x = V_S \cosh\gamma x - I_S Z_0 \sinh\gamma x \tag{17.15}$$

Similarly,

$$I_x = I_S \cosh\gamma x - (V_S/Z_0)\sinh\gamma x \tag{17.16}$$

The current and voltage at any point on a line may also be expressed in terms of the current and voltage at the receiving terminals.

$$V_x = V_i\,e^{\gamma x} + V_r\,e^{-\gamma x} \tag{17.17}$$
$$I_x = V_i\,e^{\gamma x}/Z_0 - (V_r\,e^{-\gamma x}/Z_0) \tag{17.18}$$

Alternatively, at the receiving-end terminals $x = 0$ and $V_L = V_i + V_r$. Following the same steps as before gives:

$$V_x = V_L \cosh\gamma x + I_L Z_0 \sinh\gamma x \tag{17.19}$$
$$I_x = I_L \cosh\gamma x + (V_L/Z_0)\sinh\gamma x \tag{17.20}$$

Input impedance of a mismatched line

The impedance at any point on a line is the ratio of total voltage to total current at that point. Thus the input impedance of a line is the ratio V_S/I_S. Therefore

$$Z_S = V_S/I_S = [V_i(1 + \rho_v\,e^{-2\gamma l})]/[I_i(1 + \rho_i\,e^{-2\gamma l})]$$
$$Z_S = [Z_0(1 + \rho_v\,e^{-2\gamma l})]/(1 + \rho_i\,e^{-2\gamma l})$$
$$= [Z_0(1 + \rho_v\,e^{-2\gamma l})]/(1 - \rho_v\,e^{-2\gamma l}) \tag{17.21}$$

since $\rho_i = -\rho_v$.

Equation (17.21) is often written in another form. Re-writing by substituting for ρ_V.

$$Z_S = Z_0\left[\frac{Z_L + Z_0 + (Z_L - Z_0)e^{-2\gamma l}}{Z_L + Z_0 - (Z_L - Z_0)e^{-2\gamma l}}\right]$$

$$= Z_0\left[\frac{(Z_L + Z_0)e^{\gamma l} + (Z_L - Z_0)e^{-\gamma l}}{(Z_L + Z_0)e^{\gamma l} - (Z_L - Z_0)e^{-\gamma l}}\right]$$

$$= Z_0\left[\frac{Z_L(e^{\gamma l} + e^{-\gamma l}) + Z_0(e^{\gamma l} - e^{-\gamma l})}{Z_0(e^{\gamma l} + e^{-\gamma l}) + Z_L(e^{\gamma l} - e^{-\gamma l})}\right]$$

$$Z_S = Z_0\left[\frac{Z_L\cosh\gamma l + Z_0\sinh\gamma l}{Z_0\cosh\gamma l + Z_L\sinh\gamma l}\right] \qquad (17.22)$$

Example 17.5

A line has a characteristic impedance of 600 Ω and is $\lambda/2$ long at a particular frequency. If the loss of the line is 6 dB calculate its input impedance when the load is 2000 ohms.

Solution

6 dB is very nearly a current, or voltage, ratio of 2:1 and the overall phase shift is 180°. The voltage reflection coefficient is

$$\rho_V = (2000 - 600)/(2000 + 600) = 0.54$$

Therefore $\rho_i = 0.54\angle 180°$. The incident and reflected voltages and currents at each end of the line have been marked on Fig. 17.8. This method of laying out the calculations on a mismatched line is recommended.

Total sending-end voltage

$$V_S = V_i + 0.135\ V_i = 1.135\ V_i$$

Total sending-end current

$$I_S = I_i - 0.135 I_i = 0.865 I_i$$

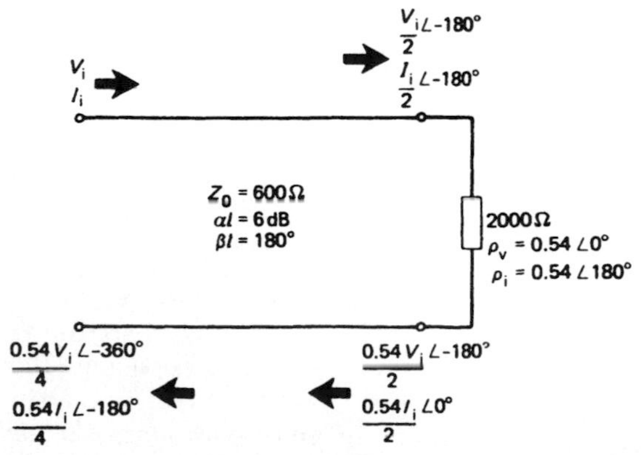

Fig. 17.8

Therefore, the input impedance of the line is

$$Z_S = V_S/I_S = (1.135V_i)/(0.865I_i) = 1.312 \times 600 = 787\ \Omega \quad (\textit{Ans.})$$

Example 17.6

A line of characteristic impedance $150\angle 0°\ \Omega$ is terminated in a resistor of 450 Ω. The line is 2.25 wavelengths long and has an attenuation of 3 dB. If there is an incident voltage of 1 V at the sending end calculate (*a*) the amplitudes of the reflected current and voltage at the sending end, and (*b*) the input impedance of the line.

Solution

$\rho_v = (450 - 150)/(450 + 150) = 0.5$ and $\rho_i = -0.5$. $\alpha l = 3$ dB = 0.345 N = $1/\sqrt{2}$ voltage ratio, and $\beta l = \pi/2$ radians.

(*a*) Total sending-end voltage

$$V_S = V_i(1 + 0.5 \times 0.5\angle 180°) = 0.75\ \text{V} \quad (\textit{Ans.})$$

Total sending-end current

$$I_S = I_i(1 - 0.5 \times 0.5\angle 180°) = 1.15 \times (1/150) = 8.33\ \text{mA} \quad (\textit{Ans.})$$

(*b*) *Method A:*

$$Z_S = V_S/I_S = (0.75\ V_I)/(1.25I_i) = (150 \times 0.75)/1125$$
$$= 90\ \Omega \quad (\textit{Ans.})$$

Method B: From equation (17.21),

$2\alpha l = 6$ dB $= 0.69$ N and $2\beta l = \pi$ radians.

$$Z_S = 150[(1 + e^{-(0.69+j\pi)})/(1 - e^{-(0.69+j\pi)})] = 90\ \Omega \quad (\textit{Ans.})$$

Method C: From equation (17.22),

$$Z_S = 150\left[\frac{450\cosh(0.345 + j\pi/2) + 150\sinh(0.345 + j\pi/2)}{150\cosh(0.345 + j\pi/2) + 450\sinh(0.345 + j\pi/2)}\right]$$

$$\cosh(0.345 + j\pi/2) = \cosh 0.345 \cos \pi/2 + j\sinh 0.345 \sin \pi/2$$
$$= j\sinh 0.345 = j[e^{0.345} - e^{-0.345}]/2 = j0.352$$

$$\sinh(0.345 + j\pi/2) = \sinh 0.345 \cos \pi/2 + j\cosh\ 0.345\ \sin\ \pi/2$$
$$= j1.06$$

$$Z_S = 150\left[\frac{(450 \times j0.352) + (150 \times j1.06)}{(150 \times j0.352) + (450 \times j1.06)}\right] = 90\ \Omega \quad (\textit{Ans.})$$

Example 17.7

A voltage source operates at 8 MHz with an internal resistance of 300 Ω. It is connected to a load by a 7.5 m length of 600 Ω loss-free line. Use the general line equations to calculate the resistance and reactance of the load for it to dissipate the maximum possible power. The velocity of propagation is 3×10^8 m/s.

Solution
From equation (17.22),

$$Z_S Z_0 \cosh \gamma l + Z_S Z_L \sinh \gamma l = Z_0 Z_L \cosh \gamma l + Z_0^2 \sinh \gamma l$$
$$Z_L(Z_S \sinh \gamma l - Z_0 \cosh \gamma l) = Z_0^2 \sinh \gamma l - Z_S Z_0 \cosh \gamma l,$$

or

$$Z_L = Z_0 \left[\frac{Z_0 \sinh \gamma l - Z_S \cosh \gamma l}{Z_S \sinh \gamma l - Z_0 \cosh \gamma l}\right] \tag{17.23}$$

$\lambda = 37.5$ m, hence the line is 7.5/37.5= 0.2λ long. Thus $\beta 1 = 72°$, and $\alpha l \approx 0$.

$$Z_L = 600 \left[\frac{(600 \times \mathrm{j} \sin 72°) - (300 \cos 72°)}{(300 \times \mathrm{j} \sin 72°) - (600 \cos 72°)}\right]$$
$$= 933 - \mathrm{j}411\ \Omega$$

This is the output impedance of the line. For a matched line the load impedance should be equal to this but for maximum power transfer the load impedance must be the conjugate of the output impedance. Therefore,

$$Z_L = 933 + \mathrm{j}411\ \Omega \quad (\textit{Ans.})$$

Line mismatched at both ends

When the source impedance as well as the load impedance is not equal to the characteristic impedance of the line, reflections will take place at both ends of the line. The reflected waves will travel back and forwards over the line until the line attenuation reduces their amplitudes to a negligibly small value. Figure 17.9 shows a line with characteristic impedance Z_0 fed by a voltage source of impedance Z_S and terminated by an impedance Z_L. The incident voltage V_S' and the many reflected voltages are also shown, where ρ_s and ρ_L are the voltage reflection coefficients at the source and at the load respectively. The total voltage at any point on the line is given by the sum to infinity of the incident and reflected voltages at that point. The voltages at any point on

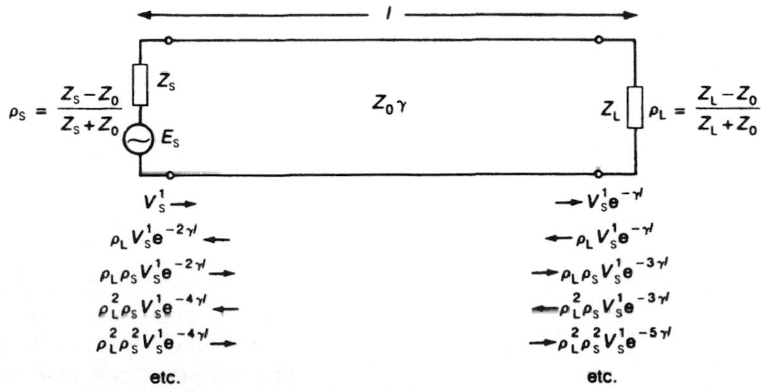

Fig. 17.9 Line mismatched at both ends

the line form a geometric progression and so their sum to infinity is equal to (initial term)/(1 – common ratio).

Sending-end voltage

Source-to-load direction

Initial term = V'_S and common ratio = $\rho_L\rho_S e^{-2\gamma l}$. Hence,

$$V_1 = V'_S(1 - \rho_L\rho_S\, e^{-2\gamma l})$$

Load-to-source direction

Initial term = $\rho_L V'_S\, e^{-2\gamma l}$ and common ratio = $\rho_L\rho_S\, e^{-2\gamma l}$. Hence

$$V_2 = (\rho_L V'_S\, e^{-2\gamma l})/(1 - \rho_L\rho_S\, e^{-2\gamma l})$$

The sending-end voltage is the sum of these voltages. Thus:

$$V_S = V'_S(1 + \rho_L\, e^{-2\gamma l})/(1 - \rho_L\rho_S\, e^{-2\gamma l}) \qquad (17.24)$$

Load voltage

Source-to-load direction

Initial value = $V'_S\, e^{-2\gamma l}$ and common ratio = $\rho_L\rho_S\, e^{-2\gamma l}$. Hence

$$V_1 = V'_S\, e^{-2\gamma l}/(1 - \rho_L\rho_S\, e^{-2\gamma l})$$

Load-to-source direction

Initial value = $\rho_L \mathrm{V}'_S\, e^{-\gamma l}$ and common ratio = $\rho_L\rho_S\, e^{-2\gamma l}$. Hence,

$$V_2 = (\rho_L V'_S\, e^{-\gamma l})/(1 - \rho_L\rho_S\, e^{-2\gamma l})$$

The load voltage is the sum of V_1 and V_2, i.e.

$$V_L = [V'_S\, e^{-\gamma l}(1 + \rho_L)]/(1 - \rho_L\rho_S\, e^{-2\gamma l}) \qquad (17.25)$$

Example 17.8

A line has a characteristic impedance of 50 Ω and negligible loss and is $\lambda/2$ in length. It is supplied by a voltage source of e.m.f. 6 V and internal resistance 100 Ω and is terminated by a load of resistance 72 Ω. Calculate the load voltage.

Solution

$$\rho_S = (100 - 50)/(100 + 50) = 1/3;$$
$$\rho_L = (72 - 50)/(72 + 50) = 0.18$$
$$V'_S = (6 \times 50)/(50 + 100) = 2 \text{ V}$$
$$e^{-\gamma l} = \angle 180° \quad \text{and} \quad e^{-2\gamma l} = \angle 360° = \angle 0°$$

Therefore,

$$V_L = [2\angle 180°(1 + 0.18)]/[1 - (1/3 \times 0.18\angle 0°)]$$
$$= -2.36/0.94 = 2.51 \text{ V} \quad (\textit{Ans.})$$

Low-loss lines

A *low-loss line* has an attenuation αl that is small enough to have $\alpha l \approx 1$ and $\sinh \alpha l \approx \alpha l$ If the load terminals are short circuited the input impedance is, from equation (17.22), $Z_S = Z_0 \tanh \gamma l$. With $\cosh \alpha l \approx 1$ and $\sinh \alpha l \approx \alpha l$ this equation becomes

$$Z_S = Z_0 \left[\frac{\sinh(\alpha + j\beta)l}{\cosh(\alpha + j\beta)l}\right]$$

$$= Z_0 \left[\frac{\sinh \alpha l \ \cosh j\beta l + \sinh j\beta l \ \cosh \alpha l}{\cosh \alpha l \ \cosh j\beta l + \sinh \alpha l \ \sinh j\beta l}\right]$$

$$= Z_0 \left[\frac{\sinh \alpha l \ \cos \beta l + j \cosh \alpha l \sin \beta l}{\cos \alpha l \ \cos \beta l + j \sinh \alpha l \sin \beta l}\right],$$

$$\text{or } Z_S \approx Z_0 \left[\frac{\alpha l \cos \beta l + j \sin \beta l}{\cos \beta l + j\alpha l \sin \beta l}\right] \qquad (17.26)$$

λ/4 Length of line

When $l = \lambda/4$, $\cos\beta l = 0$ and $\sin\beta l = 1$. Then

$$Z_S = Z_0(j/j\alpha l) = Z_0/\alpha l \qquad (17.27)$$

λ/2 Length of line

When $l = \lambda/2$, $\cos\beta l = -1$ and $\sin\beta l = 0$. Now

$$Z_S = Z_0(-\alpha l/-1) = Z_0 \alpha l \qquad (17.28)$$

Example 17.9

A 50 Ω low-loss line has a loss of 1.5 dB at a certain frequency. Calculate its input impedance when the far end terminals are short circuited and the length of the line is (*a*) $\lambda/4$ and (*b*) $\lambda/2$.

Solution

(*a*) $Z_S = 50/(1.5/8.686) \approx 290\ \Omega$ (*Ans.*)
(*b*) $Z_S = 50 \times 1.5/8.686 \approx 8.6\ \Omega$ (*Ans.*)

Loss-free lines

A line whose attenuation is very small is often said to be a loss-free line. When the loss of a line is small enough to be neglected

$$\alpha l \simeq 0 \quad \text{and} \quad \gamma l = \mathrm{j}\beta l$$

Then

$$\cosh \gamma l = \cosh \mathrm{j}\beta l = \cos \beta l$$
$$\sinh \gamma l = \sinh \mathrm{j}\beta l = \mathrm{j} \sin \beta l$$

The use of these approximations considerably simplifies the application of equation (19.22).

There are several special cases of importance.

Short-circuited line

If the far-end terminals of a line are short circuited, $Z_L = 0$. Equation (17.22) then becomes

$$Z_S = Z_0[(Z_0 \sinh \gamma l)]/(Z_0 \cosh \gamma l)$$
$$= Z_0[(\mathrm{j}Z_0 \sinh \beta l)]/(Z_0 \cos \beta l)$$
$$Z_S = \mathrm{j}Z_0 \tan \beta l \qquad (17.29)$$

This means that the input impedance of a loss-free line short circuited at its output terminals is a pure reactance whose magnitude and sign depend upon the electrical length of the line. Such a line can therefore be used to simulate either an inductor or a capacitor or, if a resonant length ($\lambda/2$ or $\lambda/4$) is employed, a series- or parallel-tuned circuit.

Example 17.10

A length of 50 Ω loss-free short-circuited line is to be used to simulate an inductance of 30 nH at a frequency of 300 MHz. Calculate the necessary length of the line.

Solution

The required inductive reactance is

$$2\pi \times 3 \times 10^8 \times 30 \times 10^{-9} = \mathrm{j}56.55\ \Omega$$

Therefore, from equation (17.29)

$$\mathrm{j}56.55 = \mathrm{j}50 \tan \beta l \qquad 1.13 = \tan \beta l = \tan(2\pi l/\lambda)$$

Therefore, $2\pi l/\lambda = 48.5° = 0.846$ radians and so

$$l = 0.846/2\pi\lambda = 0.135\lambda \quad (\textit{Ans.})$$

But $\lambda = (3 \times 10^8)/(3 \times 10^8) = 1$ m and hence $l = 13.5$ cm (*Ans.*)

Open-circuited line

Now $Z_L = \infty$ and equation (17.22) becomes

$$Z_S = -jZ_0 \cot \beta l \tag{17.30}$$

The open-circuited line is not often used to simulate a component because at very high frequencies an open-circuited line will tend to radiate energy.

$\lambda/4$ length of line

When the length of a loss-free line is exactly $\lambda/4$, $\gamma l = j\beta l = j\pi/2$. Then,

$$\cosh \gamma l = \cos \beta l = \cos \pi/2 = 0 \quad \text{and}$$
$$\sinh \gamma l = j \sin \beta l = j \sin \pi/2 = j$$

Substituting these values into equation (17.22) gives

$$Z_S = Z_0(jZ_0/jZ_L) = Z_0^2/Z_L \tag{17.31}$$

This result can just as easily be obtained from equation (17.21). Thus

$$Z_S = Z_0[(1 + \rho_v e^{-2j\beta l})/(1 - \rho_v e^{-2j\beta l})]$$
$$= Z_0[(1 + \rho_v \angle 180°)/(1 - \rho_v \angle 180°)]$$
$$= Z_0[1 - (Z_L - Z_0)/(Z_L + Z_0)]/$$
$$[1 + (Z_L - Z_0)/(Z_L + Z_0)]$$
$$= Z_0(Z_0/Z_L)$$

or $\quad Z_S = Z_0^2/Z_L$ (17.31) (again)

This result means that a $\lambda/4$ section of loss-free line can be employed to match two impedances together. The necessary characteristic impedance for the $\lambda/4$ matching section can be obtained from equation (17.32).

$$Z_0 = \sqrt{(Z_S Z_L)} \tag{17.32}$$

The device is often known as a $\lambda/4$ transformer.

Example 17.11

A low-loss transmission line of characteristic impedance 600 ohms is to be used to connect a 600 Ω source to a 300 Ω load. Calculate the required characteristic impedance for the $\lambda/4$ line matching section.

Solution
From equation (17.32)

$$Z_0 = \sqrt{(600 \times 300)} = 424\ \Omega \quad (Ans.)$$

The arrangement is shown by Fig. 17.10.

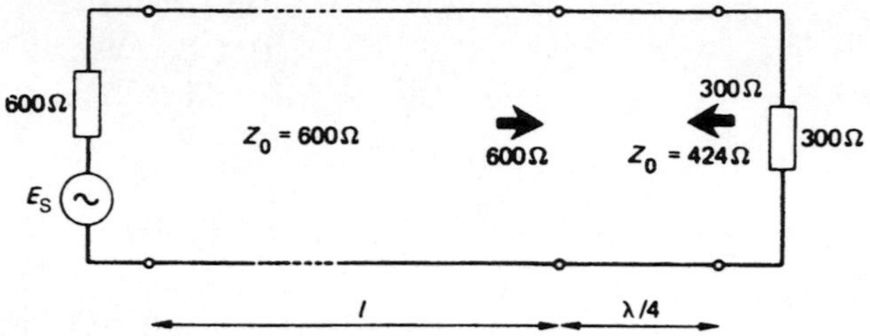

Fig. 17.10

λ\2 Length of line

For this length of loss-free line

$$\cosh \gamma l = \cos \beta l = \cos \pi = -1$$
$$\sinh \gamma l = \mathrm{j} \sin \beta l = \mathrm{j} \sin \pi = 0$$

so $$Z_S = Z_0(-Z_L/-Z_0) = Z_L \qquad (17.33)$$

Thus, the input impedance of a $\lambda/2$ length of loss-free line is equal to the load impedance. This means that the line will act like a transformer with a 1 : 1 turns ratio.

λ/8 length of line

A $\lambda/8$ length of loss-free line introduces a phase lag of 45° and hence $\cos \beta l = \sin \beta l = 1/\sqrt{2}$ and, from equation (17.22)

$$Z_S = Z_0[(Z_L/\sqrt{2} + \mathrm{j}Z_0/\sqrt{2})/(Z_0/\sqrt{2} + \mathrm{j}Z_L/\sqrt{2})]$$
$$Z_S = Z_0 \angle \tan^{-1}[(Z_0/Z_L) - (Z_L/Z_0)] \qquad (17.34)$$

Thus the input impedance of a $\lambda/8$ length of loss-free line is equal to the characteristic impedance of the $\lambda/8$ section.

Example 17.12

A 60 m length of loss-free 400 Ω line is supplied by a 100 Ω, 15 MHz voltage source of internal impedance 100 + j100 Ω. The far-end terminals of the line are open circuited and connected across the line are two 400 Ω resistors, R_1 and R_2. R_1 is connected 15 m from the end of the line and R_2 is connected 20 m from the receiving-end terminals. Determine the current flowing in each resistor.

Solution
Wavelength $\lambda = (3 \times 10^8)/(15 \times 10^6) = 20$ m. The line is hence 3λ in length and the resistors are connected 0.75λ and 1λ from the end of the line.

R_1 is $3\lambda/4$ from the end of the line and hence the line impedance at this point is $Z_S = 400^2/\infty = 0\ \Omega$. This means that R_1 is effectively short circuited. Therefore,

Current in $R_1 = 0$ (*Ans.*)

R_2 is connected $\lambda/4$ from the short circuit, or λ from the open circuit and hence the line impedance at that point is $Z_S = 400^2/0 = \infty$. This means that R_2 is effectively in parallel with an open circuit and so the line is correctly terminated at this point.

$$I_S = I_{R_2} = 100/(100 + 400 + j100) = 0.192 - 0.038$$
$$= 0.196\angle{-11.2^\circ}\ \text{A}$$

The phase shift from the sending-end terminals to R_2 is 0° and hence $I_{R_2} = 0.196\angle{-11.2^\circ}$ A (*Ans.*)

Standing waves

When reflections occur at the mismatched load of a transmission line, both incident and reflected current and voltage waves propagate along the line. At any point on the line the total voltage or current is the phasor sum of the incident and reflected waves at that point.

If the r.m.s. values of the total voltage and current are plotted to a base of distance from the load, *standing waves* are obtained. Examples of these have already been plotted for low-loss open- and short-circuited lines (see Figs 17.2 and 17.3).

When the voltage and current reflection coefficients have a magnitude other than unity, not all of the incident energy is reflected by the load. The maximum voltage V_{max} on the line occurs each time the incident and reflected voltages are in phase with one another. The minimum voltage V_{min} occurs when the incident and reflected voltages are in anti-phase with one another. Therefore, for a low-loss line

$$V_{max} = V_i(1 + |\rho_v|) \qquad V_{min} = V_i(1 - |\rho_v|) \qquad (17.35)$$

Only the magnitude of ρ_v appears in equation (17.35) because its angle is a factor in determining *where* the two waves are in phase, or in anti-phase, with one another.

Often the points of maximum voltage are known as *antinodes* and the points where the minimum voltage occurs are called *nodes*.

The ratio of maximum voltage to minimum voltage *or* the ratio of minimum voltage to maximum voltage is known as the *voltage standing-wave ratio* or the VSWR. The symbol for VSWR is *S*.

Both definitions of the VSWR are in common use, although the former is perhaps the more popular. No confusion should result since the VSWR is always either greater than, or less than, unity depending upon which of the two definitions is employed. Consider as an example a loss-free line having a characteristic impedance of 50 Ω that is terminated by a resistance of 150 Ω. Suppose a voltage source at the sending end of the line supplies an incident voltage of 10 V.

The voltage reflection coefficient is

$$\rho_v = (150 - 50)/(150 + 50) = 0.5\angle 0^\circ$$

The incident voltage is reflected by the load with zero phase change and so the maximum voltage occurs at the load, and at $\lambda/2$ intervals from the load. The maximum voltage is $10(1 + 0.5) = 15$ V. The minimum voltage occurs at *odd* multiples of $\lambda/4$ from the load and is equal to $10(1 - 0.5) = 5$ V.

The voltage standing-wave pattern for this line is shown in Fig. 17.11. The incident and reflected currents are $10/50 = 0.2$ A and $0.2 \times 0.5 = 0.1$ A respectively, Hence, $I_{max} = 300$ mA and $I_{min} = 100$ mA. The maximum current occurs at the same points as the minimum voltage and vice versa; this is because $\rho_i = -\rho_v$.

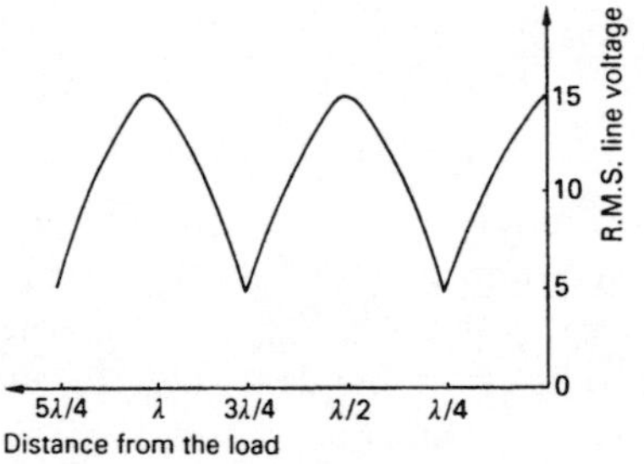

Fig. 17.11

The impedance at any point on the line is the ratio of total voltage to total current at that point. This means that the impedance of the line will vary both above and below the characteristic impedance at different points on the line.

At the load

$$Z_{max} = V_{max}/I_{min} = 15/0.1 = 150\ \Omega$$

$\lambda/4$ from the load

$$Z_{min} = V_{min}/I_{max} = 5/0.3 = 16.67\ \Omega$$

Note that from equation (17.31) $Z_{min} = 50^2/150 = 16.67$ Ω. Figure 17.11 shows that the maximum and minimum values of the line impedance repeat at $\lambda/2$ intervals along the line.

$$\text{VSWR} = S = V_{max}/V_{min} = 15/5 = 3$$

Relationship between VSWR and voltage reflection coefficient

$$\text{VSWR} = S = V_{max}/V_{min} = [V_i(1 + |\rho_v|)]/[V_i(1 - |\rho_v|)]$$

$$= (1 + |\rho_v|)/(1 - |\rho_v|) \qquad (17.36)$$

$$|\rho_v| = (S - 1)/(S + 1) \qquad (17.37)$$

Example 17.13

A low-loss line of characteristic impedance 1000 Ω is terminated in a load of 800 − j100 Ω. Calculate the VSWR on the line.

Solution

$$\rho_v = [(800 - j100) - 1000]/[(800 - j100) + 1000]$$
$$= (-200 - j100)/(1800 - j100)$$

or $|\rho_v| = 0.124$ Therefore,

$$S = (1 + 0.124)/(1 - 0.124) = 1.283 \quad (Ans.)$$

The presence of a standing wave on a line which is used to transmit energy from one point to another is undesirable for a number of reasons. These reasons are as follows:

- Maximum power is transferred from a source to a load when the load impedance is equal to the characteristic impedance. When a load mismatch exists some of the incident power is reflected at the load and so the transfer efficiency is reduced.

  The power in the incident wave is V_i^2/Z_0 and the power reflected by the load is V_r^2/Z_0. Therefore, the power P_L dissipated in the load is

  $$P_L = (V_i^2 - V_r^2)/Z_0 = [(V_i + V_r)(V_i - V_r)]/Z_0$$
  $$= (V_{max}V_{min})/Z_0$$
  $$P_L = V_{max}^2/(SZ_0) \tag{17.38}$$

  If the ratio P_L/P_i is calculated for various values of S it will be found that the ratio remains fairly constant at approximately unity, provided S is not higher than about 1.5.
- The power reflected by a mismatched load will propagate, in the form of voltage and current waves, towards the sending end of the line. The waves will be attenuated as they travel and so the total line loss is increased.
- At a voltage maximum the line voltage may be anything up to twice as great as the incident voltage. Since the breakdown voltage of the dielectric between the conductors of a line must not be exceeded, this factor limits the maximum possible peak value of the incident voltage and hence limits the maximum power that the line is able to transmit.

Measurement of VSWR

The VSWR on a mismatched line can be determined by measuring the maximum and minimum voltages on the line. Often the measurement is carried out using an instrument known as a *standing-wave indicator*. The measurement of VSWR not only shows up the presence of any reflections on a line but it also offers a way of determining the value of an unknown load impedance.

The measurement procedure is briefly as follows. The VSWR is measured and the distance in wavelengths from the load to the nearest voltage minimum is found. Referring to Fig. 17.12, suppose that the VSWR on the line is S. Then

$$|\rho_v| = (S-1)/(S+1)$$

Suppose that the first voltage minimum is at a distance of x from the load. For this point to be a voltage minimum the incident and reflected phasors must be in *anti-phase* with one another. This means that the total phase lag between V_i and V_r, i.e. $(\beta x + \theta + \beta x)$, must be equal to π radians. Hence,

$$\theta = \pi - 2\beta x \qquad (17.39)$$

Once the value of $\rho_v \angle\theta$ has been found it must be substituted into equation (17.5) in order to calculate the value of Z_L.

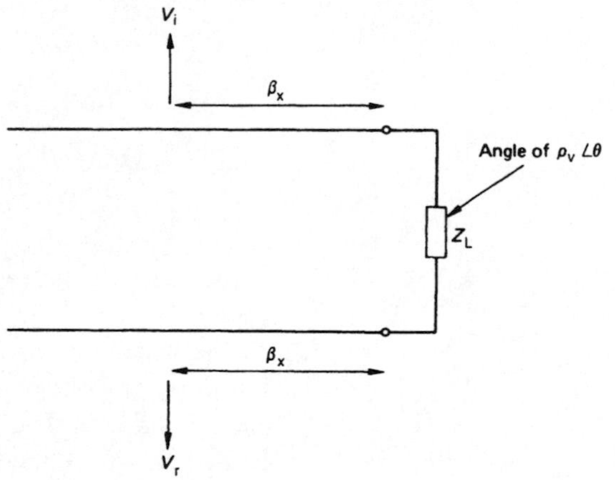

Fig. 17.12

Example 17.14

In a measurement of the VSWR on a loss-free line it was found that the ratio V_{max}/V_{min} was 5 and the first voltage minimum from the load occurred 20 cm from the load. If the characteristic impedance of the line is 50 Ω and the frequency of the test oscillator is 300 MHz calculate the magnitude and angle of the load impedance.

Solution

From equation (17.37)

$$|\rho_v| = 4/6 = 0.67 \qquad \lambda = (3\times10^8)/(3\times10^8) = 1 \text{ m}$$

Hence, 20 cm $= 0.2\lambda$ and $\beta x = 0.4\pi$. From equation (17.39)

$$\angle\theta = \pi - 0.8\pi = 0.2\pi = 36^\circ \text{ lag}$$

Therefore, from equation (17.5)

$$0.67\angle{-36^\circ} = (Z_L - 50)/(Z_L + 50)$$
$$Z_L = 75 - j108\ \Omega \quad (Ans.)$$

The calculation of Z_L using equation (17.5) is rather lengthy and tedious and it is usual to employ a graphical aid, known as a *Smith Chart* (p. 347), to greatly simplify the problem. If, however, the load impedance is known to be purely resistive an easier method becomes available. When $R_L > R_0$ and $S > 1$

$$\rho_v = (R_L - R_0)/(R_L + R_0)$$

so that

$$S = \frac{1 + (R_L - R_0)/(R_L + R_0)}{1 - (R_L - R_0)/(R_L + R_0)}$$
$$= R_L/R_0$$

Hence

$$R_L = SR_0 \tag{17.40}$$

When $R_L < R_0$,

$$S = \frac{1 - (R_L - R_0)/(R_L + R_0)}{1 + (R_L - R_0)/(R_L + R_0)}$$
$$= R_0/R_L$$

Hence

$$R_L = R_0/S \tag{17.41}$$

Impedance matching on lines

In nearly all cases when a transmission line is used to convey energy from one point to another, the requirement is for the load to be matched, or nearly matched, to the line. Whenever possible the load impedance is selected to satisfy this requirement. Very often, however, this is not possible and in such instances some kind of matching device is used.

A $\lambda/4$ length of loss-free line can be used to match a line to its mismatched load, (see equation 17.32). Alternately, *stub matching* may be employed The impedance and hence the admittance of a mismatched line varies as the distance from the load is increased. At some particular distance x from the load the admittance of the line will be equal to the characteristic admittance ($Y_0 = 1/Z_0$) in parallel with some value of susceptance. Thus, at this particular point in the line $Y_{in} = Y_0 \pm jB$. If the susceptance can be cancelled out by another susceptance of equal magnitude but opposite sign, the resulting total admittance would be equal to Y_0. The line would then be matched at this point.

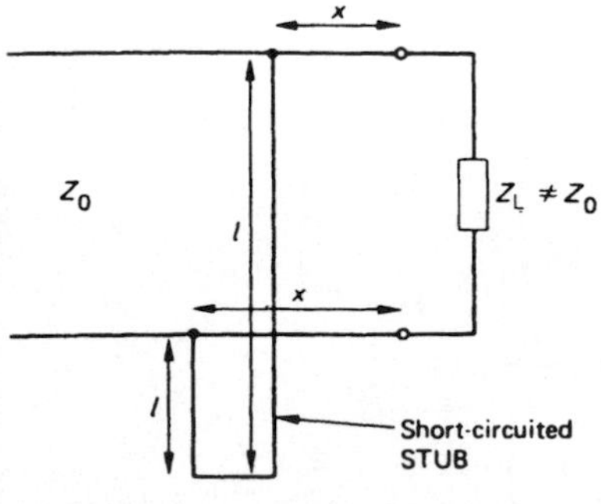

Fig. 17.13 Stub matching

The necessary shunt admittance is provided by a short-circuited line of length l – known as a *stub* – connected in parallel with the line at a distance x from the load (see Fig. 17.13). The design of a stub matching system is best performed using a Smith Chart.

Digital signals on lines

When a voltage step V supplied by a matched source is first applied to a line of characteristic impedance R_0 the current flowing into the line is $I_S = V/2R_0$ and the incident voltage is $V_S = V/2$. The current and voltage steps travel along the line until they reach the load. If the load resistance is not equal to the characteristic resistance of the line some, or all, of the incident voltage and current steps will be reflected and will travel back towards the sending end of the line.

- *Open-circuited line.* When the current step reaches the open circuit it is totally reflected with the opposite polarity so that the total current at the open circuit is zero. The voltage step is also reflected but with zero polarity change so that the voltage at the open circuit is twice as large as the incident value. Hence the line current is reduced to zero and the line voltage is increased to V volts as the reflected steps travel towards the source. The waveforms on the line are shown by Fig. 17. 14. When the voltage and current steps arrive back at the matched source they are completely absorbed and there are no further reflections. Then the current and voltage on the line are everywhere equal to 0 amps and to V volts respectively.
- *Short-circuited line.* Again, the incident current and voltage steps are $V/2R_0$ amps and $V/2$ volts respectively. At the short circuit the current step is totally reflected with no change in polarity and the voltage step is reflected with its polarity reversed This means that the current flowing in the short circuit is double the incident current and the voltage across the short circuit is zero. The reflected current and voltage

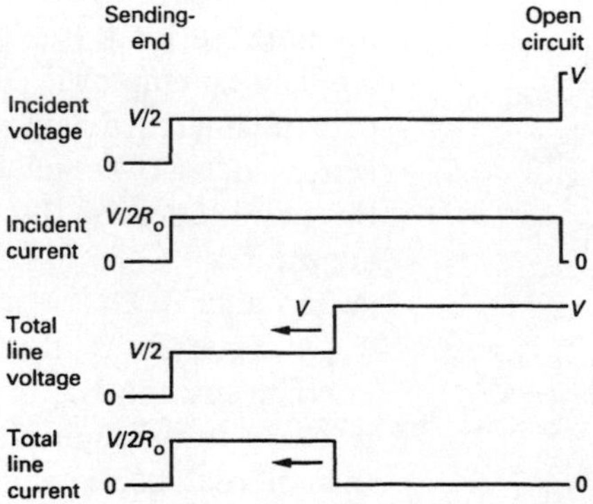

Fig.17.14

steps travel back towards the sending end of the line and as they propagate they increase the line current to V/R_0 and reduce the line voltage to zero. The waveforms on the short-circuited line are shown by Fig. 17.15. When the current and voltage steps arrive at the sending end of the line they are completely absolved by the matched source resistance. Then the line voltage is at all points zero and the line current is everywhere equal to V/R_0.

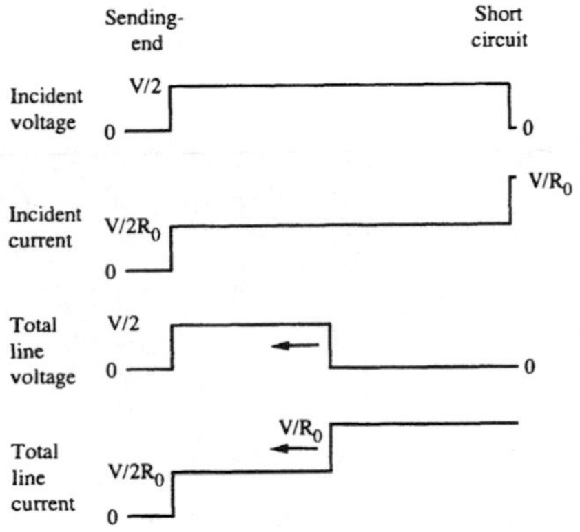

Fig. 17.15

- *Mismatched line*. If the line is neither short circuited nor open circuited but terminated in a mismatched load resistance, the fraction of the incident current and incident voltage steps that are reflected will be less than unity. If, for example, $\rho_V = 0.5$ then the maximum line voltage on the line will be $1.5 \times V/2 = 0.75\ V$ and the minimum current will be $V/4R_0$.
- *Mismatch at both ends*. If a digital line is mismatched at the source as well as at the load reflections will occur at both ends of the line. Consider a line of characteristic resistance R_0 that is supplied by a voltage source V of internal resistance $R_0/2$

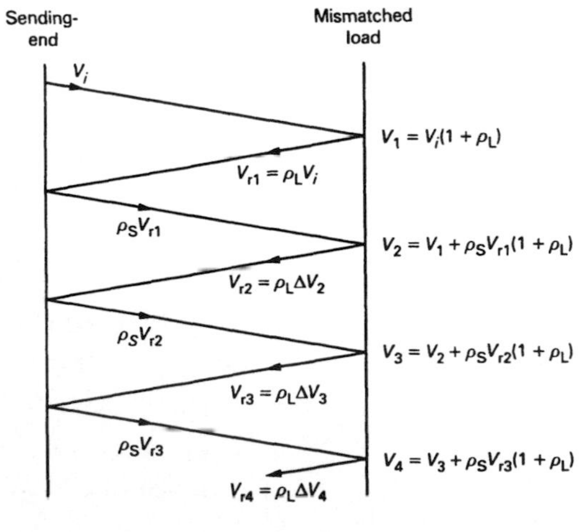

Fig. 17.16

and terminated by a load of resistance $2R_0$. The source voltage reflection coefficient is $\rho_S = (R_0/2 - R_0)/(R_0/2 + R_0) = -1/3$, and the load voltage reflection coefficient is $\rho_I = (2R_0 - R_0)/(2R_0 + R_0) = 1/3$. The incident current is $V/(R_0/2 + R_0) = (2V)/(3R_0)$, and the incident voltage is $(2VR_0)/(3R_0) = 2V/3$. The incident and reflected voltages on the line are shown by the lattice diagram of Fig. 17.16. The total voltage at each end of the line is obtained by summing the incident and reflected voltages at each end of the line.

Voltage surges on power lines

When a voltage surge that is travelling along a power line arrives at the junction of two lines of different characteristic impedances some of the incident voltage and current is reflected while the remainder which is not reflected is transmitted into the second cable.

Example 17.15

A 25 kV voltage surge travels in a line of *surge impedance* 500 Ω and arrives at the junction of two lines of surge impedances 700 Ω and 200 Ω. Find the voltages and currents in each line immediately after the surge reaches the junction.

Solution
Impedance at junction = (700 × 200)/900 = 155 Ω. Therefore the reflected voltage is 25[(155 − 500)/(155 + 500)] = −13.17 kV. Transmitted voltage = 25 − 13.17 = 11.83 kV (*Ans*).

Current in 700 Ω line = $(11.83 \times 10^3)/700 = 16.96$ A (*Ans.*)

Current in 200 Ω line = 16.96 × 700/200 = 59.15 A (*Ans.*)

Current in 500 Ω line = 16.96 + 59.36 = 76.05 A (*Ans.*)

Example 17.16

Two lines of characteristic impedance 400 Ω and 600 Ω are connected in cascade. A 100 kV voltage surge travels from the 400 Ω line into the 600 Ω line. Calculate the voltages and currents in the two lines shortly after reflections have occurred at the junction.

Solution

$\rho_v = (600 - 400)/(600 + 400) = 0.2$

Reflected voltage = 0.2 × 100 = 20 kV (*Ans.*)

Transmitted voltage = 100 + 20 = 120 kV (*Ans.*)

Reflected current = $(20 \times 10^3)/400 = 50$ A (*Ans.*)

Transmitted current = $(120 \times 10^3)/600 = 200$ A (*Ans.*)

Repeated reflections

Often a power line consists of the tandem connection of two or more cables of different surge impedances and reflections will occur at each junction along the line. Consider a line made up from three cables of surge impedances R_{01}, R_{02} and R_{03}. A voltage surge travelling along the line is first partially reflected from junction A between R_{01} and R_{02} and then from junction B between R_{02} and R_{03}. A pulse reflected from junction B will be reflected again when it reaches junction A after a time delay equal to l_1/v_p seconds, where l_1 is the length of the middle cable.

Example 17.17

A power line made up from three cables of surge resistances 100 Ω, 200 Ω and 150 Ω has a 150 kV pulse enter the first cable. Determine the voltages on the line once the surge has reached the third cable.

Solution

At junction A the reflected wave is 150[(200 − 100)]/300 = 50 kV. This reflected voltage increases the voltage on the first line to 200 kV. A 200 kV surge voltage is transmitted into cable two and at junction B the reflected wave is

$$200[(150-200)/(150+200)] = -29 \text{ kV}$$

This reflected voltage reduces the voltage on the middle line to 171 kV as it travels back towards the sending end of the line. The voltage transmitted into the third cable is 171 kV.

When the wave reflected from junction B arrives at junction A it will be re-reflected; the re-reflected voltage being

$$-29[(100-200)/(100+200)] = 9.67 \text{ kV}$$

This voltage (*i*) increases the voltage on the middle line to $200 - 29 + 9.67 = 180.67$ kV as it propagates, and (*ii*) increases the voltage on the third line to 180.67 kV. The situation at this point is shown by Fig. 17.17. The process is repeated over and over until eventually the line voltages settle down to a final value, e.g the voltage on the middle line is somewhere in between 171 and 180.67 kV.

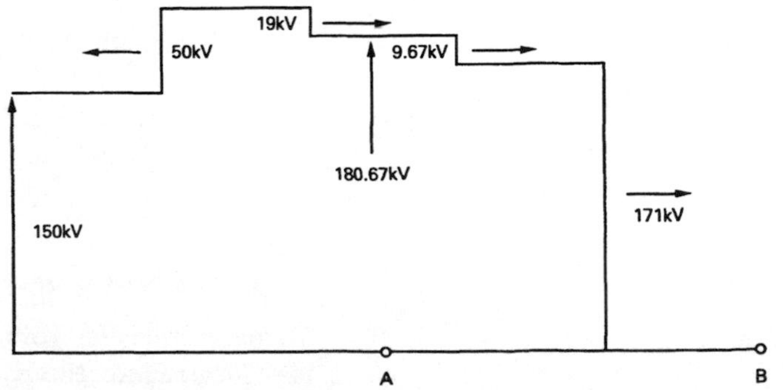

Fig. 17.17

The Smith chart

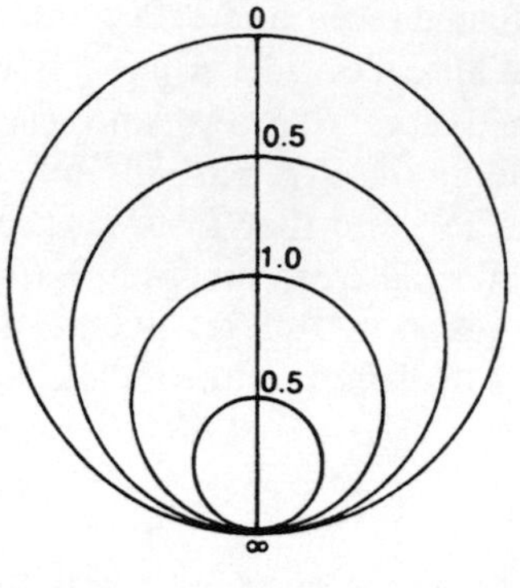

(a)

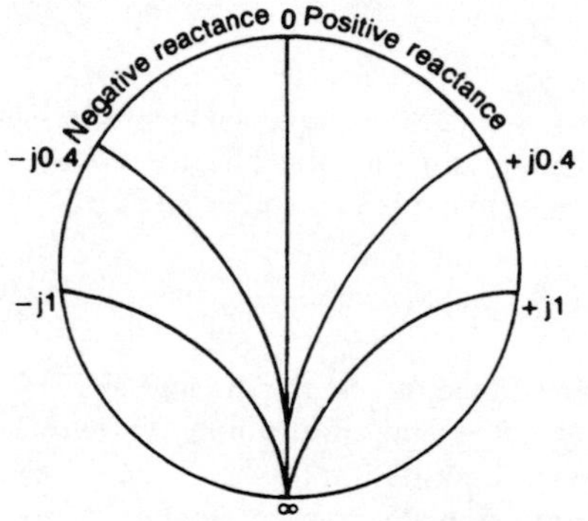

(b)

Fig. 17.18 (a) Real parts and (b) imaginary parts of normalized impedance Z/Z_0

The Smith chart is a plot of normalized impedance against the magnitude and angle of voltage reflection coefficient. It can be used in the solution of many problems involving transmission lines and waveguides, and r.f. circuits since its use can often greatly simplify a problem. The Smith chart consists of: (*a*) a real axis with values which vary from zero to infinity, with unity in the centre; (*b*) a series of circles centred on the real axis; and (*c*) a series of arcs of circles that start from the infinity point on the real axis. This is shown by Figs 17.18(*a*) and (*b*). The circles represent the real parts of the normalized impedances i.e. R/Z_0, and the arcs represent the imaginary parts of the normalized impedances, i.e. $\pm jX/Z_0$.

Figure 17.19 shows a full Smith chart. In addition to the circles and the arcs of circles representing normalized resistance and reactance the edge of the chart is marked with scales of (*a*) angle of reflection coefficient (in degrees), and (*b*) distance (in wavelengths). Movement around the edge of the chart in the *clockwise* direction corresponds to movement along the line towards the source; conversely, anti-clockwise movement around the chart represents movement along the line towards the load. It should be noticed that a complete circle around the edge of the chart represents a movement of one-half wavelength ($\lambda/2$) along the line.

Any value of normalized impedance can be located on the chart. Suppose that $Z = 100 + j100\ \Omega$ and $Z_0 = 50\ \Omega$. The normalized impedance is then $z = (100 + j100)/50 = 2 + j2$; this is shown plotted in Fig. 17.19 by the point marked as A. Similarly, an impedance $10 - j20\ \Omega$ is normalized to $(10 - j20)/50 = 0.2 - j0.4$ and is represented on the chart by the point B. Admittances can also be plotted on the Smith chart; the admittance must first be normalized by dividing it by the characteristic admittance Y_0 of the line. Thus, if $Y = 0.02 - j0.03\ S$ and $Z_0 = 50\ \Omega$ then

$$y = (0.02 - j0.03)/0.02 = 1 - j1.5$$

and is plotted on the chart as the point C.

Use of the Smith chart

Essentially, the Smith chart deals with lines of negligible loss; the effect of any line attenuation can be taken into account by the use of a separate scale; this is not considered here.

Voltage reflection coefficient

To determine the voltage reflection coefficient produced by a load impedance the impedance must first be normalized and

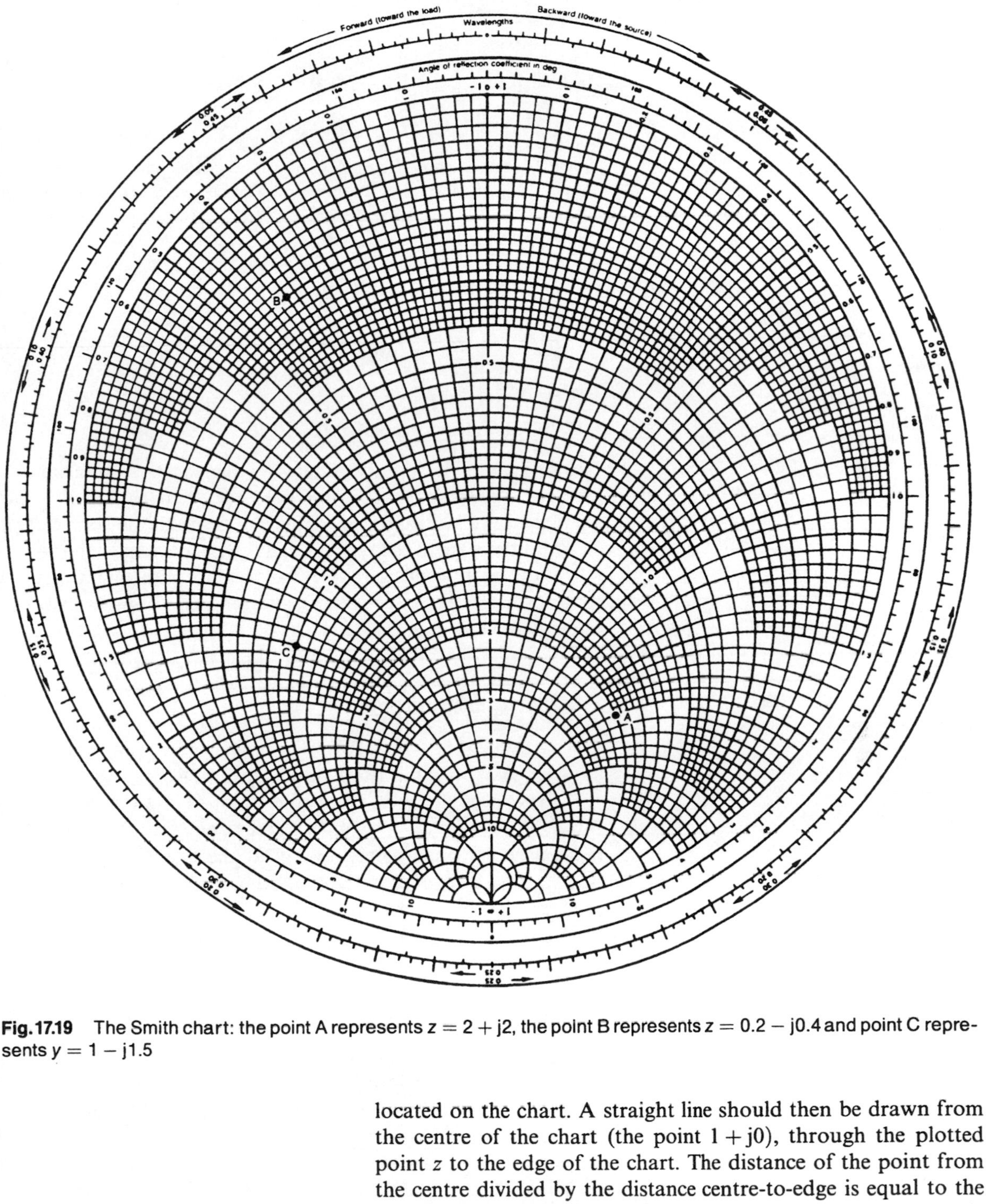

Fig. 17.19 The Smith chart: the point A represents $z = 2 + j2$, the point B represents $z = 0.2 - j0.4$ and point C represents $y = 1 - j1.5$

located on the chart. A straight line should then be drawn from the centre of the chart (the point $1 + j0$), through the plotted point z to the edge of the chart. The distance of the point from the centre divided by the distance centre-to-edge is equal to the magnitude of the voltage reflection coefficient. The phase angle of the voltage reflection coefficient is read from the scale at the edge of the chart.

Example 17.18

Calculate the load voltage reflection coefficient of a line having $Z_0 = 50\ \Omega$ and $Z_L = 100 + j20\ \Omega$.

Solution
The load impedance $Z_L = 100 + j20\ \Omega$ and the characteristic impedance $Z_0 = 50\ \Omega$ so that $z_L = 2 + j0.4$. This point is plotted on the Smith chart as shown by Fig. 17.20. The line drawn from the point $1 + j0$ through z_L passes through the voltage reflection coefficient angle scale at 14°. The distance from the point $(1 + j0)$ to z_L is 30.6 mm and the distance from $(1 + j0)$ to the edge of the chart is 85 mm and so

$$|\rho_v| = 30.6/85 = 0.36$$

Therefore,

$$\text{Voltage reflection coefficient} = 0.36\angle 14^\circ \quad (\textit{Ans.})$$

The procedure is slightly different when a load admittance is involved. First, locate the normalized admittance y_L on the chart and then draw a straight line from y_L, through the centre of the chart, to the edge of the chart. Then the magnitude of the voltage reflection coefficient $|\rho_v|$ is equal to the ratio (distance y_L to centre)/(distance edge to centre), and the angle of y_L is read from the scale at the edge of the chart.

Voltage standing-wave ratio

To calculate the VSWR on a line locate the normalized impedance z_L (or the normalized admittance y_L) on the chart and draw a circle, centred on the point $(1 + j0)$, which passes through the point z_L. The VSWR is then equal to the value of the real axis at the point where it is cut by the circle. Two values will be obtained: one greater than, and the other less than, unity and each will be the reciprocal of the other. Referring to Fig. 17.20 the VSWR is about 2.1 or 0.47.

Input impedance of a length of line

To determine the input impedance of a length of loss-free line locate the normalized load impedance on the chart and draw an arc of a circle, centred on the centre of the chart, i.e. the point $(1 + j0)$, moving clockwise. The length of the arc, measured on the wavelength scale, should be equal to the electrical length of the line. The normalized input impedance of the line is then given by the location of the end of the arc. The procedure can be reversed if the input impedance is known and the load impedance is to be determined.

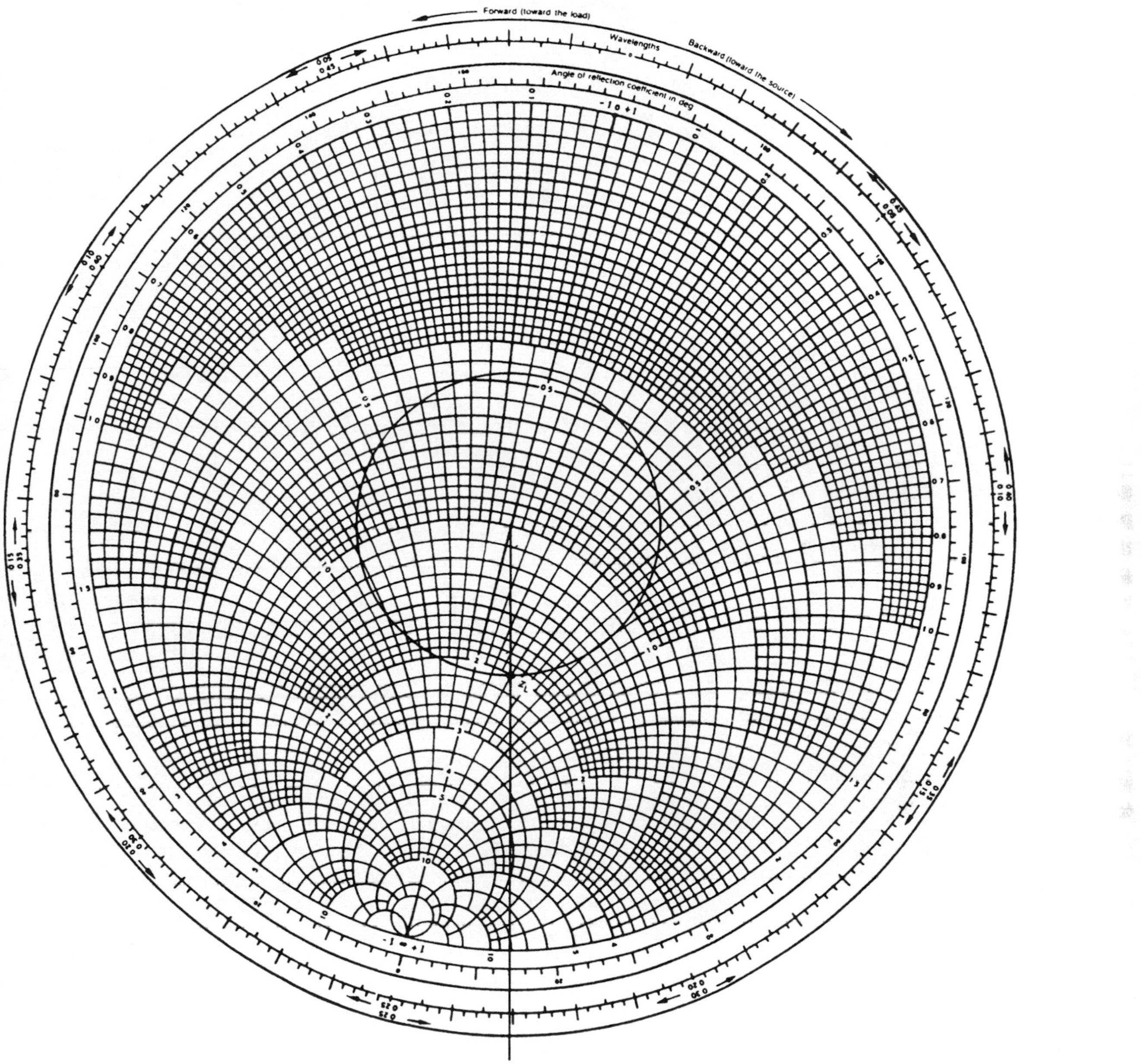

Fig. 17.20 Use of the Smith chart to determine the voltage reflection coefficient

Example 17.9

A 50 Ω line has a load impedance of 20 – j20 Ω. Use the Smith chart to find (*a*) the voltage reflection coefficient, (*b*) the VSWR, (*c*) the input impedance of a 0.2 λ length of this line and (*d*) the lengths of line that have a purely resistive input impedance and the values of these resistances.

Solution

The normalized load impedance is

$$z_L = (20 - j20)/50 = 0.4 - j0.4$$

and this is plotted on the chart shown in Fig. 17.21.

(*a*) $|\rho_v| = (42.5 \text{ mm})/(85 \text{ mm}) = 0.5$, $\angle\rho_v = -131°$

Therefore, $\rho_v = 0.5 \angle -131°$ (*Ans.*)

(*b*) $S = 3$ or 0.33.

(*c*) Travelling around the $S = 3$ circle a distance of 0.2λ from z_L towards the source gives $z_{in} = 0.65 + j0.85$. Therefore,

$$Z_{in} = (0.65 + j0.85)50 = 32.5 + j42.5 \ \Omega \quad (Ans.)$$

(*d*) Travelling from z_L towards the source 0.068λ gives $z_{in} = 0.33 + j0$, and hence $z_{in} = 16.67 \ \Omega$ (*Ans.*)

Travelling from z_L towards the source $(0.068 + 0.25)\lambda = 0.318\lambda$ makes $z_{in} = 3 + j0$ and $Z_{in} = 150 \ \Omega$. (*Ans.*)

Simulation of a component

When a length of loss-free line is used to simulate an inductance or a capacitance it is nearly always short circuited at the load terminals. Then $Z_L = z_L = 0$ and the input reactance of the line can be found by moving clockwise around the outside of the chart.

Example 17.20

A length of loss-free short-circuited line has $Z_0 = 50 \ \Omega$ and is to simulate (*a*) an inductive reactance and (*b*) a capacitive reactance of 35 Ω at 600 MHz. Calculate the necessary lengths of line.

Solution

(*a*) $x_{in} = j35/50 = j0.7$. From the Smith chart this is a distance of 0.0962λ from the top of the real axis. At 600 MHz $\lambda = 0.5$ m and so the length needed is 4.81 cm (*Ans.*)

(*b*) $x_{in} = -j35/50 = -j0.7$. Now the length needed is 0.3063λ or 15.13 cm (*Ans.*)

Determination of an unknown impedance

If both the length of the line and its input impedance are known the method previously described can be used. If not, the procedure to be adopted is as follows.

(*a*) Measure the VSWR with the unknown load connected to the line and note the position of any voltage minimum.

(*b*) Remove the load from the line and short circuit the load terminals. This will cause the noted position of the voltage

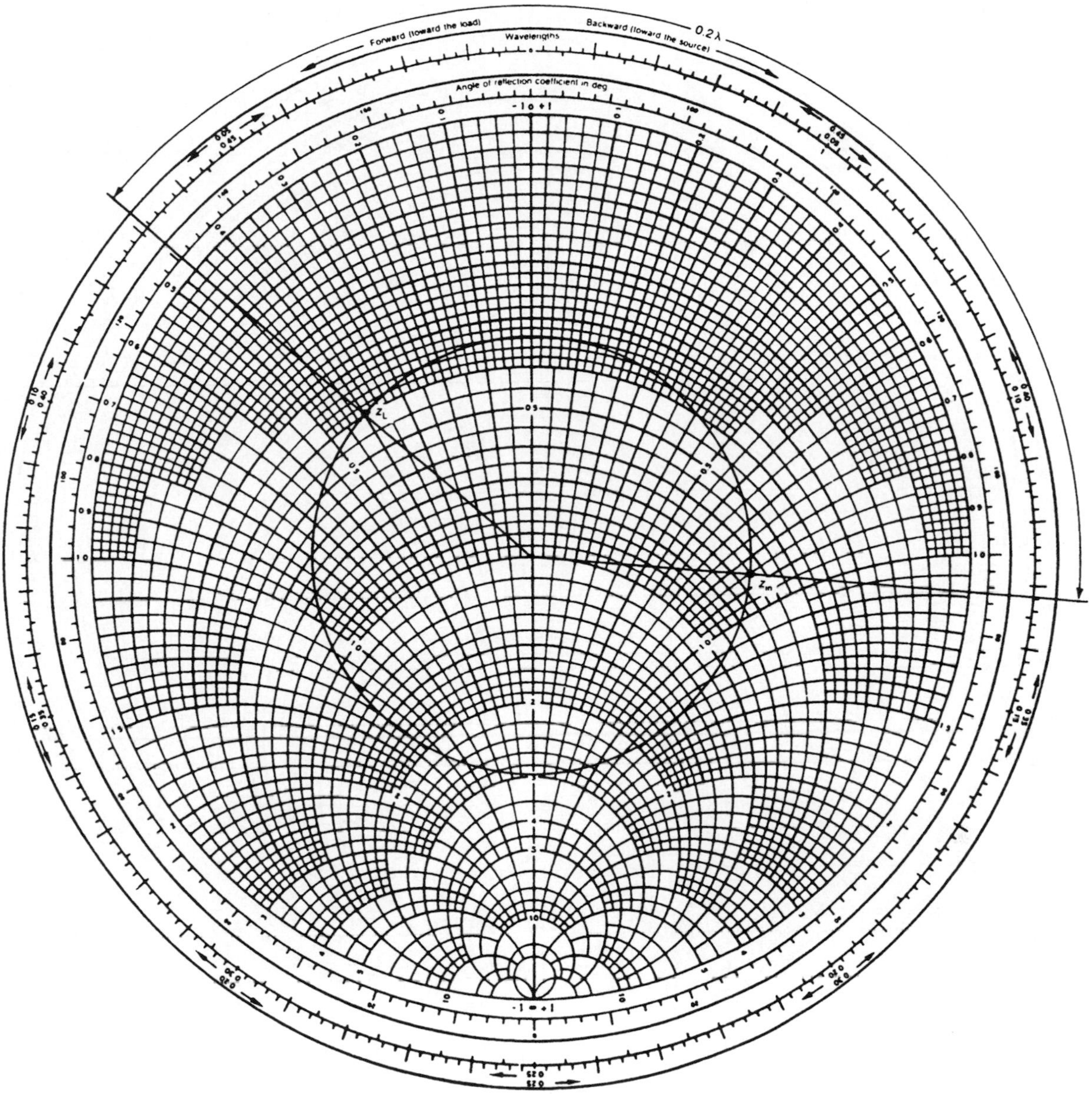

Fig. 17.21 Calculation of the VSWR and input impedance of a line

minimum to shift to a new point (which is less than $\lambda/4$ away). Note this new position.

(*c*) Measure the distance in centimetres between two adjacent voltage minima – this corresponds to one-half a wavelength on the line.

(*d*) Draw the VSWR circle.

(*e*) Starting from the point where the VSWR circle cuts the real axis at a value less than unity, move around the VSWR circle

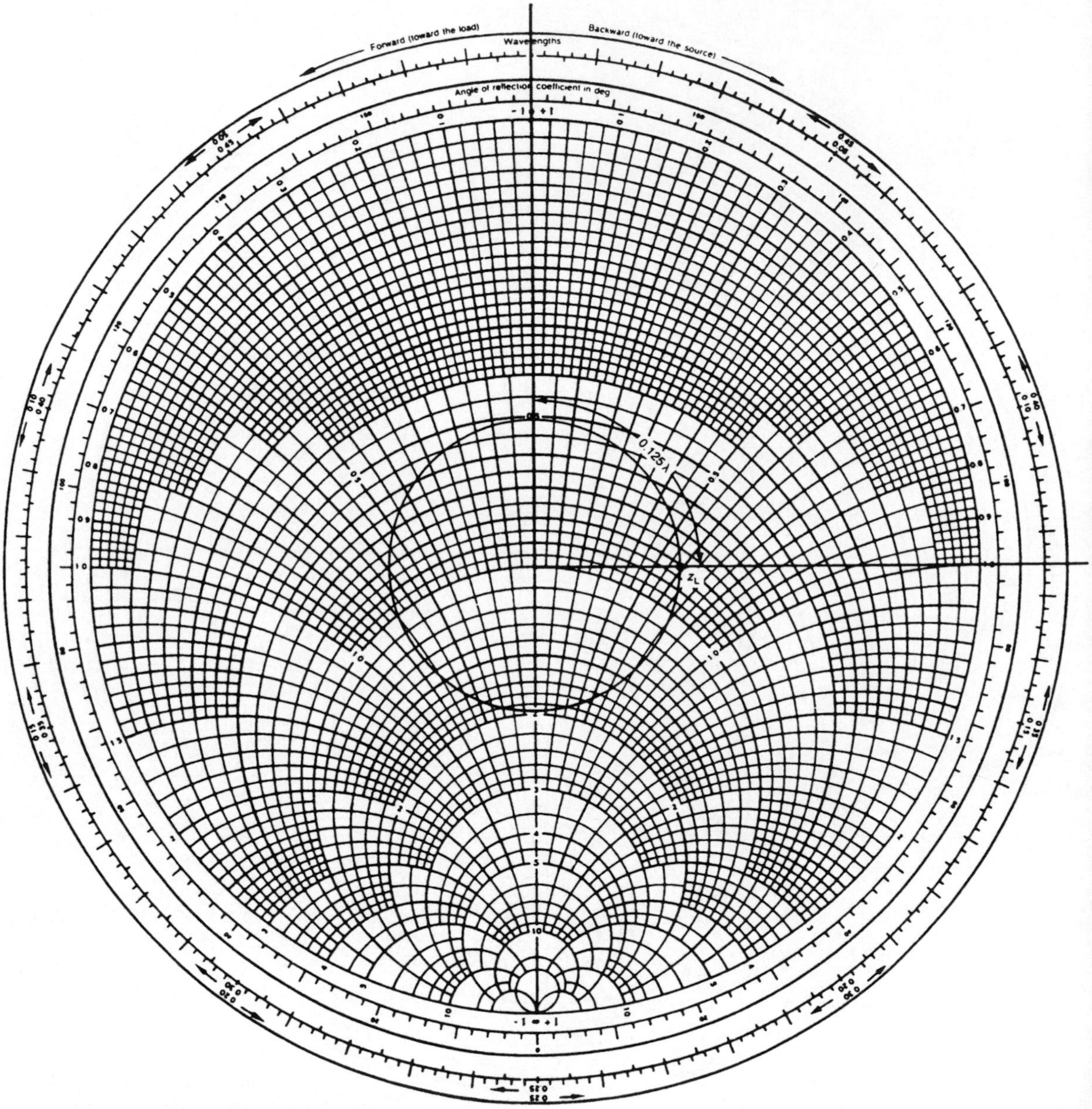

Fig. 17.22

a distance equal to the distance moved by the voltage minimum in (*b*) and in the *same* direction. The point reached is the normalized load impedance.

If an unknown load admittance is to be determined follow the same procedure but start on the Smith chart at the point where the VSWR circle cuts the real axis at a value greater than unity.

Example 17.21

A 50 Ω line has a VSWR of 2 when an unknown load impedance is connected to its output terminals. Adjacent voltage minima are found to be 30 cm apart.

When the unknown load is removed from the line and is replaced by a short circuit the voltage minima moves by 7.5 cm towards the source. Calculate the value of the unknown load impedance.

Solution

See Fig. 17.22. The VSWR = 2 circle has been drawn. 30 cm = $\lambda/2$ so that 7.5 cm = 0.125λ. Moving around the VSWR circle for this distance in the clockwise direction gives $z_L = 0.8 + j0.58$. Therefore

$$Z_L = 50(0.8 + j0.58) = 40 + j29\ \Omega \quad (Ans.)$$

Exercises 17

17.1 A loss-free line is short circuited at its output terminals and is to be used to provide a reactance of –j160 Ω at a frequency of 120 MHz. If the characteristic impedance of the line is 300 Ω calculate the length of line needed.

17.2 A loss-free transmission line has a characteristic impedance of 300 Ω and a load impedance of (200 – j300) Ω. Calculate the VSWR on the line and the impedance of the line at a distance of $\lambda/2$ from the load.

17.3 A loss-free line has a characteristic impedance of 450 Ω and a load impedance of (150 + j120) Ω. Calculate the voltage reflection coefficient and the VSWR.

17.4 A transmission line has a characteristic impedance of 600 Ω, 6 dB loss, and is 1.3 wavelengths long. If the line is terminated in an impedance of (400 – j350) Ω calculate its input impedance.

17.5 A loss-free line has a characteristic impedance of 600 Ω. When the load impedance is 200 Ω the voltage across the load is 12 V. Calculate (*a*) the voltage reflection coefficient, (*b*) the VSWR, (*c*) the voltage at a distance of $\lambda/4$ from the load and (*d*) the voltage at a distance of $\lambda/2$ from the load.

17.6 A transmission line has a characteristic impedance of 1000 Ω, 6 dB loss and is $\lambda/2$ long at a particular frequency. Calculate the input impedance of the line when the far-end terminals are short circuited.

17.7 A transmission line has a characteristic impedance of 300 Ω, negligible loss, is $\lambda/4$ long and is terminated by a load impedance of (300 – j100) Ω. Calculate the input impedance of the line.

17.8 A 2λ length of transmission line has a characteristic impedance of 50 Ω and is connected to a source of e.m.f. 12 V and internal impedance 50 Ω. If the load impedance is 100 Ω calculate (*a*) the voltage across the load and (*b*) the voltage across the input terminals if the line loss is 3 dB.

17.9 A transmission line has a characteristic impedance of 600 Ω and is correctly terminated at its far end. A resistor of 300 Ω is connected across the line $\lambda/2$ from the load. Calculate the voltage reflection coefficient at this point. If the line is 3λ long and its losses are negligible calculate the input impedance of the line.

17.10 Derive expressions for the input impedance of (*a*) an open-circuited line that is $\lambda/4$ long and (*b*) a short-circuited line that is $\lambda/2$ long. A line has a loss of 3 dB per $\lambda/4$ length and a characteristic impedance of 75 Ω. Calculate its input impedance when the line is (*i*) $\lambda/4$, (*ii*) $\lambda/2$ long and is short circuited at its output terminals.

17.11 A 100 m length of line has a characteristic impedance of 100 Ω and negligible loss. Its far-end terminals are open circuited and the line is connected to a 5 MHz voltage source of e.m.f. 50 V and internal impedance 100 Ω. Resistors of 100 Ω are connected across the line at distances of 15 m and 30 m from the far end. Calculate the currents in the resistors.

17.12 A transmission line has a characteristic impedance of 50 Ω, 3 dB loss, and is $3\lambda/4$ long at a particular frequency. The line is supplied by a voltage source of e.m.f. 10 V and internal resistance 50 Ω and it is terminated by a 100 Ω resistor. Calculate (*a*) the voltage across the load resistor and (*b*) the voltage across the sending-end terminals.

17.13 A $\lambda/8$ length of $100\angle 0°$ Ω loss-free line is to have an input impedance with a phase angle of 45°. Determine the necessary load resistance.

17.14 A 10 km length of line has its far-end terminals short circuited Calculate the ratio (sent current)/(received current) if the line has an attenuation coefficient of 0.2 N/km and a phase change coefficent of 0.1 rad/km.

17.15 A $\lambda/4$ loss-free line is terminated in a 60 Ω resistor. When 1 V is applied to the sending-end terminals the input current is 10 mA. Calculate the characteristic impedance of the line.

17.16 A transmission line is 10 km long and has a propagation coefficient of $(0.11 + j0.157)$ per km and a characteristic impedance of 900 Ω. The source has an e.m.f. of 50 V and negligible internal impedance and the load has an impedance of 600 Ω. Calculate the values of the load current and voltage.

17.17 A transmission line is $\lambda/4$ long and has an attenuation of 6 dB and a characteristic impedance of 100 Ω. Calculate the total voltage across the sending-end terminals when the load impedance is $75\angle 90°$ Ω and the voltage across the load is $250\angle 0°$ V.

17.18 A loss-free short-circuited line has a characteristic impedance of $90\angle 0°$ Ω. Calculate the length required to simulate an inductor of 0.286 μH at 100 MHz.

17.19 Two long 600 Ω loss-free lines are connected together by a length of loss-free line of capacitance 30 pF/m and inductance 4.8 μH/m. A short voltage pulse of amplitude 500 V enters the first line and arrives at their junction. Calculate (*a*) the characteristic impedance of the connecting line, (*b*) the phase velocity, (*c*) the reflection and transmission coefficients at each junction and (*d*) the magnitude of (*i*) the first three reflected pulses inside the connecting line, (*ii*) the first three pulses transmitted into the third line.

17.20 A 50 Ω coaxial line is terminated by an unknown impedance. If a VSWR of 2 is produced with a voltage minimum at a distance of 0.375 λ from the termination, calculate the value of the unknown impedance.

17.21 A 75 Ω solid dielectic coaxial line is terminated by an unknown impedance. At 200 MHz a VSWR of 1.9 and a voltage antinode 32.2 cm from the termination appear on the line. If the dielectric is loss-free and has $\varepsilon_r = 2.5$, calculate the value of the unknown impedance.

17.22 A transmission line is $\lambda/4$ long, has a loss of 3 dB, and a characteristic impedance of $100\angle 0°$ Ω. Calculate the sending-end voltage when the line is terminated by an impedance of $50\angle 90°$ Ω and the voltage across the receiving terminals is $200\angle 0°$ V.

17.23 A line with $Z_0 = 100$ Ω is 10.56λ long and is connected to another line of $Z_0 = 200$ Ω that is $\lambda/4$ long The output terminals of this line is connected to a third line of $Z_0 = 400$ Ω and length 5λ. This line is terminated by a 400 Ω resistor. Calculate the impedances at (*a*) the junction of the 200 Ω and 400 Ω lines, (*b*) the middle of the 200 Ω line, (*c*) at the junction of the 100 Ω and 200 Ω lines and (*d*) at a point in the 100 Ω line $9\lambda/16$ from the junction with the 200 Ω line.

17.24 A line has $Z_0 = 600\angle 0°$ Ω and is terminated by the impedance $650 - j50$ Ω. Calculate (*a*) the voltage reflection coefficient, and (*b*) the VSWR.

17.25 A loss-free line has $Z_0 = 600\angle 0°$ Ω and is used to connect an aerial of $300\angle 0°$ Ω impedance to a radio transmitter of output impedance $600\angle 0°$ Ω at 90 MHz. Determine length and characteristic impedance of the necessary $\lambda/4$ matching section.

17.26 A line has an attenuation of 1.68 dB per wavelength and a characteristic impedance of 75 Ω. It is connected between a voltage source of 75 Ω resistance and a load of 150 Ω. Calculate the impedance of the line if it is (*a*) $\lambda/2$, (*b*) $\lambda/4$ long. Re-calculate both answers assuming zero line losses and hence explain one effect of line loss.

17.27 A 50 Ω line is $\lambda/4$ long and has 3 dB loss with negligible contribution from the leakage *G*. (*a*) Calculate its input impedance. (*b*) Calculate the input impedance if the frequency is increased by 400%.

17.28 A line of surge resistance 80 Ω is connected to another line of surge resistance 700 Ω. A voltage step of 1.2 kV travels along the 80 Ω line. Find the voltage and current in the two lines immediately after the step reaches the junction of the two lines.

17.29 A 10 kV voltage pulse enters a line of $R_0 = 400$ Ω. This line is connected to another line of $R_0 = 50$ Ω and this other line, in turn, is connected to a third line of $R_0 = 400$ Ω. Calculate the magnitude of (*a*) the first voltage pulse, and (*b*) the second pulse sent into the third line.

17.30 A line of characteristic impedance 100 Ω is terminated by two parallel-connected lines of $R_0 = 600$ Ω and $R_0 = 1000$ Ω. If a voltage step of 1 kV travels along the cable calculate the voltage and current in each cable immediately after the wave reaches the junction.

18 Complex waveforms

All the calculations carried out so far in this book have been based upon the assumption that the signal waveform was sinusoidal. Very often, however, the signal waveform is *not* sinusoidal; such as rectangular waveforms, ramp waveforms and sawtooth waveforms. Many other voltages are initially of sinusoidal waveform but, at some stage, have been applied to a *non-linear device* and have had their wave-form altered. Examples of *non-linearity* in electrical and electronic circuits are many. Sometimes non-linearity is useful and it is used, for example, for such purposes as modulation, detection, and mixing, but often its presence is undesirable since it may produce, for example, distortion in an audio-frequency amplifier.

Synthesis of repetitive waveforms

Odd harmonics

Figure 18.1(a) shows a sinusoidal voltage $v = V_m \sin \omega t$, of peak value V_m and frequency $\omega/2\pi$, and another, smaller voltage at three times the frequency, i.e. at $3\omega/2\pi$ Hz. The first voltage is known as the *fundamental* and the other voltage is the *third harmonic*. At time $t = 0$, the two voltages are in phase with one another and the waveform produced by summing their instantaneous values is shown by the dotted line. The resultant waveform is clearly non-sinusoidal.

If the fifth harmonic, also with zero phase angle at time $t = 0$, is also added to the fundamental, the resultant waveform becomes an even better approximation to a square wave (Fig. 18.1(b)). If, first, the seventh, then the ninth, and then further higher-order odd harmonics are also added, the resultant waveform will become progressively nearer and nearer to the truly square waveform.

The equation for the resultant waveform of Fig. 18.1(a) is

$$v = V_1 \sin \omega t + V_3 \sin 3\omega t \qquad (18.1)$$

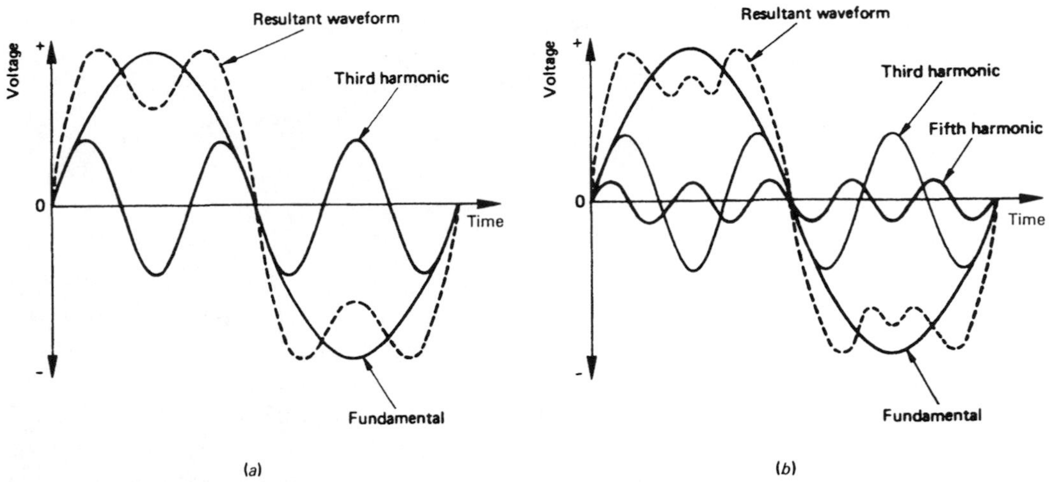

Fig. 18.1 Synthesis of a square waveform

When the fifth harmonic is added (Fig. 18.1(*b*)),

$$v = V_1 \sin \omega t + V_3 \sin 3\omega t + V_5 \sin 5\omega t \tag{18.2}$$

where V_1, V_3 and V_5 are the amplitudes of the fundamental and the third and fifth harmonics respectively, and ω is 2π times the fundamental frequency.

Similarly, when higher-order odd harmonics are also added to the fundamental

$$v = V_1 \sin \omega t + V_3 \sin 3\omega t + V_5 \sin 5\omega t + V_n \sin n\omega t \tag{18.3}$$

$$= \sum_{n=1}^{n=\infty} \mathrm{V}_n \sin n\omega t \tag{18.4}$$

Equations (18.1) through to (18.4) provide an indication that the

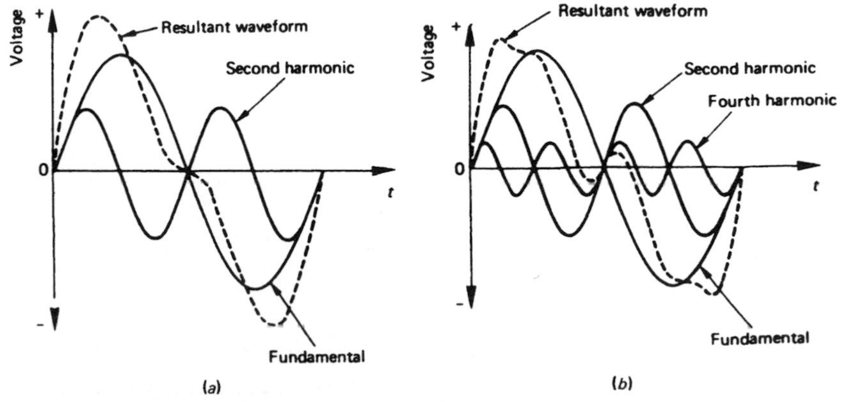

Fig. 18.2 Fundamental plus even harmonics

reverse process to the synthesis of a waveform is also possible. The reverse process is, of course, the analysis of a non-sinusoidal waveform to determine its component frequencies.

Even harmonics

If the second harmonic, and then the fourth harmonic also, are added to the fundamental, a different resultant waveshape is obtained, as is shown by Figs 18.2(a) and (b). The effect of a phase shift, at time $t = 0$, of the second and/or third harmonic is to alter the resultant waveform.

Figures 18.1 and 18.2 show complex waveforms which have the fundamental and harmonic components in phase at time $t = 0$. If the harmonic component has some other phase at $t = 0$, e.g. the waveform is $v = V_1 \sin \omega t + V_3 \sin(3\omega t + 30°)$, the resultant waveform will be different. Figure 18.3 gives one example: a fundamental with both third and fifth harmonic components with both harmonics lagging the fundamental by $30°$ at time $t = 0$.

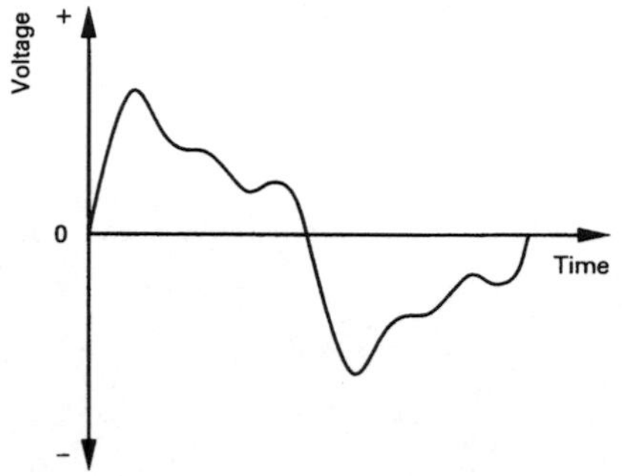

Fig. 18.3 Resultant waveform of $v = V_1 \sin \omega t + V_3 \sin(3\omega t - 30°) + V_5 \sin(5\omega t - 30°)$

Analysis of repetitive waveforms

Any repetitive waveform consists of the sum of a fundamental and a number of harmonics, in general of both even- and odd-order. Most repetitive non-sinusoidal waveforms can be analysed by the use of a mathematical process known as the *Fourier Series*. Essentially this theorem states that a repetitive waveform can be represented by an equation of the form

$$v = A_0 + A_1 \cos \omega t + A_2 \cos 2\omega t + A_3 \cos 3\omega t \cdots + B_1 \sin \omega t + B_2 \sin 2\omega t + B_3 \sin 3\omega t \cdots \qquad (18.5)$$

or

$$v = A_0 + \sum_{n=1}^{n=\infty} (A_n \cos n\omega t + B_n \sin n\omega t) \qquad (18.6)$$

where
$$A_0 = \frac{1}{T}\int_0^T f(t)\, \mathrm{d}t \qquad (18.7)$$

$$A_n = \frac{2}{T}\int_0^T f(t) \cos n\omega t\, \mathrm{d}t \qquad (18.8)$$

$$B_n = \frac{2}{T}\int_0^T f(t) \sin n\omega t\, \mathrm{d}t \qquad (18.9)$$

and T is the periodic time of the waveform.

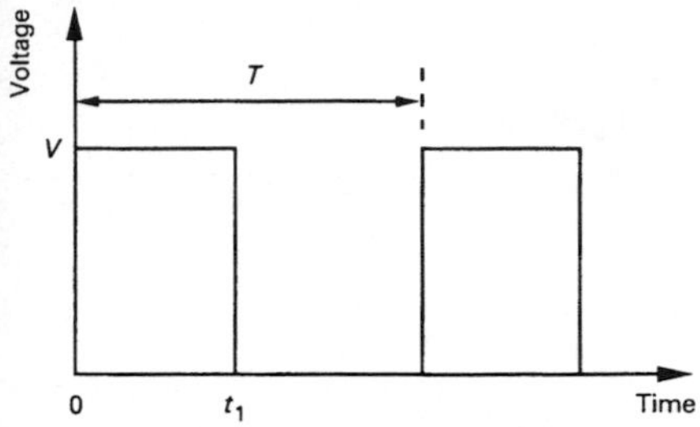

Fig. 18.4 Rectangular waveform

Rectangular waveform

Referring to Fig 18.4,

$$A_0 = (1/T)\int_0^T \mathrm{V}\,\mathrm{d}t = (1/T)\int_0^{t_1} \mathrm{V}\,\mathrm{d}t$$
$$= (1/T)[Vt]_0^{t_1} = Vt_1/T$$
$$A_n = (2/T)\int_0^T \mathrm{V}\cos n\omega t\,\mathrm{d}t = (2V/T)\int_0^{t_1} \cos n\omega t\,\mathrm{d}t$$
$$= (2V/T)\int_0^{t_1} \cos[(2\pi nt)/T]\,\mathrm{d}t$$
$$= (2V/T)[(T/2\pi n)\sin(2\pi nt/T)]_0^{t_1}$$
$$= (V/\pi n)\sin(2\pi nt_1/T)$$

When $n = 1$, $A_1 = (V/\pi)\sin(2\pi t_1/T)$; when $n = 2$, $A_2 = (V/2\pi)\sin(4\pi t_1/T)$; when $n = 3$, $A_3 = (V/3\pi)\sin(6\pi t_1/T)$, and so on.

$$B_n = (2/T)\int_0^T V\sin n\omega t\,\mathrm{d}t = (2V/T)\int_0^{t_1} \sin(2\pi nt/T)$$
$$= (-2V/T)[(T/2\pi n)\cos(2\pi nt/T)]_0^{t_1}$$
$$= (-V/\pi n)[\cos(2\pi nt_1/T) - 1]$$

when $n = 1$, $B_1 = (V/\pi)[1 - \cos(2\pi t_1/T)]$;
when $n = 2$, $B_2 = (V/2\pi)[1 - \cos(4\pi t_1/T)]$;
when $n = 3$, $B_3 = (V/3\pi)[1 - \cos(6\pi t_1/T)]$; etc.

Square waveform

For a square wave $t_1 = T/2$ and $A_0 = V/2$.

$$A_1 = (V/\pi)\sin(2\pi T/2T) = 0,$$
$$A_2 = (V/2\pi)\sin(4\pi T/2T) = 0,$$
$$A_3 = (V/3\pi)\sin(6\pi T/2T) = 0.$$
$$B_1 = (V/\pi)[1 - \cos(2\pi T/2T)]$$
$$= (V/\pi)[1 - \cos\pi] = 2V/\pi,$$
$$B_2 = (V/2\pi)[1 - \cos(4\pi T/2T)]$$
$$= (V/2\pi)[1 - \cos 2\pi] = 0,$$
$$B_3 = (V/3\pi)[1 - \cos(6\pi T/2T)]$$
$$= (V/3\pi)[1 - \cos 3\pi] = 2V/3\pi;\ \text{etc.}$$

Hence,

$$v = V/2 + (2V/\pi)[\sin\omega t + (\sin 3\omega t)/3 + (\sin 5\omega t)/5 + \cdots] \qquad (18.10)$$

If the rectangular wave is symmetrical about the zero voltage axis then the A_0 term is equal to zero. If the voltage of the waveform is $\pm V$ volts then the expression for the instantaneous voltage is

$$v = (4V/\pi)[\sin \omega t + (\sin 3\omega t)/3 + (\sin 5\omega t)/5 + \cdots] \tag{18.11}$$

Half-wave rectified waveform

For the half-wave rectified waveform, $f(t) = V \sin \omega t$ from $t = 0$ to $t = T/2$, and $f(t) = 0$ from $T/2$ to T. Hence

$$A_0 = (V/T) \int_0^{T/2} \sin \omega t \, dt = (V/\pi)[(-\cos \omega t)/\omega]_0^{T/2}$$

$$= V/\pi.$$

$$A_n = (2/T) \int_0^{T/2} V \sin \omega t \cos n\omega t \, dt + \int_{T/2}^{T} 0 \, dt$$

$$= (2V/T) \int_0^{T/2} [\sin(n+1)\omega t - \sin(n-1)\omega t] \, dt$$

$$= \frac{2V}{T}\left[\frac{-\cos(n+1)\omega t}{(n+1)\omega} + \frac{\cos(n-1)\omega t}{(n-1)\omega}\right]_0^{T/2}$$

$A_n = 0$ when n is odd.

$$A_n = (2V/\omega T)[1/(n+1) - 1/(n-1)$$

$$= (2V/\omega T)\{[(n-1) - (n+1)]/[(n+1)(n-1)]\}$$

$$= (-V/T)/[\pi(n+1)(n-1)] \text{ when } n \text{ is even.}$$

$$B_n = (2/T) \int_0^{T/2} \left[V \sin \omega t \sin n\omega t \, dt + \int_{T/2}^{T} 0 \, dt\right]$$

$$= (V/T) \int_0^{T/2} [\cos(n-1)\omega t - \cos(n+1)\omega t] \, dt$$

$$= \frac{V}{T}\left[\frac{\sin(n-1)\omega t}{n-1} - \frac{\sin(n+1)\omega t}{n+1}\right]_0^{T/2}$$

When $n = 0$, $B_n = 0$, when $n = 1$, $B_n = V/2$, when $n = 3$, $B_n = 0$, etc. Therefore,

$$v = (V/\pi)[1 + (\pi \sin \omega t)/2 - (2\cos 2\omega t)/3 - (2\cos 4\omega t)/15 - (2\cos 6\omega t)/35 + \cdots] \tag{18.12}$$

Triangular waveform

The Fourier series for a triangular wave can be derived in the same way as for the rectangular and half-wave rectified waveforms. From $t = 0$ to $t = T/2\ f(t) = kf$, and from $t = T/2$ to $Tf(t) = -kf$. Alternatively, if it realized that the differential of a triangular waveform is a square waveform the series may be obtained by integrating equation (18.11) and multiplying the result by $\pm(2\omega/\pi)$. Thus, for the triangular wave,

$$\begin{aligned} v &= (2\omega/\pi) \int \{(4/\pi)[\sin \omega t + (\sin 3\omega t)/3 \\ &\qquad + (\sin 5\omega t)/5 + \cdots]\}\ \mathrm{d}t \\ &= (-8/\pi^2)[\cos \omega t + (\cos 3\omega t)/9 + (\cos 5\omega t)/25 + \cdots] \end{aligned} \tag{18.13}$$

Example 18.1

A square current wave of amplitude ± 1 A is passed through a pure capacitor. Obtain an expression for the voltage across the capacitor and deduce its waveform.

Solution
$C = q/v_C = i\ \mathrm{d}t/v_C$, or $v_C = (1/C)i\ \mathrm{d}t$. Integrating equation (18.10) and multiplying by $(1/C)$ gives

$$v_C = (-4/\pi\omega C)[\cos \omega t + (\cos 3\omega t)/9 + (\cos 5\omega t)/25 + \cdots] \quad (Ans.)$$

Thus the capacitor voltage is of approximate triangular waveform. (*Ans.*)

Example 18.2

A symmetrical triangular current wave of amplitude ± 1 A is passed through a pure inductance. Determine an expression for the voltage across the inductance and deduce its waveform.

Solution
Voltage across inductance $v_L = L\ \mathrm{d}i/\mathrm{d}t$. Differentiating equation (18.13) and multiplying by L gives;

$$v_L = (-8L/\pi^2)[\sin \omega t + (\sin 3\omega t)/3 + (\sin 5\omega t)/5 + \cdots] \qquad (Ans.)$$

The inductance voltage is of approximate square waveshape. (*Ans.*)

Sawtooth waveform

$$v = (2V/\pi)[\sin \omega t - (\sin 2\omega t)/2 + (\sin 3\omega t)/3 - \cdots] \tag{18.14}$$

Full-wave rectified waveform

$$v = (2V/\pi) + (4V/3\pi)\cos\,\omega t - (4V/15\pi)\cos\,2\omega t + \cdots) \qquad (18.15)$$

Tabular method for determining the Fourier series of a waveform

The Fourier series representing a non-sinusoidal waveform can also be derived using a tabular method. The waveform to be analysed must have one periodic time divided up into n equal-width parts. Next erect an ordinate at the mid-point of each part. The number of ordinates must be larger than the order of the highest harmonic whose Fourier coefficient is to be determined. Then, if $y_1,\ y_2,\ y_3$, etc., is the value of each ordinate,

$$A_0 = (y_1 + y_2 + y_3 + y_4 + \cdots + y_n)/n \qquad (18.16)$$

$$A_n = 2(y_1 \sin n\omega t_1 + y_2 \sin n\omega t_2 + y_3 \sin n\omega t_3 + \cdots + y_n \sin n\omega t_n) \qquad (18.17)$$

$$B_n = 2(y_1 \cos n\omega t_1 + y_2 \cos n\omega t_2 + y_3 \cos n\omega t_3 + \cdots + y_n \cos n\omega t_n \qquad (18.18)$$

Example 18.3

The instantaneous voltages of a complex wave are given in Table 18.1. Determine the equation for the voltage if it is known to consist of a fundamental component and its third harmonic.

Table 18.1

$\theta°$	v	$\sin\theta$	$v\sin\theta$	$\sin 3\theta$	$v\sin 3\theta$	$\cos\theta$	$v\cos\theta$	$\cos 3\theta$	$v\cos 3\theta$
0	−17.32	0	0	0	0	1	−17.32	1	−17.32
20	16.88	0.342	5.77	0.866	14.63	0.924	15.85	0.5	8.44
40	64.28	0.643	41.35	0.866	55.65	0.766	49.2	−0.5	−32.14
60	103.92	0.866	89.9	0	0	0.5	51.95	−1	−103
80	115.8	0.985	114	−0.866	−100	0.174	20.1	−0.5	−57.9
100	94.48	0.985	93	−0.866	−81.8	−0.174	−16.4	0.5	47.24
120	69.28	0.866	60	0	0	−0.5	−34.65	1	69.28
140	49.96	0.643	30.2	0.866	40.6	0.766	−36	0.5	23.48
160	30.2	0.342	10.3	0.866	26.15	−0.94	−28.35	−0.5	15.1

Solution

The sum of the $v\sin\theta$ column is 444.55, of the $v\sin 3\theta$ column is −44.62, of the $v\cos\theta$ column is 4.98 and of the $v\cos 3\theta$ column is −77.94. Therefore,

$$A_1 = (2 \times 444.55)/9 = 98.8; \quad A_3 = (2 \times -44.62)/9 = -9.9;$$

$$B_1 = (2 \times 4.98)/9 = 1.11 \quad \text{and} \quad B_3 = (2 \times -77.94)/9 = -17.3.$$

Therefore,

$$v = 98.8 \sin\theta + 1.11 \cos\theta - 9.9 \sin 3\theta - 17.3 \cos 3\theta$$
$$= 98.8 \sin\theta + 19.93 \sin(3\theta - 119.8°) \quad (\textit{Ans.})$$

Example 18.4

Table 18.2 gives the instantaneous values of a non-sinusoidal current wave. Determine the amplitude of the fundamental frequency component.

Table 18.2

ωt	0°	30°	60°	90°	120°	150°	180°	210°	240°	270°	300°	330°
i (mA)	74	98	92	60	24	2	6	18	24	20	22	42

Solution
For each value of ωt calculate the values of $i \sin\omega t$ and $i \cos\omega t$. The figures obtained are given in Table 18.3.

Table 18.3

ωt	0°	30°	60°	90°	120°	150°	180°	210°	240°	270°	300°	330°
$i \sin \omega t$	0	49	79.7	60	20.8	1	0	−9	−20.8	−20	−19.1	−21
$i \cos \omega t$	74	84.9	46	0	−12	−1.7	−6	−15.6	−12	0	11	36.4

Average $i \sin\omega t = 120.6/12 = 10.05$. Average $i \cos\omega t = 200.9/12 = 16.74$. Hence

$$A_1 = 2 \times 10.05 = 20.1 \quad \text{and} \quad B_1 = 2 \times 16.74 = 33.48$$

Therefore,

$$I_1 = \sqrt{(20.1^2 + 33.48^2)} = 39.1 \text{ mA} \quad (\textit{Ans.})$$

Symmetrical waveforms

For many waveforms it is possible to recognize that only *either* cosine *or* sine terms are present and then a considerable reduction in the work involved in its analysis can be achieved.

An *odd function* is one for which $f(t) = -f(-t)$, such as $\sin\omega t$. Such a function passes through the zero voltage axis at the chosen origin and will only contain sine terms. Some examples of odd functions are given in Figs 18.3(a), (b) and (c). Figure 18.5(a) shows a sawtooth waveform that is clearly symmetrical either side of the origin ($t = 0$). Figures 18.5(b) and (c) show two different rectangular waveforms for which $f(t) = -f(-t)$. Both of these waveforms can be made into *even* functions instead by a suitable change of the origin. Lastly, Fig. 18.5(d) shows a

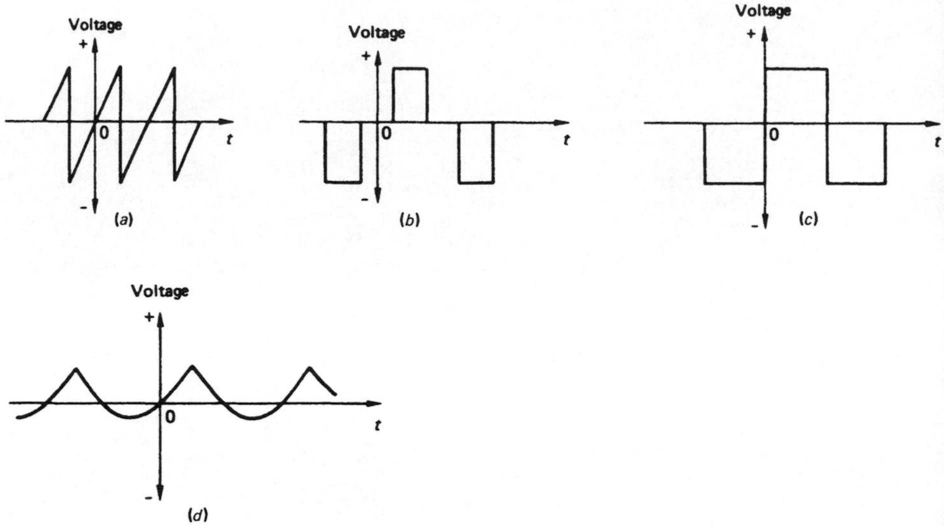

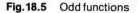

Fig. 18.5 Odd functions

waveform which, although it passes through the origin, is *not* symmetrical about it. This means that this waveform is not an odd function.

An odd function has all its A_n coefficients. including A_0, equal to zero.

An *even function* is one for which $f(t) = f(-t)$ and which contains only cosine terms. Some examples are given in Figs 18.6(*a*), (*b*) and (*c*). Note that the waveforms (*a*) and (*c*) appear in Fig. 18.5 also and have both been made into odd functions by a suitable choice of the origin, i.e. making the origin at the middle of a pulse.

An even function has all its B_n coefficients equal to zero.

It is also often possible to recognize whether a waveform contains any even or any odd harmonics. Any complex wave whose positive and negative half-cycles are identical, i.e. $f(t) = -f(t + T/2)$, will not contain any even harmonics. Some examples of waveforms that consist of a fundamental frequency plus a number of *odd* harmonics are shown in Figs 18.7(*a*) and (*b*).

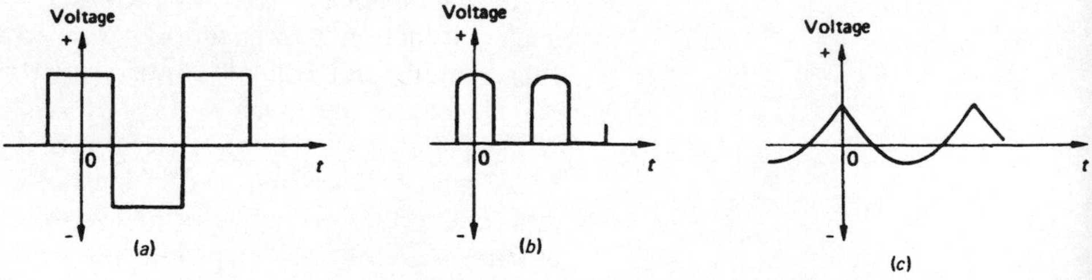

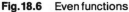

Fig. 18.6 Even functions

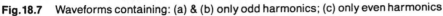

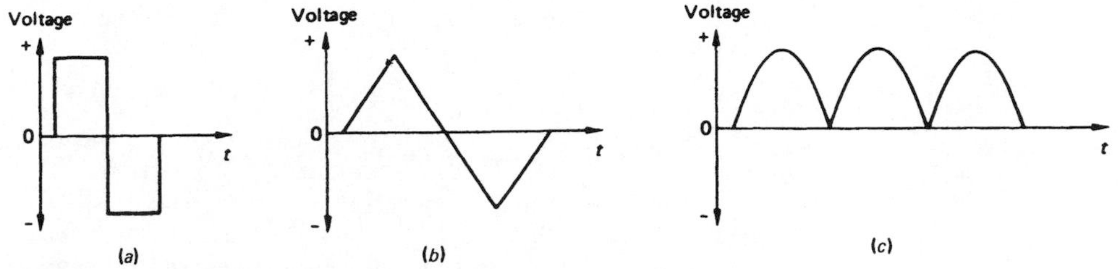

Fig. 18.7 Waveforms containing: (a) & (b) only odd harmonics; (c) only even harmonics

Other waveforms satisfy the equation $f(t) = f(t + T/2)$ and these are wave-forms that repeat themselves after each half-cycle of the fundamental frequency. Such waveforms consist of a fundamental plus a number of *even* harmonics. No odd harmonics are present in such a waveform. An example of such a waveform is given in Fig. 18.7(c)

R.M.S. value of a complex wave

The. r.m.s. value of a complex wave is the square root of the mean, or average, value of the square of the voltage. Thus,

$$V = \sqrt{\left[(1/2\pi)\int_0^{2\pi} v^2 \, d\omega t\right]} \tag{18.19}$$

Consider the complex wave

$$v = V_{DC} + V_{m1}\sin\omega t = V_{m2}\sin 2\omega t$$

The r.m.s. voltage of this wave is

$$V = \sqrt{[(1/2\pi)} \int_0^{2\pi} (V_{DC} + V_{m1}\sin\omega t + V_{m2}\sin 2\omega t)\, d\omega t]$$

$$= \sqrt{[(1/2\pi)}\int_0^{2\pi} (V_{DC}^2 + V_{m1}^2\sin^2\omega t + V_{m2}^2\sin^2 2\omega t$$

$$+\ 2V_{DC}V_{m1}\sin\omega t + 2V_{DC}V_{m2}\sin 2\omega t$$

$$+\ 2V_{m1}V_{m2}\sin\omega t\sin 2\omega t)]\, d\omega t$$

Now

$$V_{m1}^2\sin^2\omega t = (V_{m1}^2/2)(1 - \cos 2\omega t)$$

and this will have a mean value over a cycle of the fundamental frequency of $V_{m1}^2/2$. Similarly, the mean value of $V_{m2}^2\sin^2 2\omega t$ is $V_{m2}^2/2$. Further, the mean value of both $\sin\omega t$ and $\sin 2\omega t$ over a complete cycle is also zero. Lastly, the mean value of the product of two sine waves at different frequencies over a complete cycle is also zero. Therefore,

$$V = \sqrt{[(1/2\pi) \int_0^{2\pi} (V_{DC}^2 + V_{m1}^2/2 + V_{m2}^2/2)]\, d\omega t}$$

$$= \sqrt{\{(1/2\pi)[V_{DC}^2 \omega t + (V_{m1}^2/2)\, \omega t + (V_{m2}^2/2)\, \omega t]_0^{2\pi}\}}$$

$$V = \sqrt{[V_{DC}^2 + (V_{m1}^2/2) + (V_{m2}^2/2)]} \qquad (18.20)$$

$V_{m1}^2/2$ is the square of the r.m.s. voltage of the fundamental frequency component and $V_{m2}^2/2$ is the square of the r.m.s. value of the second harmonic component. Thus, equation (18.20) can be re-written as

$$V = \sqrt{(V_{DC}^2 + V_1^2 + V_2^2)} \qquad (18.21)$$

where V_1 and V_2 are respectively the r.m.s. values of the fundamental and the second harmonic frequency components.

Should any of the harmonic components of a complex wave have some phase angle other than zero, relative to the fundamental at time $t = 0$, it will have no effect upon the r.m.s. value of the wave and equation (18.21) will still apply.

Example 18.5

Calculate the r.m.s. value of the complex wave whose instantaneous voltage is given by

$$v = 4 \sin 5000t + 1 \sin 10000t + 0.5 \sin 15000t$$

Solution
From equation (18.21),

$$V = \sqrt{[(4/\sqrt{2})^2 + (1/\sqrt{2})^2 + (0.5/\sqrt{2})^2]} = 2.94 \text{ V} \quad (Ans.)$$

When a complex wave is applied to a circuit the current that flows will consist of a number of components at the same frequencies as the components in the voltage wave. The magnitude and angle of each component of the current will be equal to the corresponding voltage divided by the impedance of the circuit at that frequency. Thus, if the voltage

$$v = V_1 \sin \omega t + V_2 \sin 2\omega t$$

is applied to a circuit, the current that flows will be equal to

$$i = (V_1/Z_1) \sin \omega t + (V_2/Z_2) \sin 2\omega t$$

where Z_1 is the impedance of the circuit at frequency $\omega/2\pi$ and Z_2 is the impedance of the circuit at frequency $2\omega/2\pi$.

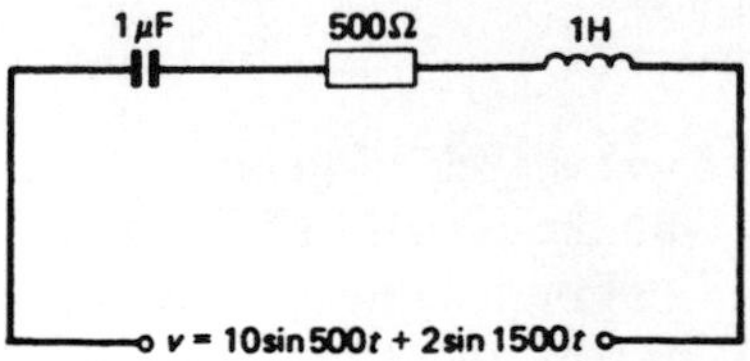

Fig.18.8

Example 18.6

Determine the expression for the current flowing in the circuit given in Fig. 18.8.

Solution
At the fundamental frequency of $500/2\pi$ Hz the reactance of the capacitor is

$$X_{C1} = 1/(500 \times 10^{-6}) = -j2000\ \Omega$$

The inductive reactance is $X_{L1} = j500\ \Omega$. Therefore, at this frequency the impedance of the circuit is

$$Z_1 = 500 - j1500 = 1581\angle{-71.6°}\ \Omega$$

At the third harmonic frequency

$$X_{C3} = -j2000/3 = -j666.7\ \Omega \quad \text{and} \quad X_{L3} = j500 \times 3 = j1500\ \Omega$$

Therefore the impedance of the circuit at the third harmonic is

$$Z_3 = 500 + j833.3 = 971.8\angle 59°\ \Omega$$

The current flowing in the circuit is

$$i = (10 \sin 500t)/(1581\angle{-71.6°}) + (2 \sin 1500t)/(971.8\angle 59°)$$
$$i = 6.33 \sin(500t + 71.6°) + 2.06 \sin(1500t - 59°)\ \text{mA} \quad (\textit{Ans.})$$

Example 18.7

A 10 Ω resistor and a capacitor whose reactance is 10 Ω at a frequency of 5000 rad/s are connected in series and the series circuit is connected in parallel with an inductor. The inductor has 10 Ω resistance and 10 Ω reactance at a frequency of 5000 rad/s. The voltage $v = 20 \sin 5000t + 10 \sin 10000t$ V is applied to the parallel circuit. (*a*) Obtain an expression for the current supplied to the circuit. (*b*) Calculate r.m.s. value of this current.

Solution
(a) At $\omega = 5000$ rad/s, the impedance Z of the circuit is

$$Z = (10 + j10)(10 - j10)/20 = 10 + j0\ \Omega.$$

Hence the fundamental frequency current is $I_f = 20/10 = 2$ A. At $\omega = 10000$ rad/s,

$$Z = (10 + j20(10 - j5)/(20 + j15) = 10 + j0\ \Omega$$

Hence second harmonic current = 10/10 = 1 A. The expression for the circuit current is

$$i = 2 \sin 5000t + 1 \sin 10000t\ \text{A}\ (\textit{Ans.})$$

(*b*) $I = \sqrt{[(2/\sqrt{2})^2 + (1/\sqrt{2})^2]} = 1.58$ A (*Ans.*)

Example 18.8

Obtain an expression for the ratio V_2/V_1 for the circuit given in Fig. 18.9. If $C = 0.2\ \mu$F, $L = 0.2$ H and $v_1 = 1 \sin 3000t + 0.5 \sin 6000t + 0.25 \sin 9000t$ V, calculate the amplitudes of the components of V_2.

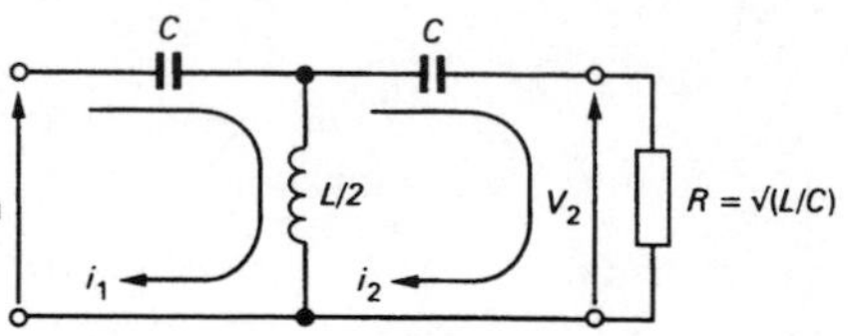

Fig. 18.9

Solution
From Fig. 18.9, $V_1 = i_1[(j\omega L/2) + (1/j\omega C)] - i_2(j\omega L/2)$ and $0 = i_2[(j\omega L/2) + (1/j\omega C) + \sqrt{(L/C)}] - i_1(j\omega L/2)$.
Also, $V_2 = i_2\sqrt{(L/C)}$. Hence,

$$i_1 = i_2[1 - 2/(\omega^2 LC) - j[2/\omega\sqrt{(LC)}]$$

$$V_1 = i_2\left[\left(\frac{j\omega L}{2} + \frac{1}{j\omega C}\right)\left(1 - \frac{2}{\omega^2 LC} - \frac{j2}{\omega\sqrt{(L/C})}\right) - \frac{j\omega L}{2}\right]$$

$$\frac{V_2}{V_1} = \sqrt{\frac{L}{C}}\left[\frac{1}{(j\omega L/2 + 1/j\omega C)(1 - 2/\omega^2 LC - j2/\omega\sqrt{(LC)}) - j\omega L/2\}}\right]$$

$$= 1/\{1 - [2/(\omega^2 LC)] - j2/[\omega\sqrt{(LC)}][1 - 1/(\omega^2 LC)]\}$$

$$|V_2/V_1| = 1/\sqrt{[1 - (4/(\omega^4 L^2 C^2)) + (4/(\omega^6 L^3 C^3))]} \quad (Ans.)$$

At $\omega = 3000$, $4/(\omega^6 L^3 C^3) = 85.7$ and $4/(\omega^4 L^2 C^2) = 30.13$

$$|V_2/V_1| = (1 - 30.13 + 85.7) = 0.13$$

At $\omega = 6000$, $4/(\omega^6 L^3 C^3) = 1.35$ and $4/(\omega^4 L^2 C^2) = 1.93$

$$|V_2/V_1| = 1/\sqrt{(1 - 1.93 + 1.35)} = 1.54$$

At $\omega = 9000$, $4/(\omega^6 L^3 C^3) = 0.12$ and $4/(\omega^4 L^2 C^2) = 0.38$

$$|V_2/V_1| = 1/\sqrt{(1 - 0.38 + 0.12)} = 1.16$$

Hence, the components of V_2 are: $0.13 \times 1 \sin 3000t$, $1.54 \times 0.5 \sin 6000t$ and $1.16 \times 0.25 \sin 9000t$. Therefore

$$v_2 = 0.13 \sin 3000t + 0.77 \sin 6000t + 0.29 \sin 9000t \quad (Ans.)$$

Power and power factor

When a complex wave v is applied to a circuit and causes a current i to flow, the instantaneous power dissipated is $P = iv$ watts. The average or *mean power* is

$$P = (1/2\pi)\int_0^{2\pi} iv \, d\omega t \tag{18.22}$$

$$P = (1/2\pi)\int_0^{2\pi} [I_1 \sin(\omega t + \theta_1) + I_2 \sin(2\omega t + \theta_2) + \cdots]$$

$$\times [V_1 \sin \omega t + V_2 \sin \omega t + \cdots] \, d\omega t$$

or $$P = V_1 I_1 \cos\theta_1 + V_2 I_2 \cos\theta_2 + \cdots \tag{18.23}$$

since most of the terms are equal to zero.

This result means that the total power dissipated in a circuit is the algebraic sum of the powers dissipated at each component frequency – including any d.c. term that may be present.

The *power factor* is not $\cos\theta$ (which θ would be used?) but is given by

$$\text{Power factor} = \text{watts/volt-amperes} \quad (18.24)$$

$$= (V_1 I_1 \cos\theta_1 + V_2 I_2 \cos\theta_2 + \cdots)/VI \quad (18.25)$$

If the equivalent series resistance of the circuit is R_S, the total power dissipated in the circuit is $P = I^2 R_S$. If the equivalent parallel resistance of the circuit is R_p the power dissipated is $P = V^2/R_p$.

Example 18.9

The voltage $v = 170 \sin \omega t + 10 \sin 2\omega t + 15 \sin 3\omega t$ V is applied to a circuit. The resulting current is $i = 100 \sin(\omega t - 30.4°) + 20 \sin(2\omega t - 94.6°) + 14 \sin(3\omega t - 100°)$ mA. Calculate (*a*) the power dissipated in the circuit, (*b*) the r.m.s. values of the applied voltage and the circuit current and (*c*) the power factor of the circuit.

Solution

(*a*) $P = [(170 \times 100)/(2 \times 10^3)] \cos(30.4°)$

$+ [(10 \times 20)/(2 \times 10^3)] \cos(94.6°)$

$+ [(15 \times 14)/(2 \times 10^3)] \cos(100°)$

$= 7.3314 - 8.02 \times 10^{-3} - 0.0182 = 7.305$ W (*Ans.*)

(*b*) $V = \sqrt{[(170^2/2) + (10^2/2) + (15^2/2)]} = 121$ V (*Ans.*)

$I = \sqrt{[(100^2/2 + (20^2/2) + (14^2/2)]} = 72.8$ mA (*Ans.*)

(*c*) Power factor $= 7.305/(121 \times 72.8 \times 10^{-3}) = 0.83$ (*Ans.*)

Example 18.10

A 20 μF capacitor is connected in series with a 50 mH inductor of resistance 25 Ω. The voltage $v = 200 \sin 300t + 40 \sin 1500t$ V is applied across the circuit. (*a*) Obtain an expression for the current taken from the supply. (*b*) Calculate the r.m.s. value of the current. (*c*) Calculate the power supplied to the circuit.

Solution

(*a*) $X_{L1} = 15\ \Omega$, $X_{C1} = 166.7\ \Omega$

$|Z_1| = \sqrt{[25^2 + (15 - 166.7)^2]} = 153.8\ \Omega$

$\theta_1 = \tan^{-1}[(15 - 166.7)/25] = -80.6°$

$|I_1| = 200/153.8 = 1.3$ A

$$X_{L5} = 5 \times 15 = 75\ \Omega, \quad X_{C5} = 166.7/5 = 33.3\ \Omega$$

$$|Z_5| = \sqrt{(25^2 + 41.7^2)} = 48.6\ \Omega.\ \theta_5 = -59.1°$$

$$|I_5| = 40/48.62 = 0.823\ \text{A. Therefore,}$$

$$i = 1.3\sin(300t + 80.6°) + 0.823\sin(1500t - 59.1°)\ \text{A} \quad (\textit{Ans.})$$

(*b*) $I = \sqrt{[(1.3^2/2) + (0.823^2/2)]} = 1.09$ A (*Ans.*)

(*c*) $P = 1.09^2 \times 25 = 29.6$ W (*Ans.*)

Example 18.11

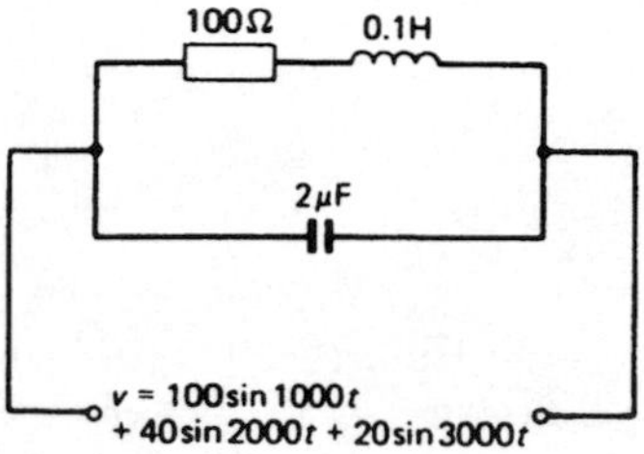

Fig. 18.10

Determine an expression for the current taken from the voltage source in the circuit of Fig. 18.10. Also calculate the power dissipated in the circuit and the power factor.

Solution

At $\omega = 1000$: $X_{L1} = 100\ \Omega$ and $X_{C1} = 500\ \Omega$

$\omega = 2000$: $X_{L2} = 200\ \Omega$ and $X_{C2} = 250\ \Omega$

$\omega = 3000$: $X_{L3} = 300\ \Omega$ and $X_{C3} = 166.67\ \Omega$

The impedance of the circuit at $\omega = 1000$ is

$$Z_1 = [(100 + j100)(-j500)]/(100 - j400) = 171.5\angle 31°\ \Omega$$

At $\omega = 2000$: $Z_2 = [(100 + j200)(-j250)]/(100 - j50) = 500\angle 0°\ \Omega$

At $\omega = 3000$: $Z_3 = [(100 + j300)(-j166.67)]/(100 + j133.33) = 316.2\angle -71.5°\ \Omega$

Therefore,

$$|I_1| = |V_1|/Z_1 = 100/171.5 = 0.583\ \text{A}$$

$$|I_2| = |V_2|/Z_2 = 40/500 = 0.08\ \text{A}$$

$$|I_3| = |V_3|/Z_3 = 20/316.2 = 0.063\ \text{A}$$

The required expression for the current flowing in the circuit is

$$i = 0.583\sin(1000t - 31°) + 0.08\sin(2000t) + 0.063\sin(3000t + 71.5°)\ \text{A} \quad (\textit{Ans.})$$

The power dissipated in the circuit is

$$P = [(100/\sqrt{2}) \times (0.583\cos 31°/\sqrt{2})] + [(40/\sqrt{2}) \times (0.08\cos 0°/\sqrt{2})] + [(20/\sqrt{2}) \times (0.063\cos(-71.5°)/\sqrt{2})]$$
$$= 24.98 + 1.6 + 0.2 = 26.78\ \text{W} \quad (\textit{Ans.})$$

The r.m.s. applied voltage is

$$V = \sqrt{[(100/\sqrt{2})^2 + (40/\sqrt{2})^2 + (20/\sqrt{2})^2]} = 77.46\ \text{V}$$

The r.m.s. current is

$$I = \sqrt{[(0.583/\sqrt{2})^2 + (0.08/\sqrt{2})^2 + (0.063/\sqrt{2})^2]}$$
$$= 0.418 \text{ A}$$

Therefore,

$$\text{Power factor} = 26.78/(77.46 \times 0.418) = 0.827 \quad (\textit{Ans.})$$

Example 18.12

A 3.18 μF capacitor is connected in parallel with a 1 kΩ resistor and the combination is connected in series with another 1 kΩ resistor. The voltage $v = 340\sin(100\pi t) + 100\sin(300\pi t + 20°)$ volts is applied across the circuit. Calculate (*a*) the power dissipated in the series resistor and (*b*) the r.m.s. voltage across the series resistor.

Solution

(*a*) At the fundamental frequency

$$X_C = 1/(100\pi \times 3.18 \times 10^{-6}) = 1000 \ \Omega$$
$$Z_1 = 1 + (1 \times -\text{j}1)/(1 - \text{j}1) \ \text{k}\Omega$$
$$= 1500 - \text{j}500 \ \Omega = 1581 \angle -18.4° \ \Omega$$

Hence,

$$|I_1| = 340/1581 = 0.215 \text{ A}$$

At the third harmonic,

$$X_C = 1000/3 \quad \text{and} \quad Z_3 = 1100 - \text{j}300 = 1140 \angle -15.3° \ \Omega$$

Hence,

$$|I_3| = 100/1140 = 0.088 \text{ A}$$

Power dissipated

$$= (0.215^2 \times 1000)/2 + (0.088^2 \times 1000)/2 = 27 \text{ W} \quad (\textit{Ans.})$$

(*b*) Voltage across 1000 Ω = $1000\sqrt{(0.215^2/2 + 0.088^2/2)}$ = 164.3 V (*Ans.*)

A.C. voltage superimposed upon d.c. voltage

Sometimes a complex wave consists of a d.c. voltage (or current) with an a.c. voltage (or current) superimposed upon it. This situation commonly arises in analogue electronic circuits. The d.c. component moves the a.c. component(s) in either the positive or the negative direction depending upon its polarity. The d.c. component of a complex waveform is treated in analysis in exactly the same way as any of the a.c. components, except, of course, its r.m.s. value is the same as the d.c. value.

Example 18.13

Determine the expression for the current flowing in the circuit given in Fig. 18.11 and calculate (*a*) the power dissipated in the circuit and (*b*) the power factor of the circuit.

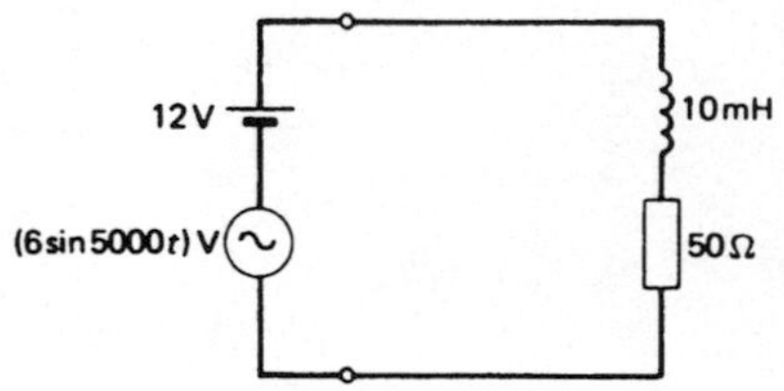

Fig. 18.11

Solution

At d.c. the impedance of the circuit is merely the resistance of 50 Ω. Therefore $I_{DC} = 12/50 = 0.24$ A.

At $\omega = 5000$ R/s, $Z = 50 + j50 = 70.7\angle 45°$ Ω and so $I_{ac} = 6/70.7 = 0.085$ A.

The d.c. power dissipated $= 0.24^2 \times 50 = 2.88$ W.

The a.c. power dissipated $= (0.085/\sqrt{2})^2 \times 50 = 0.181$ W.

The total power dissipated $= 2.88 + 0.181 = 3.061$ W (*Ans.*)

The r.m.s. voltage $V = \sqrt{[12^2 + (6/\sqrt{2})^2]} = 12.73$ V.

The r.m.s. current $I = \sqrt{[0.24^2 + (0.085/\sqrt{2})^2]} = 0.247$ A.

Therefore,

$$\text{Power factor} = 3.061/(12.73 \times 0.247) = 0.97 \quad (\textit{Ans.})$$

Mean value and form factor

Mean value

The *mean value* of a complex voltage wave is calculated using equation (18.26), i.e.

$$V_{mean} = (1/\pi)\int_0^{\pi} v \, d\omega t \qquad (18.26)$$

The value obtained for the mean value of a wave will be a function of the relative phase angles of the various harmonic components. This means that its computation may, in some cases, be a somewhat lengthy process.

Example 18.14

Calculate the mean value of the voltage

$$v = 100 \sin 314t + 15 \sin 942t + 12 \sin 1570t \text{ V}$$

Solution

$$I_{mean} = (1/\pi)\int_0^{\pi} [100 \sin \omega t + 15 \sin 3\omega t + 12 \sin 5\omega t] \, d\omega t$$

$$= (1/\pi)[-100 \cos \omega t - (15/3) \cos 3\omega t - (12/5) \sin 5\omega t]_0^{\pi}$$

$$= (1/\pi)[200 + 10 + 4.8] = 68.37 \text{ A} \quad (\textit{Ans.})$$

Form factor

The *form factor* of a complex wave is the ratio of the r.m.s. value to the mean value. The form factor of a sine wave is 1.11 and of a square wave is 1. The more removed a waveform is from sinusoidal the higher will be its form factor.

Example 18.15

Determine the form factor of the complex wave

$$v = 50\sin 200\pi t + 10\sin 400\pi t + 5\sin 600\pi t \text{ V}$$

Solution

$$V = \sqrt{[(50/\sqrt{2})^2 + (10/\sqrt{2})^2 + (5/\sqrt{2})^2]} = 36.23 \text{ V}$$

$$V_{\text{mean}} = (1/\pi)\int_0^{\pi} (50\sin\omega t + 10\sin 2\omega t + 5\sin 3\omega t)\mathrm{d}\omega t$$

$$= (1/\pi)[-50\cos\omega t - 5\cos 2\omega t - 1.67\cos 3\omega t]_0^{\pi}$$

$$= (1/\pi)(50 + 50 - 5 + 5 + 1.67 + 1.67) = 103.34/\pi$$

Therefore,

$$\text{Form factor} = (36.23 \times \pi)/103.34 = 1.1 \quad (\textit{Ans.})$$

Example 18.16

Determine the form factor of the voltage

$$v = 100\sin 314t + 15\sin 942t + 12\sin 1570t \text{ V}$$

Solution

$$V = \sqrt{[(100^2 + 15^2 + 12^2)/2]} = 72 \text{ V}.$$

$$V_{\text{mean}} = (1/\pi)\int_0^{\pi} (100\sin\omega t + 15\sin 3\omega t + 12\sin 5\omega t)\,\mathrm{d}t$$

$$= (1/\pi)[-100\cos\omega t - 5\cos 3\omega t - 2.4\cos 5\omega t]_0^{\pi}$$

$$= (1/\pi)[(100 + 5 + 2.4) - (-100 - 5 - 2.4)]$$

$$= 214.8/\pi \text{ V}$$

Therefore,

$$\text{Form factor} = 72\pi/214.8 = 1.05 \quad (\textit{Ans.})$$

Effect of harmonics on component measurement

When harmonics are present in a current or voltage waveform the measurement of an impedance will be subject to some error. The error arises because the percentage harmonic in the voltage wave will always differ from the percentage harmonic in the current wave. If the circuit is capacitive, the current wave will possess the larger harmonic content, while for an inductive circuit the voltage will have a higher percentage harmonic.

Example 18.17

The current $i = (10 \sin 314t + 3 \sin 942t)$ A flows in an inductance. At the fundamental frequency of 50 Hz the ratio reactance/resistance for the inductance is 5. Calculate the percentage error involved in measuring the impedance of the inductance as the ratio V/I.

Solution
At the fundamental frequency the impedance of the inductance is

$$Z_1 = R + j\omega L = R + j5R \quad \text{and} \quad |Z_1| = (\sqrt{26})R \ \Omega$$

At the third harmonic X_L will be j15R and so

$$Z_3 = R + j15R \qquad |Z_3| = (\sqrt{226})R \ \Omega$$

Hence, the voltage v across the inductance is

$$v = 10(\sqrt{26})R \sin 314t + 3(\sqrt{226})R \sin 942t \text{ V}$$

Therefore,

$$V = \sqrt{[(10\sqrt{26}R/\sqrt{2})^2 + (3\sqrt{226}R/\sqrt{2})^2]} = 48.14R \text{ V}$$

and $I = \sqrt{[(10/\sqrt{2})^2 + (3/\sqrt{2})^2]} = 7.38$ A

The apparent impedance of the inductance is

$$V/I = 48.14/7.38 = 6.52R$$

The true impedance of the inductance is

$$V_1/I_1 = (\sqrt{26})R = 5.1R$$

Therefore, the percentage error in the measurement of the impedance of the inductor is

$$\% \text{ error} = [(6.52 - 5.1)/5.1] \times 100 = 27.84\% \quad (\textit{Ans.})$$

Example 18.18

Determine the percentage error in measuring the impedance of a coil as voltage/current if the current waveform contains 18% third harmonic, and 5% fifth harmonic. At the fundamental frequency the ratio reactance/resistance = 12.

Solution
At $f_1, X_L = 5R$ and $Z = \sqrt{[R^2 + (12R)^2]} = R\sqrt{145}\ \Omega$. At f_3, $X_L = 36R$ and $Z_3 = \sqrt{[R^2 + (36R)^2]} = R\sqrt{1297}\ \Omega$. At f_5, $X_L = 60R$

and $Z_5 = \sqrt{[R^2 + (60R)^2]} = R\sqrt{3601}\ \Omega$. Hence,

$$i = I_1 \sin \omega t + 0.18I_1 \sin 3\omega t + 0.05I_1 \sin 5\omega t \text{ A} \quad \text{and}$$

$$v = R\sqrt{145}I_1 \sin \omega t + (R\sqrt{1297} \times 0.18I_1 \sin 3\omega t) + (R\sqrt{3601} \times 0.05I_1 \sin 5\omega t) \text{ V}$$

$$I = \sqrt{[(1^2 + 0.18^2 + 0.05^2)/2]} = 0.719I_1.$$

$$V = I_1 R\sqrt{[(145)^2/2 + (0.18\sqrt{1297})^2/2 + (0.05\sqrt{3601})^2/2]} = 9.875I_1 R.$$

Effective impedance

$$Z = (9.875I_1 R)/(0.719I_1) = 13.73R\ \Omega.$$

At the fundamental frequency, $Z = R\sqrt{145} = 12.04R\ \Omega$. Therefore,

$$\% \text{ error} = [(13.73 - 12.04)/12.04] \times 100 = 14.04\% \quad (Ans.)$$

Selective resonance

Another effect that the presence of harmonics in a waveform may have is known as *selective resonance.* This term means that a circuit containing both inductance and capacitance may resonate at any one of the harmonic frequencies. In general, this is an undesirable effect since it may result in a dangerously high voltage appearing across the capacitance at some frequency. It is particularly undesirable in power circuits where the unwanted resonant current could be very high. Selective resonance will occur at a harmonic frequency when $n\omega L = 1/n\omega C$, where n is the order of the harmonic and ω is 2π times the fundamental frequency.

Example 18.19

A complex voltage with a fundamental frequency of 50 Hz is applied to a circuit that consists of a 9.9 mH inductance in series with a 22 μF capacitor. Determine the harmonic of the fundamental frequency at which the circuit is resonant.

Solution

$n = 1/[\omega\sqrt{(LC)}] = 1/[100\pi\sqrt{(9.9 \times 10^{-3} \times 22 \times 10^{-6})}] = 7 \quad (Ans.)$

Example 18.20

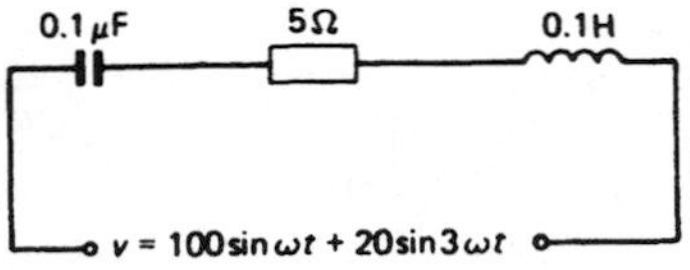

Fig. 18.12

The circuit shown in Fig. 18.12 is resonant at the third harmonic frequency of the applied voltage. Calculate (*a*) the fundamental frequency of the applied voltage, (*b*) the current flowing at the fundamental frequency and (*c*) the current flowing at the third harmonic frequency.

Solution

(*a*) $3\omega \times 0.1 = 1/(3\omega \times 0.1 \times 10^{-6})$

Fundamental frequency = 530.5 Hz (*Ans.*)

(*b*) $X_{C1} = 1/(2\pi \times 530.5 \times 10^{-7}) = -j3000\ \Omega$
$X_{L1} = 2\pi \times 530.5 \times 0.1 = j333.3\ \Omega$
$Z_1 = 5 - j2666.7\ \Omega$
$I_1 = V_1/Z_1 = 100/(5 - j2666.7) = 37.5\angle 89.9°$ mA (*Ans.*)

(*c*) $I_3 = V_3/R = 20/5 = 4$ A (*Ans.*)

Example 18.21

A complex voltage wave has a fundamental frequency of 1 kHz and contains one harmonic component. The amplitude of this harmonic is 5% of the fundamental voltage of 20 V. The voltage is applied to a circuit which consists of a 31.8 mH inductor of 4 Ω resistance in series with a variable capacitor. Resonance occurs at the harmonic frequency when the capacitor is set to 31.8 nF. Calculate (*a*) the order of the harmonic and (*b*) the ratio of the currents that flow at the harmonic and the fundamental frequencies.

Solution
(*a*) $n = 1/[(2\pi \times 1000)\sqrt{(31.8 \times 10^{-3} \times 31.8 \times 10^{-9})}] = 5$ (*Ans.*)
(*b*) At 5 kHz, the total reactance of the circuit is zero, so that $Z = R = 4\ \Omega$. Hence $I_5 = (20 \times 5)/(100 \times 4) = 0.25$ A. At 1 kHz, $X_L = 2\pi \times 1000 \times 31.8 \times 10^{-3} \approx 200\ \Omega$ and $X_C = 1/(2\pi \times 1000 \times 31.8 \times 10^{-9}) \approx 5005\ \Omega$. $I_1 = 20/[4 - j(5005 - 200)] \approx 4.162$ mA. Therefore,

Ratio $I_5/I_1 = 250/4.162 = 60$ (*Ans.*)

Exercises 18

18.1 A current $i = 120 \sin \omega t + 40 \sin 3\omega t + 15 \sin 5\omega t$ mA flows in a coil. Calculate the percentage error incurred if the impedance of the coil is calculated as the ratio of r.m.s. voltage to r.m.s. current. At the fundamental frequency $\omega/2\pi$ Hz the reactance of the coil is six times greater than the resistance of the coil.

18.2 A coil has an inductance of 30 mH and a resistance of 40 Ω and is connected in parallel with a capacitor of capacitance 8.33 μF. The circuit has the voltage

$$v = 200 \sin 10^3 t + 50 \sin 3 \times 10^3 t \text{ volts}$$

applied across it. Obtain the expression for the current taken from the source and calculate the power factor of the circuit.

18.3 The voltage $v = 200 \sin 300t + 60 \sin(900t + \frac{1}{4}\pi) + 30 \sin(1500t - \frac{1}{4}\pi)$ volts is applied across a circuit that consists of a coil in series with a capacitor. If the inductance of the coil is 100 mH and its resistance is 25 Ω calculate the capacitance value that will make the circuit resonant at the third harmonic frequency. Also find the power factor of the circuit with this value of capacitance.

18.4 The voltage $v = 100 \sin 500t + 40 \sin 1000t$ is applied across a circuit that consists of a coil in parallel with a capacitor. If the component values are, for the coil, $L = 1$ H, $R = 1000\ \Omega$; and capacitor

$C = 0.5\ \mu F$, calculate (*a*) the expression for the circuit current and (*b*) the power factor of the circuit.

18.5 For the circuit shown in Fig. 18.13 calculate (*a*) the r.m.s. voltage, (*b*) the r.m.s. current and (*c*) the power factor of the circuit.

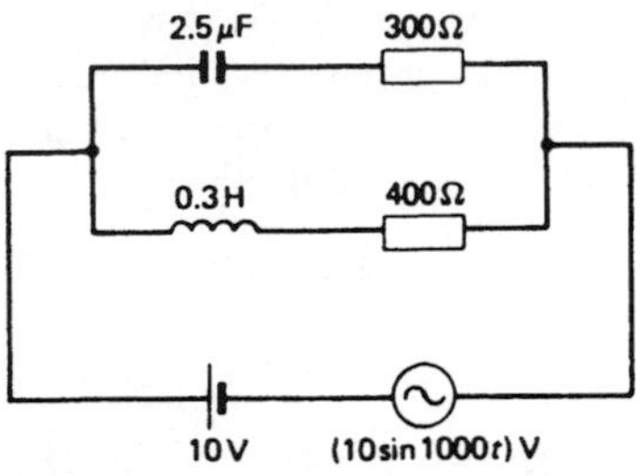

Fig. 18.13

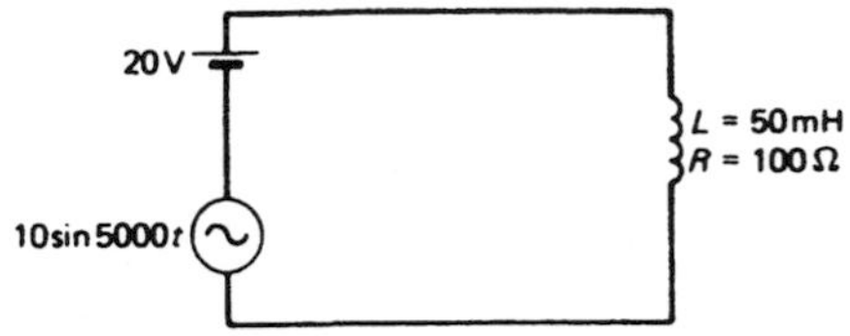

Fig. 18.14

18.6 Calculate the total power dissipated in the circuit shown in Fig. 18.14. Also find the power factor of the circuit.

18.7 Calculate the admittance of both of the series circuits given in Fig. 18.15 at frequencies of (*i*) $500/2\pi$ Hz and (*ii*) $1000/2\pi$ Hz. The two circuits are connected in parallel. Calculate their total admittance at each frequency.

If the voltage $v = 10\sin 500t + 4\sin 1000t$ volts is applied across the circuit determine the r.m s. value of the current supplied to the circuit.

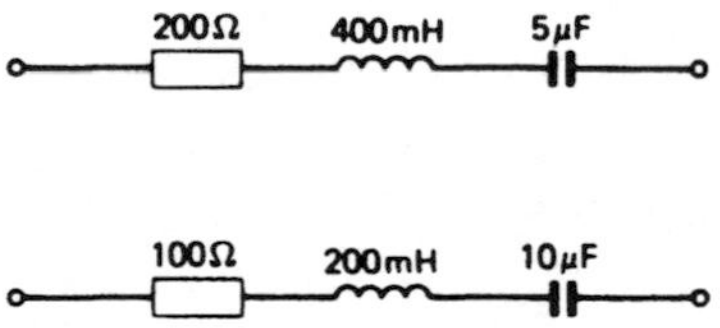

Fig. 18.15

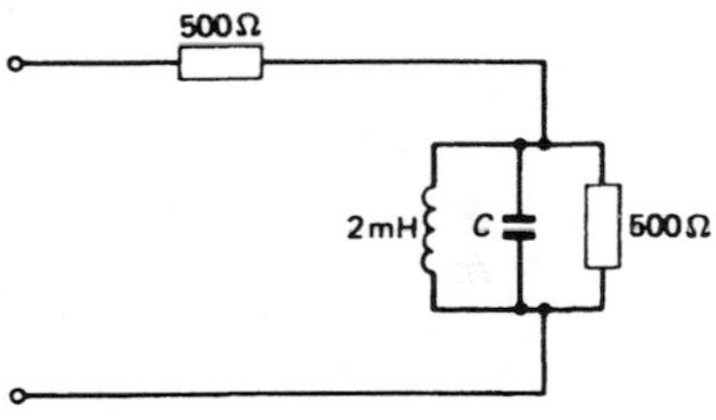

Fig. 18.16

18.8 The voltage applied to the circuit given in Fig. 18.16 is

$$v = 12\sin 4000t + \tfrac{1}{3}\sin 12000t + \tfrac{1}{5}\sin 20000t)\ \text{volts}$$

Calculate the r.m.s. value of the applied voltage. Determine the value of C that will cause the circuit to resonate at $20000/2\pi$ Hz. Calculate the peak value of the voltage across the capacitor at this frequency. Also find the impedance of the circuit at $12000/2\pi$ Hz.

18.9 The voltage $v = 1\sin \omega t - (\sin 3\omega t)/9 + (\sin 5\omega t)/25$ is applied to a circuit that consists of a 100 Ω resistor in parallel with a 2 μF capacitor. If the fundamental frequency of the applied voltage is 5000 rad/s, (*a*) determine the waveshapes of the currents in the circuit and (*b*) calculate the power dissipated in the circuit.

18.10 Calculate the r.m.s. value of the current
$i = 100\sin 314t + 15\sin 942t + 12\sin 1570t$ mA.

18.11 A 50 mH inductor of resistance 10 Ω is connected across the voltage $v = 100 \sin 100t + 200 \sin 300t$ V. Obtain an expression for the current flowing in the circuit. Calculate (*a*) the power dissipated and (*b*) the power factor.

18.12 A rectangular voltage is applied across the input terminals of a low-pass filter which has series inductances of 40 mH and shunt capacitance of (1/9) μF. The output terminals are closed in a resistance equal to $\sqrt{(L/C)}$ Ω. If the fundamental frequency is 10000 rad/s determine the waveform of the output voltage.

18.13 The voltage $v = 1000 \sin 100\pi t + 100 \sin 300\pi t$ V is applied across a 0.5 H inductor of self-resistance 100 Ω. Calculate (*a*) the r.m.s. voltage, (*b*) the r.m.s. current and (*c*) the power dissipated in the circuit.

18.14 A 100 Ω resistor is connected in series with a 2 μF capacitor and a 20 mH inductor. The voltage $v = 1 \sin 5000t + 0.5 \sin 10000t$ V is applied across the circuit. Calculate (*a*) the r.m.s. voltage, (*b*) the r.m.s. current, (*c*) the power dissipated and (*d*) the power factor.

18.15 (*a*) Calculate the r.m.s. value of the voltage

$$v = 40 + 10 \sin 3000t + 5 \cos 15000t \text{ V}$$

(*b*) Calculate the mean value of the voltage

$$v = 1 \sin 5000t + 0.4 \sin(10000t + 36°) \text{ V}$$

(*c*) Draw a fundamental wave with 25% second harmonic content with the harmonic in anti-phase at time $t = 0$. Sketch the resultant waveform.

18.16 A capacitor having a reactance of 250 Ω at a frequency of $5000/2\pi$ Hz is connected in series with an inductance whose reactance at the same frequency is 150 Ω. The voltage

$$v = 100 \sin 5000t + 20 \sin 15000t$$

is applied across the circuit. Obtain an expression for the current flowing in the circuit.

18.17 A 100 mH inductor of 1 Ω resistance is connected in series with a 0.25 μF capacitor. The voltage $v = 20 \sin \omega t + 5 \sin 3\omega t + 1 \sin 5\omega t$ is applied across the circuit. If the circuit is resonant at its third harmonic frequency, calculate (*a*) the fundamental frequency, (*b*) the r.m.s. applied voltage and (*c*) the r.m.s. current in the circuit.

18.18 The voltage $1 \sin 3000t + 0.5 \sin 6000t + 0.25 \sin 9000t$ is applied to the input terminals of a T network. The T network has series capacitors of 0.2 μF and a shunt inductance of 0.1 H and it is terminated by a resistor $R = \sqrt{(L/C)}$ Ω. Calculate the components of the output voltage of the network.

18.19 The output current of a circuit is $i = 600 + 600 \sin \omega t + 200 \sin 3\omega t$ mA. Plot one cycle of the current waveform. The third harmonic component suffers a phase lag of 180°. Plot the new waveform.

18.20 Calculate the form factor of the current $i = 2.5 \sin 157t + 0.7 \sin 471t + 0.4 \sin 785t$ A.

18.21 A circuit consists of a 50 mH inductor of resistance 50 Ω connected in series with a variable loss-free capacitor. The circuit is connected across the voltage $v = 200 \sin 500t + 50 \sin 1500t + 25 \sin 2500t$ V. The capacitor is adjusted until the circuit resonates at its third harmonic frequency. (*a*) Calculate the capacitor value. (*b*) Obtain an expression for the current.

18.22 The voltage $v = 300 \sin 300t + 60 \sin 900t$ V is applied across a circuit having two parallel branches. One branch contains a 0.1 H inductance in series with a 40 Ω resistance, and the other branch is a 20 Ω resistor in series with a 200 μF capacitor. Calculate the r.m.s. value of the current supplied to the circuit.

18.23 A variable inductance of negligible resistance is connected in series with a 10 Ω resistor and a 30 μF capacitor. The voltage $v = 2 \sin \omega t + 0.4 \sin 3\omega t + 0.1 \sin 5\omega t$ kV is applied to the circuit. The inductance is adjusted until its third harmonic resonance is obtained. Calculate the power dissipated in the circuit.

18.24 The current $i = 100 \sin 314t + 15 \sin 942t + 12 \sin 1570t$ flows in an inductance whose resistance is negligible. If the r.m.s. voltage is 75 V determine the percentage error involved in calculating the inductance value from $X = V/I$.

18.25 A loss-free 50 mH inductance is connected in parallel with a 20 μF capacitor and these components are then connected in series with a 25 Ω resistor. Calculate the power factor of the circuit when the voltage $v = 100 \sin 500t + 40 \sin 1000t + 20 \sin 1500t$ is applied across the circuit.

18.26 Determine the form factor of the voltage $v = 100 \sin 314t + 24 \sin 942t + 10 \sin 1570t$ V.

18.27 (*a*) Obtain the first three terms of the Fourier series of a square wave of amplitude 1 V. (*b*) Calculate the r.m.s. value of the terms and compare with the actual value of 1 V. Comment on the result.

18.28 The voltage $v = 100 \sin 1000t + 50 \sin(3000t - 30°)$ V is applied to a circuit consisting of a 5 Ω resistor in series with 10 Ω in parallel with j5 Ω. Calculate the power dissipated in the 5 Ω resistor.

19 Non-linear circuits

The waveforms produced by a.c. generators are usually very nearly sinusoidal but may often contain some harmonics. The harmonic content of most sinusoidal oscillators is very small indeed and usually need only be taken into account when precise measurements are carried out. Most of the harmonic content of waveforms is produced when an originally sinusoidal waveform is applied, intentionally or not, to a device having a non-linear current–voltage characteristic.

The *non-linear device* may be a non-linear resistor, a semiconductor diode, a suitably biased transistor, or the core of a transformer.

Non-linear devices

Iron core of a transformer

When a current flows in the winding of an iron-cored transformer a magnetic flux is set up in the core. The relationship between the magnetizing current and the flux set up in the core is not linear. This means that a sinusoidal magnetizing current will not produce a sinusoidal variation of the core flux.

The point is illustrated by Fig. 19.1 which shows a sinusoidal magnetizing current applied to the hysteresis loop of a core. The hysteresis loop has been drawn to exaggerate the non-linearity produced. Because the core flux is not sinusoidal the e.m.f. induced into the secondary winding will also have a non-sinusoidal waveform. Since $e = N\,d\Phi/dt$ the waveform of the secondary e.m.f. can be deduced and it is shown in Fig. 19.2. Clearly the waveform is far from sinusoidal and this shows that it is essential for a core not to be taken into saturation if waveform distortion is to be minimized. If the magnetizing force is kept small enough, fairly linear operation is possible.

It should be evident from Fig. 19.1 that if a sinusoidal core flux is needed the waveform of the magnetizing current will have to be suitably distorted. The necessary waveform can be deduced by projecting from the hysteresis loop, as has been done in Fig. 19.3.

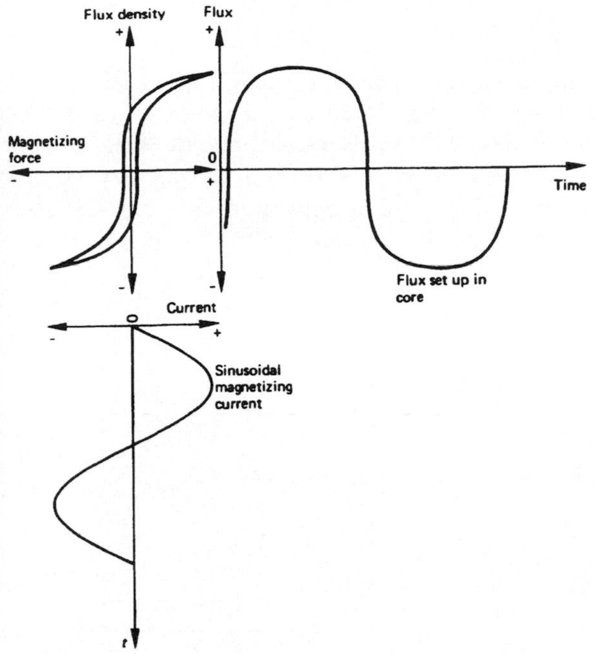

Fig. 19.1 Effect of an iron core on flux waveform

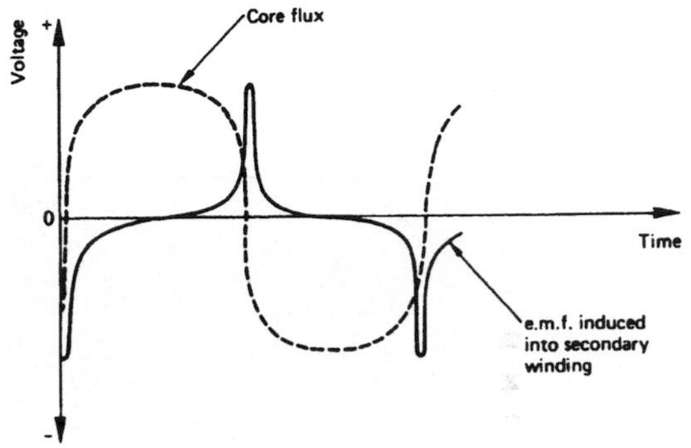

Fig. 19.2 Waveform of the secondary e.m.f. for the core flux of Fig. 19.3

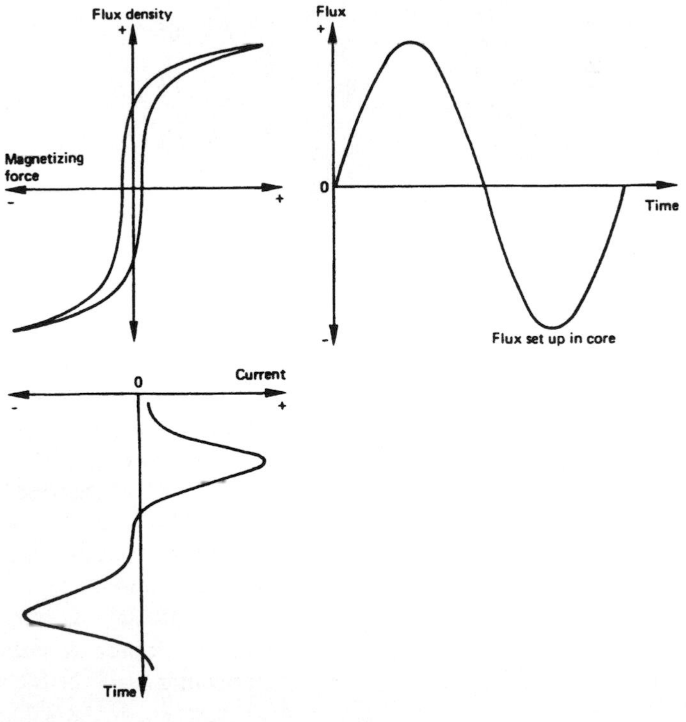

Fig. 19.3 Magnetizing current waveform needed to provide a sinusoidal flux waveform

Example 19.1

The voltage applied to the primary winding of a 50 Hz iron-cored transformer is $v = 100 \sin \omega t - 50 \cos 3\omega t$ volts. The transformer core has a cross-sectional area of 8 cm^2 and its length is 38 cm. (*a*) If there are 600 turns in the primary winding, derive an expression for the flux density in the core. (*b*) If the magnetizing force is $H = 120 \sin(\omega t - 60°) - 48 \cos(3\omega t - 20°)$ calculate the total iron loss.

Solution

(*a*) $v_p = N_p \, d\varphi/dt$ and hence

$$\int \varphi = (1/N_p) \int v_p \, dt = (1/N_p) \int (100 \sin \omega t - 50 \cos 3\omega t) \int dt$$

$$= -[(100 \cos \omega t)/\omega N_p + (50 \sin 3\omega t)/3\omega N_p].$$

$B = \varphi/A$, so that

$$B = -[(100 \cos \omega t)/\omega N_p A + (50 \sin 3\omega t)/3\omega N_p A]$$

$$= -\left[\frac{100 \cos 100\pi t}{100\pi \times 600 \times 8 \times 10^{-4}} + \frac{50 \sin 300\pi t}{300\pi \times 600 \times 8 \times 10^{-4}}\right]$$

or $B = -[0.663 \cos 100\pi t + 0.111 \sin 3\omega t]$ (*Ans.*)

(*b*) $I = Hl/N_p = [(120 \times 0.38)/600][\sin(\omega t - 60°)]$

$$- [(48 \times 0.38)/600][\cos(3\omega t - 20°)]$$

$$= 0.076 \sin(\omega t - 60°) - 0.03 \cos(3\omega t - 20°)$$

$$P = 0.5 V_1 I_1 \cos \varphi_1 + 0.5 V_3 I_3 \cos \varphi_3$$

$$= 0.5 \times 100 \times 0.076 \times 0.5 + 0.5 \times 50 \times 0.03 \times 0.94$$

$= 2.61$ W (*Ans.*)

Non-linear resistors

Non-linear resistors have been available for many years and are generally considered to be in one of two main groups. One group contains those resistors whose resistance is dependent upon temperature and these are collectively known as *thermistors*. Some thermistors have a positive temperature coefficient while others have a negative temperature coefficient. The other group of non-linear resistors contains components whose resistance is a function of the applied voltage; these are collectively known as *varistors* or as *voltage-dependent resistors*.

A VDR has a current–voltage characteristic of the form $i = av^n$, where a and n are constants for a particular component.

Diodes and transistors

The current–voltage characteristic of a semiconductor diode or the mutual characteristic of a suitably biased transistor is also non-linear. Figure 19.4, for example, shows the mutual characteristic of a bipolar transistor. If a sinusoidal voltage of 0.2 V peak is applied about a bias voltage of 0.8 V, the waveform of the collector current is clearly not sinusoidal. This means that the output current waveform will contain extra frequencies at harmonics of the input signal frequency.

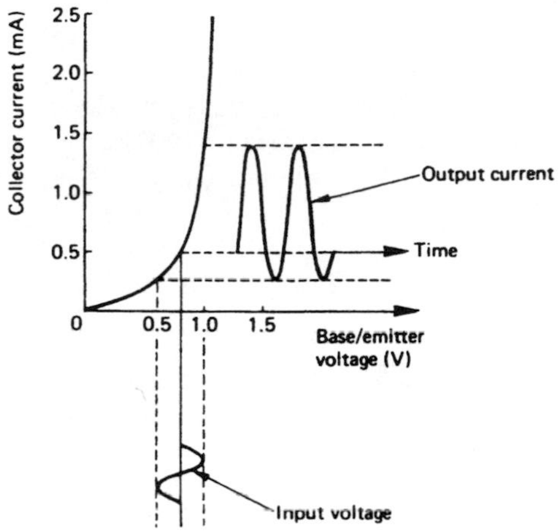

Fig. 19.4 Mutual characteristic of a bipolar transistor

Example 19.2

The current–voltage characteristic of a diode is approximately a straight line with a slope of 100 Ω passing through the point $V = 0.5$, $VI = 0$ in the forward direction, and a constant resistance of 10 kΩ in the reverse direction. The diode is connected in series with a sinusoidal voltage source of peak value 4 V and a load resistance of 500 Ω. Calculate the mean current through the diode.

Solution
Peak forward current $= (4 - 0.5)/600 = 5.83$ mA. Peak current in reverse direction $= 4/10\,500 = 0.38$ mA. The limits of forward current flow are from $\angle[\sin^{-1}(0.5/4)]$ to $\angle[180° - \sin^{-1}(0.5/4)]$, i.e. from 7.2° to 172.8°. Therefore

$$I_{\text{mean}} = \frac{1}{2\pi}\left[\int_{7.2°}^{172.8°}\left(\frac{4\sin\omega t - 0.5}{600}\right) d\omega t - \int_{\pi}^{2\pi} 0.38 \sin\omega t \; d\omega t\right]$$
$$= 1.6 \text{ mA} \quad (\textit{Ans.})$$

Determination of non-linear characteristics

The calculation of the current that flows in a non-linear device connected either in series or in parallel with a linear resistor and/or another non-linear device can be carried out either analytically or graphically. First, however, it is desirable to be able to determine an expression that will adequately describe the current–voltage characteristic of the device.

The current–voltage characteristic of most, if not all, non-linear devices can be expressed in the form of a power series:

$$i = a + bv + cv^2 + dv^3 \cdots \tag{19.1}$$

where a, b, c, etc, are constants, and v is the applied voltage. In a particular case any one, or more, of the constants may be equal to zero. In the majority of cases it is sufficient to consider only the first three terms, when the device is often said to be a 'square-law' device.

To determine the values of the three constants a, b and c, the current that flows in the device must be measured for three different values of applied voltage.

Example 19.3

Table 19.1

Applied voltage (V)	1	2	3
Current (mA)	4.2	8	13.4

Measurements of a non-linear device gave the data in Table 19.1. The characteristic may be assumed to be a square law. Determine the constants a, b, and c. Calculate (a) the mean value of the current when the applied voltage is $1 \sin \omega t$ volts and (b) the percentage second harmonic content of the current.

Solution

From equation (19.1)

$$4.2 = a + b + c \tag{19.2}$$

$$8 = a + 2b + 4c \tag{19.3}$$

$$13.4 = a + 3b + 9c \tag{19.4}$$

Solving these equations gives $a = 2$, $b = 1.4$ and $c = 0.8$. Hence the current/voltage characteristic of the device is

$$i = 2 + 1.4v + 0.8v^2 \text{ mA} \tag{19.5}$$

(a) When $v = 1 \sin \omega t$ volts,

$$i = 2 + 1.4 \sin \omega t + 0.8 \sin^2 \omega t \text{ mA}$$
$$= 2 + 1.4 \sin \omega t + 0.4 - 0.4 \cos 2\omega t$$

(A trig. identity is $\sin^2 A = (1 - \cos 2A)/2$.).

$$i = 2.4 + 1.4 \sin \omega t - 0.4 \cos 2\omega t \text{ mA}$$

Mean current = 2.4 mA (*Ans.*)

(b) % second harmonic content

$$= \frac{\text{Amplitude of second harmonic}}{\text{Amplitude of fundamental}} \times 100\%$$

$= (0.4/1.4) \times 100 = 28.57\%$ (*Ans.*)

This example shows how a second harmonic component is produced by an element with a non-linear current–voltage characteristic. If a cubic term dv^3 had been included in the problem a third harmonic term would also have made its appearance.

If two sinusoidal signals at frequencies f_1 and f_2 are applied to a square-law device, components at the sum $f_1 + f_2$ and the difference $f_1 - f_2$ would be generated. These are known as *intermodulation components*, and they are of considerable importance in telecommunication engineering.

Example 19.4

The current–voltage characteristic of a non-linear device is $i = 0.05V + 4 \times 10^{-4}V^2$ A, where V is in volts. A sinusoidal voltage of 141.42 V peak is applied to the device. (*a*) Obtain an expression for the current. (*b*) Calculate the r.m.s. value of the current waveform.

Solution

(*a*) $i = 0.05 \times 141.42 \sin \omega t + 4 \times 10^{-4} \times (141.42 \sin \omega t)^2$

$= 7.07 \sin \omega t + 4(1 - \cos 2\omega t)$

$= 4 + 7.07 \sin \omega t - 4 \cos 2\omega t\text{A}$ (*Ans.*)

(*b*) $I = \sqrt{[(4^2 + (7.07^2/2) + (4^2/2)]} = 7$ mA (*Ans.*)

Harmonic distortion

When a signal is applied to a non-linear device new frequencies are generated. The output signal then contains components at frequencies that were not present in the input signal. If these new frequencies are harmonically related to frequency of the input signal *harmonic distortion* will have occurred. Harmonic distortion of a signal waveform is usually expressed as the square root of the sum of the squares of the individual harmonic components quoted as a percentage of the fundamental frequency component. Thus;

$$\text{Percentage harmonic distortion} = \{[\sqrt{(V_{2H}^2 + V_{3H}^2 + V_{4H}^2 + \cdots)}]/V_F\} \times 100\% \qquad (19.6)$$

Example 19.5

The output voltage of a circuit is $v = 20 \sin \omega t + 5 \sin 2\omega t + 1.4 \sin 3\omega t$ V. (*a*) Calculate the power dissipated in a 600 Ω load resistor. (*b*) Calculate the percentage second harmonic distortion in (*i*) the output voltage and (*ii*) the output power.

Solution

(*a*) $V = \sqrt{(20^2/2 + 5^2/2 + 1.4^2/2)} = 14.61$ V

$P_L = 14.61^2/600 = 355.8$ mW (*Ans.*)

(*b*) (*i*) Harmonic distortion $= \{[\sqrt{(25 + 1.96)}]/20\} \times 100 = 26\%$ (*Ans.*)

(*ii*) Harmonic distortion $= [(25 + 1.96)/400] \times 100 = 6.74\%$ (*Ans.*)

Calculation of currents in circuits containing a non-linear device

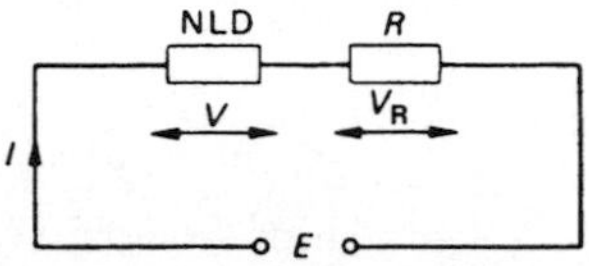

Fig. 19.5 A non-linear device in series with a linear resistor

The calculation of the current flowing in a circuit in which a non-linear device is connected in series with a linear resistor R, as in Fig. 19.5, can, in principle, be carried out either analytically or graphically. Depending upon the actual circuit and/or the non-linear characteristic one method or the other may turn out to be considerably easier in a particular case.

When a voltage E is applied to the circuit a current I will flow. The voltage V_R across the linear resistor R will then be $V_R = IR$ and the voltage V across the non-linear device will be $V = E - V_R$. Substitution of this value of V into the current–voltage characteristic of the non-linear device will then produce the value of the current I.

Alternatively, the current–voltage characteristic of the non-linear device can be plotted. The voltage across the linear resistor should then be taken as $V_R = E - V$ and the current–voltage characteristic of this component plotted on the same axes. Since $I = (E - V)/R$ this characteristic is linear and therefore only two points are needed (see Fig. 19.6). The point of intersection of the two graphs gives the current in the circuit and the voltages across each of the components as shown by the figure.

When a sinusoidal voltage is applied to a circuit similar to Fig. 19.5, a similar procedure should be followed if the peak current and/or voltages in the circuit are to be calculated.

Example 19.6

A non-linear device whose current–voltage characteristic is given by $i = 0.1V + 0.02V^2$ A is connected in series with a linear resistor of 10 Ω. A sinusoidal voltage of peak value 10 V is applied across the circuit.

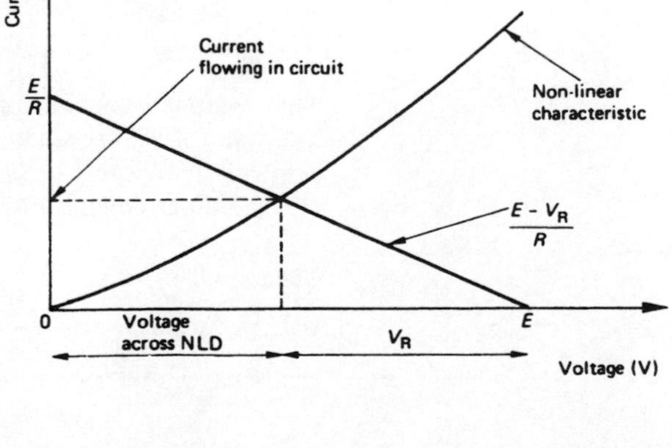

Fig. 19.6 Graphical determination of the current and voltages

Calculate (a) the peak current flowing in the circuit and (b) the voltages across each of the two components when the peak current is flowing. Use first a graphical method, and then an analytical method of solution.

Solution

Graphical method

Using the figures given in Table 19.2 the current–voltage characteristic of the non-linear device has been plotted in Fig. 19.7. The linear characteristic $(E-V)/10$ is also plotted, for the peak supply voltage of 10 V.

Table 19.2

V (volts)	1	2	3	4	5	6	7	8	9	10
$0.1V$	0.1	0.2	0.3	0.4	0.5	0.6	0.7	0.8	0.9	1.0
V^2	1	4	9	16	25	36	49	64	81	100
$0.02V^2$	0.02	0.08	0.18	0.32	0.5	0.72	0.98	1.28	1.62	2.0
I (A)	0.12	0.28	0.48	0.72	1.0	1.32	1.68	2.08	2.52	3.0

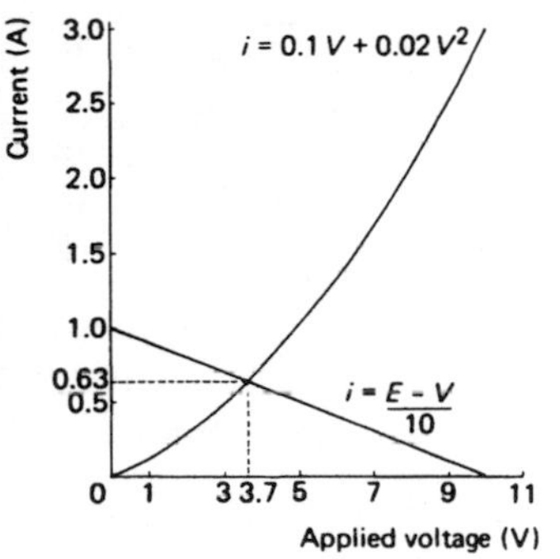

Fig.19.7

From the point of intersection of the two graphs:

(a) Peak current = 0.63 A (*Ans.*)
(b) The voltage across the non-linear device = 3.7 V (*Ans.*)
The voltage across the 10 Ω resistor = 6.3 V (*Ans.*)

Analytical method

The voltage across the non-linear device is

$$E-V=E-IR=10-10I$$

Hence

$$I=0.1(10-10I)+0.02(10-10I)^2=1-I+2-4I+2I^2$$
$$2I^2-6I+3=0$$
$$I=[6\pm\sqrt{(36-24)}]/4=2.37 \text{ A or } 0.63 \text{ A}$$

Clearly, the first result cannot be correct since it gives $IR=23.7$ V which is greater than the applied voltage. Therefore,

(a) $I=0.63$ A (*Ans.*)
(b) $V=6.3$ V (*Ans.*)
$V=10-6.3=3.7$ V (*Ans.*)

Example 19.7

A linear resistor is connected in series with a non-linear device whose current–voltage characteristic is $i=10^{-3}V^3$. A sinusoidal voltage of 2 V peak value in series with a d.c. voltage of 10 V is applied across the circuit. Calculate (a) the peak voltage that appears across each component and (b) the minimum value for the linear resistor, if the peak current is to be limited to 0.6 A.

Solution

(a) Plot a graph of current $i=10^{-3}V^3$ against voltage V over the range $V=0$ to $V=12$ V. Then plot the linear characteristic

$(12 - V)/R$ from the point $V = 12$ V with a slope such that it intersects with the non-linear characteristic at the point where $i = 0.6$ A. Projecting downwards to the horizontal (voltage) axis gives the voltage across the non-linear device as 8.3 V. Hence,

Voltage across $R = 12 - 8.3 = 3.7$ V (*Ans.*)

(*b*) $R = 3.7/0.6 = 6.17\ \Omega$ (*Ans.*)

If the waveform of the current is required, a slightly different approach is necessary. The supply voltage is always the sum of the voltages dropped across the non-linear device and the linear resistor. This fact will allow a *composite* current/voltage characteristic to be drawn. Considering again the circuit of Example 19.6, Table 19.3 has been drawn up. (The current values have been obtained from Table 19.2.)

Table 19.3

V (volts)	1	2	3	4	5	6	7	8	9	10
I (A)	0.12	0.28	0.48	0.72	1.0	1.32	1.68	2.08	2.52	3.0
IR (volts)	1.2	2.8	4.8	7.2	10.0	13.2	16.8	20.8	25.2	30.0
E (volts)	2.2	4.8	7.8	11.2	15.0	19.2	23.8	28.8	34.2	40.0

The current I is plotted against the applied voltage E in Fig. 19.8; note that as before the current that flows when E is 10 V is 0.63 A and that the characteristic has been linearized. The waveform of the current is obtained by projecting the

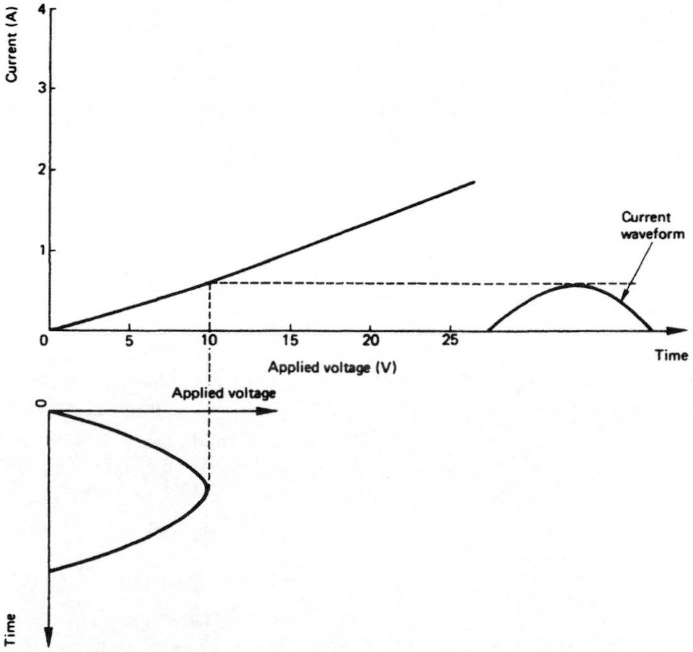

Fig. 19.8

applied voltage onto the characteristic as shown for one half-cycle only.

A similar approach can be employed when considering a parallel circuit containing a non-linear device in one of its branches. Figure 19.9 shows one possible arrangement. The total current I_T is the sum of the currents I_R and I_N flowing in the two branches of the circuit. If V is the voltage developed across the combination, then $I_R = V/R$ and I_N is given by the particular non-linear characteristic in use. Thus, in general,

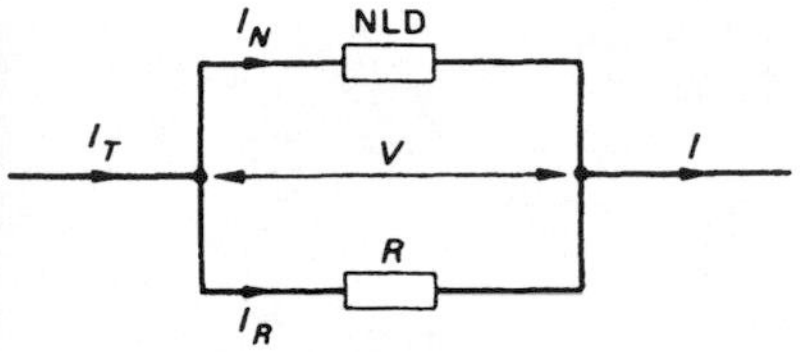

Fig. 19.9 A non-linear device in parallel with a linear resistor

$$I_T = (V/R) + a + bV + cV^2 + \cdots$$

If the total current I_T is plotted against voltage the voltage required to produce a given total current can be determined.

Example 19.8

A 100 Ω resistor is connected in parallel with a non-linear device whose $I-V$ characteristic is given by $I = 5V^2$ mA. The peak current supplied to the circuit is to be limited to 100 mA when a voltage is applied across the circuit. Calculate (a) the maximum permissible value of the applied voltage and (b) the current flowing in each component.

Solution

(a) Current in resistor $= V/100 = 10V$ mA. The total current $I_T = 10V + 5V^2$ mA, from which Table 19.4 can be obtained. This data is plotted in Fig. 19.10. From the graph it is evident that for a maximum current of 100 mA the peak value of the applied voltage must be limited to 3.6 V (*Ans.*)

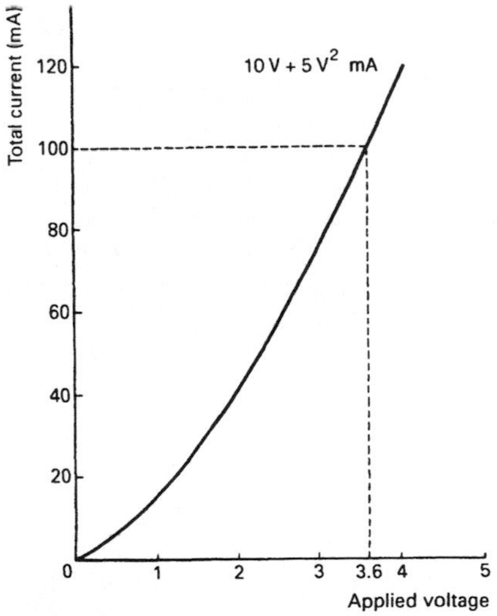

Fig. 19.10

Table 19.4

V (volts)	1	2	3	4	5
I (mA)	15	40	75	120	175

(b) With this value of applied voltage the current flowing in the linear resistor is

$3.6/100 = 36$ mA (*Ans.*)

The current flowing in the non-linear resistor is then the difference between the total current and the current in the linear resistor, i.e.

$I = 100 - 36 = 64$ mA (*Ans.*)

The peak value of the applied voltage can also be calculated using the same analytical method as before. Thus,

$100 = 10V + 5V^2$ or $V^2 + 2V - 20 = 0$

Solving this equation gives $V = 3.6$ V as before.

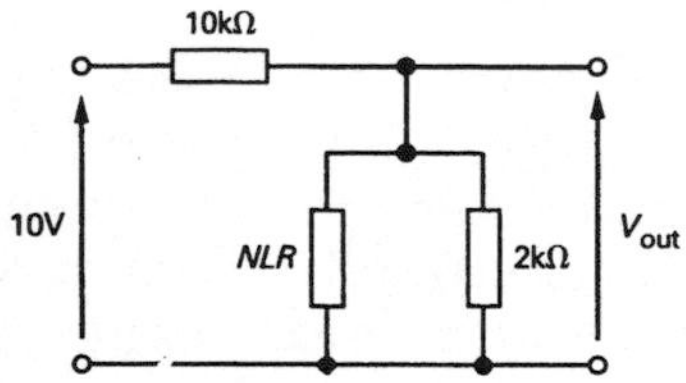

Fig. 19.11

Example 19.9

A d.c. voltage of 10 V is applied across the circuit shown in Fig. 19.11. The current/voltage characteristic of the non-linear resistor is $i = 10^{-3}V^3$ A. Calculate the output voltage of the circuit.

Solution

$$(10 - V_{out})/(10 \times 10^3) = i_{NLR} + V_{out}/2000$$
$$= 10^{-3}V_{out}^3 + V_{out}/2000$$

or $\quad 10V_{out}^3 + 6V_{out} - 10 = 0$

This equation can be solved graphically, but instead use the method of repeated estimates. Clearly, V_{out} must be somewhere in the region 0.8 to 0.9 V. Try 0.85 V: $10(0.85^3) + 6(0.85) - 10 = 1.241$. Try 0.81 V: $10(0.81^3) + 6(0.81) - 10 = 0.174$. A few more tries gives $V_{out} \approx 0.803$ V (*Ans.*)

Example 19.10

Two non-linear resistors are connected in parallel with one another. Their current–voltage characteristics are $i = 2V^3$ and $i = 4\sqrt{V}$ respectively. If a constant current of 10 A is supplied to the resistors calculate the voltage across them.

Solution

$10 = 2V^3 + 4\sqrt{V}$. Clearly, V must be less than 2. Try $V = 1.5$ V, then $i = 2 \times 1.5^3 + 4\sqrt{1.5} = 11.65$ A. Try $V = 1.4$, then $i = 2 \times 1.4^3 + 4\sqrt{1.4} = 10.22$ A. Try $V = 1.38$ V, then $i = 1.38^3 + 4\sqrt{1.38} = 9.955$ A. A few more attempts gives $V = 1.384$ V (*Ans.*)

Exercises 19

19.1 The current–voltage characteristic of a non-linear resistor is given by $i = 0.1V + 0.02V^2$ amps. This non-linear resistor is connected in series with a linear resistor of 20 Ω and a d.c. voltage source of 9 V and negligible internal resistance. Calculate the current flowing in the circuit and the voltage across each component using a graphical method.

19.2 Repeat Exercise 19.1 using an analytical method of solution.

Table 19.5

Applied voltage (V)	1	3	5
Current (mA)	1	9	25

19.3 Measurements on a non-linear device gave the data listed in Table 19.5. Assuming the device to have a square-law characteristic calculate (*a*) the mean current when a sinusoidal signal of 2 V peak value is applied, (*b*) the amplitudes of the intermodulation components when the two voltages $1 \sin \omega t$ and $1 \sin 2\omega t$ are applied.

19.4 The I–V characteristic of a non-linear circuit is $i = 4.5 + 0.5v + 0.03v^2$ mA. Calculate the r.m.s. value of the current that flows when the voltage $v = 6 \sin \omega t$ volts is applied.

19.5 A linear resistor of 10 Ω is connected in series with a non-linear resistor whose I–V characteristic is given by $i = 0.002v^3$ A. Calculate the maximum current that flows in the circuit and the voltage that is then developed across each component when a sinusoidal voltage of 10 V peak value is applied to the circuit.

19.6 A 78 Ω resistor is connected in parallel with a non-linear device whose current–voltage characteristic is given by $i = (4 \times 10^{-3} v) + (3 \times 10^{-4} v^3)$ A. Calculate the current in each component and the voltage across them when the total current flowing is 200 mA.

19.7 A non-linear device has the current–voltage characteristic $i = 10 + 3V + V^2$ mA and has the voltage $v = 2 \sin \omega t$ applied across it. (*a*) Calculate the amplitude of each component frequency in the resulting current waveform. (*b*) Calculate the percentage second harmonic distortion of the current waveform.

Table 19.6

V_{be} (V)	1	3	5
I (mA)	0.5	2	12

19.8 Measurements on a n-p-n transistor gave the data of Table 19.6. Determine the current–voltage characteristic of the transistor. The voltage applied to the transistor is $3 + 2 \sin 1000\pi t$ V. Calculate (*a*) the magnitude and frequency of each component of the current and (*b*) the percentage second harmonic distortion.

19.9 A non-linear resistor has the current–voltage characteristic $i = 0.6V + 0.1V^2$ A and it is connected in series with a 3 Ω linear resistor and a 6 V d.c. voltage supply. Calculate (*a*) the voltage across the two resistors and (*b*) the current flowing in the circuit.

19.10 The total characteristic of a bridge rectifier circuit is given by Table 19.7. Determine the waveform of the current when the rectifier is connected in series with a 3300 Ω resistor and the applied voltage is $60 \sin 100\pi t$ volts.

Table 19.7

Voltage (V)	0	6	8	10	12	14	16	18
Current (mA)	0	0	2.3	4.5	6.7	9.4	12.2	15.5

19.11 Two signals $V_1 \sin 1000t$ and $V_2 \sin 2000t$ volts are applied to a component that has a square-law I/V characteristic. Calculate the frequencies of the components of the output current.

19.12 A non-linear characteristic is given by $i = 3.8 + 1.2v + 0.6v^2$ mA. If $v = 2.5 \sin \omega t$, calculate the percentage second harmonic in the current.

19.13 Table 19.8 gives the values of flux produced in a magnetic material by various values of magnetizing current. Plot flux to a base of magnetizing current and use the graph to obtain the waveform of the current needed to produce a sinusoidal flux waveform.

Table 19.8

Magnetizing current (A)	0	0.5	0.75	1.0	1.5	2	2.5	3
Flux (Wb)	18	110	122	130	138	140	140	140
	−80	−26	0	30	80	110	128	140

19.14 List some examples of non-linearity in electrical circuits. Give both useful and undesirable examples and give an example of the use of each you consider to be useful.

19.15 A sinusoidal voltage of peak value 30 V is applied to a non-linear circuit whose I–V characteristic is given by the data of Table 19.9. Plot the half-cycle of the current that flows and hence estimate the average value of the current.

Table 19.9

Applied Voltage (V)	0	12	16	20	24	28	32	36
Current (mA)	0	0	2.6	7.0	11.4	16.8	22.4	30

19.16 A circuit consists of a 40 Ω resistor connected in parallel with a non-linear resistor having $i = 8 \times 10^{-4}V + 6 \times 10^{-5}V^3$ A. If the current supplied to the circuit is 2A find the current in each component and the voltage across it.

19.17 Measurements on a non-linear device gave the following results:

Table 19.10

Voltage (V)	−2.5	−1.0	−0.5	0.48	0.6	0.8	1.0
Current (mA)	−0.25	−0.2	−0.1	1.0	2.0	7.0	12.0

The device is connected in series with a 100 Ω linear resistor and a voltage source of 1.5 V r.m.s. Calculate the mean current.

19.18 The primary winding of a transformer has 100 turns and is connected to a voltage source of 220 V and 50 Hz. The transformer core has a cross-sectional area of 100×10^{-4} m^2 and it is 0.8 m in length. The hysteresis loop for the core material is specified by the following data:

Table 19.11

H (A/m)	−50	0	50	100	125	150	200
B (T)	0	0.65	0	0.25	0.8	0.9	1.0
			0.9	0.92	0.93	0.95	

If there are 100 turns in the primary winding determine the waveform of the magnetizing current.

19.19 A 10 Ω linear resistor is connected in series with a non-linear resistor having $i = 2 \times 10^{-3}V^3$. The voltage $10 + 7.07 \sin \omega t$ V is applied to the circuit. Calculate (*a*) the maximum current and (*b*) the voltage across each resistor when the maximum current flows.

19.20 A 12 Ω resistor is connected in series with a non-linear resistor whose current–voltage characteristic is $i = 5 \times 10^{-2}V + 4 \times 10^{-4}V^2 + 1.5 \times 10^{-6}V^3$ A. The voltage $v = 141.42 \sin \omega t$ V is applied to the circuit. Determine the two peak values of the current.

19.21 A non-linear resistor has the current–voltage characteristic $i = 0.5V^2 + V$ and is connected in series with a 2 Ω resistor and a

10 V d.c. voltage source. Calculate the voltage across the non-linear resistor.

19.22 A non-linear resistor has the characteristic $i = 0.2V^2 + V$ and is connected in series with a 4 Ω resistor. Determine the voltage across the non-linear resistor when the applied voltage is 6 V.

19.23 Measurements on a non-linear device gave the following results: Voltage 3, 6 and 9 V with corresponding currents 0.1, 6.4 and 19.9 mA. Assuming the current–voltage characteristic of the device is of the form $i = a + bV + cV^2$ mA determine the constants a, b and c. The voltage $v = 6 + 2\sin\omega t$ is applied to the device. Calculate the percentage second harmonic distortion in the current.

19.24 Two non-linear resistors are connected in parallel and supplied with a constant current of 66 mA. If the current–voltage characteristics of the two resistors are $i = V^3$ and $i = \sqrt{V}$, determine the voltage across the resistors.

20 Solution of circuits using Laplace transforms

When a voltage or a current is suddenly applied to a circuit containing capacitors and/or inductors, the currents and/or voltages produced at various points in the circuit will, in general, have two components. One of these is known as the *forced*, or *steady-state* response since it is the current or voltage that continues to exist for as long as the applied voltage or current is maintained. The other component is known as the *natural response*, or the *transient* and this will only exist for a short time after the application of the input voltage or current. When a sinusoidal voltage $v = V \sin \omega t$ is suddenly applied to a series RL circuit and the current i that flows is

$$i = I_m \sin(\omega t + \theta) + I_t \, e^{-Rt/L} \tag{20.1}$$

The first term in equation (20.1) is, of course, the current that would be determined using a.c. theory and this is the *steady-state current*. The second, transient, term is an initial current, of peak value I_t, that flows at the instant the switch is closed and then decays at a rate determined by the time constant L/R of the circuit. Clearly, after some time t the transient term will be negligibly small and then the circuit current will consist solely of the steady-state component.

A convenient method for the calculation of the complete response of a circuit involves the use of the *Laplace Transformation* and of *partial fractions*. The theory and application of Laplace transforms is normally considered in a mathematics course and hence only a brief introduction will be provided here.

Laplace transforms

The *Laplace transform* $f(S)$ of a function of time $f(t)$ is defined as

$$f(S) = \int_{-0}^{\infty} f(t)e^{-St} \, dt \tag{20.2}$$

S is a complex variable of the form $S = \alpha + j\omega$ where both α and ω have the dimensions of seconds^{-1}. The lower limit of integration is made -0 in order to include all functions that occur at time $t = 0$.

In the solution of an electrical problem it is usually not necessary to evaluate the integral given in equation (20.2) since tables giving the transforms of all the more commonly occurring waveforms are readily available. However, a few of the simpler cases will be evaluated as examples.

Unit Step Function $f(t) = 1$

$$f(S) = \int_{-0}^{\infty} e^{-St}\,dt = [-e^{-st}/S]_{-0}^{\infty} = 1/S$$

(*b*) *Unit Ramp Function* $f(t) = t$

$$f(S) = \int_{-0}^{\infty} te^{-St}\,dt = [(-e^{-St}t)/S]_{-0}^{\infty} - \int_{-0}^{\infty} (-e^{-St}/S)\,dt$$
$$= [(-e^{-St}/S^2)]_{-0}^{\infty} = 1/S^2$$

(*c*) *Exponential Delay* $f(t) = e^{-\alpha t}$

$$f(S) = \int_{-0}^{\infty} e^{-\alpha t}e^{-St}\,dt = \int_{-0}^{\infty} e^{-(S+\alpha)t}\,dt$$
$$= [(e^{-(S+\alpha)t})/[-(S+\alpha)]]_{-0}^{\infty} = 1/(S+\alpha)$$

Table 20.1 lists a number of the more commonly occurring transforms.

The unit impulse δ, transform number 1, warrants some further mention. This function, often known as the Dirac function, represents a pulse of unit area and of width δt, where $\delta t \to 0$. Since the pulse width tends to zero, the height of the pulse must tend to infinity (see Fig. 20.1). It occurs when a voltage is suddenly applied across a pure capacitor.

Fig. 20.1 The impulse function

Table 20.1 Laplace transforms

f(t)		*f(S)*
1	Unit impulse δ	1
2	Unit step	$1/S$
3	Delayed unit step	$(e^{-St})/S$
4	Rectangular pulse (of duration t)	$(1 - e^{-St})/S$
5	Unit ramp	$1/S^2$
6	Delayed unit ramp (by τ)	e^{-St}/S^2
7	Exponential delay $e^{-\alpha t}$	$1/(S+\alpha)$
8	Exponential rise $1 - e^{-\alpha t}$	$\alpha/[S(S+\alpha)^2]$
9	$t \times$ exponential decay $te^{-\alpha t}$	$1/(S+\alpha)$
10	Difference of exponentials $e^{-\alpha t} - e^{\beta t}$	$(\beta - \alpha)/[(S+\alpha)(S+\beta)]$

(*continued*)

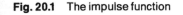

Table 20.1 *(continued)*

	$f(t)$	$f(S)$
11	Sinusoidal $\sin \omega t$	$\omega/(S^2+\omega^2)$
12	Phase advanced sinusoidal $\sin(\omega t+\theta)$	$\dfrac{\omega\cos\theta + S\sin\theta}{S^2+\omega^2}$
13	$t\sin\omega t$	$2\omega S/(S^2+\omega^2)^2$
14	$e^{-\alpha t}\sin\omega t$	$\omega/[(S+\alpha)^2+\omega^2]$
15	cosinusoidal $\cos\omega t$	$S/(S^2+\omega^2)$
16	Phase advanced cosinusoidal $\cos(\omega t+\theta)$	$\dfrac{S\cos\theta - \omega\sin\theta}{S^2+\omega^2}$
17	$t\cos\omega t$	$(S^2-\omega^2)/(S^2+\omega^2)^2$
18	$e^{-\alpha t}\cos\omega t$	$(S+\alpha)/[(S+\alpha)^2+\omega^2]$
19	$df(t)/dt = f'(t)$	$Sf(S)-f(0)$
20	$d^2f(t)/dt^2 = f''(t)$	$S^2f(S)-Sf(0)-f'(0)$
21	$\int_0^t f(t)\,dt$	$f(S)/S + f(0)/S$

Partial fractions

Before a given $f(S)$ can be transformed into the time world it is very often necessary to manipulate the equation into a form that can be recognized in Table 20.1. For this step to be carried out the use of *partial fractions* is frequently essential.

If the two equations $2/(4S+1)$ and $1/(S+4)$ are added the result is $(6S+9)/(4S^2+17S+4)$, so $2/(4S+1)$ and $1/(S+4)$ are the partial fractions of $(6S+9)/(4S^2+17S+4)$.

The problem is how the partial fractions of a given equation can be determined. Two conditions must be satisfied by an equation before it is possible to find its partial fractions. These are:

- The highest power of S in the numerator must be less than the highest power of S in the denominator. If this is not the case the numerator must be divided by the denominator to produce a remainder that does satisfy this requirement.
- The denominator must factorize.

Thus $(S+4)/(S^2+3S+2)$ can have its partial fractions found but $(S^2+4)/(S^2+3S+2)$ cannot. However, dividing the numerator of this latter equation by its denominator gives

$$\begin{array}{r|l} & 1 \\ \hline S^2+3S+2 & S^2+4 \\ & \underline{S^2+3S+2} \\ & -3S+2 \end{array} \quad = 1 + (-3S+2)/(S^2+3S+2)$$

and the partial fractions of $(-3S+2)/(S^2+3S+2)$ can be found.

Partial fractions have denominators that fall into one of a number of different types:

- *Linear factors*

  $$f(S)/[(S+a)(S+b)(S+c)] = A/(S+a) + B/(S+b) + C/(S+c)$$

  where any one of a, b or c may be zero.
- *Repeated linear factors*

  $$f(S)/(S+a)^n = A/(S+a) + B/(S+a)^2 + \cdots + N/(S+a)^n$$

- *Quadratic*

  $$f(S)/(aS^2+bS+c) = (AS+B)/(aS^2+bS+c)$$

Often it will prove necessary to modify the denominator by 'completing the square', i.e.

$$S^2+aS+b = (S+a/2)^2 + b - (a/2)^2$$

The determination of the values for A, B, C, etc. can be carried out using either of two methods or perhaps a combination of each. These two methods will be illustrated by means of the following example.

Example 20.1

Determine the partial fractions of $(S+4)/(S^2+3S+2)$.

Solution

$$\begin{aligned}(S+4)/(S^2+3S+2) &= (S+4)/[(S+2)(S+1)] \\ &\equiv A/(S+2) + B/(S+1) \\ &= [A(S+1)+B(S+2)]/[(S+2)(S+1)]\end{aligned}$$

Therefore,

$$S+4 = A(S+1) + B(S+2)$$

Method 1 Make one of the unknowns zero by suitable choice of the value of S.

(*i*) Let $S=-1$, then $-1+4=3=B(-1+2)=B$
(*ii*) Let $S=-2$, then $2=A(-2+1)=-A$ and so

$$(S+4)/(S+2)(S+1) = -2/(S+2) + 3/(S+1)$$

Method 2 Equate similar coefficients on either side of the identity. Thus

$$S+4 = AS + A + BS + 2B$$

Equating like coefficients of S:

(*i*) Constants $4 = A + 2B$
(*ii*) S $\quad 1 = A + B$

Hence, on subtracting, $3 = B$ (as before) and $4 = A + 6$ or $A = -2$ (as before).

Example 20.2

Find the partial fractions of each of the following functions of S:

(*a*) $(4+2S)/[S(S+5)]$,
(*b*) $(6+2S)/[(S+5)(S+2)]$,
(*c*) $(4+2S)/[(S^2+2S+3)(S+3)]$ and
(*d*) $(S+5)/(S+2)^2$.

Solution

(*a*) $f(S) = (4+2S)/[S(S+5)] = A/S + B/(S+5)$

$4 + 2S = AS + 5A + BS$

Equating constants: $4 = 5A$ or $A = 0.8$. Equating $S: 2 = A + B$ or $B = 2 - 0.8 = 1.2$. Hence,

$f(S) = 0.8/S + 1.2/(S+5)$ (*Ans.*)

(*b*) $f(S) = (6+2S)/[(S+5)(S+2)] = A/(S+5) + B/(S+2)$

$6 + 2S = AS + 2A + BS + 5B$

Equating constants: $6 = 2A + 5B$. Equate S: $2 = A + B$. Multiply by 2 to get $4 = 2A + 2B$ and subtract from the constants equation. Then, $2 = 3B$ or $B = 0.67$ and $A = 2 - 0.67 = 1.33$. Therefore,

$f(S) = 1.33/(S+5) + 0.67/(S+2)$ (*Ans.*)

(*c*) $(4+2S)/[(S^2+2S+3)(S+3)]$

$= (AS+B)/(S^2+2S+3) + C/(S+3)$

$4 + 2S = AS^2 + 3AS + BS + 3B + CS^2 + 2CS + 3C$

Equating constants: $4 = 3B + 3C$. Equate $S: 2 = 3A + B + 2C$. Equate $S^2: 0 = A + C$ or $A = -C$. Hence, $2 = -3C + B + 2C = B - C$. Multiply by 3 to give $6 = 3B - 3C$ and add to the constants equation. Then, $10 = 6B$ or $B = 1.67$. Also, $2 = B - C$ or $C = -0.33$ and $A = 0.33$. Therefore,

$f(S) = (0.33S + 1.67)/(S^2+2S+3) - 0.33/(S+3)$ (*Ans.*)

(*d*) $f(S) = (S+5)/(S+2)^2 = A/(S+2) + B/(S+2)^2$

$S + 5 = AS + 2A + B$

Equate constants: $5 = 2A + B$. Equate $S: 1 = A$. Hence, $B = 5 - 2 = 3$.

Therefore,

$f(S) = 1/(S+2) + 3/(S+2)^2$ (*Ans.*)

The solution of circuits

Series *RC* circuit

Step voltage

Applying Kirchhoff's voltage law to a series RC circuit gives

$$V = iR + q/C = iR + (1/C)\int_0^t i \, dt$$

If the applied voltage is a step of V volts, then taking transforms

$$V/S = i(S)R + i(S)/SC = i(S)(R + 1/SC)$$
(no. 2) (no. 21)

The complex impedance $Z(S)$ of the circuit is

$$Z(S) = V(S)/i(S) = R + 1/SC$$

This result could have been obtained by writing down the impedance of the circuit to a *sinusoidal* voltage and then replacing $j\omega$ with S throughout. This is true for any other circuit as well so that the *complex* impedance of an inductor can be written as SL, and of a capacitor as $1/SC$. The current flowing in the circuit is

$$i(S) = V/[S(R + 1/SC)]$$

This equation must be put into a form that can be recognized as one of the terms given in Table 20.1. Hence,

$$i(S) = VSC/[S(SRC + 1)] = VSC/[SRC(S + 1/CR)]$$
$$= (V/R)[1/(S + 1/CR)]$$

This can be seen to correspond with Laplace transform number 6, where $\alpha = 1/CR$. Therefore,

$$i(t) = (V/R)\, e^{-t/CR} \qquad (20.3)$$

The voltage across the capacitor is

$$v_C(S) = i(S)/SC = V/[SCR(S + 1/CR)]$$
$$= A/S + B/(S + 1/CR)$$

or $V/CR = AS + (A/CR) + BS$

Equating like coefficients:

(i) Constants: $V/CR = A/CR$ or $A = V$

(ii) S: $0 = A + B$ or $B = -A = -V$ so that

$v_C(S) = V/S - V/(S + 1/CR)$ and, from Table 20.1,

$v_C(t) = V(1 - e^{-t/CR})$

(no. 2 and no. 7) (20.4)

The method adopted for the solution of the series *RC* circuit should be followed for all other circuits. The method can be summarized by the following listed steps:

- Apply Kirchhoff's laws to the circuit and write down the equation describing the operation of the circuit.
- Re-write the equations in terms of Laplace transforms (very often the analysis can commence with this step).
- Solve the equation(s) in terms of the unknown quantities.
- Re-arrange the equations into a form that can be recognized in the available table of Laplace transforms.
- Use the appropriate parts of the table to write down the solution as a function of time.

Usually it is rearranging the equations that presents the greatest difficulty encountered in a given problem. It should be noted from Table 20.1 that none of the given transforms has any coefficient of S other than unity and that $1/S$ does not occur in either the numerator or the denominator of any term.

Ramp voltage

Now,

$$(V/S^2) = i(S)(R + 1/SC)$$

$$i(S) = V/[S^2(R + 1/SC)] = VC/[S(1 + SCR)]$$

$$= VC/[SCR(S + 1/CR)] = (V/R)/[S(S + 1/CR)]$$

Hence,

$$V/[RS(S + 1/CR)] = A/S + B/(S + 1/CR)$$

$$V/R = AS + A/CR + BS$$

Equate constants: $V/R = A/CR$ or $A = VC$
Equate S: $0 = A + B$ or $B = -A = -VC$

Hence,

$$i(S) = VC/S - VC/(S + 1/CR)$$

and $i(t) = CV(1 - e^{-t/CR})$ (20.5)

The voltage across the capacitor is

$$v_C(S) = i(S)/SC = V/S^2 - V/[S(S + 1/CR)]$$

Now

$$V/[S(S + 1/CR)] = A/S + B/(S + 1/CR)$$

$$= A(S + 1/CR) + BS$$

Let $S = 0$, then $V = A/CR$ or $A = VCR$. Let $S = -1/CR$, then $V = -B/CR$ or $B = -VCR$. Hence,

$$v_C(S) = V/S^2 - VCR/S + VCR/(S + 1/CR)$$

Therefore,

$$v_C(t) = Vt - VCR(1 - e^{-t/CR}) \tag{20.6}$$

Example 20.3

Determine an expression for the variation with time of the output voltage of the circuit given in Fig. 20.2 when the input voltage is (*a*) a 6 V step function and (*b*) a 6 V/s ramp function.

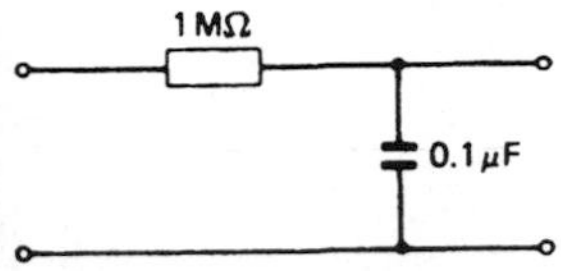

Fig. 20.2

Solution

$$V_0(S) = [V_1(S)/SC]/(R + 1/SC) = V_1(S)/(1 + SCR)$$

(*a*) $V_1(S) = 6/S$, $CR = 0.1$

$$V_0(S) = 6/[S(1 + 0.1S)] = 6/[0.1S(S + 1/0.1)]$$

$$= 60/[S(S + 10)]$$

$$60/[S(S + 10)] = A/S + B/(S + 10) = A(S + 10) + BS$$

Let $S = 0$, then $60 = 10A$ or $A = 6$. Let $S = -10$, then $60 = -10B$ or $B = -6$. Therefore

$$V_0(S) = 6/S - 6/(S + 10)$$

and $V(t) = 6(1 - e^{-10t})$ (*Ans.*)

(*b*) $V_1(S) = 6/S^2$, $CR = 0.1$

$$V_0(S) = 6/[S^2(1 + 0.1S)] = 60/[S^2(S + 10)$$

$$= (AS + B)/S^2 + C/(S + 10)$$

$$60 = AS^2 + 10AS + BS + 10B + CS^2$$

Equate constants: $60 = 10B$ or $B = 6$. Equate S: $0 = 10A + B$ or $A = -0.6$. Equate S^2: $0 = A + C$ or $C = 0.6$. Therefore,

$$V_0(S) = (6 - 0.6S)/S^2 + 0.6/(S + 10)$$

$$= 6/S^2 - 0.6/S + 0.6/(S + 10)$$

Therefore,

$$V_0(t) = 6t - 0.6(1 - e^{-10t})$$ (*Ans.*)

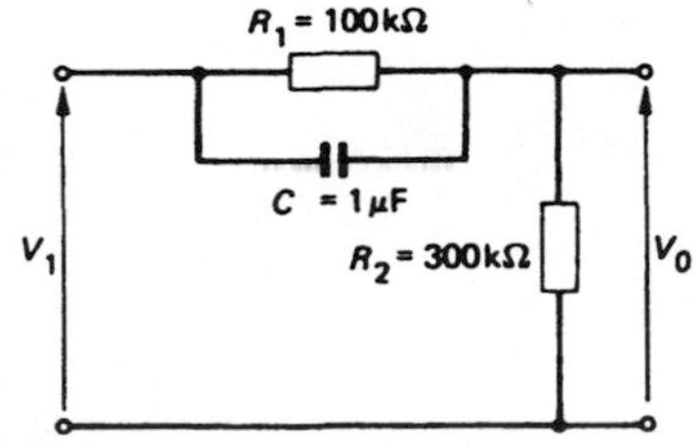

Fig. 20.3

Example 20.4

Derive an expression for the time variation of the voltage developed across the resistor R_2 in the circuit of Fig. 20.3 when the input voltage is a 4 V step function.

Solution

$$V_0(S) = [V_1(S)R_2]/\{[(R_1/SC)/(R_1+1/SC)]+R_2\}$$
$$= [V_1(S)R_2(1+SCR_1)]/[R_1+R_2+SCR_1R_2]$$

$V_1(S) = 4/S$ and hence,

$$V_0(S) = [1200\times10^3(1+0.1S)]/\{S[(400\times10^3)+(30\times10^3S)]\}$$
$$= [120(1+0.1S)]/[S(40+3S)]$$
$$= (120+12S)/[3S(S+40/3)] = (40+4S)/[S(S+40/3)]$$
$$= A/S + B/(S+40/3)$$

Therefore,

$$40+4S = AS + (40A/3) + BS$$

Equating constants: $40 = 40A/3$ or $A = 3$. Equating S: $4 = A + B$ or $B = 1$. Hence,

$$V_0(S) = 3/S + 1/(S+40/3) \quad \text{and} \quad V_0(t) = 3 + \mathrm{e}^{-40t/3} \quad (Ans.)$$

Example 20.5

A 12 MΩ resistor is connected in parallel with a 1 μF capacitor and the combination is connected in series with another 1 μF capacitor. A 6 V step is applied between the junction of the parallel components and the common line. Derive an expression for the voltage across the series capacitor. Both capacitors are initially discharged.

Solution

$$V_{out}(S) = V_{in}(S)(1/SC)/[R/(1+SCR)+1/SC]$$
$$= V_{in}(S)/[1+SCR/(1+SCR)]$$
$$= V_{in}(S)[1+SCR]/[1+2SCR]$$
$$= 6[1+S]/[S(1+2S)] = 6(1+S)/[2S(S+0.5)].$$

Hence,

$$[3(1+S)]/[S(S+0.5)] = A/S + B/(S+0.5);$$
$$3+3S = AS + 0.5A + BS$$

Equate constants: $3 = 0.5A$, or $A = 6$. Equate S: $3 = A + B$ or $B = -3$. Therefore,

$$V_{out}(S) = 6/S - 3/(S+0.5)$$

From Table 20.1,

$$V_{out}(t) = 6 - 3\,\mathrm{e}^{-0.5t} \quad (Ans.)$$

Example 20.6

An L network consists of a 4 MΩ series resistor and a 0.1 μF shunt capacitor. A 1 V step voltage is applied to the resistor. Obtain an expression for the voltage across the capacitor.

Solution

$$v_C(S) = [(1/S)(1/SC)]/(R + 1/SC) = 1/[S(1 + SCR)]$$
$$= 1/[S(1 + 0.4S)] = A/S + B/(1 + 0.4S).$$

Hence,

$$1 = A + 0.4AS + BS$$

Equate constants: $1 = A$. Equate S: $0 = 0.4A + B$ or $B = -0.4$. Therefore,

$$v_C(S) = 1/S - 0.4/(1 + 0.4S) = 1/S - 1/(S + 2.5)$$

From Table 20.1,

$$v_C(t) = 1 - e^{-2.5t} \quad (Ans.)$$

Example 20.7

Repeat Example 20.6 for a 1 V ramp input.

Solution

$$v_C(S) = 1/[S^2(1 + SCR)] = 1/[S^2(1 + 0.4S)]$$
$$= (AS + B)/S^2 + C/(1 + 0.4S)$$

Hence,

$$1 = AS + 0.4AS^2 + B + 0.4BS + CS^2$$

Equate constants: $1 = B$. Equate S: $0 = A + 0.4B$ or $A = -0.4$. Equate S^2: $0 = 0.4A + C$ or $C = 0.4^2 = 0.16$. Therefore,

$$v_C(S) = -0.4S/S^2 + 1/S^2 + 0.16/(1 + 0.4S)$$
$$= -0.4/S + 1/S^2 + 0.4/(1 + 0.4S)$$

From Table 20.1,

$$v_C(t) = 0.4 + t + 0.4\,e^{-2.5t} = t - 0.4(1 - e^{-2.5t}) \quad (Ans.)$$

Example 20.8

Repeat example 20.6 for a sinusoidal voltage input.

Solution

$$v_C(S) = V\omega/[(S^2 + \omega^2)(1+SCR)] = V\omega/[(S^2 + \omega^2)(1 + 0.4S)]$$
$$= (AS + B)/(S^2 + \omega^2) + C/(1 + 0.4S)$$

Hence

$$V\omega = AS + 0.4AS^2 + B + 0.4BS + CS^2 + C\omega^2$$

Equate constants: $V\omega = B + C\omega^2$ (20.7)

Equate S: $0 = A + 0.4B$ (20.8)

Equate S^2: $0 = 0.4A + C$ (20.9)

From equation (20.8), $B = -2.5A$ and from equation (20.9) $C = -0.4A$. Substituting into equation (20.7) gives,

$$V\omega = -2.5A - 0.4A\omega^2 = -A[2.5 + 0.4\omega^2]$$

or $A = -0.4V\omega/(1 + 0.16\omega^2)$

$$B = V\omega/(1 + 0.16\omega^2) \quad \text{and} \quad C = 0.16V\omega/(1 + 0.16\omega^2)$$

Hence,

$$[V_C(S) = [V\omega/(1 + 0.16\omega^2]\{[-0.4S/(S^2 + \omega^2)] + [1/(S^2 + \omega^2)] + [0.16/(1 + 0.4S)]\}$$

And, from Table 20.1,

$$V_c(t) = [V\omega/(1 + 0.16\omega^2)][-0.4\cos\omega t + (\sin\omega t)/\omega + (0.16/0.4)e^{-2.5t}]$$
$$= (V\sin\omega t - 0.4V\omega\cos\omega t + 0.4\,e^{-2.5t})/(1 + 0.16\omega^2)$$

(*Ans.*)

Series *RL* circuit

Step voltage

When a step voltage is applied to a series RL circuit, then $V - iR + L$ $\mathrm{d}i/\mathrm{d}t$. Taking transforms, $V/S = i(S)(R + SL)$. Therefore,

$$i(S) = V/[S(R + SL)] = (V/L)/[S(S + R/L)]$$
$$= A/S + B/(S + R/L) = AS + (AR/L) + BS$$

Equating constants: $V/L = AR/L$ or $A = V/R$. Equating S: $0 = A + B$ or $B = -A - V/R$. Hence

$$i(S) = V/RS - (V/R)/(S + R/L) \quad \text{and}$$
$$i(t) = (V/R)(1 - e^{-Rt/L}) \qquad (20.10)$$

Ramp voltage

When a ramp voltage of V volt/s is applied to the same circuit

$$V/S^2 = i(S)(R + SL) \quad \text{and}$$
$$i(S) = (V/L)/[S^2(S + R/L)]$$
$$= (AS + B)/S^2 + C/(S + R/L)$$
$$= AS^2 + ASR/L + BS + CS^2 + BR/L$$

Equating constants: $V/L = BR/L$ or $B = V/R$

Equating S: $0 = B + (AR/L) = (V/R) + (AR/L)$ or $A = -VL/R^2$

Equating S^2: $0 = A + C$, or $C = VL/R^2$. Therefore,

$$i(S) = [V/R - VSL/R^2]/S^2 + (VL/R^2)/(S + R/L)$$
$$= (V/R)/S^2 - (VL/R^2)/S + (VL/R^2)/(S + R/L)$$

and

$$i(t) = Vt/R - (VL/R^2)(1 - e^{-tR/L}) \tag{20.11}$$

The voltage $v_L(t)$ across the inductance is obtained from

$$v_L(S) = i(S)SL$$
$$= (VL/R)/S - (VL^2/R^2) - (SVL^2/R^2)/(S + R/L)$$
$$= (VL/R)/S - (VL^2/R^2)[1 - (1 - (R/L)/(S + R/L))]$$
$$= VL/SR + VL/[R(S + R/L)]$$

and

$$v_L(t) = (VL/R)(1 - e^{-Rt/L}) \tag{20.12}$$

Example 20.9

A 20 V step is applied to a 10 H inductor of self-resistance 20 Ω. Derive an expression for the current that flows in the inductor.

Solution

Kirchhoff's voltage law gives $20 = 20i + 10\,di/dt$. Taking transforms, $20/S = (20 + S)i(S)$, or $i(S) = 20/[S(20 + 20S)] = A/S + B/(20 + 10S)$. Hence,

$$20 = 20A + 10AS + BS$$

Equating constants: $20 = 20A$ or $A = 1$. Equating $S: 0 = 10A + B$, or $B = -10$. Therefore,

$$i(S) = 1/S - 10/(20 + 10S) = 1/S + 1/(S + 2)$$

From Table 20.1

$$i(t) = 1 - e^{-2t}\text{A} \quad (\textit{Ans.})$$

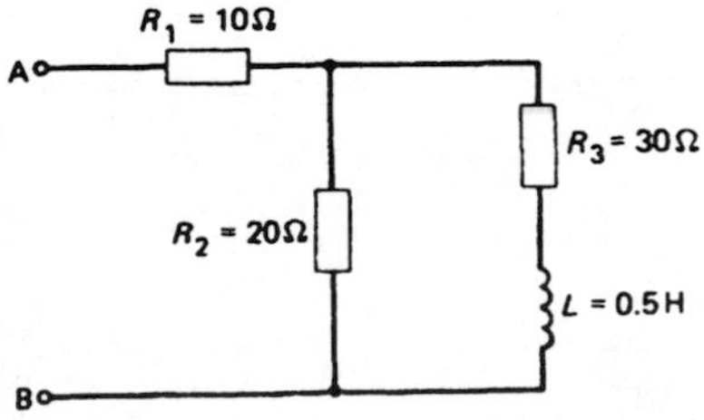

Fig. 20.4

Example 20.10

Determine an expression for the time variation of the current supplied to the circuit given in Fig. 20.4 when a 100 V step is applied to the terminals AB.

Solution

$$Z_{in}(S) = R_1 + [R_2(R_3 + SL)]/(R_2 + R_3 + SL)$$

$$= \frac{R_1R_2 + R_1R_3 + R_2R_3 + SLR_1 + SLR_2}{R_2 + R_3 + SL}$$

or $Z_{in}(S) = (1100 + 15S)/(50 + 0.5S)$

$$I_{in}(S) = V_{in}(S)/Z_{in}(S) = [100(50 + 0.5S)]/[S(1100 + 155)]$$

$$= [100(50 + 0.5S)]/[15S(S + 73.3)]$$

$$= (333.3 + 3.3S)/[S(S + 73.3)]$$

$$= A/S + B/(S + 73.3)$$

or $333.3 + 3.3S = AS + 73.3A + BS$

Equating constants: $333.3 = 73.3A$ or $A = 4.55$. Equating S: $3.3 = A + B$ or $B = -1.25$. Therefore

$$I_{in}(S) = 4.55/S - 1.25/(S + 73.3)$$

and

$$I_{in}(t) = 4.55 - 1.25\,e^{-73.3t} \quad (\textit{Ans.})$$

Sinusoidal voltage with initial phase angle

When dealing with the application of a sinusoidal voltage to a circuit it may prove easier to employ the exponential form of $\sin(\omega t + \theta)$. If the voltage $v = V_m \sin(\omega t + \theta)$ is applied to a series RL circuit then Kirchhoff's voltage law gives $V_m \sin(\omega t + \theta) = L\,di/dt + Ri$. Since $\sin(\omega t + \theta)$ is the imaginary part of $e^{j(\omega t+\theta)}$ the required solution is the imaginary part of the solution to $L\,di/dt + Ri = V_m\,e^{j(\omega t+\theta)}$. Now,

$$(R + SL)i(S) = V_m\,e^{j\theta}/(S - j\omega)$$

$$i(S) = (V_m\,e^{j\theta})/(R + SL)(S - j\omega)$$

$$= (V_m\,e^{j\theta})[L(S + R/L)(S - j\omega)]$$

$$= A/(S + R/L) + B/(S - j\omega)$$

Hence,

$$(V_m/L)e^{j\theta} = AS - j\omega A + BS + BR/L$$

Equating constants: $(V_m/L)e^{j\theta} = -j\omega A + BR/L$, and equating S: $0 = A + B$. Thus:

$$A = -(V_m\,e^{j\theta})/(R + j\omega L) \quad \text{and} \quad B = (V_m\,e^{j\theta})/(R + j\omega L)$$

Therefore,

$$i(S) = V_m\,e^{j\theta}\left[\frac{-1}{(R + j\omega L)(S + R/L)} + \frac{1}{(R + j\omega L)(S - j\omega)}\right]$$

Now, $i(t)$ is the imaginary part of $[(V_m\,e^{j\theta})/(R+j\omega L)]$ $[e^{j\omega t}-e^{-Rt/L}]$. Hence,

$$i(t) = IP\left[\frac{V_m\,e^{j\theta}}{\surd(R^2+\omega^2L^2)\tan^{-1}(\omega L/R)}\right](e^{j\omega t}-e^{-Rt/L})$$

$$= IP\{[(V_m\,e^{j(\theta-\beta)}/\surd(R^2+\omega^2L^2)][e^{j\omega t}-e^{Rt/L}]\}$$

$$\text{where } \beta = \tan^{-1}(\omega L/R)$$

$$= IP\{V_m/\surd(R^2+\omega^2L^2)[e^{j(\omega t+\theta-\beta}-e^{j(\theta-\beta)-Rt/L}]\}$$

Therefore ($\beta = \tan^{-1}(\omega L/R)$),

$$i(t) = \left[\frac{V_m}{\surd(R_2+\omega^2L^2)}\right][\sin(\omega t+\theta-\beta)-\sin(\theta-\beta)e^{-Rt/L}] \quad (20.13)$$

The worst transient occurs when $\theta-\beta = 90°$, and there is zero transient term when $\theta-\beta=0$. This means that the magnitude of the transient depends upon the instant the circuit is switched on.

Example 20.11

The voltage $v = 10\ \sin(1000+\theta)t$ V is applied to a 60 mH inductor of 10 Ω resistance. (*a*) Show that there is zero transient current if $\theta = \tan^{-1}(\omega L/R)$. (*b*) Calculate the maximum value of the current for this value of θ and the time at which it occurs.

Solution

(*a*) When $\theta = \tan^{-1}(\omega L/R) = \beta$ in equation (20.13),

$$i = [V_m/\surd(R^2+\omega^2L^2)][\sin(\omega t+\beta-\beta)-\sin(\beta-\beta)e^{-Rt/L}]$$

$$= [V_m/\surd(R^2+\omega^2L^2)]\sin\omega t$$

and there is no transient term.

(*b*) $I_{max} = 10/\surd[10^2+(1000\times 0.06)^2] = 10/60.83$

$= 164.4$ mA (*Ans.*)

The maximum value of the current occurs when $\sin\omega t = 1$, i.e. when $\omega t = \pi/2$. Hence,

$t = \pi/2000 = 1.57$ ms (*Ans.*)

Initial conditions

Capacitor with initial charge

If a capacitor has an initial charge at time $t=0$ then the voltage across the capacitor is

$$v_C(t) = v_{C0} + (1/C)\int_0^t i(t)\,dt$$

or $\quad v_C(S) = v_{C0}/S + i(S)/SC \quad (20.14)$

The current in the capacitor is $i(t) = C\,dv_C(t)/dt$ and

$$i(S) = C[Sv_C(S) - v_{C0}] \quad (20.15)$$

This equation states that the circuit is initially like a d.c. voltage source connected in series with an uncharged capacitor.

Series RC circuit

When the applied voltage is removed from the *RC* circuit and replaced by a short circuit,

$$0 = i(t)R + v_C = i(t)R + q/C$$

$$= i(t)R + (1/C)\int_0^t i(t)\,dt + v_{C0}$$

$$= i(S)[R + 1/SC] - v_{C0}/S = i(S)[R + 1/SC] - (V/S)$$

assuming the capacitor is fully charged to voltage V. Therefore,

$$i(S) = V/[S(R + 1/SC)] = V/[R(S + 1/CR)]$$

and

$$i(t) = (V/R)\,e^{-t/CR} \quad (20.16)$$

Example 20.12

A 1.5 μF capacitor is charged to have a voltage of 25 V across its terminals. The voltage source is then removed and the capacitor is connected across a 10 kΩ resistor. Obtain an expression for the current and determine its value 10 ms after the capacitor starts its discharge.

Solution

$$i(S) = (25/S)/[(10 \times 10^3 + 1)/(1.5 \times 10^{-6}S)]$$

$$= (25 \times 10^{-4})/(S + 66.67)$$

Therefore,

$$i(t) = 2.5e^{66.67t} \quad (Ans.)$$

When $t = 10 \times 10^{-3}$

$$i(t) = 2.5 \times 0.513 = 1.28 \text{ mA} \quad (Ans.)$$

Inductor with initial current

When an inductor is already carrying a current, at time $t = 0$, the voltage across the inductor is then

$$v_L(t) = L\,di(t)/dt$$

and

$$v_L(S) = L[Si(S) - i(0)] \quad (20.17)$$

The current in the inductor is then

$$i_L(t) = i_0 + (1/L)\int_0^t v_L(t)\,dt$$

or $$i(S) = i_0/S + v_L(S)/SL \qquad (20.18)$$

This equation states that the circuit is initially like a d.c. current source connected in parallel with an inductor which is not carrying a current.

Series RL circuit

When the voltage step is removed and replaced by a short circuit

$$0 = i(t)R + L\,di/dt$$
$$0 = i(S)(R + SL) - LI(0) = i(S)(R + SL) - (LV/R)$$
$$i(S) = LV/[R(R + SL)] = V/[R(S + R/L)]$$

and $$i(t) = (V/R)e^{-Rt/L} \qquad (20.19)$$

Example 22.13

A d.c. voltage of 100 V is applied to a 500 mH inductor of 50 Ω resistance. When steady-state conditions have been established the d.c. voltage source is removed and replaced by the voltage $v = 200\cos 100\pi t$ volts. Derive an expression for the subsequent current in the inductor.

Solution
Steady-state current $= i_0 = 100/50 = 2$ A, $v = iR + L\,di/dt$. Therefore,

$$200S/(S^2 + 314^2) = i(S)50 + 0.5(S\,i(S) - 2)$$
$$i(S) = [200S/(S^2 + 314^2) + 1]/[50 + 0.5S]$$
$$= [200S/(S^2 + 314^2) + 1]/[0.5(S + 100)]$$
$$= 400[A/(S + 100) + (BS + C)/(S^2 + 314^2)] + 2/(S + 100)$$
$$AS^2 + 314^2A + BS^2 + 100BS + CS + 100C = 400S$$

Solving: $A = -0.368$, $B = 0.368$ and $C = 363$. Therefore,

$$i(S) = -0.368/(S + 100) + (0.368S + 363)/(S^2 + 314^2) + 2/(S + 100)$$

and $$i(t) = 0.368\,e^{-100t} + 0.368\cos 314t + 1.156\sin 314t + 2\,e^{-100t}$$
$$= 1.213\cos(314t - 72.3°) + 2\,e^{-100t} \text{ A} \quad (\textit{Ans.})$$

Impedance functions

An impedance function is defined as $Z(S) = V(S)/I(S)$. The impedance of a capacitor can be written as $1/SC$ and the impedance of an inductor as SL. This means that the normal sinusoidal expressions can be used with S replacing $j\omega$. This step can be extended to the determination of the impedance, or the

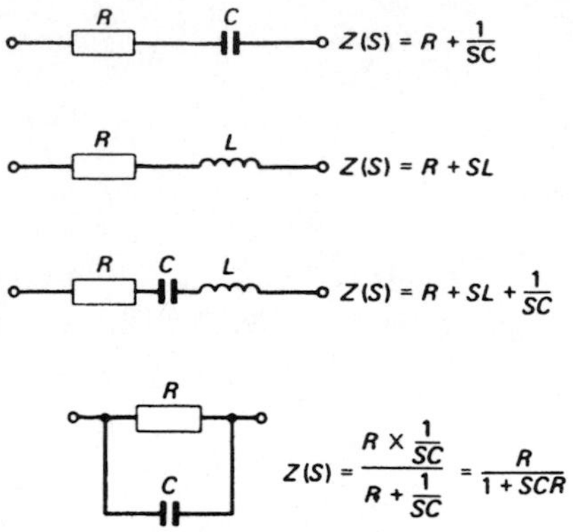

Fig. 20.5 Impedance function

admittance of any circuit *provided the initial conditions are zero*. Some examples are given in Fig. 20.5.

Example 20.14

An inductor L of self-resistance R_3 is connected in parallel with a resistor R_2 and the combination is connected in series with a third resistor R_1. (a) If $R_1 = 4\ \Omega$, $R_2 = 6\ \Omega$, $R_3 = 3\ \Omega$ and $L = 1$ H determine an expression for the complex impedance of the circuit. (b) Find the current supplied to the circuit when a 1 V step is applied.

Solution

(a) $Z(S) = R_1 + [R_2(R_3 + SL)]/[R_2 + R_3 + SL]$

$$= 4 + 6(3 + S)/(9 + S)$$

$$= 4 + (18 + 6S)/(9 + S) = (54 + 10S)/(9 + S) \quad (Ans.)$$

(b) $I(S) = V(S)/I(S) = (9 + S)/[S(10S + 54)]$

$$= A/S + B/(10S + 54)$$

Hence,

$$9 + S = 10AS + 54A + BS$$

Equate constants: $9 = 54A$ or $A = 0.167$. Equate S: $1 = 10A + B$ or $B = -0.67$. Thus,

$$I(S) = 0.167/S - 0.67/(10S + 54) = 0.167/S - 0.067/(S + 5.4)$$

Therefore,

$$I(t) = 0.167 - 0.067\ e^{-5.4t}\ \text{A} \quad (Ans.)$$

Transfer functions

The *transfer function* (T.F.) of a network is the ratio of the output quantity to the input quantity expressed using Laplace transforms. In the case of the voltage transfer function, the output terminals of the network are open circuited, while for the current transfer function the output terminals are short circuited.

Three examples are shown in Fig. 20.6.

- Figure 20.6(a)

$$V_0(S) = V_1(S)R/(R + 1/SC) = V_1(S)SCR/(1 + SCR)$$

or $\text{T.F.} = V_0(S)/V_1(S) = SCR/(1 + SCR)$

- Figure 20.6(b)

$$V_0(S) = (V_1(S)/SC)/(R + 1/SC) = V_1(S)/(1 + SCR)$$

or $\text{T.F.} = V_0(S)/V_1(S) = 1/(1 + SCR)$

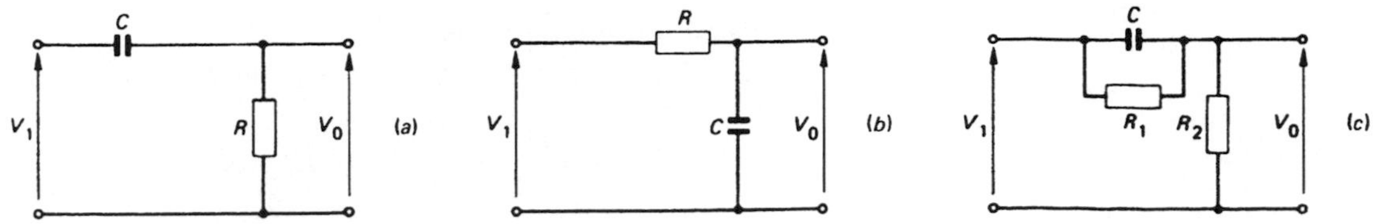

Fig. 20.6

- Figure 20.6(c)

$$V_0(S) = (V_1(S)R_2)/[R_1/(1 + SCR_1) + R_2]$$

$$= [V_1(S)R_2(1 + SCR_1)]/[R_1 + R_2(1 + SCR_1)]$$

or $\quad \text{T.F.} = V_0(S)/V_1(S)$

$$= [R_2(1 + SCR_1)]/[R_1 + R_2 + SCR_1R_2]$$

LCR circuits

Assuming that at time $t = 0$, $i = 0$ and $v_C = 0$, the application of Kirchhoff's voltage law to a series LCR circuit gives $V = L\,\mathrm{d}i/\mathrm{d}t + Ri + (1/C)\int_0^t i\,\mathrm{d}t$. Hence,

$$V(S) = I(S)[R + SL + 1/SC]$$

$$= (I(S)/SC)[S^2LC + SCR + 1]$$

and $I(S) = (V(S)SC)/[S^2LC + SCR + 1]$ (20.20)

For a step input voltage of V volts, $V(S) = V/S$ and

$$I(S) = VC/[S^2LC + SCR + 1]$$

$$= (V/L)[1/(S^2 + SR/L + 1/LC)] \qquad (20.21)$$

The roots of the denominator of equation (20.21) are

$$S = [-R/L \pm \sqrt{(R^2/L^2 - 4/LC)}]/2$$

There are three cases:

- If $R^2/L^2 < 4/LC$ the circuit is *under-damped* and will oscillate before it settles down to its steady-state value.
- If $R^2/L^2 = 4/LC$ the circuit is *critically damped* and will not oscillate. The circuit reaches its steady-state value in the shortest possible time.
- If $R^2/L^2 > 4/LC$ the circuit is *over-damped* and will not break into oscillation. The time taken for steady-state value to be reached depends upon the degree of damping.

The value of resistance that gives critical damping is obtained as $R = 2\sqrt{(L/C)}\ \Omega$. When the circuit is under-damped:

$$I(S) = (V/L)\{1/[(S + R/2L)^2 + (1/LC - R^2/4L^2)]\}$$

Let $\alpha = R/2L$ (known as the *damping ratio*) and $\omega = \sqrt{(1/LC - R^2/4L^2)}$, then

$$I(S) = (V/\omega L)\{1/[(S+\alpha)^2 + \omega^2]\} \quad \text{and}$$

$$I(t) = (V/\omega L)[e^{-at}\sin \omega t] \qquad (20.22)$$

Example 20.15

A voltage step of 20 V is applied to a series *LCR* circuit with $R = 5\ \Omega$, $L = 1$ H and $C = 0.1\ \mu$F. Determine an expression for the current that flows in the circuit.

Solution
$\alpha = R/2L = 2.5$ and $\omega = \sqrt{[1/(1 \times 0.1 \times 10^{-6}) - 25/4]} = 3162.3$ rad/*s*.
$I(t) = (20/3162.3)[e^{-25t}\sin 3162.3t] = 6.325(e^{-2.5t}\sin 3162.3t)$mA(*Ans.*)

Example 20.16

A series *LCR* circuit consists of a 100 Ω resistance, a 1 H inductance and a 1 μF capacitor. A 100 V step voltage is applied across the circuit. Derive an expression for the variation with time of the capacitor voltage.

Solution
Since $v_C = q/C$,

$$V = R\,dq/dt + L\,d^2q/dt^2 + q/C$$

or $$100 = 100\,dq/dt + d^2q/dt^2 + 10^6 q$$

Taking transforms:

$$100/S = q(S)(S^2 + 100S + 10^6),$$

$$q(S) = 100/[S(S^2 + 100S + 10^6)]$$

$$= A/S + (BS + C)/(S^2 + 100S + 10^6)$$

Hence,

$$100 = AS^2 + 100AS + 10^6 A + BS^2 + CS$$

Equating constants: $100 = 10^6 A$, or $A = 10^{-4}$. Equating S: $0 = 100A + C$ or $C = -10^{-2}$. Equating $S^2 : 0 = A + B$, or $B = -10^{-4}$.

Thus,

$$q(S) = (10^{-4}/S) - (10^{-4}S + 10^{-2})/(S^2 + 100S + 10^6)$$
$$= (10^{-4}/S) - (10^{-4}S + 10^{-2})/[(S+50)^2 + 997500]$$
$$= 10^{-4}[1/S - (S+100)/[(S+50)^2 + 997500]$$
$$= 10^{-4}\left[\frac{1}{S} - \frac{S+50}{(S+50)^2 + 997500} + \frac{50}{(S+50)^2 + 997500}\right]$$
$$= 10^{-4}\left[\frac{1}{S} - \frac{S+50}{(S+50)^2 + 997500} + \frac{998.75/19.975}{(S+50)\ 2 + 997500}\right]$$

From Table 20.1,

$$q(t) = 10^{-4}[1 - e^{-50t}\cos 998.75t + 0.05\, e^{-50t}\sin 998.75t]$$
$$= 10^{-4}[1 - e^{-50t}\sqrt{(1^2 + 0.05^2)}\sin(998.75t + \varphi)$$

where $\varphi = \tan^{-1}\ 20 = 87°$. Therefore,

$$v_C = q/C = 100[1 - e^{-50t}\sin(998.75t + 87°)]\text{V} \quad (Ans.)$$

Example 20.17

A series LCR circuit has $R = 200\ \Omega$, $L = 2$ H and $C = 40\ \mu$F. Obtain an expression for the variation with time of the capacitor voltage if the capacitor is initially discharged and a 250 V step voltage is applied to the circuit.

Solution

$V = 2\ d^2q/dt^2 + 200\ dq/dt + q/(40 \times 10^{-6})$. Hence,

$$250/S = q(S)[2S^2 + 200S + 25 \times 10^3]$$

and $q(S) = 250/[S(2S^2 + 200S + 25 \times 10^3)]$

$$= A/S + (BS + C)/(2S^2 + 200S + 25 \times 10^3)$$

Hence,

$$250 = 2AS^2 + 200AS + 25 \times 10^3A + BS^2 + CS$$

Equating constants: $250 = 25 \times 10^3A$ or $A = 0.01$. Equating S: $0 = 200A + C$ or $C = -2$. Equating $S^2: 0 = 2A + B$ or $B = -0.02$. Thus,

$$q(S) = 0.01/S - (0.02S + 2)/(2S^2 + 200S + 25\ 000)$$
$$= 0.01\{1/S - (2S + 200)/[2(S+50)^2 + 100^2]\}$$
$$= 0.01\{(1/S) - (S + 100)/[(S+50)^2 + 100^2]\}$$
$$= 0.01\left[\frac{1}{S} - \frac{S+50}{(S+50)^2 + 100^2} + \frac{0.5 \times 100}{(S+50)^2 + 100^2}\right]$$

Therefore,

$$q(t) = 0.01[1 - e^{-50t}(\cos 100t + 0.5\sin 100t)]$$
$$= 0.01[1 - 1.12\ e^{-50t}\sin(100t + 63.4°)]$$

and $v_C(t) = q/C = 250[1 - 1.12\ e^{-50t}\sin(100t + 63.4°)]$ V $(Ans.)$

Transient reflections on transmission lines

When a voltage pulse, or a steep-fronted voltage surge, arrives at the end of mismatched transmission line some of the incident energy will be reflected. If the mismatched load contains reactance a transient will exist.

Load impedance: R in parallel with C

The complex impedance of the load is $Z_L = (R/SC)/(R + 1/SC) = R/(1 + SCR)$. The reflected voltage is equal to the incident voltage times the voltage reflection coefficient (p. 323). Hence,

$$\text{Reflected voltage} = V_i(Z_L - Z_0)/(Z_L + Z_0)$$

and the voltage V_L across the load is

$$V_i[1 + (Z_L - Z_0)/(Z_L + Z_0)] = V_i[2Z_L/(Z_L + Z_0)]$$

For a step incident voltage $V_i(S) = V_i/S$ and

$$\begin{aligned} V_L(S) &= (V_i/S)\{[2R/(1 + SCR)]/[R/(1 + SCR) + Z_0]\} \\ &= (V_i/S)\{2R/[R + Z_0(1 + SCR)]\} \\ &= [(V_i/S)2R]/(R + Z_0 + SCRZ_0) \\ &= [(V_i/S)2R]/\{CRZ_0\,[S + (R + Z_0)/(CRZ_0)]\} \\ &= [(V_i/S)2/CZ_0]/[S + (R + Z_0)/CRZ_0] \\ &= A/S + B/(S + Z) \quad \text{where } Z = (R + Z_0)/CRZ_0 \end{aligned}$$

Hence,

$$2V_i/CZ_0 = AS + AZ + BS$$

Equating constants: $2V_i/CZ_0 = AZ$ or $A = 2V_iR/(R + Z_0)$.
Equating S: $0 = A + B$ or $B = -2V_iR/(R + Z_0)$. Thus,

$$V_L(S) = (2V_iR/(R + Z_0)[1/S - 1/(S + Z_0) \quad \text{and}$$

$$V_L(t) = (2V_iR/(R + Z_0)[1 - e^{-Zt}] \tag{20.23}$$

Load impedance: R in series with C

Now the voltage reflection coefficient at the load is

$$\rho_v = (R + 1/SC - Z_0)/(R + 1/SC + Z_0)$$

The voltage V_L across the load is

$$V_L = 2V_i[(R + 1/SC)/(R + 1/SC + Z_0)]$$

For an incident step voltage,

$$V_L(S) = [(2V_i/S)(1+SCR)]/[1+SC(R+Z_0)]$$

$$= \frac{2V_i}{S}\left[\frac{1+SCR}{S+1/C(R+Z_0)}\right]\left[\frac{1}{C(R+Z_0)}\right]$$

$$= \left[\frac{2V_i}{C(R+Z_0)}\right]\left[\frac{1+SCR}{S(S+1/C(R+Z_0))}\right]$$

$$= A/S + B/[S+1/C(R+Z_0)]$$

$$2V_i/C(R+Z_0) + 2V_iRS/(R+Z_0)$$
$$= AS + A/C(R+Z_0) + BS$$

Equating constants:

$$2V_i/C(R+Z_0) = A/C(R+Z_0)$$

or $A = 2V_i$. Equating S:

$$2V_iR/(R+Z_0) = A + B$$

or $\quad B = 2V_iR/(R+Z_0) - 2V_i = 2V_i[R/(R+Z_0) - 1]$

Hence,

$$V_L(S) = \frac{2V_i}{S} + \frac{2V_i[R/(R+Z_0)] - 1}{S+1/C(R+Z_0)}$$

Therefore,

$$V_L(t) = 2V_i\{1 + [R/(R+Z_0) - 1]e^{-t/C(R+Z_0)} \qquad (20.24)$$

Load impedance; inductance in series with resistance

Following the same lines as before, when a step voltage V arrives at the end of the line the current in the load is

$$I_L(S) = [2Z_0I_i/(R+Z_0)]\{(1/S) - 1/[S+(R+Z_0)/L]\}$$

and

$$I_L(t) = [2Z_0I_i/(R+Z_0)][1 - e^{-(R+Z_0)t/L}] \qquad (20.25)$$

Example 20.18

A 1 kV voltage step arrives at the end of a line whose characteristic impedance is 250 Ω. The line is terminated by a 10 mH inductance and a 9.75 kΩ resistance. Calculate (*a*) the value of the current in the load 1 μs after the arrival of the step voltage at the load and (*b*) the final value of the load current.

Solution

(*a*) From equation (20.25), $I_L(t) = 0.2[1 - e^{-10^6 t}]$A. After 1 μs,

$$I_L(t) = 0.2(1 - e^{-1}) = 0.126 \text{ A} \quad (Ans.)$$

(*b*) Final load current = 0.2 A (*Ans.*)

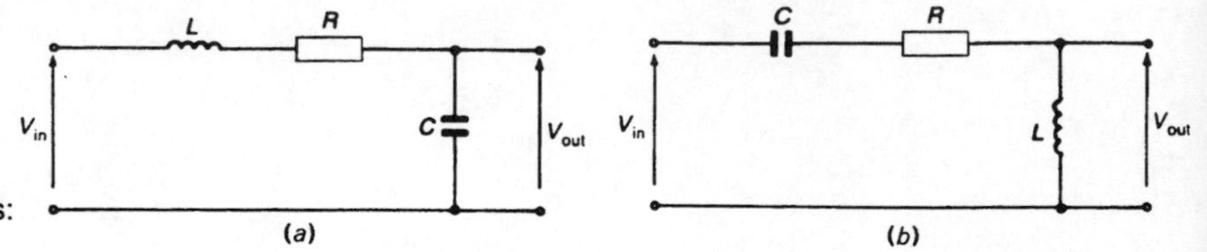

Fig. 20.7 Second-order filters: (a) low-pass; (b) high-pass

Second-order filters

Figure 22.7(*a*) shows a passive second-order low-pass filter whose selectivity and cut-off frequency are determined by the values of its inductance and capacitance. The transfer function $F(S)$ of the filter is

$$F(S) = V_{out}(S)/V_{in}(S) = (1/SC)/(SL + R + 1/SC)$$
$$= 1/(S^2LC + SCR + 1)$$
$$F(S) = (1/LC)/(S^2 + SR/L + 1/LC) \quad (20.26)$$

The final term, $1/LC$ in the denominator of equation (20.26), is the square of (2π times the resonant frequency f_0 of the circuit), i.e. $\omega_0^2 = 1/LC$. Also,

$$R/L = R\sqrt{(C/L^2C)} = [R\sqrt{(C/L)}][1/\sqrt{(LC)}] = \omega_0/Q$$

Therefore, the equation for $F(S)$ can be re-written as

$$F(S) = \omega_0^2/(S^2 + S\omega_0/Q + \omega_0^2) \quad (20.27)$$

or

$$F(S) = 1/(S^2/\omega_0^2 + S/\omega_0 Q + 1) \quad (20.28)$$

For the second-order high-pass filter of Fig. 20.7(*b*),

$$F(S) = V_{out}(S)/V_{in}(S) = SL/(SL + R + 1/SC)$$
$$= S/(S + R/L + 1/SCL) = S^2/(S^2 + SR/L + 1/LC)$$

or

$$F(S) = S^2/(S^2 + S\omega_0/Q + \omega_0^2) \quad (20.29)$$

It should be noted that the two transfer functions have the same denominator.

Active filters

The vast majority of active filters are of the *Sallen and Key* type. The general form of a Sallen and Key filter is shown in Fig. 20.8. The resistors R_3 and R_4 set the voltage gain $A_{v(f)}$ of the op-amp to the desired value.

Summing the currents at the node A:

$$(V_{in} - V_x)Y_1 + (V_{out} - V_x)Y_4 - V_{out}Y_3/A_{v(f)} = 0$$
$$V_{in}Y_1 = (Y_1 + Y_4)V_x - V_{out}Y_4 + Y_{out}V_3/A_{v(f)}$$

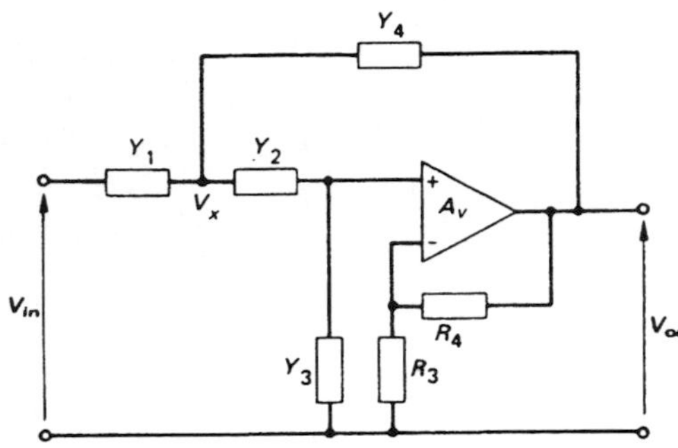

Fig. 20.8 Active filter of the Sallen and Key type

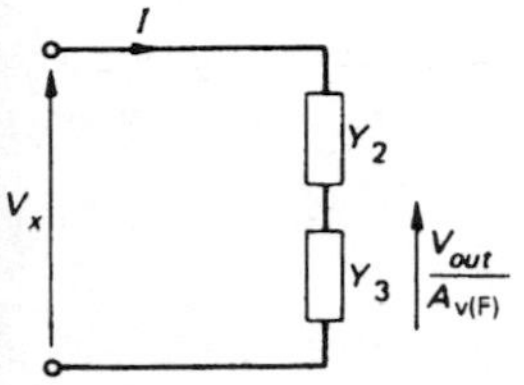

Fig. 20.9

Referring to Fig. 20.9 $i = V_{out}Y_3/A_{v(f)}$ and so

$$V_x = [V_{out}Y_3/A_{v(f)}][(Y_2 + Y_3)/Y_2Y_3]$$
$$= [V_{out}/A_{v(f)}][Y_2 + Y_3)/Y_2]$$

Hence

$$V_{in}Y_1 = \left[\frac{(Y_1 + Y_4)(Y_2 + Y_3)}{Y_2} + Y_3\right]\left(\frac{V_{out}}{A_{v(f)}}\right) - V_{out}Y_4$$

or

$$F(S) = V_{out}/V_{in}$$
$$= \frac{A_{v(f)}Y_1Y_2}{(Y_1 + Y_4)(Y_2 + Y_3) + Y_2Y_3 - A_{v(f)}Y_2Y_4)}$$
$$= \frac{A_{v(f)}Y_1Y_2}{Y_2[Y_1 + Y_4(1 - A_{v(f)})] + Y_3(Y_1 + Y_2 + Y_4)} \quad (20.30)$$

For the network to act as a low-pass or a high-pass filter the admittances Y_1, Y_2, etc. must be provided by the appropriate components (R or C) so that the transfer function has the form of equations (20.26 through to 20.29).

Low-pass filter

Since the numerator of equation (20.26) does not contain a term in S, both Y_1 and Y_2 must be provided by resistors. Equation (20.30) then reduces to

$$F(S) = \frac{A_{v(f)}G_1G_2}{G_2[G_1 + Y_4(1 - A_{v(f)})] + Y_3(G_1 + G_2 + Y_4)} \quad (20.31)$$

and from this it can be seen that for terms in both S and S^2 to appear in the denominator both Y_3 and Y_4 must be provided by capacitors. The required circuit is shown by Fig. 20.10.

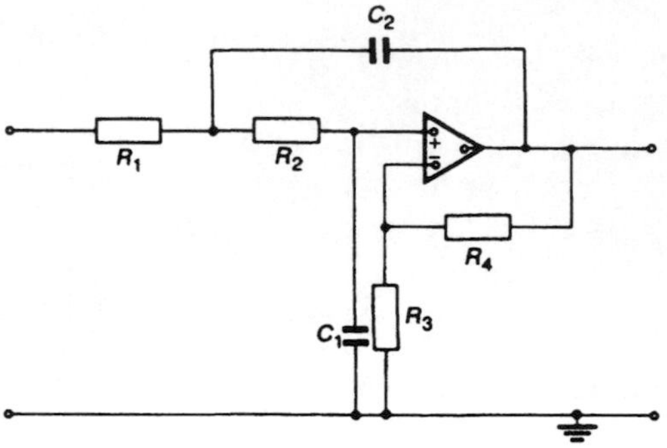

Fig. 20.10 Second-order low-pass active filter

The transfer function of Fig. 20.10 can be derived directly or by substituting into equation (20.31) with $Y_3 = SC_1$ and $Y_4 = SC_2$, thus

$$F(S) = \frac{A_{v(f)}G_1G_2}{G_1G_2 + SC_2G_2(1 - A_{v(f)}) + SC_1(G_1 + G_2 + SC_2)}$$

$$= \frac{A_{v(f)}}{1 + SC_2R_1(1 - A_{v(f)}) + SC_1(R_1 + R_2 + SC_2R_1R_2}$$

or $$F(S) = \frac{A_{v(f)}}{S^2C_1C_2R_1R_2 + S[C_2R_1(1 - A_{v(f)})C_1(R_1 + R_2)] + 1} \qquad (20.32)$$

Comparison of equation (20.32) with equation (20.28) gives

$$\omega_0 = 1/\sqrt{(R_1R_2C_1C_2)}$$

and

$$Q = \frac{\sqrt{(R_1R_2C_1C_2)}}{R_1C_2(1 - A_{v(f)}) + C_1(R_1 + R_2)} \qquad (20.33)$$

Often the design of a particular filter commences with making $R_1 = R_2$ and $C_1 = C_2$. Then

$$F(S) = A_{v(f)}/[S^2R^2C^2 + SRC(3 - A_{v(f)}) + 1] \qquad (20.34)$$

Also,

$$\omega_0 = 1/RC \quad \text{and} \quad Q = 1/(3 - A_{v(f)}) \qquad (20.35)$$

This means that the Q factor, and hence the passband, of the filter is set by the closed-loop gain of the op-amp.

High-pass filter

Comparing equations (20.29) and (20.30) it can be seen that the admittances Y_1 and Y_2 must both be capacitances in order to obtain an S^2 term in the numerator of the transfer function. Then

$$F(S) = \frac{A_{v(f)} SC_1 SC_2}{SC_2[SC_1 + Y_4(1 - A_{v(f)})] + Y_3(SC_1 + SC_2 + Y_4)} \tag{20.36}$$

For terms in S^2, S and constant to appear in the denominator, Y_3 and Y_4 must both be conductances. Hence

$$F(S) = \frac{A_{v(f)} SC_1 SC_2}{SC_2[SC_1 + G_4(1 - A_{v(f)})] + G_3(SC_1 + SC_2 + G_4)}$$

$$= \frac{A_{v(f)} S^2}{S^2 + SG_4(1 - A_{v(f)})/C_1 + SG_3(1/C_2 + /C_1) + G_3G_4/C_1C_2}$$

$$= \frac{A_{v(f)} S^2}{S^2 + S(1 - A_{v(f)})/C_1R_4 + S(1/C_2 + /C_1)/R_3 + /C_1C_2R_3R_4}$$

$$= \frac{A_{v(f)} S^2}{S^2C_1C_2R_3R_4 + S[C_2R_3(1 - A_{v(f)}) + R_4(C_1 + C_2)] + 1} \tag{20.37}$$

Comparing with equation (20.27),

$$\omega_0 = \sqrt{(C_1C_2R_3R_4)}$$

and

$$Q = \sqrt{(R_3R_4C_1C_2)}/[R_3C_2(1 - A_{v(f)}) + R_4(C_1 + C_2)] \tag{20.38}$$

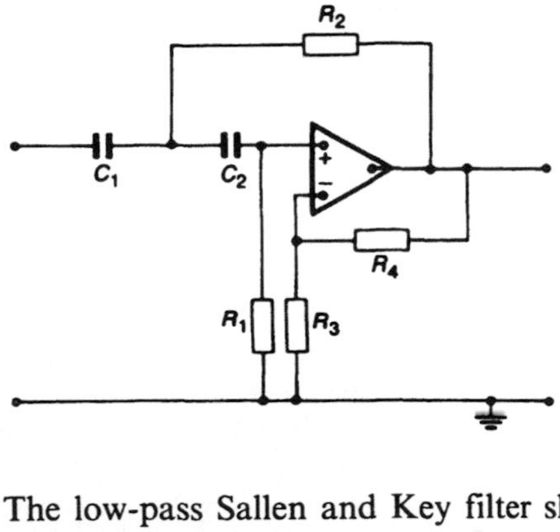

Fig. 20.11 Second-order high-pass active filter

The low-pass Sallen and Key filter shown in Fig. 20.10 can be converted into a high-pass filter by interchanging the resistors with the capacitors. Figure 20.11 shows the circuit of an aotive high-pass filter.

Further detail on active filters is contained in Electronics 4.

Exercises 20

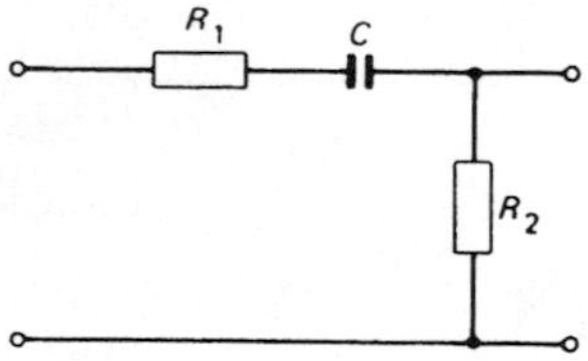

Fig. 20.12

20.1 A V volt/s ramp voltage is applied to the circuit of Fig. 20.12. Derive an expression for the subsequent variation with time of the current flowing in the circuit when the initial voltage stored in the capacitor is zero.

20.2 Repeat Exercise 20.1 for the case when the initial capacitor voltage is V/4 volts.

20.3 Derive an expression for the time variation of the voltage across the output terminals of the circuit in Fig. 20.13 if the input voltage is a 6 V step function.

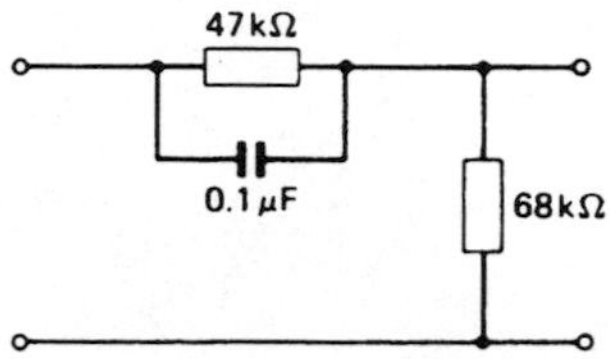

Fig. 20.13

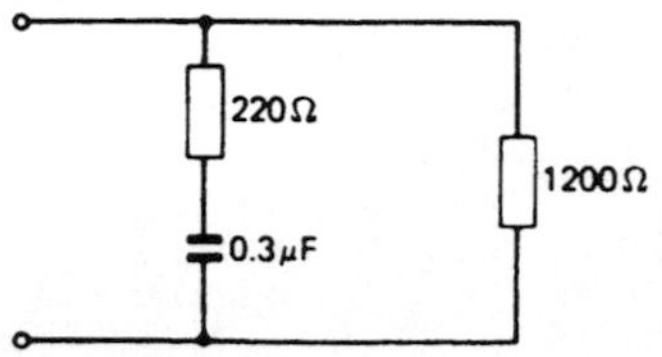

Fig. 20.14

20.4 A 1 V step voltage is applied to the circuit shown in Fig. 20.14. Derive an expression for the variation with time of the current supplied to the circuit.

20.5 Obtain an expression for the output voltage of the circuit given in Fig. 20.15 if $R_1 = 160$ kΩ, $R_2 = 350$ kΩ and $C = 0.3$ μF when the input voltage is a 6 V step.

20.6 Derive an expression for the current flowing in the 5 Ω resistor of Fig. 20.16 when the input voltage is a unity step function.

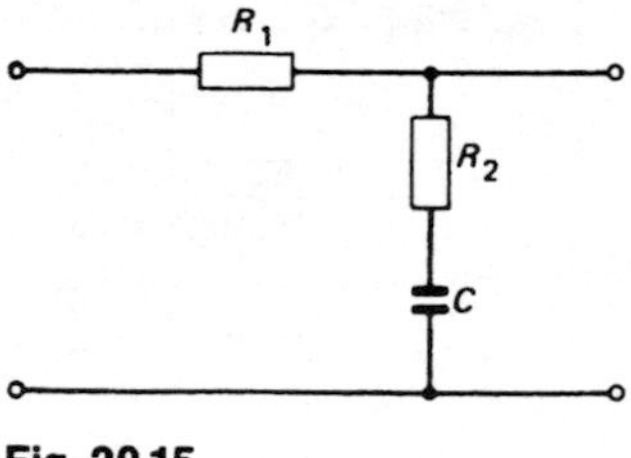

Fig. 20.15

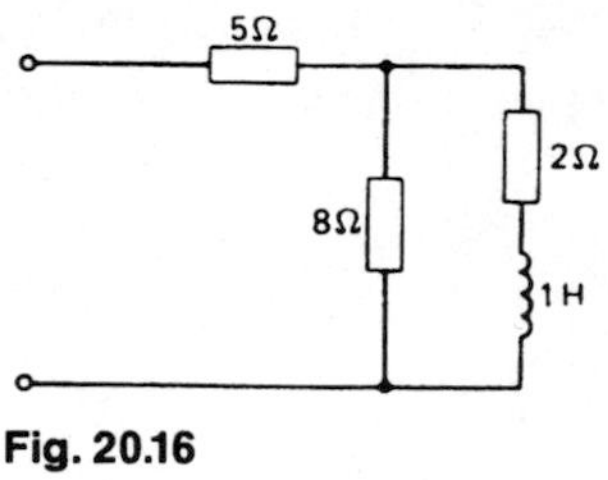

Fig. 20.16

20.7 For the circuit given in Fig. 20.16 obtain an expression for the current flowing in the inductor when the applied voltage is a 3 V/s ramp function.

20.8 A circuit consists of a resistor R_2 and a capacitor C connected in parallel with one another and with the signal path. Another resistor R_1 is connected in series with the signal path. A constant current I is suddenly caused to flow into R_1. Derive an expression for the applied voltage if the capacitor has zero charge at time $t = 0$. If $R_1 = 10$ kΩ, $R_2 = 12$ kΩ and $C = 22$ μF, calculate the applied voltage 0.25 s after the 10 mA current first flows into the circuit.

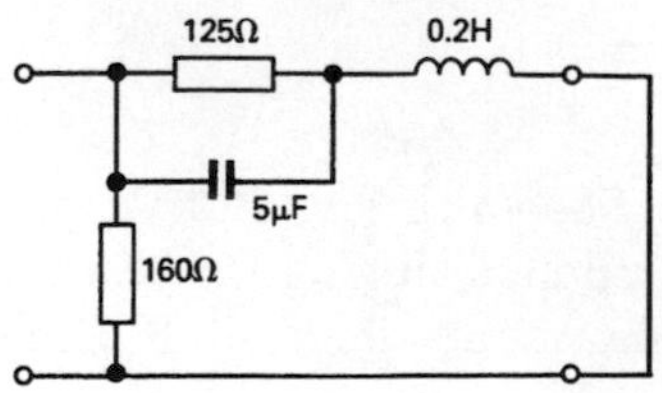

Fig. 20.17

20.9 For the circuit shown in Fig. 20.17 derive an expression for the current into the circuit when a 160 volt step is applied.

20.10 A 50 μF capacitor is connected in series with a 1 MΩ resistor and then a 10 μF capacitor is connected in parallel with the circuit. A 120 V step voltage is applied to the circuit. Obtain an expression for the current supplied to the circuit.

20.11 A 50 μF capacitor is charged to 300 V and is then suddenly connected across a 2 H inductor of 500 Ω resistance. (*a*) Derive expressions for the voltage across the capacitor and the current in the circuit. (*b*) Calculate the time at which the maximum current occurs.

20.12 An L network has a series resistor of 2 MΩ and a shunt impedance that consists of a 3 MΩ resistor in series with a 0.5 μF capacitor. A 6 V d.c. voltage source is suddenly connected across the input to the circuit and the output voltage is taken from across the shunt impedance. Derive expressions for the current into the circuit and for the output voltage. Calculate the current and the output voltage 2.5 s after the voltage is applied.

20.13 A 2 H inductance and a 1 kΩ resistance are connected in series and a 1 μF capacitor is connected in parallel with them. Derive an expression for the current into the circuit when a 100 V step voltage is applied.

20.14 A LCR circuit has an inductance of 1 H, a capacitance of 1 μF and a resistance whose value is one-half of the value which would give critical damping. If a 20 V step voltage is applied to the circuit obtain an expression for the current.

20.15 A 1.5 μF capacitor is connected in series with a 100 Ω resistor. The circuit has a step voltage of 25 V applied across it. Obtain an expression for the current and determine its value 10 ms after the application of the voltage.

20.16 A circuit consists of a capacitor and a resistor connected in series with one another and then connected in series with the parallel combination of an identical capacitor and resistor. Determine an expression for the complex impedance of the circuit.

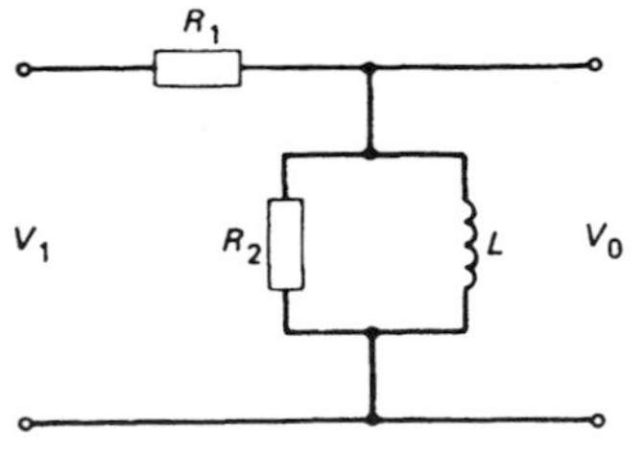

Fig. 20.18

20.17 Show that the transfer function of the circuit of Fig. 20.18 is

$$V_0(S)/V_1(S) = SR_2/\{[S + R_1R_2/L(R_1 + R_2)][R_1 + R_2]\}$$

20.18 (*a*) Find the expression for the current $i(t)$ flowing in a circuit if

$$i(S) = 6/S + (5S + 16)/[S(S + 6)]$$

(*b*) Calculate the current after 0.993s.

20.19 Find the expression for the current $i(t)$ in a circuit if

$$i(S) = 25/S + (20S + 12)/[S(S + 4)]$$

20.20 A transmission line has a characteristic impedance of 250 Ω and is terminated by an inductance of 20 mH and a resistance of 4750 Ω in series. A step voltage of 500 V arrives at the load. Determine an expression for the load current. Calculate the current 1 μs after the step arrives at the load.

20.21 A parallel tuned circuit has $L = 2$ H, $R = 3\ \Omega$ and $C = 0.1$ F and is connected in series with a 5 Ω resistor. The current $i = 6\,e^{-2t}$ flows into the 5 Ω resistor. Determine the voltage across the capacitor.

20.22 A 5 H inductor of 10 Ω resistance has a ramp voltage applied across it. The ramp rises linearly from 0 V to 50 V in 1 s. Derive an expression for the current in the circuit.

20.23 An L network with series resistor R_1 and shunt capacitor C_1 is connected in cascade with a second L network with series resistor R_2 and shunt capacitor C_2. If the input voltage is applied to R_1 determine the transfer function for the network.

20.24 A steep-fronted voltage surge of amplitude V volts travels along a line of characteristic impedance Z_0. The line is terminated by a resistor R in parallel with a capacitor C. Calculate the value of the load voltage 1.5 μs after the surge arrives at the load $V = 10$ kV, $C = 2$ nF, $R = 15$ kΩ and $Z_0 = 300\ \Omega$.

20.25 A resistor R is connected in series with the parallel combination of an inductance L and a capacitor C. A step voltage V is applied to the circuit. Once steady-state conditions have been reached both the resistor and the voltage source are removed. Derive an expression for the subsequent voltage across the capacitor.

20.26 A 1.5 V d.c. source is applied across a circuit which consists of a 200 Ω resistor in series with a 1 F capacitor and a 1200 Ω resistor in parallel. Obtain an expression for the current flowing into the circuit.

20.27 Show that for the circuit given in Fig. 20.19

$$Z(S) = \frac{R_1R_2 + R_1R_3 + R_2R_3 + SL(R_1 + R_2)}{R_2 + R_3 + SL}$$

20.28 Determine the impedance function $Z(S)$ for the circuit shown in Fig. 20.20.

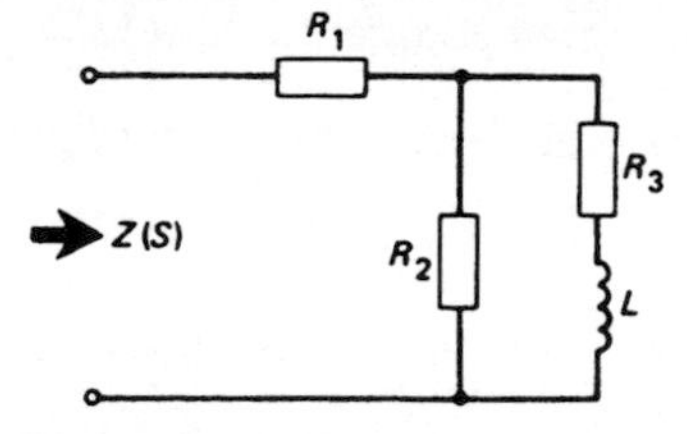

Fig. 20.19

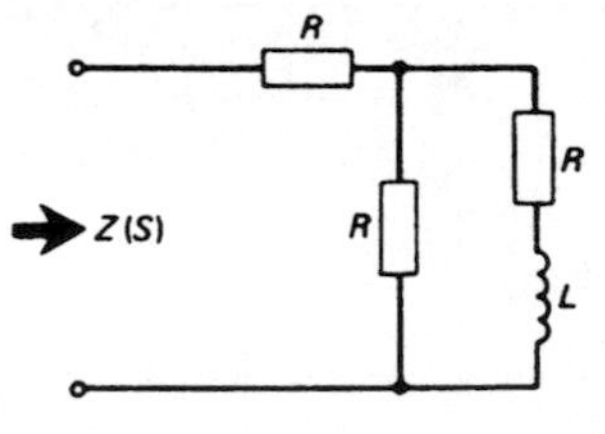

Fig. 20.20

Answers

1.1 (*a*) 60, 20; (*b*) 25, 0; (*c*) 0, -500; (*d*) 2.5, 4.33; (*e*) 1.73, -1; (*f*) -0.042, 0.9; (*g*) -18.9×10^{-6}, -41.3×10^{-6}

1.2 (*a*) 0, -1; (*b*) 1, -1; (*c*) 10, -5; (*d*) 2, -3; (*e*) 1, -1; (*f*) 1, 1; (*g*) -0.22, 0.03

1.3 (*a*) $13.42\angle 63.4°$; (*b*) $0\angle 0°$; (*c*) $6\angle 0°$; (*d*) $10.8\angle 54.9°$; (*e*) $14.9\angle 85.1°$; (*f*) $10.78\angle 109.5°$; (*g*) $2.91\angle -89.9°$

1.4 (*a*) $275.8\angle 133.5°$; (*b*) $0.02\angle 0°$; (*c*) $16.37\angle -22.2°$; (*d*) $129.06\angle -17.5°$; (*e*) $0.448\angle -112.3°$; (*f*) $22.05\angle -61.6°$; (*g*) $89.1\angle -63.9°$

1.5 (*a*) $4.2\angle 106°$; (*b*) $405.1\angle -138.3°$; (*c*) $60\angle -90°$; (*d*) $0.3\angle -68.8°$; (*e*) $0.184\angle -69.4°$; (*f*) $13.354\angle 11.5°$; (*g*) $1523\angle -13.2°$

1.6 (*a*) $1.813\angle -37°$; (*b*) $13.624\angle 40.2°$; (*c*) $2\angle 183°$; (*d*) $1\angle 57.3°$; (*e*) $0.684\angle -57.3°$; (*f*) $0.5\angle 0°$; (*g*) $1.179\angle -58°$

1.7 (*a*) $0.796\angle -12.9°$; (*b*) $45.16\angle -4.5°$; (*c*) $2.916\angle 59°$; (*d*) $1.294\angle -10.2°$; (*e*) $4.518\angle 83.2°$

1.8 (*a*) $0.058\angle 35.5°$; (*b*) $0.0762\angle -2.4°$; (*c*) $1.303\angle 67.5°$; (*d*) $3.314\angle 35.3°$

1.9 (*a*) $70.195\angle 3.7°$; (*b*) $43.196\angle -12.1°$; (*c*) $97.268\angle -64.8°$

1.10 (*a*) $9.658\angle -9.4°$; (*b*) $12.5\angle -67.2°$; (*c*) $2.976\angle 37.4°$

1.11 (*a*) (*i*) $52\angle 112.6°$; (*ii*) $52\angle 112.6°$; (*iii*) $400\angle 64°$; (*iv*) $e^{-(0.8+j0.4)}$
(*b*) (*i*) $2.685\angle 118.2°$ or $2.685\angle -61.9°$; (*ii*) $2.685\angle 28.2°$ or $2.685\angle 208.2°$; (*iii*) $4.472\angle 16°$ or $4.472\angle 196°$; (*iv*) $e^{-(0.2+j0.1)}$ or $e^{-(0.2+j3.242)}$

1.12 (*a*) $v_1 = 50\angle 35°$ V; (*b*) $v_2 = 100\angle -60°$ V; (*c*) $v_3 = 5\angle 90°$ V; (*d*) $v_4 = 10\angle 125°$ V; (*e*) $v_5 = 40\angle 30°$ V

1.13 $13\angle 52.2°$ mA

1.14 $i_4 = 16.34\angle 69.7°$ mA

1.15 $v = 15.654 \sin(1000\pi t - 8.2°)$ V

1.16 $v_S = 28.43 \sin(\omega t + 94.3°)$ V

1.17 $x = 25.045 \times 10^{-3}$, $y = 8.587 \times 10^{-3}$, $z = 6.44 \times 10^{-3}$

1.18 $x = 1.042$, $y = 0.203$, $z = -0.191$

1.19 $\begin{bmatrix} 22 & 42 \\ 24 & 46 \end{bmatrix}$

1.20 $\begin{bmatrix} 2+j6 & 30-j20 \\ -2+j6 & 10+j24 \end{bmatrix}$

1.21 $A = 36/52$, $B = 28/52$

1.22 $\begin{bmatrix} 4-j3 & 1+j2 \\ 8+j3 & -1+j4 \end{bmatrix}$

2.1 (*a*) $R = 10\ \Omega$, $L = 120/100\pi = 0.382$ H
(*b*) $R = 25\ \Omega$, $C = 1/(100\pi \times 80) = 39.8\ \mu$F
(*c*) $50\angle{-20^\circ} = 47 - j17.1$, $R = 47\ \Omega$, $C = 1/(100\pi \times 17.1) = 186.2\ \mu$F
(*d*) $120\angle 60^\circ = 60 + j103.92$, $R = 60\ \Omega$, $L = 103.92/100\pi = 0.331$ H

2.2 (*a*) $Z = 300 + j(2000 \times 0.2) = 300 + j400 = 500\angle 53.1^\circ\ \Omega$
(*b*) $I = (17\angle 0^\circ)/(500\angle 53.1^\circ) = 34\angle{-53.1^\circ}$ mA

2.3 (*a*) (*i*) $Z = 100 + j(800\pi \times 0.3) = 100 + j754 = 760.6\angle 82.5^\circ\ \Omega$
(*ii*) $X_C = 846.6\ \Omega$, $Z = 100 - j846.6 = 852.5\angle{-83.3^\circ}\ \Omega$
(*iii*) $Z = 100 + j(754 - 846.6) = 100 - j92.6 = 136.3\angle{-42.8^\circ}\ \Omega$
(*b*) (*i*) $|I| = 15/760.6 = 19.72$ mA, $i = 19.72\sin(800\pi t - 82.5^\circ)$ mA
(*ii*) $|I| = 15/852.5 = 17.6$ mA, $i = 17.6\sin(800\pi t + 83.3^\circ)$ mA
(*iii*) $|I| = 15/136.3 = 110.1$ mA, $i = 110.1\sin(800\pi t + 42.8^\circ)$ mA

2.4 (*a*) $Z = 60 + j120 = 134.2\angle 63.4^\circ\ \Omega$, $|I| = 120/134.2 = 894.2$ mA,
$i = 894.2\sin(2000\pi t - 33.4^\circ)$ mA
(*b*) $V_{Rm} = iR = 53.65\angle{-33.4^\circ}$ V, $V_R = 53.65/\angle\sqrt{2} = 37.93$ V,
$V_{Lm} = iX_L = 107.3\angle 56.6^\circ$ V, $V_L = 107.3\angle\sqrt{2} = 75.86$ V
(*c*) $V = 53.65\angle{-33.4^\circ} + 107.3\angle 56.6^\circ = (44.8 - j29.5) + (59.1 + j89.6)$
$= 103.9 + j60.1 = 120\angle 30^\circ$ V

2.5 (*a*) $Z = 560 - j120 = 572.7\angle{-12.1^\circ}\ \Omega$
(*b*) $I_{(max)} = 5/(572.7\angle{-12.1^\circ}) = 8.73\angle 12.1^\circ$ mA, $I = 6.17$ mA,
$V_R = IR = 3.46$ V, $V_{C1} = 0.123$ V, $V_{C2} = 0.617$ V

2.6 $Z = 9.31\angle 2.4^\circ\ \Omega$, $I = 150/(9.31\angle 2.4^\circ) = 16.112\angle{-2.4^\circ}$ A,
$I_L = (16.112\angle{-2.4} \times 100\angle 0^\circ)/(110 + j60) = 12.86\angle{-30.9^\circ})$ A,
$I_R = I - I_L = 16.112\angle{-2.4^\circ} - 12.86\angle{-30.9^\circ} = 5.07 + j5.95$
$= 7.82\angle 49.6^\circ$ A

2.7 (*a*) $X_C = 3183\ \Omega$, $\tan^{-1}(-3183/R) = 40^\circ$, $R = 3794\ \Omega$
(*b*) $Z = 3794 - j3183 = 4952.4\angle{-40^\circ}\ \Omega$,
$I = 100/(4952.4\angle{-40^\circ}) = 20.19\angle 40^\circ$ mA,
$i = 20.19\sin(100\pi t + 40^\circ)$ mA

2.8 $Y_1 = 1/(8 + j12) = 0.03846 - j0.05769$ S,
$Y_2 = 1/(5 + j10) = 0.04 - j0.08$ S, $Y_T = 0.07846 - j0.13769$ S
$Z = 1/Y_T = 3.124 - j5.482\ \Omega = 6.31\angle{-60.3^\circ}\ \Omega$
(*a*) $I = 100/(6.31\angle{-60.3^\circ}) = 15.85\angle 60.3$ A
$I_1 = (15.85\angle 60.3^\circ \times 11.18\angle 63.4^\circ)/(13 + j22) = 6.94\angle 64.3^\circ$ A
$I_2 = 15.85\angle 60.3^\circ - 6.94\angle 64.3^\circ = 8.94\angle 57.2^\circ$ A

2.9 $Z_1 = (10 + j82)(-j44)/(10 + j38) = 12.54 - j91.65\ \Omega$
$Z_T = 59.54 - j91.65 = 109.29\angle{-57^\circ}\ \Omega$
$I = (12\angle 0^\circ)/(109.29\angle{-57^\circ}) = 109.8\angle 57^\circ$ mA,
$I_C = (109.8\angle 57^\circ \times 82.61\angle 83^\circ)/39.29\angle 75.2^\circ) = 230.9\angle 64.8^\circ$ mA

2.10 (*a*) $V = (12\angle{-40^\circ} \times 2200\angle 0^\circ)/(5500 - j7800) = 2.766\angle 14.8^\circ$ V
(*b*) (*i*) $X_C = 3900\ \Omega$, $V = (12\angle{-40^\circ} \times 2200\angle 0^\circ)/(5500 - j3900)$
$= 3.92\angle{-4.7^\circ}$ V
(*ii*) $X_C = 15600\ \Omega$, $V = (12\angle{-40^\circ} \times 2200\angle 0^\circ)/$
$(5500 - 15600) = 1.6\angle 30.6^\circ$ V

2.11 $V_{out} = (50 \times -jX_C)/[300 + j(250 - X_C) = 25\angle\theta°$
$X_C = 0.5/[300^2 + (250 - X_C)^2]$ or $X_C = 157\ \Omega$
(*a*) At 50 Hz, $C = 1/(100\pi \times 157) = 20.27\ \mu F$
(*b*) At 5 kHz, $C = 1/(10^4\pi \times 157) = 202.7$ nF

2.12 (*a*) $Y = 1/100 + j/160 = 10 + j6.25 = 11.79\angle 32°$ mS
(*b*) $I = VY = 60\angle 0° \times 11.79 \times 10^{-3}\angle 32° = 707.4\angle 32°$ mA
(*c*) $I_r = 60 \times 0.01 = 600$ mA, $I_C = 60 \times 6.25 \times 10^{-3} = 375\angle 90°$ mA

2.13 (*a*) $Y = 1/50 + j/100 + -j/200 + j/200 = 0.02 + j0.01$ S
(*b*) $Z = 1/Y = 40 - j20\ \Omega$, $R = 40\ \Omega$, $X_C = 20\ \Omega$

2.14 (*a*) $Y = 1/(40 - j60) + 1/(52 + j30) = 22.32\angle 8.2°$ mS
(*b*) $Z = 1/Y = 44.8\angle -8.2°\ \Omega$
(*c*) $I = 100\angle 0° \times 22.32\angle 8.2° = 2.232\angle 8.2°$ A
(*d*) $I_{40} = 100/(40 - j60) = 1.387\angle 56.3°$ A,
$I_{52} = 100/(52 + j30) = 1.67\angle -30°$ A

2.15 $Y = 1/10k - j/10k + j/8k = 0.1 + j0.25$ mS,
$R_S = 0.1/(0.1^2 + 0.025^2) = 9.4\ k\Omega$,
$X_S = 0.25/(0.1^2 + 0.025^2) = 2.353\ k\Omega$

2.16 (*a*) $Z = 1 + j10 - j30/(5 - j6) = 3.951 + j7.541 = 8.513\angle 62.4°\ \Omega$,
$Y = 1/Z = 0.118\angle -62.4°$ S
(*b*) $I = (50\angle -45°)/(8.513\angle 62.4°) = 5.873\angle -107.4°$ A,
$i = 5.873\sin(\omega t + 107.4°)$ A

2.17 (*a*) $Z = 20 + (39 \times j20)/(39 + j20) = 28.12 + j15.84 = 32.27\angle 29.4°\ k\Omega$,
$Y = 1/Z = 31\angle -29.4°\ \mu S$
(*b*) $R = 28.11\ k\Omega$, $X_L = 15.84\ k\Omega$, $G = 31\cos(-29.4°) = 27\ \mu S$,
$B = 31\sin(-29.4°) = -j15.22\ \mu S$
(*c*) $I = (18\angle 0°)/(32.27 \times 10^3\angle 29.4°) = 558\angle -29.4°\ \mu A$
(*d*) $I_{39} = (558\angle -29.4°) \times 20\angle 90°)/(39 + j20) = 254.7\angle 33.4°\ \mu A$,
$I_{j20} = 558\angle -29.4° - 254.7\angle 33.4° = 304.4\angle 26.1°\ \mu A$

2.18 (*a*) $V_L = 0.5\angle -30° \times (5 + j50) = 25.125\angle 84.3°$ V
Current I_{100} in $100\ \Omega = [0.5\angle -30° \times (47 - j30)]/(147 - j30)$
$= 0.186\angle -51.1°$ A,
$V_2 = 100 \times I_{100} = 18.6\angle -51.1°$ V,
$I_{47} = 0.5\angle -30° - 0.186\angle -51.1° = 0.333\angle -18.4°$ A,
$V_3 = 47 \times I_{47} = 15.65\angle -18.4°$ V

2.19 (*a*) $Z = -j20 + [20(5 + j32)]/(25 + j32) = 20\angle -76.7°\ \Omega$,
$Y = 1/Z = 0.054\angle 76.7°$ S
(*b*) $R = 20\cos(-76.7°) = 4.6\ \Omega$, $X = 20\sin(-76.7°) = -j19.5\ \Omega$
(*c*) $G = 0.05\cos(76.7°) = 0.012$ S, $B = 0.05\sin(76.7°) = j0.049$ S
(*d*) $I = VY = 100\angle 0° \times 0.05\angle 76.7° = 5\angle 76.7°$ A
(*e*) $I_5 = (5.4\angle 76.7° \times 20)/(25 + j32) = 2.66\angle 24.7°$ A,
$I_{20} = 5.4\angle 76.7° - 2.66\angle 24.7° = 4.306\angle 105.8°$ A

2.20 $Z = [(4 + j8)(6 - j5)/(10 + j3)] + j8 = 6.64 + j8.81\ \Omega = 11.03\angle 53°\ \Omega$,
$I = 120/11.03\angle 53° = 10.88\angle -53°$ A

2.21 $jX_L = j2497.6\ \Omega$, $I = 25/(500 + j2497.6) = 9.82\angle -78.7°$ mA

2.22 $0.8\angle -46.7°$ A

2.23 $18.39\angle -8.2°\ \Omega$

2.24 $1006\angle 83.8°\ \Omega$

2.25 $1.05\angle 109.3°$ V

2.26 $6.01\angle 28.6°$ A

2.27 (*a*) $i_R = 1 \sin 5000t$ A, $i_C = 0.5 \sin(5000t + 90°)$ A
(*b*) $i = 1.118 \sin(5000t + 26.6°)$ A

2.28 (*a*) $v = 87.72 \sin(100\pi t + 27.8°)$ V
(*b*) (*i*) $R = 6.9\ \Omega$, $C = 185\ \mu F$,
(*ii*) $R = 118\ \Omega$, $L = 93.6$ mH,
(*iii*) $R = 200\ \Omega$, $C = 21.2\ \mu F$

2.29 $C = 74\ \mu F$

2.30 $Z = 79\angle -29.2°\ \Omega$

2.31 (*a*) $Y = 0.707\angle -45°$ S, $Z = 14.14\angle 45°\ \Omega$
(*b*) $Y = 0.943 \times 10^{-3}\angle 32°$ S, $Z = 1060\angle -32°\ \Omega$

2.32 $I_R = 1.94\angle 35.9°$ A, $I_C = 1.41\angle -54.1°$ A

2.33 $R_p = 122\ \Omega$, $X_p = j101.6\ \Omega$

2.34 (*a*) $R = 7.69\ \Omega$, $X_L = j11.54\ \Omega$
(*b*) $R = 25\ \Omega$, $X_L = j16.7\ \Omega$

2.35 (*a*) $6.384\angle -57.9°$ A; (*b*) $1.62\angle 78.3°$ A,
(*c*) $1.308\angle -74.2°$ A; (*d*) $7.93\angle 62°$ A

2.36 (*a*) $Z = 984.9\angle 24°\ \Omega$
(*b*) $i = 40.61 \sin(200t - 24°)$ mA
(*c*) $V_{150} = 6.09\angle -24°$ V, $V_{j250} = 50.76\angle 66°$ V,
$V_{-j900} = 36.55\angle -114°$ V, $V_{j300} = 12.18\angle 66°$ V,
$V_{-j250} = 10.15\angle -114°$ V, $V_{750} = 30.5\angle -24°$ V

2.37 $C = 13.5\ \mu F$

2.38 $341.4\angle -81.5°\ \Omega$

2.39 (*a*) (*i*) $Z = 40.15\angle 51.5°\ \Omega$; (*ii*) $Z = 153.08\angle -70.9°\ \Omega$
(*b*) (*i*) $Z = 135.8\angle -56.5°\ \Omega$; (*ii*) $Z = 45.24\angle 37.1°\ \Omega$
(*c*) $i - 7.516 \sin(100\omega t - 37.1°)$ A

2.40 $V = 180.28\angle -56.3°$ V, $V_R = 100\angle 0°$ V, $V_C = 150\angle -90°$ V

2.41 $I = 5.029\angle -27.4°$ A

2.42 $R_p = 40\ \Omega$, $C_p = 4.4\ \mu F$

2.43 (*a*) $0.432\angle 40.9°$ S; (*b*) 0.326 S; (*c*) j0.283 S; (*d*) $V = 3.47\angle 40.9°$ V;
(*e*) $I_{30} = 115.7\angle 40.9°$ mA

2.44 (*a*) $Z = 28.6\angle -50.6°\ \Omega$
(*b*) $I_A = 3.3\angle 86.3°$ A; $I_B = 4.47\angle 63.4°$ A; $I_C = 2\angle 90°$ A

2.45 (*a*) $I = 1.4\angle -44.3°$ A
(*b*) $I_C = 21.5\angle 77.6°$ mA
(*c*) $I_L = 1.414\angle 45°$ A
(*d*) $Z = 7.1\angle 44.3°\ \Omega$

3.1 (*a*) $P = (12/\sqrt{2})^2/50 = 1.44$ W
(*b*) Peak power $= 12^2/50 = 2.88$ W

3.2 $I = 240/(20\angle 63.3°) = 12\angle -63.3°$ A
$P = 240 \times 12 \cos(-63.3°) = 1296$ W
Apparent power $= 240 \times 12 = 2880$ VA
Reactive power $= 2880 \sin(-63.3°) = 2572$ var

3.3 Total real power $= 200 + 800 \times 0.72 + 400 \times 0.8 = 1096$ kW
Total reactive power $= 800 \times 0.69 - 400 \times 0.6 = 312$ kvar

(*a*) Total kVA $= \sqrt{(1096^2 + 312^2)} = 1139.5$ kVA,
(*b*) Power factor $= 1096/1139.5 = 0.962$

3.4 $240I^* = k + jQ$, $12k = 240I \times 0.8$, or $240I = 15k$, $\theta = \cos^{-1} 0.8 = 36.9°$ and $\sin\theta = 0.6$, $jQ = j15k \times 0.6 = j9k$, $240I^* = 12k + j9k$, $I^* = 50 + j37.5$ and $I = 50 - j37.5 = 62.5\angle{-38.9°}$ A

3.5 (*a*) $S = 400 \times 30 = 12$ kVA; (*b*) $P = 12 \times 0.72 = 8.64$ kW; (*c*) $Q = 12\sin(\cos^{-1} 0.72) = 8.33$ kvar

3.6 $Z = (82.46\angle 76°)$, $I_L = (50\angle 0°)/(82.46\angle 76°) = 0.606\angle{-76°}$ A, $P = 0.606^2 \times 20 + 50^2/56 = 52$ W

3.7 (*a*) $Z = 30 + j10 = 31.62\angle 18.4°\ \Omega$; (*b*) Power factor $= \cos 18.4° = 0.949$; (*c*) $I = (50\angle 0°)/(31.62\angle 18.4°) = 1.581\angle{-18.4°}$ A, $P = 50 \times 1.581 \times 0.949 = 75.02$ W

3.8 (*a*) $X_L = 31.4\ \Omega$, $R_p = (5^2 + 31.4^2)/5 = 202.2\ \Omega$,
$X_p = (5^2 + 31.4^2)/31.4 = j32.2\ \Omega$, $L_p = 32.2/314 = 0.103$ H,
$B = 1/j\omega L = -j0.031$ S. For unity power factor,
$B_C = +j0.031\ S = j\omega C_p$, $C_p = 0.031/314 = 98.9\ \mu F$
(*b*) $P = V^2G = (120/\sqrt{2})^2/202.2 = 35.61$ W

3.9 (*a*) $P = 100^2/56 = 178.57$ W, $Q = 0$, $S = P = 178.57$ VA
(*b*) $P = 0$, $Q = 100^2/100 = 100$ var (inductive), $S = Q = 100$ VA
(*c*) $Z = 100 - j200 = 223.61\angle{-63.4°}\ \Omega$ and $I = V/Z = 0.447\angle 63.4°$ A
$P = 0447^2 \times 100 = 19.981$ W, $Q = 0.447^2 \times 200 = 39.962$ var
$S = \sqrt{(19.981^2 + 39.962^2)} = 44.679$ VA

3.10 (*a*) Total real power $= 40 + 10 + 38 = 88$ kW
(*b*) $X_C = 10.61\ \Omega$. Total reative power $= -230^2/10.61 + 10k - 5k + 10k = 10\,014$ var (inductive).
Total apparent power $= 88.568$ kVA
(*c*) Power factor $= 88/88.568 = 0.994$

3.11 (*a*) $Z = (30 + j20)/(1 + j0.5) = 16 + j28 = 32.25\angle 60.3°\ \Omega$
(*b*) $P = 30 \times 1 + 20 \times 0.5 = 40$ W
(*c*) $Q = 20 \times 1 - 30 \times -0.5 = 35$ var
(*d*) $S = \sqrt{(40^2 + 35^2)} = 53.15$ VA

3.12 (*a*) $22\,000 = 0.883$ S, or $S = 24.915$ kVA
(*b*) $I = (24.915 \times 10^3)/240 = 103.8$ A

3.13 $P = 25 \times 10^3 \times 0.81 = 20250$ W, $\phi = \cos^{-1} 0.81 = 35.9°$, $Q = 25 \times 10^3 \sin 35.9° = 14661$ var. When power factor $= 0.93\phi = 21.6°$, $\tan 21.6° = 0.396 = (14661 - Q)/20250$, or $Q = 6642$ var. $I_C = 6642/240 = 27.675$ A, $X_C = 240/27.675 = 8.672\ \Omega$ and $C = 1/(2\pi \times 50 \times 8.672) = 367\ \mu F$

3.14 (*a*) $Z = 50 + j62.83 = 80.3\angle 51.5°\ \Omega$, $I = V/Z = 12.453\angle{-51.5°}$ A, $S = 1000 \times 12.453 = 12.453$ kVA
(*b*) Power factor $= \cos 51.5° = 0.623$
(*c*) $P = 12.453 \times 0.623 = 7.758$ kW
(*d*) $Q = 12.453 \times \sin 51.5° = 9.746$ kvar
(*e*) $Y = 1/Z = 7.75 - j9.75$ mS. With the correction capacitor $Y' = 7.75 \times 10^{-3}\ j(9.75 - \omega C)$. For a power factor of 0.92, $\phi = 23.1° = \tan^{-1}[(9.75 \times 10^{-3} - 100\pi C)]/(7.75 \times 10^{-3})$, or $C = 20.5\ \mu F$

3.15 $10\text{ mA} = I \times -j100/(100 - j100) = I(0.5 - j0.5) = 14.14\angle{-45°}$ mA
$V_{50} = 50I = 0.71\angle{-45}\ \circ$ V, $V_{100} = 100 \times 0.01 = 1$ V

$V = V_{50} + V_{100} = 1.5 - j0.5 = 1.58\angle -18.4°$
Therefore, $V = 1.58$ V
$\phi = -45° - (-18.4°) = 26.6°$
$P = 50(14.14 \times 10^{-3})^2 + 100(10 \times 10^{-3})^2 = 20$ mW
Power factor $= \cos 26.6° = 0.894$

3.16 (*a*) $I = 100/(300 + j200) = 0.277\angle -33.7°$ A
(*b*) $P = |I|^2 R = 0.277^2 \times 300 = 23.02$ W
(*c*) $Y = 1/Z = 1/(300 + j200) = 2.31 - j1.54$ mS

3.17 (*a*) $I = VY = 12(100 + j22) \times 10^{-6} = 1.23\angle 12.4°$ mA
(*b*) $P = V^2 G = 144 \times 100 \times 10^{-6} = 14.4$ mW
(*c*) Power factor $= \cos 12.4° = 0.977$
(*d*) $Z = 1/Y = 10^6/(100 + j22) = 9538 - j2098\ \Omega$

3.18 (*a*) $Y = 1/(100 - j50) + 1/(200 + j50) + 1/200 = (17.71 + j2.82) \times 10^{-3}$ S
$Z = 1/Y = 55.1 - j8.8\ \Omega$
(*b*) $I = V/Z = 100/(55.1 - j8.8) = 1.77 + j0.283 = 1.793\angle 9.1°$ A
(*c*) $P = |I|^2 R = 1.793^2 \times 55.1 = 177$ W

3.19 $Z = 50 + j1200 = 1201\angle 87.6°\ \Omega$
$Y = 1/Z = 832.6\angle -87.6°\ \mu S$
$P = V^2 G = 144 \times 832.6 \times 10^{-6} \times \cos(-87.6°) = 5$ mW

3.20 (*a*) $Z = 1/Y = 10^6/(200 - j250) = 1951.2 + j2439\ \Omega$
(*b*) $P = V^2 G = 50^2 \times 200 \times 10^{-6} = 0.5$ W
(*c*) $B_C = j\omega C = j2\pi \times 120 \times 10^3 \times 400 \times 10^{-12} = j301.6\ \mu S$
$Y = 200 - j250 + j301.6 = 200 + j51.6\ \mu S$
$Z = 10^6/(200 + j51.5) = 4688 - j1209\ \Omega = 4841\angle -14.5°\ \Omega$

3.21 (*a*) $Y = 1/Z = 1/(3500 + j5000) = 163.85\angle -55°\ \mu S$
For unity power factor the j terms must be zero, hence
$B_C = +j13.423 \times 10^{-5}$ S $= j5000C$, or $C = 26.85$ nF

3.22 $jX_L = j2497.6\ \Omega$, $I = 25/(500 + j2497.6) = 9.82\angle -78.7°$ mA

3.23 (*a*) $Z = [(40 - j80)(30 + j40)]/(70 - j40) = 52.31 + j18.46\ \Omega$
$Y = 1/Z = (17 - j6)$ mS $= 18.03\angle -19.4°$ mS
(*b*) $I = VY = 12 \times 18.03\angle -19.4° \times 10^{-3} = 216.4\angle -19.4°$ mA
(*c*) $P = V^2 G = 144 \times 17 \times 10^{-3} = 2.45$ W

3.24 $Q = 46.087$ kvar, PF $= 0.735$

3.25 140.625 kW

3.26 (*a*) $S = 4938.27$ VA; (*b*) $Q = 2896$ var; (*c*) $Z = 11.66\angle 35.9°\ \Omega$

3.27 (*a*) $S = 2913.6$ VA; (*b*) $P = 1769.65$ W; (*c*) $Q = 2314.61$ var

3.28 12.3 μF

3.29 520.7 W

3.30 $P = 10$ W

3.31 (*a*) 14.14 mA; (*b*) 20 V

3.32 (*a*) $Z = 17.88\angle -7.8°\ \Omega$; (*b*) PF $= 0.991$; (*c*) $P = 3192$ W,
(*d*) $Q = 437.2$ var; (*e*) $S = 3221.8$ VA

3.33 (*a*) $I = 83.3$ mA; (*b*) 25.8°

3.34 15.94 W

3.35 (*a*) $121.25\angle -11°$ mA; (*b*) $S = 1.47 + j0.88$ VA

3.36 (*a*) $G = 5.1$ mS, $B = -0.99$ mS, $Z = 192\angle 11°\ \Omega$; (*b*) $C = 49.5$ nF

3.37 (*a*) $316\angle -71.6$ mA; (*b*) 10 W

3.38 8.1 μF

3.39 27 mW

4.1 $Z = R_2 + jωL + R_1/(1 + jωCR_1)$
$= R_2 + R_1/(1 + ω^2C^2R_1^2) + jωL - CR_1^2/(1 + ω^2C^2R_1^2)$.
At resonance, the j terms sum to zero, hence
$L = CR_1^2/(1 + ω_0^2C^2R_1^2)$ and $ω_0 = \sqrt{[(1/LC) - L/C^2R_1^2)]}$
Therefore,
$f_0 = 2037$ Hz, $Z_0 = 30 + j256 + 2000/(1 + j7.7) = 63.3\ \Omega$

4.2 $Q = f_0/B_{3dB} = 100/5 = 20$
(*a*) $|Z/R_d| = 1/2 = 1/\sqrt{[1 + 400B_{6dB}^2/(1 \times 10^{10})]}$,
$4 = 1 + 400B_{6dB}^2(1 \times 10^{10})$
and $B_{6dB} = \sqrt{[(3 \times 10^{10})/400]} = 8660$ Hz
(*b*) $|Z/R_d| = 1/3.162 = 1/\sqrt{[1 + 400B_{10dB}^2/(1 \times 10^{10})]}$,
$B_{10dB} = \sqrt{[(9 \times 10^{10})400]} = 15$ kHz
(*c*) $|Z/R_d| = 1/10$ and $B_{20dB} = 49.75$ kHz

4.3 $LC \times L/C = L^2 = 2 \times 10^{-10} \times 5000$ and $L = 1$ mH
$C = (1 \times 10^{-3})/5000 = 0.2\ \mu$F
$f_0 = 1/2\pi\sqrt{(2 \times 10^{-10})} = 11.254$ kHz
$Q = (1/R)(\sqrt{L/C}) = (1/10)(\sqrt{5000}) = 7.07$
$B_{3dB} = f_0/Q = (11.254 \times 10^3)/7.07 = 1592$ Hz
Similarly for circuits (*b*) and (*c*):
(*b*) $L = 1$ mH, $C = 0.2\ \mu$F, $f_0 = 11.254$ kHz, $Q = 3.54$ and
$B_{3dB} = 3184$ Hz
(*c*) $L = 0.5$ mH, $C = 0.4\ \mu$F, $f_0 = 11.254$ kHz, $Q = 3.54$ and
$B_{3dB} = 3184$ Hz. Overall, $L_T = 2.5$ mH, $C_T = 0.08\ \mu$F, $R_T = 40\ \Omega$,
$f_0 = 11.254$ kHz, $Q = 4.42$ and $B_{3dB} = 2546$ Hz

4.4 (*a*) $f_0 = 1/2\pi\sqrt{(50 \times 10^{-3} \times 600 \times 10^{-12})} = 29.058$ kHz
(*b*) $X_C = -j9129\ \Omega$. $I_C = j25/9129 = j2.74$ mA
(*c*) $Q = Q_L = 80$
(*d*) $R_d = Qω_0L = 730.3$ kΩ, $R_{d(eff)} = 730.3\|250 = 186.24$ kΩ
$R = ω_0L/Q = 114\ \Omega$
$Z_L = R + jω_0L = 114 + j9129\ \Omega$
$I_L = 25/(114 + j9129) = 3.42 \times 10^{-5} - j2.74 \times 10^{-3}$ A
$I = I_L + I_C = 3.42 \times 10^{-5}$ A
$V = 3.42 \times 10^{-5} \times 186.24 \times 10^3 = 6.37$ V

4.5 $B_{3dB} = (5 \times 10^6)/63 = 79365$ Hz,
$f_1 = 5$ MHz $- 79365/2 = 4960.318$ kHz,
$f_2 = 5$ MHz $+ 79365/2 = 5039.683$ kHz

4.6 Coupled resistance $= 10^2 \times 4700 = 470$ kΩ,
$R_{d(eff)} = (1 \times 0.47)/1.47$ M$\Omega = 319.73$ kΩ,
$Q_{eff} = 200/(1 + 1/0.47) = 63.95$

4.7 (*a*) $C = 1/[4\pi^2 \times (24 \times 10^3)^2 \times 10 \times 10^{-3}] = 4.4$ nF
(*b*) $Q = f_0/B_{3dB} = (24 \times 10^3)/(3 \times 10^3) = 8$
$8 = (1/R)\sqrt{(10 \times 10^{-3})/(4.4 \times 10^{-9})}$ and $R = 188.5\ \Omega$
(*c*) If a 4.7 nF capacitor is used, $f_0 = (1/[2\pi\sqrt{(10 \times 10^{-3} \times 4.7 \times 10^{-9})}]$
$= 23.215$ kHz,
% error $= [(23.215 - 24)/24] \times 100 = -3.27\%$

4.8 (*a*) $f_0 = 59.941$ kHz
(*b*) (*i*) $Z = R = 22\ \Omega$;
(*ii*) 25% below resonance $= 0.75 \times 59.941 = 44.956$ kHz
Then, $X_L = j4237\ \Omega$ and $X_C = -j7532\ \Omega$,
$Z = 22 + j4237 - j7532 = 22 - j3295 = 3295\angle{-89.6°}\ \Omega$
(*iii*) 5% above resonance $f = 1.05 \times 59.941 = 62.938$ kHz. Then
$Z = 22 + j552 = 552\angle 87.7°\ \Omega$
(*c*) $Q = (1/22)\sqrt{(15 \times 10^{-3})/(470 \times 10^{-12})} = 256.8$
(*d*) $V = 12/256.8 = 46.73$ mV

4.9 (*a*) $f_0 = 1779$ Hz
(*b*) $I_R = 15/100 = 0.15$ A
$I_L = 15/X_L = -j0.335$ A
$I_C = 15/X_C = j0.335$ A
(*c*) $I_T = 0.15$ A
(*d*) At f_0, $Z = 100\angle 0°\ \Omega$. At $0.1f_0 = 177.9$ Hz,
$Z = 100 - j442.8\ \Omega = 454\angle{-77.3°}\ \Omega$. At $0.5f_0 = 889.5$ Hz,
$Z = 100 - j67.1 = 120.4\angle{-33.9°}\ \Omega$. At $2f_0$, $Z = 120\angle 33.9°\ \Omega$.
At $10f_0$, $Z = 454\angle 77.3°\ \Omega$

4.10 (*a*) $f_0 = (1/2\pi\sqrt{LC})\sqrt{(1 - R^2C/L)} = 10.597$ kHz
(*b*) $f_a = 10.61\sqrt{(1 - R^2C/4L)} = 10.607$ kHz
(*c*) $Q = (1/R)\sqrt{(L/C)} = 20$
(*d*) $R_d = L/CR = 20\ \text{k}\Omega$. At f_a, $X_L = j999.7\ \Omega$ and $X_C = -j1000.3\ \Omega$,
$Z_{max} = [(50 + j999.7)(-j1000.3)]/(50 + j999.7 - j1000.3)$
$= 20.023\angle{-2.2°}\ \text{k}\Omega$

4.11 (*a*) $f_0 = 2997$ Hz; (*b*) $I_0 = 24/12 = 2$ A;
(*c*) $Q = (1/12)\sqrt{(60 \times 10^{-3})/(47 \times 10^{-9})} = 94.2$;
(*d*) $V_C = 94.2 \times 24 = 2261$ V; (*e*) $i = 2\sin(18831t)$ A

4.12 (*a*) $V_C = 2 \times 45 = 90$ V
(*b*) $5000\pi = 1/\sqrt{LC}$, $L = 1/(25 \times 10^6 \times 47 \times 10^{-9}\pi^2) = 86.2$ mH,
$R = (5000\pi \times 86.2 \times 10^{-3})/45 = 30.1\ \Omega$
(*c*) $Q = (1/80.1)\sqrt{(86.2 \times 10^{-3})}\sqrt{(47 \times 10^{-9})} = 16.9$

4.13 (*a*) $f_0 = 503$ Hz
(*b*) $Q = 31.6$
(*c*) $f_1 = [503\sqrt{(1/31.6^2 + 4)} - 503/31.6]/2 = 495$ Hz
$f_2 = [503\sqrt{(1/31.6^2 + 4)} + 503/31.6]/2 = 511$ Hz

4.14 $f_0 = 1.073$ MHz, $R_d = Q\omega_0 L = 53.935\ \text{k}\Omega$. At 99.2% of f_0,
$Z = R_d/[1 + jQ(1 - f_0^2/f^2)] = 53.935/[1 + j80(1 - 1/0.992^2)]$
$= 53.935/(1 - j1.296) = 20.128 + j26.086 = 32.949\angle 52.4°\ \text{k}\Omega$
$B_{3dB} = f_0/Q = (1.073 \times 10^6)/80 = 13.413$ kHz

4.15 (*a*) (*i*) $f_0 = 2729$ Hz; (*ii*) $f_0 = 2729$ Hz
(*b*) (*i*) $Q = 60$; (*ii*) $Q = 35.3$
(*c*) (*i*) $Z = R = 14.3\ \Omega$; (*ii*) $Z = 24.3\ \Omega$

4.16 (*a*) $C = 5.6$ pF; (b) $R_d = 188.5\ \text{k}\Omega$
(*c*) $B_{3dB} = 150$ kHz, $B_{6dB} = 259.8$ kHz, $B_{20dB} = 1.493$ MHz

4.17 (*a*) $f_0 = 313.3$ Hz; (*b*) $f_a = 318.3$ Hz; (*c*) $f_1 = 305.6$ Hz, $f_2 = 331$ Hz.
(*d*) $R_d = 6250\ \Omega$; (*e*) $Z_{max} = 6270\ \Omega$

4.18 (*a*) $f_s = 225.079$ kHz, (*b*) $f_p = 238.732$ kHz; (*c*) $Q = 707$

4.19 (*a*) $L = 19.9$ mH; (*b*) $C = 318$ nF; (*c*) $B_{3dB} = 80$ Hz; (*d*) $f_1 = 1960$ Hz,
$f_2 = 2040$ Hz; (*e*) $71\angle{-45°}$ mA

4.20 (*a*) $f_0 = 14.342$ kHz; (*b*) $Q = 8.1$

4.21 (*a*) $L = 0.21$ H; (*b*) $C = 190.5$ nF; (*c*) 769 Hz, 824 Hz

4.22 (*a*) $f_0 = 53$ Hz, $R_d = 909\ \Omega$; (*b*) $Q = 6.76$; (*c*) $I_L = 742.4$ mA, $I_C = 737$ mA

4.23 $f_0 = 712$ Hz, $Z = 2083\ \Omega$

4.24 (*a*) $C = 6$ nF, $Q = 32.4$; (*b*) $C = 79$ nF, $Q = 32.4$

4.25 (*a*) 20 kΩ; (*b*) 79.577 kHz; (*c*) 1

4.26 1125 Hz, 56.55 Ω, 23.5

4.27 353.4 Ω, 2278 Ω

4.28 (*a*) 2373 Hz; (*b*) 55.91 kΩ; (*c*) 25; (*d*) 54.22 kΩ

4.29 (*a*) 142.9 kHz; (*b*) 50 mA; (*c*) 50

4.30 (*a*) 405 pF; (*b*) 24.1 Ω; (*c*) 32.6

4.31 (*a*) 29.058 kHz; (*b*) 32.9 μA; (*c*) 3.29 mA

4.32 (*a*) 253 pF; (*b*) 12.22 kΩ; (*c*) 38.9

5.1 With node D as the reference.
Node A: $5 = V_A/(10 + j20) + (V_A - V_C)/5$,
$5 = (0.22 - j0.04)V_A - 0.2V_C$
Node B: $(V_A - V_B)/10 = V_B/j20$, $0 = -0.1V_A + V_B(0.1 - j0.05)$
Node C: $(V_A - V_C)/5 = V_C/10 + V_C/(-j20)$,
$0 = V_A(-0.2) + V_C(0.3 + j0.05)$
Solving, $V_B = 38.43\angle{-14.4^\circ}$ V and $V_C = 28.27\angle - 50.5^\circ$ V
$V_{BC} = V_B - V_C = (37.23 - j9.56) - (17.98 - j21.81) = 19.25 + j12.25$
$= 22.82\angle 32.5^\circ$ V

5.2 Loop ABEF: $5 = i_1(10 + j5) + i_2(12 - j6)$
Loop BCDE: $8.66 + j5 = (i_1 - i_2)(8 + j8) - i_2(12 - j6)$
$= i_1(8 + j8) + i_2(-20 - j2)$
Hence, $i_1 = 0.425\angle{-5.1^\circ}$ A, $i_2 = 0.168\angle 235.8^\circ$ A and
$i_1 - i_2 = 0.55\angle 10.5^\circ$ A

5.3 Loop ABGH: $40 = i_1(5 - j10) + j10i_2$
Loop BCFG: $0 = i_2(15 + j10 - j10) - i_1(-j10) - i_3(5 + j10)$
$= j10i_l + 15i_2 + i_3(-5 - j10)$
Loop CDEF: $0 = i_3(5 + j10 - j10) - i_2(5 + j10) = i_2(-5 - j10) + 5i_3$
Solving the three simultaneous equations: $i_L = 1.4\angle 52.1^\circ$ A

5.4 $V_{in} = R(i_1 - i_3) + (i_2 - i_3)/j\omega C$, $(r + j\omega L)i_2 + (i_2 - i_3)/j\omega C$
$+ (i_2 - i_1)/j\omega C = 0$
Therefore $(r + j\omega L + 2/j\omega C)i_2 - (i_1 + i_3)/j\omega C = 0$. When $V_{out} = 0$,
$i_3 = 0$ and hence $Ri_1 + i_2/j\omega C = 0$ or $i_2 = -j\omega CRi_1$ and
$i_1 = -i_2/j\omega CR$. Now $(r + j\omega L + 2/j\omega C - 1/\omega^2C^2R)i_2 = 0$. Hence,
$r = 1/\omega^2C^2R$ and $L = 2/\omega^2C$

5.5 (*a*) (*i*) With the 15 V source short circuit, $Z = 12\ \Omega$ and
$I_s = 24/12 = 2$ A,
$I_{L1} = [2 \times (-j10)]/[5 + j5 - j10] = 2 - j2 = 2.828\angle - 45^\circ$ A
(*ii*) With 24 V source short circuit, $Z = 1.622 - j9.729\ \Omega$ and
$I_s = (15\angle 90^\circ)/(9.862\angle - 80.5^\circ) = 1.52\angle 170.5^\circ$ A
$I_{L2} = [(1.52\angle 170.5^\circ) \times 2]/(7 + j5) = -0.25 + j0.25$. Therefore
$I_L = I_{L1} - I_{L2} = 2.25 - j2.25 = 3.18\angle - 45^\circ$ A

(*b*) $(24 - V_A)/2 = V_A/(5 + j5) + (V_A + j15)/(-j10)$ or $V_A = 22.5$ V
$I_L = 22.5/(7.07\angle 45°) = 3.18\angle -45°$ A

5.6 *V_2 suppressed*:
$Z_L = 50(20 + j100)/(70 + j100) = 38.26 + j16.78\ \Omega$
$I_s = (100 - j100)/(68.26 + j46.78) = 0.3137 - j1.68$ A
$I_1 = (0.3137 - j1.68)50/(70 + j100) = -0.49 - j0.5$ A
V_1 suppressed:
$Z_L = [(30 + j30)(20 + j100)]/(50 + j130) = 17.938 + j25.36\ \Omega$
$I_s = (100 + j100)/(67.94 + j25.36) = 1.774 + j0.81$ A
$I_2 = (1.774 + j0.81)(30 + j30)/(50 + j130) = 0.594$ A
$I = I_1 + I_2 = -0.49 - j0.5 + 0.594 = 0.51\angle -78.2°$

5.7 $10\angle 20° = 9.4 + j3.42$ V. $120\angle 45° = 84.85 + j84.85$ V
With the 120 V source suppressed:
$Z_L = 4 + j5 + (8 - j10)(10 + j6)/(18 - j4) = 12.024 + j3.894\ \Omega$
$I_1 = (9.4 + j3.42)/(12.024 + j3.894) = 0.791 + j0.028$ A
$I_2 = (0.791 + j0.028)(8 - j10)/(18 - j4) = 0.44 - j0.329$ A
$I_3 = I_1 - I_2 = 0.351 + j0.357$ A
With the 10 V source suppressed:
$Z_L = 10 + j6 + (4 + j5)(8 - j10)/(12 - j5) = 15.822 + j8.426\ \Omega$
$I_A = (84.85 + j84.85)/(15.822 + j8.426) = 6.403 + j1.953$ A
$I_B = (6.403 + j1.953)(8 - j10)/(12 - j5) = 6.455 - j1.348$ A
$I_C = I_A - I_B = -0.052 + j3.301$ A
Current in $(4 + j5)\ \Omega = I_1 - I_B = -5.664 + j1.376 = 5.83\angle 166.4°$ A
Current in $(10 + j6)\ \Omega = I_2 - I_A = 6.38\angle -159°$ A
Current in $(8 - j10)\ \Omega = I_3 + I_C = 3.67\angle -85.3°$

5.8 Let I_1 flow in $(30 + j30)\ \Omega$, $I_2 = 3\angle 45°$ in Z_3, and $(I_1 - I_2)$ in $-j20\ \Omega$
Mesh A: $100 + j100 = I_1(30 + j30) + I_2 Z_3$ (1)
Mesh B: $60 = -(I_1 - I_2) \times (-j20) + I_2 Z_3$ (2)
(1) − (2): $40 + j100 = I_1(30 + j30 - j20) + j20 I_2$
But $I_2 = 3\angle 45° = 2.12 + j2.12$ A
Hence, $40 + j100 = I_1(30 + j30) - 42.42 + j42.42$
$I_1 = (82.42 + j57.58)/(30 + j10) = 3.05 + j0.9 = 3.18\angle 16.4°$
Substitute into equation (1);
$100 + j100 = (3.05 + j0.9)(30 + j30) + Z_3(2.12 + j2.12)$
and $Z_3 = 4 - j12.74 = 13.35\angle -72.6°\ \Omega$
$I_1 - I_2 = (3.05 + j0.9) - (2.12 + j2.12) = 0.93 - j1.22$
$= 1.53\angle -52.7°$ A

5.9 $10 - j10 = j20I_1 - j15I_1 + j15I_2 = j5I_1 + j15I_2$ (3)
$j10 = j15I_2 + j20I_2 + j15I_1 = +j15I_1 + j5I_2$ (4)
From equation (4), $j5I_2 = j10 - 15I_1$ and $I_2 = 2 - 3I_1$
Substituting into equation (3), $10 - j10 = -j40I_1 + j30$
$I_1 = (-10 + j40)/j40 = 1 + j0.25$ A

5.10 $(10 - V_A)/j20 - [(V_A - j10)/-j15] - V_A/j20 = 0$
$0.67 + j0.5 = j0.033V_A$, $V_A = (0.67 + j0.5)/j0.033 = 15.15 - j20.3$ V
$I_1 = [10 - (15.15 - j20.3)]/j20 = 1.015 + j0.25$ A

5.11 $V_C = 140\angle -68.2°$ V

5.12 $E = 9.7$ V

5.13 $I = 0.2$ A

5.14 $I = 1.14\angle -90°$ A

5.17 $I_{12} = 0.55$ A

5.18 $V = 0.615\angle 40.6^\circ$ V

5.19 0.25 A

5.20 $I_1 = 2.49\angle -96.7^\circ$ A, $I_2 = 1.18\angle -28.9^\circ$ A

5.21 $I = 1.5\angle 90^\circ$ A

5.22 $V = 22.8\angle -165^\circ$ V

5.23 $I = 22.8\angle 15.3^\circ$ A

5.24 $I = 57.78\angle -8.5^\circ$ mA

6.1 $V_{oc} = (10 \times 1000)/1100 = 9.091$ V
$Z_{oc} = (100 \times 1000)/1100 - j200 = 90.91 - j200\ \Omega$
Hence load impedance $= 90.91 + j200\ \Omega$
$I = 9.091/181.2$, $P = I^2 \times 90.91 = 227.3$ mW

6.2 Magnitude of source impedance $= \sqrt{(500^2 + 30^2)} = 500.9\ \Omega$
Turns ratio $n = \sqrt{(500.9/8)} = 7.913 : 1$
$I = 15/[1000 + j30 + (7.913^2 \times 8)] = 14.97 - j0.45$ mA, $|I| = 14.98$ mA
$P = (14.98 \times 10^{-3})^2 \times 500 = 112.2$ mW

6.3 $V_{oc} = (20 \times -j1000)/(600 - j1000) = 14.71 - j8.82\ \text{V} = 17.15\angle 31^\circ$ V
$Z_{oc} = (600 \times -j1000)/(600 - j1000) = 441.2 - j264.7\ \Omega$
$I = (14.71 - j8.82)/(491.2 - j204.7) = 32.2\angle -8.4^\circ$ mA
$Z_L = 50 + j60 = 78.1\angle 50.2^\circ\ \Omega$, $V_L = IZ_L = 2.51\angle 41.8^\circ$

6.4 (*a*) $Z_{oc} = 100 + j50\ \Omega$, so $Z_L = 100 - j50\ \Omega$
$C = 1/(50 \times 10^4) = 2\ \mu$F. Hence $R_L = 100\ \Omega$ and $C_L = 2\ \mu$F
(*b*) $Y_L = 1/Z_L = 8 + j4$ mS, $R_P = 1/(8 \times 10^{-3}) = 125\ \Omega$
$X_P = 1/(4 \times 10^{-3}) = 250\ \Omega = 1/(1 \times 10^4 C_P)$ and $C_P = 0.4\ \mu$F

6.5 $X_L = j60\ \Omega$, $X_C = -j20\ \Omega$
$V_{oc} = j600/(100 + j60) = 2.647 + j4.412$ V
$Z_{oc} = j6000/(100 + j60) = 26.47 + j44.12\ \Omega$
Hence, $I = (2.647 + j4.412)/(126.47 + j24.12) = 26.62 + j29.81$ mA
$|I| = 39.97$ mA, $P_L = [(39.97 \times 10^{-3})/\sqrt{2}]^2 \times 100 = 79.8$ mW

6.6 $Z_{oc} = -j5 + (10 \times 5)/(10 + 5) = 3.33 - j5\ \text{k}\Omega$,
$V_{oc} = (50 \times 10)/(5 + 10) = 33.3$ V
Hence $I_L = 33.3/(3.33 - j5 + 20) = 1.37 + j0.29 = 1.4\angle 12^\circ$ mA
$P = (1.4 \times 10^{-3})^2 \times 20 \times 10^3 = 39.2$ mW

6.7 (*a*) Load removed: $i = (50 - 25)/(5 + 2) = 3.571$ A, and
$V_{oc} = 50 - (3.571 \times 5) = 32.15$ V
$Z_{oc} = (5 \times 2)/7 = 1.43\ \Omega$,
$i_L = 32.14/(1.43 + 2 + j40) = 0.684 - j0.798 = 0.8\angle -85.1^\circ$ A
(*b*) Short circuit the 25 V source:
$Z' = 5 + 2(2 + j40)/(4 + j40) = 7\ \Omega$, $I'_s = 50/7 = 7.143$ A, and
$I'_L = 7.143 \times 2/(4 + j40) = 0.0354 - j0.354$ A
Short circuit the 50 V source:
$Z'' = 2 + 5(2 + j40)/(7 + j40) = 6.89 + j0.61\ \Omega = 6.92\angle 5^\circ\ \Omega$
$I''_s = 25/(6.92\angle 5^\circ) = 3.613\angle -5^\circ$ A and
$I''_L = 3.61\angle -5^\circ \times 5/(7 + j40) = 0.445\angle -88.5^\circ$
$I_L = I'_L + I''_L = 0.073 - j0.797 = 0.8\angle -84.8^\circ$ A
(*c*) $(50 - V_A)/5 = V_A/(2 + j40) + (V_A - 25)/2$ or $V_A \approx 32.14$ V
and $I_L = 32.14/(2 + j40) = 0.8\angle -87.1^\circ$ A

6.8 $Z_{oc} = 600(50 - j500)/(650 - j500) = 252 - j267.6\ \Omega$,
$V_{oc} = (6 \times 600)/(650 - j500) = 3.48 + j2.68$ V

$I_L = (3.48 + j2.68)/(282 - j67.7) = 0.118\angle 85.4°$ A and
$P = 0.118^2 \times 30 = 0.418$ W

6.9 $V_{oc} = (10 \times 100)/150 - (10 \times 75)/225 = 3.33$ V,
$R_{oc} = (75 \times 150)/225 + (50 \times 100)/150 = 83.33\ \Omega$
$I_{200} = 3.33/(83.33 + 200) = 11.75$ mA

6.10 Open circuit the current source. Then
$Z_{oc} = (22 + j10)(1 + j8)/(23 + j18) = 2.36 + j6.24\ \Omega$
With the output short circuited
$I_{sc} = 1\angle 60° \times (2 + j10)/(22 + j10) = 0.422\angle 114.3°$ A
Hence
$I_L = [(0.433\angle 114.3°) \times (2.36 + j6.24)]/(12.36 + j6.24) = 0.203\angle 156.8°$ A

6.11 $Z_{oc} = 23.8 + j4.92\ \Omega$, $V_{oc} = 9.73\angle 36.7°$ V. Hence
$R_L = 23.8\ \Omega$ and $C = 1/(10^4 \times 4.9) = 20.4\ \mu F$
$I = (9.73\angle 36.7°)/(2 \times 23.8) = 0.204\angle 36.7°$ and
$P_L = 0.204^2 \times 23.8 = 0.99$ W

6.12 (*a*) $92.6\angle -90°$ mA, j2.7 kΩ; (*b*) $2\angle -180°$ V, 3 kΩ

6.13 $R = 500\ \Omega$, $C = 36.8$ nF

6.14 $0.91\angle 12.3\ \Omega$ A

6.15 3.045 V

6.16 (*a*) 5Ω; (*b*) 8 W

6.17 $n = 3.52 :: 1$, $P = 15$ mW

6.19 0.15 W

6.20 3.36 : 1

7.1 Refer to Fig. 7.14 and apply the star/delta transform to the 10, 20 and 50 Ω resistors.
$Z_1 = (10 \times 50)/(10 + 20 + 50) = 500/80 = 6.25\ \Omega$,
$Z_2 = 200/80 = 2.5\ \Omega$, and
$Z_3 = 1000/80 = 12.5\ \Omega$.
Redraw the circuit and then apply Thevenin's theorem to the left of the 10 Ω resistor.
$V_{oc} = (12.5 \times 6.25)/(2.5 + 6.25) = 8.93$ V,
$R_{oc} = 42.5 + (2.5 \times 6.25)/8.75 = 44.29\ \Omega$
Apply Thevenin again including the 10 Ω resistor,
$V_{oc} = (8.93 \times 10)/54.29 = 1.645$ V
and $R_{oc} = 442.9/54.29 = 8.2\ \Omega$
Therefore $P_L = (1.645/16.4)^2 \times 8.2 = 82.5$ mW

7.2 3.33 Ω each

7.3 30 Ω each

7.4 $(37.1 + j21.8)\ \Omega$; $(42.5 + j12.5)\ \Omega$; $(55 - j30)\ \Omega$

7.5 (*a*) Convert the star network 2k, 2k, 3k into delta. $Z_A = 8$ kΩ, $Z_B = 8$ kΩ and $Z_C = 5.33$ kΩ. The 8 kΩ resistances are now in series with the 1 kΩ resistors. Hence
$Z_{in} = 600\ \Omega + (5.33 \times 19.5)/(5.33 + 19.5)\ k\Omega = 4786\ \Omega$,
$I_s = 9/486 = 1.88$ mA
(*b*) $I_L = (1.88 \times 5.33)/24.83 = 0.404$ mA
$V_L = 1500 \times 0.404 \times 10^{-3} = 0.61$ V

7.6 Replace the upper delta by the equivalent star.
$Z_1 = (j10 \times -j10)/(10 + j10 - j10) = 10\ \Omega$
$Z_2 = (j10 \times 10)/10 = j10\ \Omega$, $Z_3 = (-j10 \times 10)/10 = -j10\ \Omega$
Input impedance $= 10 + [(j10 + j20)(-j10 - j8)]/(j10 + j20 - j10 - j8) = 10 - j45\ \Omega$, $I = 6/(10 - j45) = 0.13\angle 77.6°$ A

7.7 $Z_1 = 240\angle 53.1°\ \Omega$, $Z_2 = 360\angle -36.9°\ \Omega$, $Z_3 = 120\angle 143.1°\ \Omega$

7.8 $246\angle 29.2°$ mA

7.9 $I_{50} = 40$ mA

7.10 $C = 2\ \mu$F, $R = 5\ \Omega$

7.11 3.43 V

7.12 $24.6\angle 54.8°\ \Omega$

7.14 $G = 2 \times 10^{-6}$ S. $B = -j10^{-4}$S

7.15 159 Hz

7.16 10 MΩ

7.18 318 Hz

8.1 (*a*) $(R_x + j\omega L_x)(R_3 + 1/j\omega C_3) = R_2R_4$,
$R_x + j\omega L_x = R_2R_4/(R_3 - j/\omega C_3) =$
$R_2R_4(R_3 + j/\omega C_3)(R_3^2 + 1/\omega^2 C_3^2)$,
$R_x = R_2R_3R_4/(R_3^2 + 1/\omega^2 C_3^2)$ and
$\omega L_x = (R_2R_4/\omega C_3)/(R_3^2 + 1/\omega^2 C_3^2)$.
(*b*) (*i*) $R_x = 482.4\ \Omega$; (*ii*) $\omega L_x = 644.1\ \Omega$ and $L_x = 51.26$ mH; (*iii*) $Q = 1.34$

8.2 $R_x = 1$ kΩ, $L_x = 2.5 \times 10^{-3}(2 \times 110 + 1200) = 3.55$ H

8.3 When $L = 200$ mH and $C = 200$ nF the bridge is balanced at 5000 rad/s since $\omega L = 1/\omega C = 1000\ \Omega$. Hence all the current in the impedance must be second harmonic. Applying Kirchhoff's laws to the circuit gives:
$0 = i_1 1800 - i_2 600 - i600$ or $0 = 3i_1 - i_2 - i$. Also,
$0 = -600i + i_2(1800 + j1500) - 600i$ or $-0 = i_1 + (3 + j2.5)i_2 - i$. But $i_1 - i_2 = 10$ mA, so $i = 2i_1 + 10 = 30 + 2i_2$.
$0 = 4(i_1 - i_2) - j2.5i_2$. Solving, $i_2 = -j16$ mA, $i_1 = 10 - j16$ mA and $i = 30 - j32$ mA
Therefore, $V_2 = 1200i - 600(i_1 + i_2)$ mV $= 35.7\angle -32.6°$ V
When $L = 100$ mH and $C = 100$ nF the bridge balances at the second harmonic. Hence the 20 mA current in the impedance is all fundamental. Solving as above gives $V_f = 71.2\angle 32.6°$ V.
Therefore % second harmonic $= (35.6/71.2) \times 100 = 50\%$.

8.4 The bridge has equal ratio arms. Therefore, $C_2 = C_1 = 10$ nF
The power factor $= 1/\sqrt{[1 + (2\pi \times 1000 \times \times 10^{-9} \times 50)^2]} = 3.14 \times 10^{-3}$
(*a*) Voltage across $C_2 = 1$ V
(*b*) $i \approx 2/(2.01Z)$. Thus
$V \approx 1 - Z \times 2/(2.01Z) \approx 1 - (1 - 0.005) = 5$ mV

8.5 $C_x = M/R_1R_3$, $M = C_xR_1R_3$, $R_x = (L - M)/M - R_4$,
$R_xM = L - M - R_4M$
$M(1 + R_x + R_4) = L$ and $M = L/(1 + R_x + R_4)$
(*a*) $R_3 = R_4 = 120\ \Omega$, $M = 120C_xR_1$, so $L = 120C_xR_1(1 + R_x + 120)$
$k = M/L = (120C_xR_1)/[120C_xR_1(1 + R_x + 120) = 1/(R_x + 121)$
(*b*) $R_3 = R_4 = 1200\ \Omega$ and $k = 1/(R_x + 1201)$

8.6 At balance, $Z = (C_1R_4/C_2 - R_3) + (1/j\omega C_1)$
$[(R_4C_1/C_2)(1/R_1) - C_1/C_3]$
With $Z = 0$, $R_4C_1/C_2 = R_1 = 3500$ and $3500/3500 = 1 = C_1/C_3$
Hence $Z = (3500 - R_3) + (1/j\omega C_1)(3500/R_1 - 1)$. When $R_1 = R_3 = 3000\ \Omega$
$Z = (3500 - 3000) + (1/j\omega C)(3500/3000 - 1) = 500 - j500 \times 0.167$
$= 500 - j83.5\ \Omega$

8.9 (*a*) $r_x = 2.16\ \Omega$, $C_x = 0.78\ \mu F$; (*b*) $PF = 0.013$

8.10 (*a*) $R = 3398\ \Omega$, $C = 123$ nF; (*b*) $R = 3892\ \Omega$, $C = 15.6$ nF

8.11 $L_x = 45$ mH, $r_x = 11.83\ \Omega$, $Q = 23.9$

8.12 $r_x = 160\ \Omega$, $C_x = 11.5$ nF

8.13 $r_x = 2.92\ \Omega$, $L_x = 24.2$ mH

8.14 $r_x = 30.56\ k\Omega$, $C_x = 18$ nF

8.15 $r_x = 1.6\ \Omega$, $L_x = 2$ mH

8.16 298–920 Hz

9.1 $R = R_S + (\omega_p^2 L^2 R_p)/(R_p^2 + \omega^2 L^2)$, and $L = (\omega L R_p^2)/(R_p^2 + \omega^2 L^2)$
Substituting the given values, $R = 71.32\ \Omega$ and $X_L = 799.4\ \Omega$

9.2 (*a*) $C = 0.2\ \mu F$
$PF = 3 \times 10^{-4} = \omega C R_s$. $R_s = (3 \times 10^{-4})/(0.4 \times 10^{-6}\omega) = (750/\omega)\ \Omega$
$PF = 4.5 \times 10^{-4}$: $R_s = (4.5 \times 10^{-4})/(0.4 \times 10^{-6}\omega) = (1125/\omega)\ \Omega$
Total series resistance $R_T = (1875/\omega)\ \Omega$
Hence the overall $PF = 0.2 \times 10^{-6}\omega \times (1875/\omega)$
$= 3.75 \times 10^{-4}$
(*b*) $C = 0.8\ \mu F$
$R_p = (750 \times 1125)/(750 + 1125)\omega = (450/\omega)\ \Omega$
Overall $PF = \omega C_p R_p = \omega \times 0.8 \times 10^{-6} \times (450/\omega) = 3.6 \times 10^{-4}$

9.3 (*a*) $X_C = 1250\ \Omega$
(*b*) $R_s = (PF)/\omega C = (2.5 \times 10^{-4})/(8000 \times 0.1 \times 10^{-6}) = 0.3125\ \Omega$
(*c*) $R_p = 1/(PF \times \omega C) = 1/(2.5 \times 10^{-4} \times 8000 \times 0.1 \times 10^{-6}) = 5\ M\Omega$

9.4 (*a*) $200 \times 10^{-6} = 20^2/R_p$, $R_p = 2\ M\Omega$
(*b*) $PF = 1/(\omega C R_p) = 2.5 \times 10^{-4}$
(*c*) $Q = 1/PF = 3996$

9.5 (*a*) $C = 1/(2\pi \times 4000 \times 1200) = 33.2$ nF
(*b*) $I_C = V/X_C = 12/1200 = 10$ mA
(*c*) $P = 12 \times 10 \times 10^{-3} \times 4 \times 10^{-4} = 48\ \mu W$
$48 \times 10^{-6} = 12^2/R_p$, $R_p = 3\ M\Omega$
Or, $R_s = (PF)/\omega C = 0.48\ \Omega$

9.6 (*a*) $PF = \omega C R_s = 2\pi \times 10^6 \times 110 \times 10^{-12} \times 0.05$
$= 3.46 \times 10^{-5}$
(*b*) $Q = 1/(4.46 \times 10^{-5}) = 28902$
(*c*) $R_p = 1/(2\pi \times 10^6 \times 110 \times 10^{-12} \times 3.46 \times 10^{-5}) = 41.82\ M\Omega$

9.7 $X_L = 2500\ \Omega$. $Z = [(10 + j2500)20000]/(20010 + j2500) = 317.2 + j2459\ \Omega$
Hence $R = 317.2\ \Omega$, and $L = 2459/5000 = 0.492$ H
$P = (100 \times 10^{-3})^2 \times 317.23 = 3.172$ W

9.8 $C_1 = \varepsilon_1 A/1_1$, where $A = Bd$, and $C_2 = \varepsilon_2 A/(1_2 - 1_2)$
$V_2 = [V\varepsilon_1 A/1_1] \times 1/[\varepsilon_1 A/1_1) + (\varepsilon_2 A/(1_2 - 1_1)$
$= V\varepsilon_1(1_2 - 1_1)/[\varepsilon_1(1_2 - 1) + \varepsilon_2 1_1$

Substituting values;
$V_2 = [3500 \times 4(10-4)]/[4(10-4) + (6 \times 4)] = 1750$ V. The voltage gradient $= 1750/(6 \times 10^{-3}) = 291.7$ kV/m
Also, $V_1 = 3500 - 1750 = 1750$ V
The voltage gradient $= 1750/(4 \times 10^{-3}) = 437.5$ kV/m

9.9 $400 = k_1 \times 60^{-0.6} \times 6600^{1.6}$. Therefore, $k_1 = 3.6 \times 10^{-3}$
At 50 Hz, $P_H = 3.6 \times 10^{-3} \times 6600^{1.6} \times 50^{-0.6} = 444.8$ W
At 50 Hz, $P_c = 180 \times (50/60)^2 = 125$ W
Total iron losses $= 444.8 + 125 = 569.8$ W

9.10 $220 = k_1 50^{(1-x)} V^x$, and $k_1 = 220/[50^{(1-x)} V^x]$
$160 = k_2 \times 50^2 \times V^2$, $k_2 = 160/(50^2 V^2)$
At the new voltage
$220 = [220/50^{(1-x)} V^x] \times 50^{(1-x)} \times 0.75^x V^x$
$\quad + [160\ (50 \times 0.75 \times V)^2/(50^2 V^2)]$
$220 = 220 \times 0.75^x + 160 \times 0.75^2$, $0.75^x = 0.591$, and $x = 1.828$.
$P_H = [(220/50^{-0.828} V^{1.828}) \times 60^{-0.828} \times 0.7 V^{1.828}]$
$\quad + [(160/50^2 V^2) \times 60^2 \times (0.75 V^2)] = 241.4$ W

9.11 $125 = \eta \times 50 \times 1.5^{1.6}$, $\eta \approx 1.31$. When $f = 25$ Hz and $B = 1.1$ T, hysteresis loss $= 1.31 \times 25 \times 1.1^{1.6} = 38.15$ W

9.12 $B_{max} = (12 \times 10^{-3})/(100 \times 10^{-4}) = 1.2$ T, $120 = \eta \times 50 \times 1.2^{1.6}$ or $\eta = 1.793$
At 100 Hz and 10 mWb, $B_{max} = (10 \times 10^{-3})/(100 \times 10^{-4}) = 1$ T and hysteresis loss $= 1.793 \times 100 \times 1^{1.6} = 179.3$ W

9.13 $P_H = 28 \times (100/50) = 56$ W,
$P_E = 18 \times (100/50)^2 = 72$ W. Total loss $= 128$ W

9.14 (*a*) 10 W; (*b*) 50 W

10.1 (*a*) $100 = (500 \times 10^3 \times L_p)/5$, or $L_p = L_s = 1$ mH
(*b*) $C_p = C_s = 1/[(500 \times 10^3)^2 \times (1 \times 10^{-3})] = 4$ nF
(*c*) $R_s = (500 \times 10^3 \times 1 \times 10^{-3})/50 = 10\ \Omega$
$E_p = 0.05 \times 15 = 0.75 V = \omega M I_p = 500 \times 10^3 M \times 0.1$
or $M = 15\ \mu H$
(*d*) $k = M/\sqrt{L_p L_s} = (15 \times 10^{-6})/(1 \times 10^{-3}) = 0.015$
(*e*) $k_{crit} = 1/\sqrt{Q_p Q_s} = 1/\sqrt{(100 \times 50)} = 0.014$

10.2 $k = 0.01 = (10 \times 10^{-6})/L_p$, or $L_p = 1$ mH
$f_0 = 1/[2\pi\sqrt{(1 \times 10^{-3} \times 2.533 \times 10^{-9})}] = 100$ kHz
$\omega_0 = 2\pi f_0 = 628.3 \times 10^3$ rad/s. $\omega_0 M = 6.283\ \Omega$
$\omega_0 L_s = 628.3\ \Omega$
$Z_{p(eff)} = R_{p(eff)} = 10 + (6.283^2 \times 10)/100 = 13.95\ \Omega$

10.3 $M = 0.2\sqrt{(160 \times 10^{-6} \times 125 \times 10^{-6})} = 28.28\ \mu H$
$R_p = (2\pi \times 85 \times 10^3 \times 160 \times 10^3)/80 = 1.07\ \Omega$
$R_s = (2\pi \times 85 \times 10^3 \times 125 \times 10^{-6})/40 = 1.67\ \Omega$
$\omega M = 2\pi \times 85 \times 10^3 \times 28.28 \times 10^{-6} = 15.1\ \Omega$
$\omega L_p = 2\pi \times 85 \times 10^3 \times 160 \times 10^{-6} = 85.45\ \Omega$
$\omega L_s = 85.45 \times 125/160 = 66.76\ \Omega$
$Z_{p(eff)} = 1.07 + j85.45 + 15.1^2(1.67 - j66.76)/(1.67^2 + 66.76^2)$
$= 1.2 + j82\ \Omega$
$I_p = 1/82 = 12.19$ mA
$Q_{(eff)} = 82/1.2 = 68.42$

10.4 $100 = 2\pi f_0 \times 20 \times 10^{-3}, f_0 = 7957.8$ Hz
$C_p = 1/(4\pi^2 \times 7957.8^2 \times 20 \times 10^{-3}) = 0.02\ \mu F$
$C_s = 4C_p = 0.08\ \mu F$
$Q_s = (2\pi \times 7957.8 \times 5 \times 10^{-3})/5 = 50.\ k_{crit} = 1/\surd(50 \times 100) = 0.014$
$M = 0.014\surd(20 \times 5) = 0.14$ mH.
$\omega M = 2\pi \times 7957.8 \times 0.14 \times 10^{-3} = 7\ \Omega$
Therefore, $R_{p(eff)} = 10 + (7^2/5) = 19.8\ \Omega$

10.5 $Q_p = 100 = 1/\omega_0 C_p R_p$, so $R_p = 10\ \Omega$
$Q_s = 50 = 1/\omega_0 C_s R_s$, so $R_s = 20\ \Omega$.
$\omega_0 M = 10^6 \times 10 \times 10^{-6} = 10\ \Omega$
$R_{p(eff)} = 10 + (100/20) = 15\ \Omega.\ I_p = 1/15$ A
$E_s = 10/15$V and $V_s = 10/15 \times 50 = 33.33$V

10.6 (*a*) $f_0 = 1/[2\pi\surd(300 \times 10^{-6} \times 1.5 \times 10^{-9})] = 237.25$ kHz
$\omega_0 M = 2\pi f_0 \times 50 \times 10^{-6} = 74.53\ \Omega$
$R_{p(eff)} = 100 + 74.53^2/100 = 155.6\ \Omega$
$I_p = 12/155.6 = 77.1$ mA
(*b*) $k = (50 \times 10^{-6})/(300 \times 10^{-6}) = 0.167$
(*c*) $E_s = 74.53 \times 77.1 \times 10^{-3} = 5.75$V. $I_s = 5.75/100 = 57.5$ mA

10.7 $E = 2.25 = 0.5L \times 25$ or $L = 0.18$H. This inductance is equal to sum of the individual values $\pm$ 2M. Therefore, $0.18 = 0.5 - 2M$ and $M = 0.16$ H

10.8 (*a*) $I_s = 0.25\angle 10.5°$ mA; (*b*) $I_p = 0.5\angle -52.9°$ mA; (*c*) $P = 0.3\ \mu$W

10.9 2250 Hz

10.10 (*a*) 0.012; (*b*) 5.5 μH; (*c*) 84.85; (*d*) 38.4 pF

10.11 $(19.2 + j2480)\ \Omega$

10.12 480 mA, 7.24V

10.13 $V_C = 8.28$V

10.14 0.6°

10.15 9

10.16 104 pF, 44 W

10.17 63 mV

10.18 0.05, $1584\angle 89.6°\ \Omega$

11.1 $R_{p(eff)} = 2.5 + (3200/400)^2 \times 0.2 = 15.3\ \Omega$
$X_{p(eff)} = (3200/400)^2 \times 0.3 = 19.2\ \Omega$
Full-load primary current = (10000)/3200 = 3.125 A
Therefore $V_p = 3.125(15.3 + j19.2) = 47.813 + j60$ and $|V_p| = 76.72$ V

11.2 (*a*) $\eta = (100 \times 10^3 \times 1)/(100 \times 10^3 + 1000 + 1500) = 97.56\%$
(*b*) On half load the copper losses are reduced to
$(1/2)^2 \times 1500 = 375$ W
$\eta = (50 \times 10^3)/(50 \times 10^3 + 1000 + 375) = 97.32\%$
(*c*) For maximum efficiency copper losses = iron losses, i.e. copper loss = 1000 W
$I_{s(max)} = (100 \times 10^3)/230 = 434.78$ A
$434.78^2 R_{s(eff)} = 1500$ W, $I_s R_{s(eff)} = 1000$ W
Therefore, $(434.78/I_s)^2 = 1.5$, and $I_s = 434.78/\surd 1.5 = 355$ A
(*d*) $\eta_{(max)} = (355 \times 230)(355 \times 230 + 1000 + 1000) = 97.61\%$

11.3 $R_{s(eff)} = 0.01 + (230/3200)^2 \times 5 = 0.036\ \Omega$
$X_{s(eff)} = 25(230/3200)^2 = 0.129\ \Omega$
(*a*) $(20 \times 10^3)/230 = 86.96\text{V}$
per cent regulation $= [(86.96 \times 0.036 \times 1) + 0]230] \times 100 = 1.36\%$ or 3.13V
(*b*) per cent regulation $= [(86.96 \times 0.036 \times 0.82)$
$+(86.96 \times 0.129 \times 0.57)]/230 \times 100\% = 3.9\%$ or 8.96 V

11.4 $400 = 385 + I_s R_{s(eff)} \cos\phi_1 + I_s X_{s(eff)} \sin\phi_1$ (1)
$400 = 401.6 + I_s R_{s(eff)} \cos\phi_2 - I_s X_{s(eff)} \sin\phi_2$ (2)
From equation (1), $15 = 25R_{s(eff)} \times 0.8 + 25X_{s(eff)} \times 0.6$
$15 = 20R_{s(eff)} + 15X_{s(eff)}$ (3)
Substitute into equation (2),
$-1.6 = 25 \times 0.62R_{s(eff)} - 25 \times 0.785X_{s(eff)}$
$-1.6 = 15.5R_{s(eff)} - 19.625X_{s(eff)}$ (4)
From equation (3), $R_{s(eff)} = (15 - 15X_{s(eff)})/20 = 0.75(1 - X_{s(eff)})$
Substitute into equation (4),
$-1.6 = 11.625 - 11.625X_{s(eff)} - 19.625X_{s(eff)}$
Hence, $X_{s(eff)} = 0.423\ \Omega$
And so $R_{s(eff)} = 0.75(1 - 0.423) = 0.433\ \Omega$

11.5 $R_{s(eff)} = 0.12 + (1100/6600)^2 \times 3.8 = 0.256\ \Omega$
$X_{s(eff)} = 0.26 + (1100/6600)^2 \times 8.8 = 0.504\ \Omega$
$I_{FL} = 10000/1100 = 9.091$ A
(*a*) per cent regulation
$= [(9.091 \times 0.256 \times 0.86) + (9.091 \times 0.504 \times 0.51)]/1100 \times 100\%$
$= 0.394\%$
(*b*) per cent regulation $= [(9.091 \times 0.256 \times 1)/1100] \times 100\% = 0.212\%$

11.6 For maximum efficiency $I_s^2 R_{s(eff)} = P_c = 3000$ W
$I_{FL} = (100 \times 10^3)/330 = 303$ A, $0.9\ I_{FL} = 272.73$ A
Therefore, $R_{s(eff)} = 3000/(272.73)^2 = 0.04\ \Omega$
$303 = 100/\sqrt{(0.04^2 + X_s^2)}$, $3.03^2 = 0.04^2 + X_s^2$ or $X_s = 0.328\ \Omega$

11.7 (*a*) $350 = k_1 \times 60^{-0.6} \times 6600^{1.6}$, $k_1 = 350 \times 60^{0.6} \times 6600^{-1.6}$
At 50Hz, $P_H = (350 \times 60^{0.6} \times 6600^{-1.6}) \times 50^{-0.6} \times 6600^{1.6} = 390.5$ W
Also, $150 = k_2 \times 6600^2 \times 60^2$, $k_2 = 150/(6600 \times 60)^2$
At 50Hz, $P_c = 150/(6600 \times 60)^2 \times 6600^2 \times 50^2 = 104.2$ W
Therefore $P_1 = 390.5 + 104.2 = 494.7$ W
(*b*) $I_{FL} = (100 \times 10^3)/330 = 303$ A. $0.9I_{FL} = 272.73$ A
For maximum efficiency
$390.5 = 272.73^2 R_{s(eff)}$ or $R_{s(eff)} = 390.5/272.73^2 = 5.25 \times 10^{-3}\ \Omega$
$I_{FL} = 303 = 100/\sqrt{(5.25^2 \times 10^{-6} + X_{s(eff)}^2)}$
$0.33 = \sqrt{(27.56 \times 10^{-6} + X_{s(eff)}^2)}$, and
$X_{s(eff)} = \sqrt{(0.109 - 27.56 \times 10^{-6})} = 0.33\ \Omega$
Referred to the primary circuit,
$R_{p(eff)} = 5.25 \times 10^{-3} \times (6600/330)^2 = 2.1\ \Omega$
$X_{p(eff)} = 0.33 \times 400 = 132\ \Omega$

11.8 $R_{s(eff)} = 0.25 + (400/2000)^2 \times 5 = 0.45\ \Omega$
$X_{s(eff)} = 1.2 + (400/2000)^2 \times 10 = 1.6\ \Omega$. $I_{FL} = 10000/400 = 25$ A
(*a*) $V_s = 400 - (25 \times 0.45 \times 0.8) + (25 \times 1.6 \times 0.6) = 367$ V
(*b*) $V_s = 400 - (25 \times 0.45 \times 0.6) + (25 \times 1.6 \times 0.8) = 425.25$ V

11.9 When short circuit, $Z_p = 8.2/2 = 4.1\ \Omega$, $\cos\beta = 8/(8.2 \times 2) = 0.488$
$\beta = 60.8°$, $\sin\beta = 0.873$, $R_{p(eff)} = 4.1 \times 0.488 = 2\ \Omega$

$X_{p(eff)} = 4.1 \times 0.873 = 3.58\ \Omega$
When open circuit, $I_c = 0.7 \times 0.468 = 0.328$ A, $I_m = 0.7 \times 0.884 = 0.619$ A, $R_c = V_p/I_c = 220/0.328 = 670.7\ \Omega$
$X_m = 220/0.619 = 355.4\ \Omega = 100\pi\ L_p$, $L_p = 1.13$ H

11.10 $R_{p(eff)} = 9.2 + (6600/330)^2 \times 0.1 = 49.2\ \Omega$
$X_{p(eff)} = 30 + (400 \times 0.3) = 150\ \Omega$
$I_{FL} = 10000/330 = 30.3$ A
$R_{s(eff)} = 49.2/400 = 0.123\ \Omega$, $X_{s(eff)} = 150/400 = 0.375\ \Omega$
(*a*) per cent regulation $= [(30.3 \times 0.123 \times 0.8) + (30.3 \times 0.375 \times 0.6)]/330 \times 100 = 2.97\%$
(*b*) per cent regulation $= [(30.3 \times 0.123 \times 0.8) - (30.3 \times 0.375 \times 0.6)]/330 \times 100 = -1.15\%$

11.11 1 kW, 17 kW

11.12 5.753 Ω

11.13 0.38 Ω, 1.61 Ω

11.14 592.1 W

11.15 1.187 W

11.16 (*a*) $0.894\angle{-63.4^\circ}$ A; (*b*) $2.828\angle{-45^\circ}$ A

11.17 (*a*) 352.4 V; (*b*) 5.31%

11.18 (*a*) 18.52 kVA; (*b*) (*i*) 97.27%, (*ii*) 96.61%

11.19 $1.975\angle{-45.6^\circ}$ A

11.20 223.2 V

11.21 19.92 V

12.1 The capacitance is 60 pF in 0.4 m and hence the capacitance per metre is 60/0.4 = 150 pF. Now $C = 150 \times 10^{-12} = 2\pi\varepsilon/\log_e(R/r)$ and $E_{(max)} = 3.5 \times 10^6 = (235 \times 10^3)/[r\ \log_e(R/r)]$.
Dividing the second equation by the first gives:
$(3.5 \times 10^6)/(150 \times 10^{-12}) = (235 \times 10^3)/r\ \log_e(R/r) \times [\log_e(R/r)]/(2\pi \times 8.85 \times 10^{-12})$
Hence $9.929 \times 10^{10} = 1/(5.5606 \times 10^{-11} r)$ and $r = 181$ mm
From the first equation,
$\log_e(R/181) = (2\pi \times 8.85 \times 10^{-12})/(150 \times 10^{-12}) = 0.3707$,
$R = 181 \times 1.449 = 262$ mm

12.2 $V_{rR} = -(q/2\pi\varepsilon r)\ dr = (q/2\pi\varepsilon)\log_e(R/r)$
$V_{xr} = -(q/2\pi\varepsilon x)dx = -V_{rR}/r\log_e(R/r)dx$
$= V_{rR}[\log_e(R/x)/\log_e(R/r)]$

12.3 $W = (\varepsilon A/2d) \times (E^2 d^2)$
$= (5 \times 8.85 \times 10^{-12} \times 20 \times 10^{-4} \times 1000^2 \times 25 \times 10^{-4})/(2 \times 5 \times 10^{-4} \times 25 \times 10^{-4})$
$= 0.89\ \mu$J
Also, $0.89 \times 10^{-6} = 0.5 \times 1000^2 C$ or
$C = (0.89 \times 10^{-6})/(0.5 \times 1000^2) = 1.78$ pF

12.4 (*a*) $C = \pi\varepsilon/\log_e[(15 - 0.75)/0.75] = 9.44$ pF/m
(*b*) $L = [\mu_0/\pi][\log_e(d - r)/r + \mu_0/4\pi$
$= [(4 \times 10^{-7})\log_e 19] + 10^{-7} = 1.28$ H/m

12.5 $C = 2\pi\varepsilon/\log_e(R/r) = (2\pi \times 8.85 \times 10^{-12} \times 2.5)/\log_e(0.5/0.15) = 115.5$ pF/m

12.6 $C = 2\pi\varepsilon/\log_e(R/r) = Q/V$, $V = Q/C = [Q \log_e(R/r)]/2\pi\varepsilon$
$V/I = R = \rho\log_e(R/r)/2\pi\rho = [9 \times 10^{12} \times \log_e(22/6)]/2\pi\rho$
$= 1.86 \times 10^{12}$ Ω/m

12.8 At midway $x = R/2$. Hence
$V = 100 \log_e[(2)/(3)] = 100 \log_e 0.67$ and $V = 40.55$ kV

12.9 17.28×10^6 A/m^2

12.11 Current inside conductor $= Ix^2/r^2$, $H = (Ix^2/r^2)(1/2\pi x)$,
$B = \mu_0 Ix/(2\pi r^2)$. Flux through cylindrical shell of radial thickness and 1 m length is

$$(\mu_0 Ix/2\pi r^2)\int_0^r \mathrm{d}x = (\mu_0 I/2\pi r^2)\int_0^r x\,\mathrm{d}x.$$

This flux only links with x^2/r^2 of the conductor so
linkages $= (\mu_0 I/2\pi r^2)(x^2/r^2) \times \mathrm{d}x = (\mu_0 I/2\pi r^4)x^3\,\mathrm{d}x$
$= \mu_0 I/8\pi$ per conductor $= \mu_0 I/4\pi$ in total

12.12 $r = 7.07$ mm, $R = 19.22$ mm

12.13 $d = 11.16$ cm

12.14 $V = 32.86$ kV

12.15 0.93 μH

12.16 (*a*) $H = 3.5$A, $B = 44 \times 10^{-7}$ T; (*b*) $H = 1.75$ A, $B = 22 \times 10^{-7}$ T

12.17 8.3 V

12.18 135 pF

13.1 $NI = I = 160 = Hl = H \times 45\pi \times 10^{-2}$, $H = 160/(45\pi \times 10^{-2})$;
$B = \mu_0\mu_r H$ and
$F = 2B^2A/2(\mu_0 = 2 \times [4\pi \times 10^{-7} \times 400 \times 160/(45\pi \times 10^{-2})]^2$
$\times (50 \times 10^{-4})/(2 \times 4\pi \times 10^{-7}) = 12.88$ N

13.2 $F = (I^2\,\mathrm{d}L/\mathrm{d}l)/2$ and hence $\mathrm{d}L/\mathrm{d}l$ is required where $1 = 2\pi R$.
$\mathrm{d}L/\mathrm{d}l = \mathrm{d}L/\mathrm{d}R \times \mathrm{d}R/\mathrm{d}l$
$= (\mu_0)[\log_{10}(R/0.015) + R(1/0.4343 \times 0.015/R \times 1/0.015)] \times (1/2\pi)$
$= (\mu_0/2\pi)[\log_{10}(8 \times 10^{-3})/0.015 + 1/0.4343]$
$= 2 \times 10^{-7}(-0.273 + 2.303) = 4.06 \times 10^{-7}$
Peak force $= 0.5 \times 125^2 \times 4.06 \times 10^{-7} = 3.17$ mN

13.3 $F = B^2A/2\mu_0$, $B = \sqrt{(2 \times 4\pi \times 10^{-7} \times 4.8)/(1.4 \times 10^{-4})} = 0.294$ T
$H = 0.294/(4\pi \times 10^{-7})$
$I = Hl/N = 0.294/(4\pi \times 10^{-7}) \times 0.6 \times 10^{-5} = 1.4$ A

13.4 Full-load current $= 2000/240 = 8.33$ A
Armature loss $= 8.33^2 \times 1 = 69.4$ W
Then $-2000 + W_M = 69.4 + 200 + 100$, $W_M = 2369.4$ W
The efficiency $= (2000/2369.4) \times 100\% = 84.4\%$.

13.5 $2.9 = CV^2/2$, $20 \times 10^{-3} = VC$, $C = (20 \times 10^{-3})/V$. Hence,
$V = 5.8/(20 \times 10^{-3}) = 290$ V
$C = (20 \times 10^{-3})/290 = 69$ μF
Total capacitance $= 79$ μF. The charge is unchanged so
$20 \times 10^{-3} = 79 \times 10^{-6}$V, or V $= 253.2$ V

13.6 m.m.f. $= NI = 1500 \times 4 = 6000$ A
(*a*) $H_{(\mathrm{min})} = NI/l = 6000/(2 \times 10^{-2}) = 3 \times 10^5$ A/m
$B_{(\mathrm{min})} = \mu_0 H = 4\pi \times 10^{-7} \times 3 \times 10^5 = 0.377$ T
$H_{(\mathrm{max})} = 6000/(0.76 \times 10^{-2}) = 7.895 \times 10^5$ A/m

$B_{(max)} = 4\pi \times 10^{-7} \times 7.895 \times 10^5 = 0.992$ T

(b) $L = N\phi/I$

$\phi_{(min)} = 0.377 \times \pi \times (2.1 \times 10^{-2})^2$, and hence

$L_{(min)} = 0.377 \times \pi \times (2.1 \times 10^{-2})^2 \times 1500/4 = 195.9$ mH

$L_{(max)} = 0.992/0.377 \times 195.9 = 515$ mH

(c) Distance moved $x = 2 - 0.76 = 1.24$ cm

$F = (I^2\ \mathrm{d}L/\mathrm{d}x)/2$

$= 8 \times [(515.9 - 15.9) \times 10^{-3}]/(1.24 \times 10^{-2}) = 205.9$ N

Work done $= Fx = 205.9 \times 1.24 \times 10^{-2} = 2.55$ J

13.7 Force $= B^2A/2\mu_0$ per air gap. There are two air gaps so the total force is B^2A/μ_0 N

m.m.f. $= 0.536 \times 200 = 107.2$ A, $H = 107.2/0.76 = 141$ A/m

$B = \mu_0\mu_r H = 700 \times 4\pi \times 10^{-7} \times 141 = 0.124$ T

$F = (0.124^2 \times 4 \times 10^{-4})/(4\pi \times 10^{-7}) = 4.89$ N

13.8 Let change in air gap be x when spring is not strained. Then $l = l_0 - x$, or $x = l_0 - l$

Flux $\phi = \left(\frac{Ni}{2l}\right)/A\mu_0) = NiA\mu_0/2l$

Inductance $L = N\phi/I = N^2A\mu_0/[2(l_0 - x)]$. Hence,

$\mathrm{d}L/\mathrm{d}x = -N^2A\mu_0/[2(l_0 - x)] = -L/(l_0 - x)$

At any time t, electrical force = mechanical force, so

$(I^2\,\mathrm{d}L/\mathrm{d}x)/2 = -kx$. $(I^2/2)[-L/(l_0 - x)] = -kx$

And, $LI^2/2 = kx(l_0 - x)$ = energy in magnetic field

When armature stops, $W_E = W_M + W_F$

$$W_M = \int_0^{l_0 - l_1' = x_1} kx\,\mathrm{d}x = k(l_0 - l_1)^2/2$$

At start, $x = 0$, and $W_{F1} = 0$. At end $x = x_1 = l_0 - l_1$

$W_{F2} = k(l_0 - l_1)(l_0 - l_0 + l_1) = kl_1(l_0 - l_1)$

$W_E = k(l_0 - l_1)^2/2 + kl_1(l_0 - l_1) = k(l_0^2 - l_1^2)/2$

13.9 $W = QV/2$, $1.56 = (1.8 \times 10^{-3})V/2$ or $V = 1733$ V

$1.8 \times 10^{-3} = 1733C$, or $C = 1.04\ \mu$F

Total capacitance $= 0.51\ \mu$F

Total voltage $= Q/C = (1.8 \times 10^{-3})/(0.51 \times 10^{-6}) = 3529$ V

Voltage across $1.04\ \mu$F capacitor $= (3529 \times 1)/2.04 = 1730$ V

Voltage across each $2\ \mu$F capacitor $= (3529 - 1730)/2 = 900$ V

13.10 $S_i = 0.3/(600 \times 4\pi \times 10^{-7} \times 0.6 \times 10^{-4}) = 6.632 \times 10^6$

$S_a = (0.3 \times 10^{-2})/(4\pi \times 10^{-7} \times 0.6 \times 10^{-4}) = 39.789 \times 10^6$

$S = S_i + S_a = 46.421 \times 10^6$, m.m.f. $= 10^4 \times 0.05 = 500$ A,

$\varphi = 500/(46.421 \times 10^6) = 10.77\ \mu$Wb,

$B = (10.77 \times 10^{-6})/(0.6 \times 10^{-4}) = 0.18$ T

Pull $= (0.18^2 \times 0.6 \times 10^{-4})/(2 \times 4\pi \times 10^{-7}) = 0.77$ N

13.11 Flux density in air gaps $= B$ and in magnetic material $= 1.2B$. Then $400 \times 3 = (1.2B \times 0.6)/(1000\ \mu_0) + (2B \times 1 \times 10^{-3})/\mu_0$. Hence

$B = 0.55$ T

$W = (0.55^2 \times 2 \times 10^{-3})/(8\pi \times 10^{-7}) = 240.72$ N.

Mass lifted $= 240.72/9.81 = 24.54$ kg

13.12 $1600 = 0.75B/(4\pi \times 10^{-7} \times 1800) + (1.8B \times 10^{-3})/(4\pi \times 10^{-7})$ or

$B = 0.91$ T

$W = 0.91^2/(2 \times 4\pi \times 10^{-7} \times 1800) = 183$ J/m^3.

Volume of material $= 0.75 \times 1000 \times 10^{-6}$ m^2

Energy stored $= 183 \times 0.75 \times 10^{-3} = 0.137$ J

13.13 0.345 J

13.14 0.1 J

13.15 5 H

13.16 8 J

13.17 0.9 W

13.18 (*a*) 300 V; (*b*) 0.075 J

13.19 (*a*) 0; (*b*) 2.75 J; (*c*) 11 J

13.20 2.64 mJ, 2.88 mJ

13.21 99.5 A

13.22 (*a*) 1.97 H, 0.66 H; (*b*) 4.09 J

13.23 (*a*) 9.47 mH, 28.42 mH; (*b*) 85.27 mJ; (*c*) 8.53 N

13.24 1.35 A

14.1 $A = 1 + Z_1 Y_2 = 1 + (10 + j10)/(5 - j15) = 0.6 + j0.8 = 1\angle 53.1°$
$B = Z_3(1 + Z_1 Y_2) + Z_1 = (20 + j5)(0.6 + j0.8) + 10 + j10$
$= 18 + j29 = 34.13\angle 58.2°\ \Omega$
$C = Y_2 = 1/(5 - j15) = 0.02 + j0.06 = 0.063\angle 71.6°\ \text{S}$
$D = 1 + Y_2 Z_3 = 1 + (20 + j5)/(5 - j15) = 1.1 + j1.3 = 1.7\angle 49.8°$

14.2 (*a*) $Z_L = \infty$. $Z_{in} = A/C = 1\angle 30°/0.02\angle -6.8° = 50\angle 36.8°\ \Omega$
(*b*) $Z_L = 0$. $Z_{in} = B/D = 150\angle 50°/3\angle 0° = 50\angle 50°\ \Omega$

14.3 (*a*) $V_{out}/V_{in} = Z_L/(AZ_L + B)$
$= 600/(14.14 \times 600\angle 0° + 50\angle 60°) = 600/(8484 + 25 + j43.3)$
$= 0.071 - j3.6 \times 10^{-4} = 0.071$
(*b*) $Z_{in} = (AZ_L + B)/(CZ_L + D)$
$= (14.14 \times 600 + 50\angle 60°)/(4\angle -60° + 14.14) = 1.77 + j3.06\ \Omega$

14.4 (*a*) With $V_2 = 0$, $I_1 = V_1/100$, $y_{11} = 1/100 = 0.01$ S
$I_2 = V_1/100$, $y_{21} = 0.01$ S
With $V_1 = 0$, $I_2 = V_2/[(100 \times 500)/600] = V_2/83.33$
$y_{22} = 1/83.33 = 0.012$ S
$I_1 = I_2 \times 500/600 = (V_2/83.33)(5/6) = 0.01 V_2$, $y_{12} = 0.01$ S
(*b*) With $V_2 = 0$,
$h_{11} = V_1/I_1 = 1/y_{11} = 100\ \Omega$,
$h_{21} = I_2/I_1 = (V_1/100)/V_1/100) = 1$
With $I_1 = 0$,
$h_{12} = V_1/V_2 = 1$,
$h_{22} = I_2/V_2$, $I_2 = V_2/500$, $h_{22} = 1/500 = 2$ mS

14.5 Matrix is

$$(1/0.5)\begin{bmatrix} -(1000 \times 0.015 - 0.5^2) & -1000 \\ -0.015 & -1 \end{bmatrix} = \begin{bmatrix} 29.5 & -2000\ \Omega \\ -0.03\ \text{S} & -2 \end{bmatrix}$$

14.6 (*a*) $\begin{bmatrix} 1 + 120/330 & 220(1 + 120/330) + 120 \\ 1/330 & 1 + 220/330 \end{bmatrix}$

$$= \begin{bmatrix} 1.364 & 420\ \Omega \\ 3.03 \times 10^{-3}\ \text{S} & 1.67 \end{bmatrix}$$

(*b*) $V_2/V_1 = 1200/(1.364 \times 1200 + 420) = 0.58$

14.7 Network A: $A = 1 + 100/200 = 1.5$. $B = (100 \times 1.5) + 100 = 250\ \Omega$
$C = 1/200 = 5 \times 10^{-3}$ S. $D = 1 + 100/200 = 1/5$
Network B: $A = 1 + 200/100 = 3$. $B = (200 \times 3) + 200 = 800\ \Omega$
$C = 1/100 = 0.01$ S. $D = 1 + 200/100 = 3$
The overall matrix is

$$\begin{bmatrix} 1.5 & 250 \\ 5 \times 10^{-3} & 1.5 \end{bmatrix} \begin{bmatrix} 3 & 800 \\ 0.01 & 3 \end{bmatrix} = \begin{bmatrix} 7 & 1950\ \Omega \\ 0.03\ \text{S} & 8.5 \end{bmatrix}$$

14.8 (*a*) $$\begin{bmatrix} 1 + 400/\text{j}400 & -\text{j}300(1-\text{j}) + 400 \\ 1/\text{j}400 & 1 + -\text{j}300/\text{j}400 \end{bmatrix}$$
$$= \begin{bmatrix} 1 - \text{j} & 100 - \text{j}300\ \Omega \\ -\text{j}2.5 \times 10^{-3}\ \text{S} & 0.25 \end{bmatrix}$$

(*b*) $V_2/V_1 = 500/[500(1-\text{j}) + 100 - \text{j}300]$
$= 500/1000\angle -53.1° = 0.5\angle 53.1°$
Therefore, $V_{in} = 100/0.5\angle 53.1° = 200\angle -53.1°$ V

14.9 $$\begin{bmatrix} 1 + \dfrac{250 + \text{j}350}{-\text{j}180} & (250 + \text{j}350)\left[1 + \dfrac{250 + \text{j}350}{-\text{j}180}\right] + 250 + \text{j}350 \\ \dfrac{1}{-\text{j}180} & 1 + \dfrac{250 + \text{j}350}{-\text{j}180} \end{bmatrix}$$
$$= \begin{bmatrix} -0.944 + \text{j}1.39 & -472.5 + \text{j}1033\ \Omega \\ \text{j}5.56 \times 10^{-3}\ \text{S} & -0.944 + \text{j}1.39 \end{bmatrix}$$

14.10 $$\begin{bmatrix} 1 + 330/-\text{j}400 & 330(1 + 300/-\text{j}400) + 330 \\ 1/-\text{j}400 & 1 + 330/-\text{j}400 \end{bmatrix}$$
$$= \begin{bmatrix} 1 + \text{j}0.825 & 660 + \text{j}272.25 \\ \text{j}2.5 \times 10^{-3}\ \text{S} & 1 + \text{j}0.825 \end{bmatrix}$$
$Z_L = 100/(50 \times 10^{-3}) = 2000\ \Omega$
$100 = 2000\ \text{V}_{in}/[(1 + \text{j}0.825)2000 + 660 + \text{j}272.25]$
$= 2000\ \text{V}_{in}/(2660 + \text{j}1922.25)$
$V_{in} = 100(2660 + \text{j}1922.25)/2000 = 164.1\angle 35.9°$ V
$Z_{in} = [(1 + \text{j}0.825)2000 + 660 + \text{j}272.25]$
$/[(\text{j}2.5 \times 10^{-3} \times 2000) + 1 + \text{j}0.825]$
$= 396.7 - \text{j}388.6 = 555.3\angle -44.4°\ \Omega$
$I_{in} = V_{in}/Z_{in} = 164.1\angle 35.9°/555.3\angle -44.4° = 0.296\angle 80.3°$ A

14.11 For the l.h.s. T network:

$$\begin{bmatrix} 1 + 500 \times 4 \times 10^{-3} & 600(1 + 500 \times 4 \times 10^{-3}) + 500 \\ 4 \times 10^{-3} & 1 + 600 \times 10^{-3} \end{bmatrix}$$
$$= \begin{bmatrix} 3 & 2300\ \Omega \\ 4 \times 10^{-3}\ \text{S} & 3.4 \end{bmatrix}$$

Hence the overall matrix is

$$\begin{bmatrix} 3 & 2300 \\ 4 \times 10^{-3} & 3.4 \end{bmatrix} \begin{bmatrix} 1 & 0 \\ 2.5 \times 10^{-3} & 1 \end{bmatrix} = \begin{bmatrix} 8.75 & 2300\ \Omega \\ 12.5 \times 10^{-3}\ \text{S} & 3.4 \end{bmatrix}$$

$Z_L = 25/(16 \times 10^{-3}) = 1562.5\ \Omega$
Hence, $25 = 1562.5V_1/[(8.75 \times 1562.5) + 2300]$ and $V_1 = 255.6$ V

14.12 3 V

14.14 $\begin{bmatrix} 50 & 40 \\ 10 & 40 \end{bmatrix}$

14.16 20 S

14.17 $\begin{bmatrix} 10^{-3}\ \text{S} & -1 \times 19^{-9}\ \text{S} \\ 120 \times 10^{-3}\ \text{S} & 4 \times 10^{-4}\ \text{S} \end{bmatrix}$

14.18 $\begin{bmatrix} 3.5 & 120 \\ 0.094 & 3.5 \end{bmatrix}$

14.19 $\begin{bmatrix} 1 & 100\angle 60^\circ \\ 1.2 \times 10\angle 90^\circ\ \text{S} & 1 \end{bmatrix}$

14.22 $A = 1.6\angle 0^\circ$, $B = 855\angle -69^\circ\ \Omega$, $C = 2$ mS, $D = \sqrt{2}\angle -45^\circ$

14.23 $A = 1.07\angle 10.8^\circ$, $B = 206\angle 76^\circ\ \Omega$, $C = 1.53\angle 3.8^\circ$ mS, $D = 1.03\angle 5.6^\circ$

14.24 $A = 1.45$, $B = \text{j}57\ \Omega$, $C = \text{j}0.015$ S $D = 0.1$

15.1 Image impedances: $Z_{oc} = 130\ \Omega$, $Z_{sc} = 20 + (35 \times 110)/145 = 46.55\ \Omega$
$Z_A = \sqrt{(130 \times 46.55)} = 77.75\ \Omega$
$Z_{oc} = 145\ \Omega$, $Z_{sc} = 35 + (20 \times 110)/130 = 51.92\ \Omega$
$Z_B = \sqrt{(145 \times 51.92)} = 86.77\ \Omega$
Image transfer coefficient: $I_S = E_S/(2 \times 77.5)$, $V_S = E_s/2$
$I_R = (E_S/155)[110/(145 + 86.77)] = 3.062 \times 10^{-3} E_S$
$V_R = I_R Z_B = 3.062 \times 10^{-3} E_S \times 86.77 = 0.266 E_S$
$\gamma = (1/2)\log_e[(E_S/155) \times (E_S/2)]/(3.063 \times 10^{-3} \times 0.266 E_S^2]$
$= (1/2)\log_e(3.96) = 0.688\text{N} = 6$ dB
Insertion loss: $I_1 = E_S/(77.75 + 86.77) = E_s/164.52$
$I_2 = I_R = 3.062 \times 10^{-3} E_S$
Insertion loss $= 20\ \log_{10}[1/(164.52 \times 3.062 \times 10^{-3})]$
$= 20\ \log_{10} 1.985 = 6$ dB

15.2 $Z_2 = \sqrt{[1650(1100 - 550)]} = 952.63\ \Omega$
$Z_1 = 1100 - 952.63 = 147.37\ \Omega$
$Z_3 = 1650 - 952.63 = 697.37\ \Omega$

15.3 Image impedances: $Z_{oc} = 1400\ \Omega$,
$Z\ sc = 400 + (800 \times 1000)/1800 = 844.4\ \Omega$
$Z_A = \sqrt{(1400 \times 844.4)} = 1087.3\ \Omega$
$Z_{oc} = 1800\ \Omega$, $Z_{sc} = 800 + (400 \times 1000)/1400 = 1085.7\ \Omega$
$Z_B = \sqrt{(1800 \times 1085.7)} = 1398\ \Omega$
Iterative impedances: $R_A = 400 + 1000(800 + R_A)/(1800 + R_A)$
$1800 R_A + R_A^2 = 720 \times 10^3 + 400 R_A + 800 \times 10^3 + 1000 R_A$
$R_A^2 + 400 R_A - 1.52 \times 10^6 = 0$
$R_A = -400 \pm \sqrt{[400^2 + (4 \times 1.52 \times 10^6)]}$
$R_A = 1049\ \Omega$
The negative value of 1449 Ω is the iterative impedance R_B in the other direction.
Insertion loss: $I_1 = E_S/(1049 + 1449) = 4 \times 10^{-4} E_S$
$I_2 = (E_S/2498) \times 1000/(1000 + 800 + 1449) = 1.232 \times 10^{-4} E_S$
Insertion loss $= 20 \log_{10}(4/1.232) = 10.2$ dB

15.4 (*a*) $Z_0 = \sqrt{[(2000^2/4) + (2000 \times 5600)]} = 3493\ \Omega$
(*b*) $8 = \log_e[1 + 2000/(2 \times 5600) + 3493/5600]$
$= \log_e 1.802 = 0.59$ N
$I_R = 10 \times e^{-5\times0.59} = 0.523$ mA

15.5 $Z_{oc} = 340\ \Omega$, $Z_{sc} = 120 + (220 \times 105)/325 = 191.08\ \Omega$
$Z_A = \sqrt{(340 \times 191.08)} = 254.9\ \Omega$
$Z_{oc} = 325\ \Omega$, $Z_{sc} = 105 + (220 \times 120)/340 = 182.65\ \Omega$
$Z_B = \sqrt{(325 \times 182.65)} = 243.6\ \Omega$
$I_1 = E_S/(254.9 + 243.6) = 2 \times 10^{-3} E_S$
$I_{in} = E_S/(254.9 + 254.9) = 1.962 \times 10^{-3} E_S$
$I_2 = (1.962 \times 10^{-3}) \times 220/(220 + 105 + 243.6) = 7.59 \times 10^{-4} E_S$
Insertion loss $= 20\ \log_{10}[(2 \times 10^{-3} E_S)/(7.59 \times 10^{-4} E_S)] = 8.4$ dB

15.6 (*a*) $Z_{oc} = 5600\ \Omega$, $Z_{sc} = (5600 \times 7800)/(5600 + 7800) = 3259.7\ \Omega$
$Z_A = \sqrt{(5600 \times 3259.7)} = 4272.5\ \Omega$
$Z_{oc} = 5600 + 7800 = 13400\ \Omega$, $Z_{sc} = 7800\ \Omega$
$Z_B = \sqrt{(7800 \times 13400)} = 10223.5\ \Omega$
$I_1 = E_S/(4272.5 + 10223.5) = 6.9 \times 10^{-5} E_S$
$I_{in} = E_s/(2 \times 4272.5) = 1.17 \times 10^{-4} E_S$
$I_2 = (1.17 \times 10^{-4} E_s) \times 5600/(5600 + 7800 + 10223.5)$
$= 2.774 \times 10^{-5} E_S$
Insertion loss $= 20\ \log_{10}(6.9/2.774) = 7.91$ dB
(*b*) $R_A = [5600(7800 + R_A)]/(5600 + 7800 + R_A)$
$13\,400 R_A + R_A^2 = 4.368 \times 10^7 + 5600 R_A$
$R_A^2 + 7800 R_A - 4.368 \times 10^7 = 0$
$R_A = [-7800 \pm \sqrt{(7800^2 + (4 \times 4.368 \times 10^7))}]/2$
$= (-7800 \pm 15348)/2$
Therefore, $R_A = 3774\ \Omega$ and $R_B = 11574\ \Omega$
$I_1 = E_S/(3774 + 11574) = 6.52 \times 10^{-5} E_S$
$I_{in} = E_S/(2 \times 3774) = 1.325 \times 10^{-4} E_S$
$I_2 = (1.325 \times 10^{-4}) E_S \times 5600/(5600 + 7800 + 11574)$
$= 2.97 \times 10^{-5} E_S$
Insertion loss $= 20\ \log_{10}(6.52/2.97) = 6.8$ dB

15.7 $Z_0 = \sqrt{[400^2/4 + (400 \times 380)]} = 438.2\ \Omega$
$\gamma = \log_e[1 + 400/760 + 438.2/380] = 0.986$ N
For four sections $4\gamma = 3.944\ N = 34.26$ dB

15.8 $Z_{oc} = 4500\ \Omega$, $Z_{sc} = 1200\ \Omega$
$Z_A = \sqrt{(4500 \times 1200)} = 2324\ \Omega$
$Z_{oc} = 3300\ \Omega$, $Z_{sc} = (200 \times 3300)/4500 = 880\ \Omega$
$Z_B = \sqrt{(3300 \times 880)} = 1704\ \Omega$
$I_1 = E_S/(2324 + 1704) = 2.483 \times 10^{-4}$
$I_{in} = E_S/4648 = 2.152 \times 10^{-4}\ E_s$
$I_2 = (2.152 \times 10^{-4} E_s) \times 3300/(3300 + 1704) = 1.419 \times 10^{-4} E_S$
Insertion loss $= 20\ \log_{10}(2.483/1.419) = 4.86$ dB

15.9 $1200 = Z_1 + Z_2\ Z_1 = 1200 - Z_2$
$480 = Z_1 + (Z_2 Z_3)/(Z_2 + Z_3) = 1200 - Z_2 + (Z_2 Z_3)/(Z_2 + Z_3)$
$Z_2^2 = 720\ Z_2 + 720\ Z_3$ (1)
$560 = Z_3 + (Z_1 Z_2)/(Z_1 + Z_2)$
$= Z_3 + (1200 - Z_2) Z_2/(1200 - Z_2 + Z_2)$
$Z_2^2 = 1200\ Z_2 + 1200\ Z_3 - 6.72 \times 10^5$ (2)
Equating equations (1) and (2) gives:
$720\ Z_2 + 720\ Z_3 = 1200\ Z_2 + 1200\ Z_3 - 6.72 \times 10^5$

$Z_3 = 1400 - Z_2$
Substitute into equation (1) to give:
$Z_2^2 = 720\ Z_2 + 720\ (1400 - Z_2) = 720\ Z_2 + 1.008 \times 10^6 - 720\ Z_2$
or $Z_2 = 1004\ \Omega$
$Z_1 = 1200 - 1004 = 196\ \Omega$, $Z_3 = 1400 - 1004 = 396\ \Omega$

15.10 15 dB is a voltage ratio of 5.62
$R_1 = 140(5.62 - 1)/6.62 = 97.7\ \Omega$
$R_2 = (280 \times 5.62)/(5.62^2 - 1) = 51.45\ \Omega$

15.11
$$\text{I.L.} = 20\left[\log_{10}\left(\frac{1600}{2\sqrt{(1000\times 600)}}\right) + \log_{10}\left(\frac{1250}{2\sqrt{(750+500)}}\right) + 0.6514\right.$$
$$\left. + \left[\log_{10}\left[1 - \left(\frac{400}{1600}\right)\left(\frac{250}{1250}\right)e^{-3}\right] - \log_{10}\left(\frac{1100}{2\sqrt{(600\times 500)}}\right)\right]\right]$$

$= 20[\log_{10} 1.033 + \log_{10} 1.021 + 0.6514 + \log_{10} 0.998 - \log_{10} 1.004]$
$= 13.44$ dB

15.12 (*a*) $600 = 480 + R(480 + 600)/(R + 480 + 600)$, $R = 135\ \Omega$
(*b*) Direct current $= E/(600 + 600)$
Current with network in
$= (E/1200) \times 135/(135 + 480 + 600) = 135E/(1200 \times 1215)S$
Insertion loss $= 20\ \log_{10}[(1200 \times 1215)/(1200 \times 135)] = 19.1$ dB
(*c*) With 300 Ω load, direct current $= E/900$. Current with network
$= 135E/(1080 \times 915 + 135 \times 780)$
Insertion loss
$= 20\ \log_{10}[1080 \times 915 + 135 \times 780)/(135 \times 900) = 19.2$ dB

15.13 $\omega_{co} = 2/\sqrt{(LC)} = 1000R/S$. $\alpha = 2\cosh^{-1}(2500/1000) = 2\cosh^{-1}\ 2.5$. $e^{\alpha/2} + e^{-\alpha/2} = 5$. Multiply by $e^{\alpha/2}$ to get $e^{\alpha} - 5e^{\alpha/2} + 1 = 0$. This is of the form $x^2 - 5x + 1 = 0$ and so $x = [5 \pm \sqrt{(25 - 4)}]/2 = 4.79$ or 0.21. Hence $\alpha/2 = 1.567$ or -1.561, the latter value would mean that the filter had a gain so $\alpha = 3.134$ nepers $= 27.2$ dB

15.14 Constant k: $600 = \sqrt{(L/C)}$ and $25000 = 2/\pi\sqrt{LC}$. Hence $C = 133.3$ nF and $L = 48$ mH, $m = \sqrt{[1 - (25000/32500)]} = 0.48$

15.15 $R_A = 20 + 120R_A/(120 + R_A)$, $R_B = 120(20 + R_B)/(120 + 20 + R_B)$. Solving, $R_A = 60\ \Omega$ and $R_B = 40\ \Omega$

15.16 (*a*) 350 Ω; (*b*) No; (*c*) 8.57 mA; (*d*) 16.14 dB; (*e*) 16.14 dB

15.17 (*a*) 2500 Ω, 1250 Ω; (*b*) 1.73 N

15.18 $L = 153\ \mu H$, $C = 3.7$ nF

15.19 75 Ω, 90 Ω, 6 dB

15.22 391 Ω, 8.36 dB

15.23 4.76 dB

15.24 (*a*) 1090 Ω; (*b*) 1400 Ω; (*c*) 9 dB

15.25 (*a*) 600 Ω, 1950 Ω; (*b*) 56.3°; (*c*) 2

15.26 (*a*) $A = D = 1.5$, $B = 500$, $C = 1/400$; (*b*) $Z_0 = 447\ \Omega$;

(*c*) $\begin{bmatrix} 3.5 & 1500 \\ 3/400 & 3.5 \end{bmatrix}$

15.27 (*a*) 95.4 Ω, 110 Ω; (*b*) 231.5 μH, 22 nF

15.28 $Z_0 = \sqrt{(L_1/C_1)}[\sqrt{(1-\omega^2 L_1 C_1)}][\sqrt{(1-(C_1+C_2))}/(\omega^2 L_1 C_1 C_2)]$

15.30 (*a*) 1080 Ω, 732 Ω, (*b*) 17.2 dB

15.31 17.8 dB

15.32 7.23 dB

15.33 (*a*) 235 Ω, 16.2 Ω; (*b*) 22.5°

16.1 (*a*) $V_s = 10/2 = 5$ V, $3.5 = 5e^{-0.05l}$
$\log_e 0.7 = -0.3567 = -0.05l$, $l = 7.134$ km
(*b*) $2\pi = (\pi/10)l$, $l = 20$ km
(*c*) $|I_s| = 5/600$ A, $|I_r| = (5/600)\ e^{-(0.05\times 22)} = 2.774$ mA
$P_L = (2.774 \times 10^{-3})^2 \times 600 \times \cos(-30°) = 4$ mW

16.2 (*a*) $|I_s| = 12/550$ A, $|I_r| = (12/550)\ e^{-(0.015\times 20)} = 16.163$ mA
$P_L = (16.163 \times 10^{-3})^2 \times 550 \cos(-25°) = 130.2$ mW
(*b*) $\beta l = (\pi/60) \times 20 = \pi/3 = 60°$

16.3 (*a*) $I_s = I_r = 20$ mA, $P_L = (20 \times 10^{-3})^2 \times 500 = 200$ mW
(*b*) $v_p = \omega/\beta = (2\pi \times 30 \times 10^6)/(\pi/3) = 1.8 \times 10^8$ m/s

16.4 $I_s = 10/600\angle -30°$ A
$I_r = (10/600)\ e^{-(0.1\times 12)}\angle[-30° - (90° \times 12)] = 5.02\angle -1110°$ mA
$= 5.02\angle -30°$ mA

16.5 $\omega L = 5000 \times 16 \times 10^{-3} = 80\ \Omega$
$\omega C = 5000 \times 0.045 \times 10^{-6} = 2.25 \times 10^{-4}$ S
$Z_o = \sqrt{[(R + j\omega L)/j\omega C]} = 100\sqrt{[20 + j80)/j2.25]} = 605\angle -7°\ \Omega$
$\gamma = \sqrt{[(R + j\omega L)/j\omega C]} = 10^{-2}\sqrt{(20 + j80)j2.25} = 0.1362\angle 83°$
$= 0.0166 + j0.1352$. 0.1352 radians = 7.7°
Therefore, $I_s = 12/(500 + 600\angle -7°) = 11\angle 3.8°$ mA
$V_s = 605\angle -7° \times 0.011\angle 3.8° = 6.655\angle -3.2°$ V
$I_r = 11\ e^{-0.166}\angle(3.2° - 77°) = 9.3\angle -73.8°$ mA (w.r.t. V_s)

16.6 (*a*) $\lambda = (3 \times 10^8)/(1 \times 10^8) = 3$ m
(*b*) $P_L = 100/75 = 1.33$ W
(*c*) $v_p = 3 \times 10^8 = \omega/\beta$, $\beta = (2\pi \times 10^8)/(3 \times 10^8) = 2\pi/3$ rad/km
$\beta l = 120°$, $I_s = 10/75 = 133.3$ mA, $I_l = 133.3\angle -120°$ mA

16.7 (*a*) $V_s = 6/2 = 3$ V
(*b*) $V_r = 3$ V
(*c*) $P_L = 9/50 = 180$ mW
(*d*) $\beta = 2\pi/\lambda = 2\pi/200 = \pi/100$ rad/km

16.8 (*a*) $\omega L = 10000 \times 1 \times 10^{-3} = 10\ \Omega$,
$\omega C = 10000 \times 0.05 \times 10^{-6} = 5 \times 10^{-4}$ S,
$Z_o = \sqrt{[(25 + j10)/(j5 \times 10^{-4})]} = 100\sqrt{[(26.93\angle 21.8°)/5\angle(90°)]}$
$= 232\angle -34.1°\ \Omega$
(*b*) $\gamma = \sqrt{[(25 + j10)(j5 \times 10^{-4})]} = 10^{-2}(26.93\angle 21.8° \times 5\angle 90°)$
$= 0.116\angle 58.5°$
$\alpha = 0.116 \cos 58.5° = 0.06$ N/km
(*c*) $\beta = 0.116 \sin 58.5° = 0.1$ rad/km

16.9 (*a*) 12 dB $= 20 \log_{10}[(3/1800)/I_r]$
$3.98 = (3/1800)/I_r$ or $I_r = (3/1800)/3.98 = 418.8\mu$A
$\beta l = 1.24$ rad $= 71°$ and so I_r lags V_s by $28° - 71° = 43°$
(*b*) $\beta l = (71/360)\lambda = 0.197\lambda$

16.10 $\omega_L = 18 \times 0.6 = 10.8\ \Omega$, $\omega C = 18 \times 0.055 \times 10^{-3} = 9.9 \times 10^{-4}$ S
$\gamma = \sqrt{(28 + j10.8)(j9.9 \times 10^{-4})} = 0.172\angle 55.6°$
$\beta = 0.172 \sin\ 55.6° = 0.142$ rad/km
$v_p = \omega/\beta = 18000/0.142 = 126.76 \times 10^3$ km/s

16.12 $Z_1 = 106 + j22\ \Omega$, $Z_2 = -668 + j279\ \Omega$

16.13 (*a*) $244\angle -28.8°$ A; (*b*) $96.34\ \angle 4.8°$ kV; (*c*) 0.914

16.14 $A = D = 0.992\angle 0.2°$, $B = 42.8\angle 65.3°$, $C = 4.17 \times 10^{-4}\angle 90°$ S,

16.15 (*a*) 3.16 W; (*b*) $-240°$, $-220°$

16.16 $160\angle -10°\ \Omega$, $0.233 + j0.119$ per km

16.17 (*a*) 2.564 W; (*b*) 2.338×10^8 m/s

16.18 193 mW, 25.71 dB

16.19 $A = D = 1.1\angle 7°$, $B = 87.15\angle 0°\ \Omega$, $C = 3.87 \times 10^{-3}\angle 60°$ S

16.20 (*a*) $254\angle -38.6°\ \Omega$; (*b*) $0.18\angle 51.8°$; (*c*) 0.11 N/km; (*d*) 0.14 rad/km

16.22 (*a*) $139\angle 3°$ V phase and 240 V line; (*b*) $350.9\angle -40.8°$ A; (*c*) 94.75%

16.23 $A = D = 0.88\angle 0°$, $B = 239\angle 75.6°\ \Omega$, $C = 9.56 \times 10^{-4}\angle 100°$ S

16.24 $166\angle -17.5°\ \Omega$

16.25 (*a*) $324 + j128\ \Omega$; (*b*) $54.7\angle -21.6°$ A; (*c*) 29.94 kV; (*d*) 3.1°

16.26 (*a*) $Z_0 = 785\angle -6.5°\,\Omega$; (*b*) $\gamma = 0.19\angle 80°$ per km

17.1 $-j160 = j300\ \tan(2\pi l/\lambda)$, $2\pi/\lambda = \tan^{-1}(-16/30) = 152° = 2.653$ rad
$l = (2.653/2\pi)\lambda = 0.422\lambda$, $\lambda = 2.5$ m and so
$l = 0.422 \times 2.5 = 1.056$ m

17.2 $\rho_v = [(200 - j300) - 300]/(500 + j300) = 0.542\angle -139.4°$
$S = 1.542/(1 - 0.542) = 3.37$
$Z_s = 200 - j300 = 360.6\angle -56.3°\ \Omega$

17.3 $\rho_v = [(150 + j120) - 450]/(600 + j120) = 0.528\angle 146.9°$
$S = 1.528/(1 - 0.528) = 3.24$

17.4 6 dB is a voltage ratio of 1/2, $1.3\lambda = 0.3 \times 360° = 108°$
$\rho_v = [(400 - j350) - 600]/[400 - j350 + 600] = 0.38\angle -100.5°$ and
$\rho_i = 0.38\angle 79°$
Hence, $V'_s \rightarrow V'_s/2\angle -108°$, $\quad I_s \rightarrow I'_s 2\angle -108°$
$0.095V'_s\angle -316.5° \leftarrow 0.19V'_s\angle -208°$,
$0.095I'_s\angle -136.5° \leftarrow 0.19I'_s\angle -28.5°$
$V_s = V'_s + 0.095\ V'_s \cos(-316.5°) - j0.095\ V'_s \sin(-316.5°)$
$= 1.069\ V'_s\angle 3.5°$ V
$I_s = I'_s[1 + 0.095\cos(-136.5°) + j0.095\sin(-136.5°)]$
$= 0.935I'_s\angle -4°$ A
$Z_s = V_s/I_s = [1.069\angle 3.5°/0.935\angle -4°]600 = 686\angle 7.5°\ \Omega$

17.5 (*a*) $\rho_v = (200 - 600)/(200 + 600) = -0.5$
(*b*) $S = 1.5/0.5 = 3$
(*c*) $12 = V'_s(1 - 0.5) = 0.5\ V'_s$ and so $V'_s = 24$ V
At distance $\lambda/4$ from the load the incident voltage is 90° in advance and the reflected voltage is 90° lagging. Hence the two voltages are in phase with one another and the total voltage = 36 V.
(*d*) At distance $\lambda/2$ from the load the two voltages are in antiphase, so the total voltage is 12 V.

17.6 6 db = voltage ratio of 2 $\lambda/2 = 180°$, $\rho_v = -1$ and $\rho_i = +1$
$V'_s \to V'_s/2\angle-180°$, $I'_s \to I'_s/2\angle-180°$
$V'_s/4/-180° \leftarrow V'_s/2\angle 0°$, $I'_s/4/-360° \leftarrow I'_s/2\angle-180°$
$V_s = 0.75\ V'_s$ and $I_s = 1.25\ I'_s$
$Z_s = V_s/I_s = (0.75/1.25) \times 1000 = 600\ \Omega$

17.7 $\lambda/4 = 90°$, $\rho_v = [(300 - j100) - 300]/(600 - j100) = 0.164\angle-80.5°$
$V'_s \to V'_s\angle-90°$, $I'_s \to I'_s\angle-90°$
$0.164\angle-260.5°\ V'_s \leftarrow 0.164\angle-170.5°\ V'_s$,
$0.164\angle-80.5°\ I'_s \leftarrow 0.164\angle 9.5°\ I'_s$
$V_s = V'_s[1 + 0.164\cos(-260.5°) + j0.164\sin(-260.5°)]$
$= 0.986\angle 9.5°\ V'_s$
$I_s = I'_s[1 + 0.164\cos(-80.5°) + j0.164\sin(-80.5°)]$
$= 1.04\angle-9°\ I'_s$
$Z_s = V_s/I_s = [(0.986\angle 9.5°)/1.04\angle-9°)] \times 300 = 284.4\angle 18.5°\ \Omega$

17.8 $\rho_v = (100 - 50)/150 = 1/3$. Line loss $= 1/\sqrt{2}$
$V'_s = 6V \to 6/\sqrt{2}\angle 0°$, $2/2\angle 0° \to 2/\sqrt{2}\angle 0°$
(*a*) $V_L = 6/\sqrt{2} + 2/\sqrt{2} = 5.66$ V
(*b*) $V_S = 6 + 1 = 7$ V

17.9 At the point $Z_{(eff)} = (600 \times 300)/900 = 200\ \Omega$
$\rho_v = (200 - 600)/(200 + 600) = -0.5$
Length of line to the point $= 2.5\lambda$, $\beta l = 180°$
$V'_s \to -V'_s$, $I'_s \to -I'_s$
$-V'_s/2 \leftarrow +V'_s/2$, $+I'_s/2 \leftarrow -I'_s/2$
$V_s = V'_s(1 - 0.5) = V'_s/2$
$I_s = I'_s(1 + 0.5) = 3I'_s/2$
$Z_s = V_s/I_s = (1/3) \times 600 = 200\ \Omega$

17.10 $\rho_v = (600 - 900)/1500 = -1/5$, $\rho_i = +1/5$
$V'_s = 50$ V, $I'_s = 50/900 = 55.56$ mA
$50\ e^{-1.1} = 16.64$ V, $V_L = 16.64 - 16.64/5 = 13.312$ V
$55.56\ e^{-1.1} = 18.49$ mA, $I_L = 18.49 + 18.49/5 = 22.19$ mA

17.11 $\lambda = (3 \times 10^8)/(5 \times 10^6) = 60$ m. Therefore, 15 m $= \lambda/4$ and 30 m $= \lambda/2$. At the 15 m point the 100 Ω resistor is effectively in parallel with $Z_s = 100^2/\infty = 0$ and hence the total resistance at this point is zero. There is zero current in the furthest 100 Ω resistor. At the 30 m point the line resistance = 100 Ω and hence the total resistance at this point is 50 Ω. Current in resistor = 50/100 = 0.5 A.

17.12 $V'_s = 10/2 = 5$ V. At receive end of line $V' = (5/\sqrt{2})\angle-270°$ V, $\rho_v = (100 - 50)/(100 + 50) = (1/3)\angle 0°$. Reflected voltage $= (5/3\sqrt{2})\angle-270°$ V and load voltage $V_L = (5/\sqrt{2}) + (5/3\sqrt{2}) = 4.714$ V. Reflected voltage at sending-end of line $= [5/(3\sqrt{2} \times \sqrt{2})] = (5/6)\angle-180°$ V and sending-end voltage $V_S = 5 - 5/6 = 4.17$ V.

17.13 $45° = \angle[\tan^{-1}(100/R_L - R_L/100)]$, $1 = 100/R_L - R_L/100$. Hence $R_L = 61.8\ \Omega$

17.14 Incident current at receive end $= I'_S\ e^{-(2+j1)}$. Reflected current $= I'_S\ e^{-(2+j1)}$. Reflected current at sending end $= I'_S\ e^{-(4+j2)}$. Total sending-end current I_S is
$I'_S[1 + e^{-(4+j2)}] = I'_S[1 + 0.0183\angle-114.6°]$
$= I'_S[0.992 - j0.167] = I'_S[1\angle-9.6°]$
$I_R = 2I'_S[e^{-(2+j1)}] = 2I'_S[0.135\angle 57.3°] = 0.27\angle 57.3°$.

Ratio $|I_S|/|I_R| = 1/0.27 = 3.704$

17.15 $Z_S = 1/(10 \times 10^{-3}) = 100\ \Omega = Z_0^2/600$. Hence, $Z_0 = \surd(100 \times 600) = 245\ \Omega$

17.16 13.32 V, 22.2 mA

17.17 $455.3\angle -24°$ V

17.18 0.53

17.19 (*a*) $Z_0 = 400\ \Omega$,
(*b*) $v_p = 83.33 \times 10^6$,
(*c*) $\rho_{AC} = -0.2$, $T_{AC} = 0.8$, $\rho_{BC} = 0.2$, $T_{BC} = 1.2$,
(*d*) (*i*) 80 V, 16 V, 3.2 V,
(*ii*) 480 V, 19.2 V, 0.77 V

17.20 $50\angle 36.9°\ \Omega$

17.21 $58.7\angle 32.5°\ \Omega$

17.22 $V_S = 430\angle 171°$ V

17.23 (*a*) 400 Ω; (*b*) $200\angle -37°\ \Omega$; (*c*) 100 Ω; (*d*) 100 Ω

17.24 (*a*) $\rho_v = 0.06\angle -42.7°$; (*b*) $S = 1.12$

17.25 $1 = 0.833$ m, $Z_0 = 424\ \Omega$

17.26 (*a*) 132 Ω, 40 Ω; (*b*) 150 Ω, 37.5 Ω

17.27 (*a*) 35.7 Ω; (*b*) 59.1 Ω

17.28 2.154 kV, 3.08 A

17.29 (*a*) 3.95 kV; (*b*) 22.4 kV

17.30 1578 V, 4.22 A, 1578 V, 2.63 A, 1578 V, 1.578 A

18.1 $Z_1 = R + j\omega L = R + j6R$, $|Z_1| = \surd(37)R\ \Omega$
$Z_3 = R + j18R$, $|Z_3| = \surd(325)R\ \Omega$
$Z_5 = R + j30R$, $|Z_5| = \surd(901)R\ \Omega$
$v_L = 120\surd 37 R \sin \omega t + 400\surd 325 R \sin 3\omega t + 15\surd 901 R \sin 5\omega t$ mV
$= 0.73R \sin \omega t + 0.72R \sin 3\omega t + 0.45R \sin 5\omega t$ V
$V_L = R\surd[(0.73/\surd 2)^2 + (0.72/\surd 2)^2 + (0.45/\surd 2)^2]$
$= 0.792R$ V
$I = \surd[(0.12/\surd 2)^2 + (0.04/\surd 2)^2 + (0.015/\surd 2)^2] = 0.09$ A
Apparent impedance $= V/I = 0.792\ R/0.09 = 8.8R\ \Omega$
True impedance $= \surd(37)\ R = 6.083R\ \Omega$
% error $= (8.8 - 6.083)/(6.083 \times 100\% = +44.7\%$

18.2 At $\omega = 1000$ rad/s $\omega L = 30\ \Omega$ and $1/\omega C = 120\ \Omega$
At $\omega = 3000$ rad/s $3\omega L = 90\ \Omega$ and $1/3\omega C = 40\ \Omega$
$Z_1 = (40 + j30)(-j120)/(40 - j90) = 60.92\angle 12.9°\ \Omega$
$Z_3 = (40 + j90)(-j40)/(40 + j50) = 61.52\angle -75.3°\ \Omega$
$|i_1| = 200/60.92 = 3.282$ A and $|i_3| = 50/61.52 = 0.813$ A
Therefore, $i = 3.283 \sin(1000t - 12.9°) + 0.813 \sin(300_0 t + 75.3°)$ A
Power factor $= P/VI$, $P = |I_L|^2 R$
$I_{L1} = 200/(40 + j30)$, $|I_{L1}| = 200/50 = 4$ A and $I_{L1(r.m.s.)} = 2.828$ A
$P_1 = 2.828^2 \times 40 = 319.9$ W
$I_{L3} = 50/(40 + j90)$, $|I_{L3}| = 50/98.5 = 0.508$ A and
$I_{L3(r.m.s.)} = 0.359$ A
$P_3 = 0.359^2 \times 40 = 5.16$ W
Total power $P = 319.9 + 5.16 = 325.06$ W

$V = \sqrt{[(200/\sqrt{2})^2 + (50/\sqrt{2})^2]} = 145.75$ V
$I = \sqrt{[(3.283/\sqrt{2})^2 + (0.813/\sqrt{2})^2]} = 2.39$ A
Therefore power factor $= 325.06/(145.75 \times 2.39) = 0.933$

18.3 For third harmonic resonance $900 \times 100 \times 10^{-3} = (1/900C)$
or $C = 12.35\mu$F
$\omega L_1 = 300 \times 0.1 = 30\ \Omega$, $1/\omega C_1 = 1/(300 \times 12.35 \times 10^{-6}) = 270\ \Omega$
$Z_1 = 25 + j30 - j270 = 25 - j240 = 241.3\angle -84.1°\ \Omega$
$I_1 = 200/241.3 = 0.829$ A $= 0.586$ A r.m.s.
$Z_3 = 25 + j90 - j90 = 25\ \Omega$
$I_3 = 60/25 = 2.4$ A $= 1.7$ A r.m.s.
$Z_5 = 25 + j150 - j54 = 25 + j96 = 99.2\angle 75.4°\ \Omega$
$I_5 = 30/99.2 = 0.3$ A $= 0.214$ A r.m.s.
$V = \sqrt{[(200/\sqrt{2})^2 + (60/\sqrt{2})^2 + (30/\sqrt{2})^2]} = 149.1$ V
$I = \sqrt{[0.586^2 + 1.7^2 + 0.214^2)} = 1.81$ A
$P = 1.81^2 \times 25 = 81.9$ W
Power factor $= 81.9/(1.81 \times 149.1) = 0.303$

18.4 (*a*) At $\omega = 500$ rad/s, $\omega L = 500\ \Omega$ and $1/\omega C = 4000\ \Omega$
At $\omega = 1000$ rad/s, $\omega L = 1000\ \Omega$ and $1/\omega C = 2000\ \Omega$
$Z_1 = [(1000 + j500)(-j4000)]/(100 - j3500) = 1232\angle 10.8°\ \Omega$
$I_1 = 100/1232\angle 10.8° = 81.17\angle -10.8°$ mA
$Z_2 = [(1000 + j1000)(-j2000)]/(1000 - j1000) = 2000\ \Omega$
$I_2 = 40/2000 = 20$ mA
Therefore, $i = 81.17 \sin(500t - 10.8°) + 20 \sin 1000t$ mA
(*b*) $V = \sqrt{[(100/\sqrt{2})^2 + (40/\sqrt{2})^2]} = 76.15$ V
$I = \sqrt{[(81.17/\sqrt{2})^2 + (20/\sqrt{2})^2]} = 59.1$ mA
At 500 rad/s, $i_L = 70.7/(1000 + j500)$, $|i_L| = 70.7/1118 = 63.24$ mA
$P_1 = (63.24 \times 10^{-3})^2 \times 1000 = 4$ W
At 1000 rad/s, $i_L = 28.28/(1000 + j1000)$,
$|i_L| = 28.28/1414.2 = 20$ mA
$P_2 = (20 \times 10^{-3})^2 \times 1000 = 0.4$ W
Total power dissipated = 4.4 W and
Power factor $= 4.4/(76.15 \times 59.1 \times 10^{-3}) = 0.978$

18.5 (*a*) $V = \sqrt{[10^2 + (10/\sqrt{2})^2]} = 12.25$ V
(*b*) $I_{DC} = 10/400 = 25$ mA
$\omega L = 1000 \times 0.3 = 300\ \Omega$, $1/\omega C = 1/(1000 \times 2.5 \times 10^{-6}) = 400\ \Omega$
$I_{acC} = 7.07/(300 + j400) = 14.14\angle 53.1°$ mA
$I_{acL} = 7.07/(400 + j300) = 14.14\angle -36.9°$ mA
Adding, the total a.c. current $= 20\angle 8.1°$ mA
$I = \sqrt{(25^2 + 20^2)} = 32$ mA
(*c*) A.C. power dissipated $= (14.14 \times 10^{-3})^2 \times 700 = 140$ mW
D.C. power dissipated $= (25 \times 10^{-3})^2 \times 400 = 250$ mW
Therefore power factor $= 390/(12.25 \times 32) = 0.995$

18.6 $V = \sqrt{[20^2 + (10/\sqrt{2})^2]} = 21.21$ V
$I_{DC} = 20/100 = 0.2$ A, $\omega L = 5000 \times 50 \times 10^{-3} = 250\ \Omega$
$|Z_{ac}| = \sqrt{(100^2 + 250^2)} = 269.3\ \Omega$
$I_{ac} = 7.07/269.3 = 26.25$ mA, $I = \sqrt{(200^2 + 26.25^2)} = 201.7$ mA
$P_{DC} = 0.2^2 \times 100 = 4$ W
$P_{ac} = (26.25 \times 10^{-3})^2 \times 100 = 69$ mW
Total power dissipated = 4.069 W
Power factor $= 4.069/(21.21 \times 0.2017) = 0.95$

18.7 For circuit A: $\omega_1 L = 500 \times 0.4 = 200\ \Omega$, $1/\omega_1 C = 1/(500 \times 5 \times 10^{-6})$

$= 400\ \Omega$. And $\omega_2 L = 400\ \Omega$ and $1/\omega_2 C = 200\ \Omega$
$Z_1 = 200 - \mathrm{j}200\ \Omega$, $Y_1 = 1/Z_1 = 2.5 + \mathrm{j}2.5$ mS
$Z_2 = 200 + \mathrm{j}200\ \Omega$, $Y_2 = 2.5 - \mathrm{j}2.5$ mS
Similarly for circuit B, $Y_1 = 5 + \mathrm{j}5$ mS and $Y_2 = 5 - \mathrm{j}5$ mS
When connected in parallel the total admittance of the circuit at frequency ω_1 is $Y_{t1} = 7.5 + \mathrm{j}7.5$ mS and at frequency ω_2 is $Y_{t2} = 7.5 - \mathrm{j}7.5$ mS
$I_1 = 10 \times Y_1 = 75 + \mathrm{j}75$ mA $= 106\angle 45°$ mA and $I_2 = 30 - \mathrm{j}30$ mA $= 42.43\angle -45°$ mA
$I = \sqrt{[(106/\sqrt{2})^2 + (42.43/\sqrt{2})^2]} = 80.7$ mA

18.8 $V = \sqrt{[(12/\sqrt{2})^2 + (4/\sqrt{2})^2 + (2.4/\sqrt{2})^2]} = 9.1$ V
For resonance, $20000 \times 2 \times 10^{-3} = 1/(20000C)$
or $C = 1.25\ \mu$F
At this frequency, $V_C = 2.4 \times 500/1000 = 1.2$ V
At $\omega = 12000$, $1/\omega L = 41.67$ mS and
$\omega C = 15$ mS. Also $G = 2$ mS
$Y_p = (2 + \mathrm{j}15 - \mathrm{j}41.67) \times 10^{-3} = 2 - \mathrm{j}26.67$ mS
$Z_p = 1/Y_p = 2.8 + \mathrm{j}37.3\ \Omega$
Therefore, $Z = 502.8 + \mathrm{j}37.3 = 504.2\angle 4.4°\ \Omega$

18.9 Voltage is triangular. Capacitor current is $i = C\,\mathrm{d}v/\mathrm{d}t = (\pi^2/8)/(\pi/2\omega) = \pi\omega/4$ from $\omega t = 0$ to $\omega t = \pi/2$ and $i = -\pi\omega/4$ from $\omega t = \pi/2$ to $3\pi/2$. Hence current is of square waveshape with amplitude of $\omega\pi C/4 = 5000 \times 2 \times 10^{-6} \times \pi/4$ $=$ 7.85 mA. Current in resistor is triangular with amplitude $(\pi^2/8)/100 = 12.34$ mA. Hence the total current is of parabolic shape. When $\omega t = 0$, $i_R = 0$, when $\omega t = \pi/2$ $i_R = 10\pi^2/8$ mA and $P = (10\pi^2/8)^2 \times 10^{-6} \times 100 = 15.22$ mW. The area under the parabola has a mean value of 1/3 so $P = 15.22/3 = 5.07$ mW.

18.10 $I = \sqrt{[(100^2 + 15^2 + 12^2)/2]} = 72$ A

18.11 (*a*) $Z_f = 10 + \mathrm{j}5 = 11.18\angle 26.6°\ \Omega$,
$Z_3 = 10 + \mathrm{j}15 = 18.03\angle 56.3°\ \Omega$,
$|I_f| = 100/11.18 = 8.94$ A, $|i_3| = 200/18.03 = 11.09$ A,
$i = 8.94\sin(100t - 26.6°) + 11.09\sin(300 - 56.3°)$ A
(*b*) $I = 10.07$ A, $P = 10.07^2 \times 10 = 10.14.7$ W
(*c*) Power factor $= 1014/(158.1 \times 10.07) = 0.64$

18.12 At $\omega = 10000 X_C = -\mathrm{j}900\ \Omega$ and $X_L = \mathrm{j}400\ \Omega$,
$R = \sqrt{(0.04 \times 9 \times 10^6)} = 600\ \Omega$
$V_S = i_1(\mathrm{j}400 - \mathrm{j}900) + \mathrm{j}900 i_2$, $0 = i_2(600 + \mathrm{j}400 - \mathrm{j}900) + \mathrm{j}900 i_1$
or $i_1 = (0.56 + \mathrm{j}0.67)i_2$
Hence, $V_S = [(0.56 + \mathrm{j}0.67)(-\mathrm{j}500) + \mathrm{j}900]i_2 = 704.72\angle 61.6°\, i_2$
$V_R = i_2 R = 600 i_2$, $|V_R/V_S| = 600/704.72 = 0.851$
At $\omega = 30000 X_L = 1200\ \Omega$ and $X_C = -\mathrm{j}300\ \Omega$,
$V_S = 2545.6\angle -135°\, i_2$ and $V_R = 600 i_2$
$|V_R/V_S| = 600/2545.6 = 0.236$. Therefore
ratio $= (0.236/3)/0.851 = 0.092$.
So output wave is approximately sinusoidal.

18.13 (*a*) $V = \sqrt{[(10^6 + 10^4)]} = 710$ V. (*b*) At 100π radians,
$X_L = 157\ \Omega$ and $Z = 186\ \Omega$, so that $i = 710/186 = 3.82$ A
At 300π radians, $X_L = 471\ \Omega$ and $Z = 482\ \Omega$,
giving $i = 0.147$ A, $I = \sqrt{(3.82^2 + 0.147^2)} \approx 3.82$ A,

(c) $P = 3.82^2 \times 100 = 1459$ W

18.14 At $\omega = 5000$, $X_C = -j100\ \Omega$ and $X_L = j100\ \Omega$,
$I_f = 1/100 = 10$ mA. At $\omega = 10000 X_C = -j50\ \Omega$ and $X_L = j200\ \Omega$,
$Z = 100 + j150 = 180.3\angle 56.3°$, $I_3 = 2.78\angle -56.3°$ mA,
(a) $V = \sqrt{[(1+.25)/2]} = 0.79$ V, (b) $I = \sqrt{(7.07^2 + 1.96^2)} = 7.32$ mA,
(c) $P = 7.32^2 \times 10^{-6} \times 100 = 5.39$ mW.
(d) Power factor $= 5.59/(0.79 \times 7.32) = 0.93$

18.15 (a) 40.77 V; (b) 0.267 V

18.16 $i = 1 \sin(5000t + \pi/2) + 0.055 \sin(15000t - \pi/2)$A

18.17 (a) 336 Hz; (b) 14.59 V; (c) 3.54 A

18.18 $v = 0.133 \sin 3000t + 0.772 \sin 6000t + 0.291 \sin 9000t$ V

18.20 1.035

18.21 (a) 8.8 μF,
(b) $i = 0.96 \sin(500t + 76°) + 1 \sin 1500t + 0.26 \sin(2500t - 57°)$A

18.22 10.23 A

18.23 8.05 kW

18.24 22%

18.25 0.56

18.26 1.045

18.27 (a) 0.96 V; (b) $\approx 4\%$

18.28 423 W

19.1 $V = 9 - 20I$. Therefore $I = 0.1(9 - 20I) + 0.02(9 - 20I)^2$,
$8I^2 - 10.2I + 2.52 = 0$
$I = 10.2 \pm \sqrt{[104.04 - (4 \times 8 \times 2.52)]}/16 = 10.2 \pm 4.84/16$
I is either $15.04/16 = 0.94$ A or it is $5.36/16 = 0.335$ A. If $I = 0.94$ A the voltage across the 20 Ω resistor would be 18.8 V which cannot be.
Therefore $I = 0.335$ A
Voltage across the 20 Ω resistor $= 20 \times 0.335 = 6.7$ V
Voltage across NLR $= 9 - 6.7 = 2.3$ V

19.3 $1 = a + b + c$, $9 = a + 3b + 9c$ and $25 = a + 5b + 25c$. Solving these equations gives $a = b = 0$ and $c = 1$. Hence $i = v^2$
(a) When $v = 2 \sin \omega t$, $i = 4 \sin^2 \omega t = 2 - 2 \cos \omega t$ mA
Therefore the mean current = 2 mA
(b) When $v = 1 \sin \omega t + 1 \sin 2\omega t$
$i = 1 \sin^2 \omega t + 2 \sin \omega t \sin 2\omega t + 1 \sin^2 2\omega t$
The middle term contains the intermodulation products and expanding it gives $1 \cos \omega t - 1 \cos 3\omega t$.
Therefore IM products = 1 mA

19.4 $i = 4.5 + 3 \sin \omega t + 1.08 \sin^2 \omega t$
$= 4.5 + 3 \sin \omega t + 0.54 - 0.54 \cos 2\omega t$
$I = \sqrt{[5.04^2 + (3/\sqrt{2})^2 + (0.54/\sqrt{2})^2]} = 5.48$ mA

19.5 Plot the non-linear characteristic.
Draw the 10 Ω load line.
From the point of intersection $I_{max} = 410$ mA
Maximum voltage across 10 Ω resistor = 4.1 V
Voltage across NLR $= \sqrt[3]{(0.41/0.002)} = \sqrt[3]{205} = 5.9$ V

19.6 This problem can also be solved graphically using a similar method to that used in Exercise 19.5.
Alternatively, $200 = V/78 + 4 \times 10^{-3}v + 3 \times 10^{-4}v^3$ A
$200\text{ mA} = (16.8 \times 10^{-3})v + (3 \times 10^{-4})v^3$
$(3 \times 10^{-4})v^3 + (16.8 \times 10^{-3})v - 0.2 = 0$
Solving: $v = 6.63$ V
$I_{78} = 6.63/78 = 85$ mA
$I_{\text{NLR}} = 200 - 85 = 115$ mA

19.7 $i = 10 + 3V + V^2\text{ mA} = 10 + 3 \times 2\sin\omega t + (2\sin\omega t)^2$
$= 12 + 6\sin\omega t - 2\cos 2\omega t$ mA
Percentage distortion $= (2 \times 100)/6 = 33.33\%$

19.8 $0.5 = a - b + c$, $2 = a - 3b + 9c$ and $12 = a - 5b + 25c$. Solving, $a = 2.933$, $b = 3.5$ and $c = 1.063$. (*a*) 0 Hz, 2 mA; 500 Hz, 5.756 mA, 1000 Hz, 2.126 mA.
(*b*) Distortion $= [(2.126 \times 100)/5.756] \times 100 = 36.9\%$

19.9 $i = 0.6V + 0.1V^2 = (6 - V)/3$, or $V = 1.8$ V,
$i = 0.6 \times 1.8 + 0.1 \times 1.8^2 = 1.4$ A

19.11 159.2 Hz, 318.3 Hz, 477.5 Hz, 636.6 Hz

19.12 62.5%

19.15 8.5 mA

19.16 0.7 A, 1.3 A, 27.8 V

19.17 3.2 A

19.19 (*a*) 0.93 A; (*b*) 9.3 V, 7.7 V

19.20 6 A, 2.67 A

19.21 2 V

19.22 4.96 V, 1.04 V

19.23 1, 1.5, 0.4, 12.12%

19.24 4 V

20.1 $I(S) = V(S)/[R_1 + R_2 + 1/SC] = V/[S^2(R_1 + R_2 + 1/SC)]$
$= VC/S[SC(R_1 + R_2) + 1] = [VC/C(R_1 + R_2)]/S[S + 1/C(R_1 + R_2)]$
$I(t) = VC[1 - e^{-t/C(R_1+R_2)}]$ A

20.2 $V(S) = I(S)[R_1 + R_2 + 1/SC] - V/4S$
$I(S) = [V/S^2 - V/4S]/[R_1 + R_2 + 1/SC]$

The left-hand side of this equation gives the same result as 20.1. The right-hand side gives
$V/4S(R_1 + R_2 + 1/SC) = (V/4)/[S(R_1 + R_2) + 1/C]$
$= [V/4(R_1 + R_2)]/[S + 1/C(R_1 + R_2)]$
$I(t) = CV(1 - e^{-t/C(R_1+R_2)}) - (1/4)(R_1 + R_2)e^{-t/C(R_1+R_2)}$ A

20.3 Let $47\ k\Omega = R_1$ and $68\ k\Omega = R_2$. Then,
$V_0(S) = [6R_2/S]/[R_2 + R_1/(1 + SCR_1)]$
$= [6R_2(1 + SCR_1)]/[S(SCR_1R_2 + R_1 + R_2)]$
$= (6R_2/CR_1R_2)/S[S + (R_1 + R_2)/CR_1R_2]$
$+ (6SCR_1R_2/CR_1R_2)/S[S + (R_1 + R_2)/CR_1R_2]$
$6/CR_1 = 1277, \quad (R_1 + R_2)/CR_1R_2 = 360$
Hence, $V_0(S) = 1277/[S(S + 360)] + 6/(S + 360)$
$V(t) = (1277/360)[1 - e^{-360t}] + 6\,e^{-360t}$
$= 3.55 + 2.45\,e^{-360t}$ V

20.4 There is a steady current of $1/1200 = 0.833$ mA in the 1200 Ω resistor.
In the capacitive branch of the circuit:
$I(S) = (1/S)/[R + 1/SC] = SC/S[1 + SCR] = C/[1 + SCR]$
$= (1/R)/[S + 1/CR]$
Therefore, $I_c(t) = 4.545\ e^{-1.52}$ mA
Total current $I(t) = 0.833 + 4.55\ e^{-1.52\times 10^{-4}t}$ mA

20.5 $V_0(S) = V_{in}(S)[R_2 + 1/SC]/[R_1 + R_2 + 1/SC]$
$= 6/S[1 + SC(R_1 + R_2)] + 6\ CR_2/[1 + SC(R_1 + R_2)]$
Now $1/C(R_1 + R_2) = 6.54$. Hence,
$V_0(S) = (6/6.54)/S[S + 6.54] + [6CR_2/6.54]/[S + 6.54]$
And $V_0(t) = 6(1 - e^{-6.54t}) + (6 \times 0.686)\ e^{-6.54t}$
$= 6 - 1.88\ e^{-6.54t}$ V

20.6 $Z(S) = 5 + 8(2 + S)/(10 + S) = (66 + 13S)/(10 + S)$
$I(S) = V(S)/I(S) = [S + 10]/[S(13S + 66)]$
$= 1/(13S + 66) + 10/S(13S + 66)$
$= (1/13)/(S + 5.08) + (10/13)/(S(S + 5.08))$
$I(t) = 76.92\ e^{-5.08t} + (769.2/5.08)(1 - e^{-5.08t})$ mA
$= 151.4 - 74.5\ e^{-5.08t}$ mA

20.7 $I(S) = (3/S^2)[(10 + S)/(66 + 13S)]$
$= [(3/13)]/[S(S + 66/13)] + [(30/13)]/[S^2(S + 66/13)]$
$= [0.23/S(S + 5.08)] + [2.3/S^2(S + 5.08)]$
The first term gives $(0.23/5.08)(1 - e^{-5.08t}) = 0.045(1 - e^{-5.08t})$ A
The second term gives
$I(S) = (AS + B)/S^2 + C/(S + 5.08)$
$= AS^2 + 5.08S + BS + 5.08B + CS^2$
Equating constants: $2.3 = 5.08B$ or $B = 0.453$
Equating S:$0 = 5.08A + B$ or $A = -0.089$
Equating S^2:$0 = A + C$ or $C = 0.089$
Therefore, $I(S)' = -0.089/S + 0.453/S^2 + 0.089/(S + 5.08)$
$I(t)' = 0.453t - 0.089(1 - e^{-5.08t})$. Adding the two currents gives
$I(t) = 0.453t - 0.044(1 - e^{-5.08t})$ A

20.8 $Z(S) = R_1 + R_2/(1 + SCR_2) = (R_1 + R_2 + SCR_1R_2)/(1 + SCR_2)$
$V(S) = (I/S)](R_1 + R_2 + SCR_1R_2)/(1 + SCR_2)]$
$= (I/S)\{[R_1 + R_2 + SCR_1R_2]/[CR_2(S + 1/CR_2)]\}$
$= IR_1[S + (R_1 + R_2)/CR_1R_2]/[S(S + 1/CR_2)]V(t)$
$= I(R_1 + R_2[1 - e^{-t/CR_2}]$
Substituting values: $V(t) = 146.6$ V

20.9 $Z_1(S) = [125 \times 10^6/5S]/[125 + 10^6/5S] = (2 \times 10^5)/(S + 1600)$
$Z_s(S) = 0.2S + (2 \times 10^5)/(S + 1600)$
$= (0.2S^2 + 1600S + 10^6)/(S + 1600)$
$Z_T(S) = (160 \times Z_s)/(160 + Z_s)$
$= [160(S^2 + 1600S + 10^6)]/(S^2 + 2400 + 2.28 \times 10^6)\ I(S)$
$= (160/S)/Z_T = (S^2 + 2400S + 2.28 \times 10^6)/$
$[S(S + 800)^2 + (6 \times 10^2)^2]$
and $I(t) = 2.28 + 1.33\ e^{-800t} \sin(600t + 73.6°)$ V

20.10 $Z(S) = [(10^5/S)(10^6 + 10^5/5S)]/[10^5/S + 10^6 + 10^5/5S]$
$= [5 \times 10^6S + 10^5]/[S(5 + 50S + 1)] = [10^5(S + 0.02)]/[S(S + 0.12)]$
$I(S) = (120/S)[S(S + 0.12)]/[10^5(S + 0.02)]$
$= 120 \times 10^{-5}[1 + 0.1/(S + 0.02)]$ and
$I(t) = 120 \times 10^{-5}$ [unit impulse $+ 0.1\ e^{-0.02t}$]

20.11 $L\,\mathrm{d}i/\mathrm{d}t + Ri + (1/C)\int_0^k i\,\mathrm{d}t = 0$. $L\,\mathrm{d}^2q/\mathrm{d}t^2 + R\,\mathrm{d}q/\mathrm{d}t + q/C = 0$
Hence, $2\,\mathrm{d}^2q/\mathrm{d}t^2 + 500\,\mathrm{d}q/\mathrm{d}t + 2 \times 10^{-4}q = 0$
$S^2q(S) - Sq(0) - \mathrm{d}q/\mathrm{d}t(0) + 250Sq(S) - 250q(0) + 10^4q(S) = 0$
$q(S)(S^2 + 250S + 10^4) = Sq(0) + \mathrm{d}q/\mathrm{d}t(0) + 250q(0)$.
When $t = 0$, $q = Cv = 50 \times 10^{-6} \times 300 = 15 \times 10^{-3} = q(0)$.
Also, $\mathrm{d}q/\mathrm{d}t(0) = i(0) = 0$. Therefore,
$q(s)(S + 50)(S + 200) = 15 \times 10^{-3}(S + 250)$ and
$q(S) = (15 \times 10^{-3})[(S + 200) + 50]/[(S + 50)(S + 200)]$
$= (15 \times 10^{-3})[1/(S + 50) + (50/150)][(200 - 50)/(S + 200)(S + 50)]$.
Hence, $q(t) = 5 \times 10^{-3}[4\,\mathrm{e}^{-50t} - \mathrm{e}^{-200t}]$
Therefore, $v_C = q/C = 100(4\,\mathrm{e}^{-50t} - \mathrm{e}^{-200t})$ V
$i = \mathrm{d}q/\mathrm{d}t = \mathrm{e}^{-200t} - e^{-50t}$ A. For maximum current $4 = \mathrm{e}^{150t}$ or
$t = 9.25$ ms

20.12 $I(S) = \{(6/10^6S)/[2 + 3 + (1/0.5S)]\} = (6/10^6)\{1/[S(5 + 2/S)]\}$
$= [6/(5 \times 10^6)][1/(S + 2/5)]$ and $I(t) = 1.2 \times 10^{-6}\,\mathrm{e}^{-0.4t}$ A
$v_{\text{out}} = 6 - iR_2 = 6 - (2 \times 10^6)(1.2 \times 10^{-6}\,\mathrm{e}^{-0.4t}) = 6 - 2.4\,\mathrm{e}^{-0.4t}$
After 2.5 s, $i = 1.2\,\mathrm{e}^{-1} = 0.44\ \mu$A
$v_{\text{out}} = 6 - 2.4\,\mathrm{e}^{-0.4t} = 5.12$ V

20.13 $Z(S) = (1000 + 2S)(10^6/S)/(1000 + 2S + 10^6/S)$
$= [(1000 + 2S) \times 10^6]/[1000S + 2S^2 + 10^6]$
$= [(S + 500) \times 10^6]/(S^2 + 500S + 5 \times 10^5)$
Hence $I(S) = (100/S)[S^2 + 500S + 5 \times 10^5]/[(S + 500) \times 10^6]$
$= 10^{-4}[S^2 + 500S + 5 \times 10^5]/[S(S + 500)]$
$= 10^{-4}\{S/(S + 500) + 500/(S + 500) + (5 \times 10^5)/[S(S + 500)]\}$
$= 10^{-4}\{1 + (5 \times 10^5)/[S(S + 500)]\}$
$= 10^{-4}\{1 + (5 \times 10^5)/500[500/S(S + 500)]\}$
Therefore, $I(t) = 10^{-4}$ [unit impulse] $+ 0.1(1 - \mathrm{e}^{-500t})$

20.14 $R_{\text{crit}} = 2\sqrt{[1/(1 \times 10^{-6})]} = 2000\ \Omega$, $\alpha = R/2L = 500$,
$\omega = \sqrt{[(1/10^{-6}) - (10^6/4)]} = 866$ rad/s
$I(S) = 20[(1/(S + 500)^2 + 866^2]$ and $I(t) = 23.1\,\mathrm{e}^{-500t}\sin 866t$ mA

20.15 $i(S) = (25/S)/(1.5S + 100) = (25/1.5)/[S(S + 100/1.5)]$
$= 0.25\{(100/1.5)/[S(S + 100/1.5)]\}$
Therefore, $i(t) = 0.25(1 - \mathrm{e}^{-66.7t})$
When $t = 10 \times 10^{-3}$, $i(t) = 0.25(1 - 0.513) = 121.65$ mA

20.18 (*b*) 8.676 mA

20.20 $i = 0.2[1 - \mathrm{e}^{-t/4}]$ A, 44 mA

20.21 $V_C = -5\,\mathrm{e}^{-2t}$ V

20.22 $i = 5t - 2.5(1 - \mathrm{e}^{-2t})$

20.23 Transfer function $= 1/[1 + S(R_1C_1 + R_1C_1 + R_2C_2) + S^2R_1R_2C_1C_2]$

20.24 18.1 kV

20.25 $v = (V/R)[\sqrt{(L/C)}\sin(t\sqrt{(LC)})]$

20.26 $i = 5 + 2.5\,\mathrm{e}^{-0.015t}$ mA

Index